Lecture Notes in Nanoscale Science and Technology

Volume 41

Lecture Notes in Nanoscale Science and Technology (LNNST) aims to report latest developments in nanoscale science and technology research and teaching – quickly, informally and at a high level. Through publication, LNNST commits to serve the open communication of scientific and technological advances in the creation and use of objects at the nanometer scale, crossing the boundaries of physics, materials science, biology, chemistry, and engineering. Certainly, while historically the mysteries in each of the sciences have been very different, they have all required a relentless step-by-step pursuit to uncover the answer to a challenging scientific question, but recently many of the answers have brought questions that lie at the boundaries between the life sciences and the physical sciences and between what is fundamental and what is application. This is no accident since recent research in the physical and life sciences have each independently cut a path to the edge of their disciplines. As both paths intersect one may ask if transport of material in a cell is biology or is it physics? This intersection of curiosity makes us realize that nanoscience and technology crosses many if not all disciplines. It is this market that the proposed series of lecture notes targets.

Gang Wei · Jianming Zhang
Editors

Cellulose-Based Hydrogels and Aerogels

Synthesis, Functionalization, Sustainable Applications

Springer

Editors
Gang Wei
Qingdao University of Science and Technology
Qingdao, China

Jianming Zhang
Qingdao University of Science and Technology
Qingdao, China

ISSN 2195-2159 ISSN 2195-2167 (electronic)
Lecture Notes in Nanoscale Science and Technology
ISBN 978-3-032-03007-8 ISBN 978-3-032-03008-5 (eBook)
https://doi.org/10.1007/978-3-032-03008-5

This Springer imprint is published by the registered company Springer Nature Switzerland AG
The registered company address is: Gewerbestrasse 11, 6330 Cham, Switzerland

Contents

Chapter 1
Synthesis Methods and Properties of Cellulose-Based Hydrogels

Chenxi Du and Gang Wei

Abstract Cellulose-based hydrogels (CBHs) are three-dimensional polymeric networks derived from natural, nanostructured, or chemically modified cellulose, offering outstanding biocompatibility, biodegradability, and tunable physicochemical properties. Owing to their intrinsic hydrophilicity and structural versatility, CBHs have garnered increasing attention for the use in biomedical, tissue engineering, sensors, agricultural, and environmental applications. This chapter presents a comprehensive overview of the sources and types of cellulose, followed by detailed discussions on the synthesis strategies of CBHs, including physical, chemical, and radiation-induced cross-linking methods. Meanwhile, the fundamental properties of CBHs, including mechanical strength, swelling behavior, porosity, and biodegradability, are examined thoroughly. Emphasis is placed on how the synthesis parameters and structural design affect these properties and their implications for specific applications. This chapter aims to provide foundational knowledge and design insights for the development of advanced, sustainable hydrogel systems from cellulose-based materials.

Keywords Cellulose-based hydrogels · Cross-linking strategies · Nanocellulose · Functional regulation · Mechanical properties

1.1 Introduction

Cellulose, a biomass derived from renewable resources, is widely distributed in plant cell walls and certain microorganisms, and represents the most abundant biopolymer in nature [1]. Cellulose is a linear polymer composed of β-D-glucose units,

---

C. Du
College of Chemistry and Chemical Engineering, Qingdao University, Qingdao, China

C. Du · G. Wei (✉)
School of Polymer Science and Engineering, Qingdao University of Science and Technology, Qingdao, China
e-mail: wei-lab@qust.edu.cn

G. Wei and J. Zhang (eds.), *Cellulose-Based Hydrogels and Aerogels*, Lecture Notes in Nanoscale Science and Technology 41,
https://doi.org/10.1007/978-3-032-03008-5_1

characterized by excellent mechanical strength and significant hydrophilicity [2]. Owing to its excellent biocompatibility, environmental friendliness, and biodegradability, cellulose has emerged as a focal point in recent research on green and sustainable materials. With the advancement of material innovation, cellulose is no longer confined to traditional industrial applications [3, 4]. Through chemical modification or physical treatments, cellulose has been transformed into an emerging platform for the development of multifunctional materials [5]. Among these, cellulose hydrogels have garnered widespread attention due to their unique three-dimensional (3D) network structure as well as outstanding water absorption and swelling properties.

Cellulose hydrogels have polymeric network structures formed from cellulose or its derivatives, cross-linked with agents or reagents in aqueous media [6]. They exhibit excellent plasticity, flexibility, and biocompatibility, making them widely applicable in medicine, food science, agriculture, and environmental engineering. Based on their origin and composition, cellulose-based hydrogels (CBHs) can be classified into three categories: natural cellulose hydrogels, cellulose derivative hydrogels, and composite hydrogels [7, 8]. Natural cellulose hydrogels are typically prepared through cross-linking of plant-derived or bacterial cellulose (BC). However, because natural cellulose exhibits poor solubility in water and conventional organic solvents, dissolution typically requires the use of ionic liquids, ammonia/alkaline systems, or N-methylmorpholine N-oxide (NMMO) [9, 10]. Cellulose derivatives have found broader application in hydrogel synthesis due to their improved solubility and chemical reactivity. For instance, derivatives like carboxymethyl cellulose (CMC) and hydroxyethyl cellulose (HEC) are soluble in water and mild solvents, eliminating the need for harsh dissolution conditions [11]. In recent years, researchers have integrated cellulose with other polymeric materials to endow hydrogels with enhanced functional properties, including stimuli-responsiveness, self-healing, and electrical conductivity, facilitating their applications in smart materials and advanced biomedical technologies [12].

Compared with other commonly used hydrogel matrices, such as polyvinyl alcohol (PVA), polyacrylamide, and gelatin. CBHs exhibit tremendous potential in both biomedical and industrial applications owing to their unique structure and properties. Their high hydrophilicity and 3D network architecture endow them with excellent water retention, controlled release capabilities, and a high surface area, making them ideal platforms for applications such as drug delivery, tissue engineering, and wound dressings. In drug delivery, CBHs can respond to environmental stimuli such as pH, temperature, or targeting ligands by undergoing structural changes or degradation, enabling sustained or targeted drug release. Derivatives such as CMC and HEC can form stable complexes with drug molecules, thereby enhancing drug stability and bioavailability [13, 14]. Furthermore, by incorporating functional materials such as chitosan, gelatin, or PVA, the mechanical properties and release behavior of CBHs can be tailored to meet diverse therapeutic requirements [15]. In the field of tissue engineering, CBHs exhibit excellent biocompatibility and a structure resembling the extracellular matrix, providing favorable support for the cell adhesion and proliferation [16]. Studies have shown that porous scaffolds constructed from nanocellulose can mimic natural tissue environments, facilitating cell migration and new tissue formation [17]. Moreover, CBHs can encapsulate stem cells and growth factors to further enhance tissue regeneration [18].

In industrial applications, CBHs are widely used as food thickeners, cosmetic carriers, agricultural water-retaining agents, and environmental remediation materials. For example, in the food industry, their excellent gel-forming ability and stability make them ideal thickening and structural-modifying agents. In agriculture, CBHs can enhance soil water retention and provide controlled release of fertilizers and pesticides, thereby improving resource utilization. In environmental purification, cellulose hydrogels exhibit high adsorption capacity and tunable degradability, making them effective for removing heavy metal ions, dyes, and organic pollutants [19]. Additionally, with the rise of smart materials and wearable technologies, research on conductive CBHs as flexible sensors, bioelectrodes, and energy storage devices has grown rapidly, driving their development in cutting-edge fields such as electronic skin, human–machine interfaces, and bioelectronics. In terms of functional expansion, traditional cellulose hydrogels generally possess singular properties, such as water absorption or controlled release, which limits their applicability in multi-functional or stimuli-responsive systems. Although recent studies have introduced conductive fillers, thermo-responsive monomers, and self-healing agents to endow new functionalities, challenges remain in system complexity, component stability, and long-term functionality, particularly regarding their reliability in clinical and industrial settings [20].

This chapter not only highlights the diversity and adaptability of CBH material design but also underscores the importance of adopting a structure–property–function synergy perspective in future research. By doing so, it is possible to develop predictive and application-oriented hydrogel systems with engineered compatibility. The following contents will delve into the relationships between the microstructural features, physicochemical properties, and application-driven requirements of CBHs, thereby uncovering the intrinsic connections between material characteristics and functional performance.

1.2 Sources and Types of Cellulose for CBH Formation

Unlike other polysaccharides, cellulose is composed of β-D-glucose units linked by β-1,4-glycosidic bonds. In contrast, starch and glycogen are composed of α-D-glucose units linked by α-1,4-glycosidic bonds, forming helical structures [21]. The unique glycosidic linkage orientation in cellulose results in molecular chains that extend linearly in a zigzag conformation. This allows adjacent chains to pack closely, forming dense crystalline regions [22]. Moreover, the hydrogen bonding network between chains aligns the cellulose molecules into compact and rigid microfibrils. This molecular structure endows cellulose with excellent mechanical strength and chemical stability, making it a vital matrix for hydrogel materials. Depending on their source and processing method, cellulose materials used in hydrogels can be classified into three main categories: natural cellulose, nanocellulose, and chemically modified cellulose [23].

1.2.1 Natural Sources of Cellulose

Natural cellulose is primarily derived from sources such as plants, bacteria, algae, and agricultural waste. Among these, plants are the most abundant source of cellulose, including cotton, wood, straw, bamboo, and bagasse, as illustrated in Fig. 1.1a. Cotton contains over 90% pure cellulose, making it an ideal raw material for laboratory-scale production of high-purity cellulose. In contrast, wood-derived cellulose exists within a lignocellulosic complex and requires processes such as alkaline treatment, bleaching, and cooking to remove lignin and hemicellulose [24]. Therefore, plant-derived cellulose typically undergoes mechanical grinding, acid hydrolysis, or enzymatic treatment to break down its dense structure, allowing it to disperse and form gel networks. These treatment methods can modulate the crystallinity and degree of polymerization of cellulose, thereby influencing the structure and performance of the resulting hydrogels.

In addition to plant sources, certain bacteria can synthesize highly pure and impurity-free cellulose, commonly referred to as BC. For instance, Gram-positive Lactococcus, *Komagataeibacter diospyri*, and *Leifsonia sp.* can produce cellulose via extracellular synthesis under suitable fermentation conditions (Fig. 1.1b). The resulting fibers are finer in diameter, possess higher crystallinity, and exhibit superior mechanical strength and film-forming capabilities. Since bacterial cellulose requires no complex purification and avoids chemical reagent residues, it offers enhanced biocompatibility. This makes it highly valuable for biomedical applications such as tissue engineering, artificial skin, and wound dressings [25]. However, BC production is limited by bacterial strains, culture media, and cultivation methods, hindering large-scale production and increasing costs.

Additionally, agricultural wastes such as corn stalks, wheat straw, rice husks, and coconut shells are important sustainable sources of cellulose. Through green separation and extraction techniques, cellulose components can be effectively recovered, reducing environmental pollution and providing low-cost raw materials for eco-friendly hydrogel fabrication. However, further optimization of extraction processes is needed to reduce impurity interference. Natural cellulose sources are diverse, each with distinct advantages suitable for various hydrogel applications. Nevertheless, due to differences in origin, notable variations exist in extraction efficiency, purity control, and functional adaptability, necessitating tailored processing strategies.

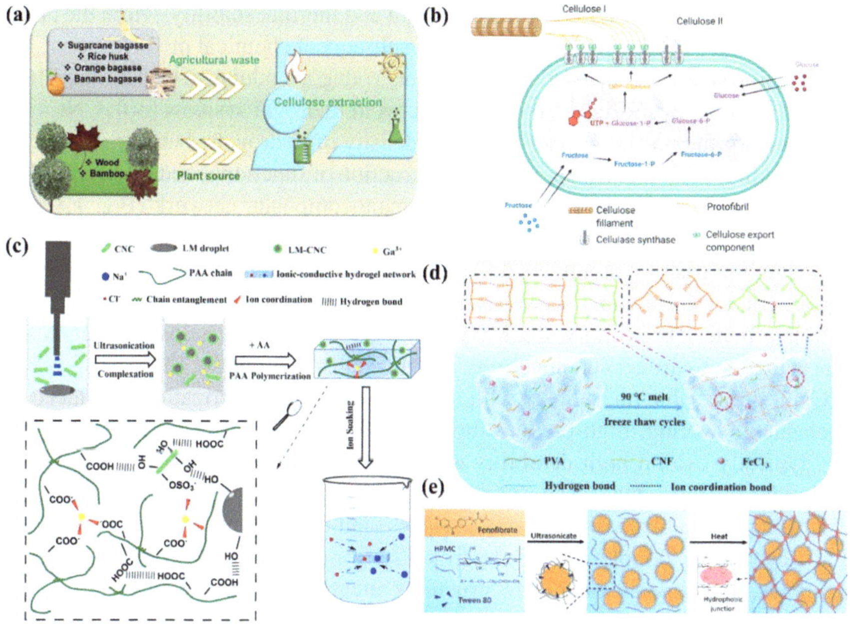

Fig. 1.1 Sources and types of cellulose: **a** Natural source of cellulose. Reproduced with the permission from Ref. [45]. **b** The synthesis process of BC. Reproduced with the permission from Ref. [46]. **c** Steps and mechanism of preparing hydrogels using CNC and LMNPs. Reproduced with the permission from Ref. [32]. **d** Synthesis process and mechanism of PVA@CNF-$FeCl_3$ hydrogel. Reproduced with the permission from Ref. [33]. **e** Preparation process of HPMC thermo-gel emulsion. Reproduced with the permission from Ref. [44]

1.2.2 Nanocellulose

Although natural cellulose provides an abundant and cost-effective source for hydrogel preparation, its high crystallinity and dense fibrous structure limit its dispersibility and gelation capability in hydrogel formation [26]. In contrast, nanocellulose that produced by chemical, mechanical, or microbial techniques to "microstructure" natural cellulose offers significantly enhanced morphology, size control, surface activity, and mechanical properties, thus exhibiting superior gel-forming capabilities [27]. Nanocellulose is generally categorized into cellulose nanocrystals (CNCs), cellulose nanofibers (CNFs), and bacterial nanocellulose (BNC) [28]. These materials typically possess nanoscale dimensions, high specific surface area, excellent mechanical strength, and good dispersibility, endowing CBHs with novel functionalities and structural benefits.

CNC is highly crystalline nanostructures extracted from natural cellulose by removing amorphous regions through acid hydrolysis and other methods, and they generally exhibit rod-like or needle-like morphologies [29]. Its highly crystalline domains impart excellent mechanical strength and thermal stability, while the abundance of surface hydroxyl groups facilitates subsequent chemical modifications. The high specific surface area, tensile strength, and Young's modulus of CNCs make them ideal reinforcing agents or structural templates in hydrogel systems [30]. CNCs not only improve the mechanical strength and density of hydrogels, but also enable the stable loading of nanoparticles and the construction of functionalized interfaces [31]. For example, Rahmani et al. constructed a nanocomposite network (Fig. 1.1c) by physically integrating CNCs with liquid metal nanoparticles (LMNPs), significantly enhancing the fracture toughness and strain-sensing capabilities of the hydrogel. In this system, CNCs not only served as mechanical support scaffolds, but also formed hydrogen-bonded networks with LMNPs via surface hydroxyl groups, synergistically improving the hydrogel's conductivity and mechanical strength [32].

CNFs are flexible microfibers with high degrees of polymerization, produced by mechanical shearing, high-pressure homogenization, or enzymatic treatment, and they typically exhibit high flexibility and lower crystallinity. Owing to their flexibility, adhesiveness, and strong water absorption capacity, CNFs are widely used for fabricating injectable hydrogels, bioadhesives, and flexible sensors. For instance, Li et al. developed a double-network hydrogel sensor (Fig. 1.1d) by integrating CNFs with PVA, significantly enhancing structural integrity and energy dissipation capability [33]. In this system, CNFs acted as uniform micro-scale support scaffolds and formed hydrogen bonds that synergistically enhance both mechanical strength and fatigue resistance of hydrogels, enabling flexibility and cyclic stability under high loads. This strategy highlights the critical role of CNFs as flexible reinforcing agents in constructing high-performance hydrogels. Moreover, the CNF incorporation significantly improved the mechanical stability, anti-swelling properties, and cell-supporting capacity of hydrogels, making them suitable for the applications in bioprinting and tissue regeneration.

BNC is biosynthesized by specific bacterial strains, among which *Acetobacter xylinum* is widely used for BNC production. The BNC synthesis in bacteria can be divided into two stages: (1) intracellular polymerization of glucose monomers and (2) self-assembly of polymers into cellulose fibrils [34, 35]. As a result, BNCs are free from the impurities found in plant cellulose, offering high purity and crystallinity without requiring decontamination processes. Furthermore, their inherent biocompatibility and biodegradability make BNCs highly suitable for applications in wound dressings and tissue engineering [36].

Nanocellulose materials introduce enhanced structural tunability and performance versatility into conventional cellulose hydrogels, forming a critical foundation for the development of next-generation functional hydrogel systems.

1.2.3 Chemically Modified Cellulose

To overcome the poor solubility and low reactivity of natural cellulose, researchers have developed a variety of cellulose derivatives. These chemically modified celluloses exhibit excellent water solubility, processability, and functionalizability, making them among the most widely used materials in hydrogel research. Common chemical modifications include carboxymethylation, hydroxypropylation, methylation, and acetylation [37].

CMC is an anionic, water-soluble cellulose derivative. The introduction of carboxymethyl groups ($-CH_2COOH$) into the cellulose backbone alters its inherent hydrophobic nature. As a result, CMC dissolves in water to form a clear solution, exhibiting excellent thickening properties and ionic responsiveness [38]. In hydrogel construction, negatively charged CMC can form ionically cross-linked hydrogels with cationic polymers (e.g., chitosan) or metal ions (e.g., Ca^{2+}, Fe^{3+}), and can also participate in chemical cross-linking to establish stable networks [39]. The unique charge properties of CMC hydrogels have enabled their broad application in drug delivery, tissue engineering scaffolds, and heavy metal adsorption.

HEC is a nonionic, water-soluble cellulose derivative produced by introducing hydroxyethyl substituents ($-CH_2CH_2OH$) into the cellulose backbone. Its nonionic nature renders HEC resistant to pH and electrolyte fluctuations within the system, resulting in high stability [40]. Additionally, the incorporation of HEC into other solutions can significantly alter the system's viscosity. For example, in a study by Kokol, adding HECs with varying molecular weights and degrees of substitution into microfibrillated cellulose (MFC) suspensions significantly increased the zero-shear viscosity (η_0) of the system. This suggests that HEC reinforces the fibrillar network structure, thereby enhancing system viscosity [41]. Therefore, HEC-based hydrogels are often employed as carrier materials for stabilizing drugs, moisturizing agents, or cosmetic formulations. Due to its nonionic characteristics, HEC is less likely to interact with other electrolytes, providing excellent compatibility across various formulations.

Methyl cellulose (MC) and hydroxypropyl methyl cellulose (HPMC) have been obtained by chemically modifying natural cellulose with methyl ($-CH_3$) and hydroxypropyl ($-CH_2CH(OH)CH_3$) groups, respectively [42]. Both materials exhibited thermogelation behavior with forming gels upon heating and revert to solutions upon cooling. This property is particularly valuable in injectable hydrogels and temperature-responsive drug release systems [43]. Utilizing this characteristic, Chen et al. developed drug-loaded oral films using HPMC thermo-gel emulsions as templates, as illustrated in Fig. 1.1e. HPMC formed a gel network upon heating, enabling the entrapment of drugs and achieving drug loadings up to 63 wt% [44]. This study highlights the promising application of HPMC's thermo-responsive gelation in drug delivery systems. MC and HPMC are also widely used as food stabilizers, tablet coatings, and controlled-release agents due to their excellent biocompatibility and film-forming properties.

Chemically modified cellulose introduces abundant reactive sites and diverse functional features into hydrogel systems. By tailoring the type of substituent, degree of substitution, and molecular weight, the physicochemical properties, responsiveness, and biocompatibility of cellulose hydrogels can be precisely tuned, thereby greatly expanding their potential in biomedical, food, and environmental applications.

1.3 Synthesis Methods of CBHs

The synthesis methods of CBHs play a decisive role in determining their structural characteristics, mechanical properties, biocompatibility, and functional attributes. Different synthesis strategies endow hydrogels with distinct performance characteristics. Based on the underlying cross-linking mechanisms, current synthesis strategies of CBHs are primarily categorized into three types: physical cross-linking, chemical cross-linking, and radiation cross-linking [47, 48].

1.3.1 Physical Cross-Linking

Physical cross-linking refers to the formation of hydrogel networks without covalent bonds, relying mainly on hydrogen bonding, electrostatic interactions, and hydrophobic interactions to establish a 3D structure [49]. These methods do not require the use of chemical cross-linkers, operate under mild conditions, and offer excellent biocompatibility, making them particularly suitable for toxicity-sensitive biomedical applications such as cell scaffolds, drug delivery systems, and skin-contact dressings.

Among these, hydrogen bonding is the most commonly employed physical cross-linking mechanism for hydrogel formation due to the abundance of hydroxyl groups in cellulose. s shown in Fig. 1.2a, exposure of hydroxyl groups through solvent treatment facilitates the formation of intramolecular hydrogen bonds within a single cellulose chain and intermolecular hydrogen bonds between adjacent chains, resulting in a 3D hydrogel network. For example, using CNFs, Chen et al. regulated the hydrogen bonding between CNFs and PVA via a solvent-exchange method [50]. Water diffusion was used to enhance hydrogen bonding, resulting in hydrogels with excellent mechanical strength and water retention capabilities. This demonstrated the crucial role of hydrogen bonding in enhancing hydrogel properties. While native cellulose is electrically neutral, surface modifications (e.g., CMC) can introduce charges, allowing electrostatic interactions with oppositely charged cross-linkers to form hydrogels (Fig. 1.2b). In a study by Wan et al., positive charges were introduced onto cellulose via organic solvent treatment, enabling dynamic electrostatic interactions with negatively charged MXene nanosheets [51]. This resulted in hydrogels with exceptional stretchability and self-healing capabilities.

Compared to hydrogen bonding and electrostatic interactions, hydrophobic interactions are relatively weaker. Hydrophobic interactions are not direct attractive forces; rather, they arise from the repulsion of water molecules toward nonpolar groups, leading to an entropically driven aggregation effect that lacks directional specificity [52]. As shown in Fig. 1.2c, the introduction of hydrophobic groups into cellulose enables their aggregation through hydrophobic interactions in aqueous solutions, forming a network structure. Therefore, hydrogen bonding and electrostatic interactions typically serve as the main driving forces for network formation,

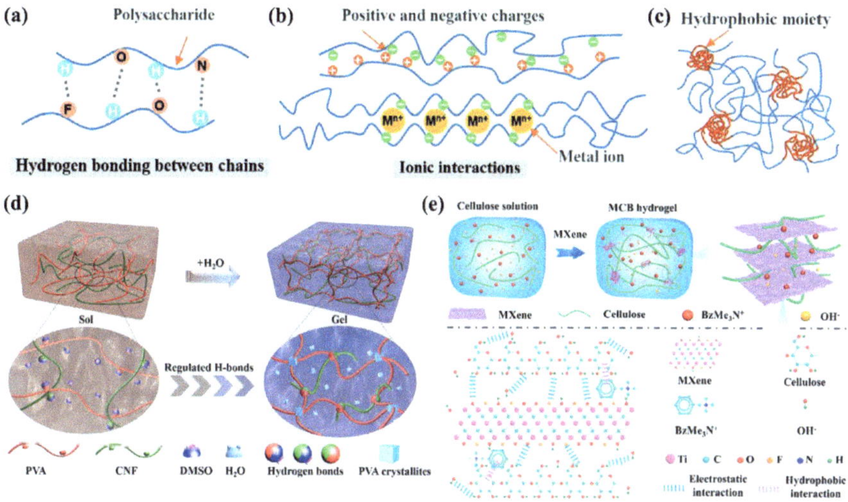

Fig. 1.2 Several physical crosslinking mechanisms and applications: **a** Hydrogen bonding, **b** electrostatic interactions and **c** hydrophobic interactions are cross-linking mechanisms. Reproduced with the permission from Ref. [53]. **d** Schematic diagram of the solvent exchange method to enhance the hydrogen bonding of gels. Reproduced with the permission from Ref. [50]. **e** Cross-linking mechanism of MXene and cellulose via electrostatic interaction. Reproduced with the permission from Ref. [51]

while hydrophobic interactions are mainly employed to impart functionalities such as thermo-responsiveness, reversibility, and flexibility. However, their ability to solely maintain network stability is limited.

Physical cross-linking offers several advantages, including the absence of toxic cross-linkers, operational simplicity, environmental friendliness, and the formation of reversible networks that exhibit good stimuli-responsiveness. However, its limitations include relatively weak mechanical properties, poor network stability, and insufficient long-term water stability, which restrict its application in load-bearing or long-term implantable materials. Overall, physical cross-linking is suitable for applications requiring high safety and flexibility. However, performance enhancements often require composite or double-network strategies.

1.3.2 Chemical Cross-Linking

Chemical cross-linking is one of the most common and effective methods for constructing CBHs. It forms stable covalent bonds between polymer chains, resulting in irreversible and mechanically robust 3D networks. Compared with physical cross-linking, chemical cross-linking offers greater structural controllability and enhanced potential for functional modification. Based on specific reaction mechanisms, chemical cross-linking can be categorized into click chemistry, free radical polymerization, and covalent cross-linking using linkers [54].

Click chemistry does not refer to a single reaction, but rather a class of chemical reactions characterized by high efficiency, strong selectivity, and minimal byproducts [55]. In the synthesis of CBHs, functionalizing cellulose with highly reactive and biocompatible click groups enables efficient, rapid, and selective cross-linking, leading to the formation of stable or reversible 3D hydrogel networks. Common click reactions include azide–alkyne cycloaddition, thiol–ene/yne reactions, and the Diels–Alder reaction [56]. These reactions proceed under mild conditions, are atom-economical, and can be carried out in aqueous media, making them especially suitable for modification and cross-linking in the presence of bioactive molecules. For example, Grilo et al. utilized a Diels–Alder click reaction to introduce furan and maleimide groups into lignin and cellulose, as illustrated in Fig. 1.3a, to prepare bio-based gel materials [57]. The resulting gels exhibited excellent thermal stability, offering a promising strategy for the development of bio-based thermosetting materials. Click chemistry offers a biocompatible and efficient synthetic strategy with more precise reaction control. However, the synthetic pathways are complex, and both the reaction cost and material expenses are relatively high, which require further optimization for clinical translation.

Free radical polymerization is an efficient strategy for constructing cellulose-based copolymer hydrogels, typically involving the addition of initiators (e.g., ammonium persulfate) and functional monomers (e.g., acrylamide, acrylic acid) to form

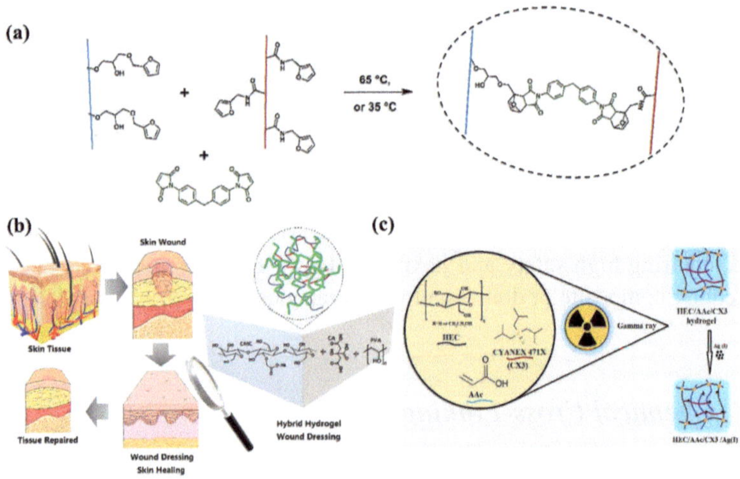

Fig. 1.3 Applications of chemically cross-linked and irradiation cross-linked hydrogels: **a** Mechanism of introducing furan and maleimide groups into lignin and cellulose. Reproduced with the permission from Ref. [57]. **b** Preparation and application of CMV/PVA polymer hydrogel. Reproduced with the permission from Ref. [61]. **c** Gelation mechanism of HEC/AAc/CX hydrogel under γ-ray irradiation. Reproduced with the permission from Ref. [62]

copolymeric networks around the cellulose backbone [58]. This method not only enhances the mechanical strength and swelling capacity of the hydrogel, but also allows for the incorporation of stimuli-responsive functional groups, broadening its applications in smart materials. For example, HEC and CMC can copolymerize with acrylamide to form hydrogels that exhibit good hydrophilicity and tunable reversible swelling behavior through monomer design [59]. Based on this principle, Blazic et al. conducted free radical polymerization of cellulose with vinyl/divinyl monomers to graft stimuli-responsive poly (dimethylaminoethyl methacrylate) onto cellulose, forming a copolymeric hydrogel [60]. The resulting hydrogel exhibited excellent swelling properties and dynamic viscoelasticity, presenting a viable strategy for hydrogel-based wound treatment applications. However, free radical polymerization may leave residual initiators and unreacted monomers, which could pose potential risks to biomedical systems and therefore require thorough safety evaluations.

Covalent cross-linking using linkers typically involves the reaction of functionalized cross-linkers (e.g., glutaraldehyde, epoxy compounds, azelaic acid, or PVA) with functional groups on the cellulose chains, forming stable covalent networks. In a study by Capanema et al., CMC and PVA formed ester bonds in the presence of citric acid (CA) as the cross-linker, resulting in a polymeric cross-linked network (Fig. 1.3b) [61]. The resulting composite hydrogels demonstrated excellent water absorption and mechanical properties, showing great potential for soft tissue engineering applications. This method offers high reaction efficiency and strong network stability, making it suitable for high drug-loading applications and the construction of supportive tissue engineering scaffolds. However, some cross-linkers, such as glutaraldehyde, exhibit cytotoxicity, necessitating strict control over reaction conditions and dosage to minimize residual toxicity and ensure biosafety. Therefore, the development of more biocompatible cross-linkers remains a critical challenge in the research of covalently cross-linked hydrogels.

Overall, chemical cross-linking enables comprehensive control over both the structural formation and functional tuning of CBHs, and its covalent networks provide enhanced mechanical stability and environmental tolerance. Future research should focus on developing green cross-linking strategies, biodegradable cross-linking systems, and multi-mechanism synergistic cross-linking technologies to advance their applications in biomedical and related fields.

1.3.3 Radiation-Induced Cross-Linking (E.G., γ-rays, UV, Electron Beam)

Radiation-induced cross-linking employs high-energy radiation (e.g., γ-rays, UV, and electron beams) to generate free radicals in cellulose macromolecules, initiating interchain cross-linking without the need for chemical crosslinkers. This approach is regarded as an additive-free and residue-free green crosslinking method. γ-rays and electron beams offer high penetration power, making them suitable for thick films or internal crosslinking, while UV radiation is typically used for surface or shallow-layer cross-linking, often in conjunction with photosensitizers [63].

Compared with conventional chemical cross-linking, radiation-induced methods eliminate the need for potentially toxic cross-linkers and can simultaneously sterilize the material, offering eco-friendliness and operational simplicity. Recent studies have demonstrated significant advances in applying radiation cross-linking to CBHs. For instance, Masry et al. used γ-ray irradiation to cross-link HEC, acrylic acid (AAc), and triisobutyl phosphine sulfide to fabricate a hydrogel capable of capturing silver ions in acidic nitric media (Fig. 1.3c) [62]. The resulting hydrogel exhibited excellent adsorption performance, highlighting the potential of radiation-induced crosslinking in environmental remediation. Despite its advantages of being clean, efficient, and additive-free, the high cost of equipment, stringent environmental requirements, and complex dosage control hinder its widespread adoption in small-scale laboratories and industrial applications. Future research should further elucidate the underlying mechanisms, optimize processing parameters, and expand its application in high-performance functional hydrogels.

1.4 Properties of CBHs

1.4.1 Mechanical Properties (Compression, Tensile, and Rheology)

CBHs with favorable mechanical properties are widely used in biomedical engineering, flexible electronics, tissue engineering, and environmental remediation. The mechanical properties of hydrogels can be categorized from multiple dimensions, collectively determining their performance in practical applications. Key mechanical parameters include compressive strength, tensile strength, and rheological behavior [64]. These properties are influenced by various factors such as the type of nanofiller, cross-linking density, cellulose type, and water content. By tuning these variables, hydrogels can be tailored to meet specific application requirements.

Incorporating nanofillers is an effective strategy to enhance the mechanical properties of hydrogels. Commonly used nanofillers include carbon-based like graphene oxide (GO) and carbon nanotubes (CNTs) and inorganic nanomaterials. The 2D sheet-like structure of GO (Fig. 1.4a) interacts with cellulose networks via hydrogen bonding and π–π interactions, suppressing crack propagation and dispersing stress, while its oxygen-containing groups (e.g., –COOH, –OH) enhance interfacial adhesion. CNT, with its 1D tubular structure (Fig. 1.4b), form 3D conductive reinforcement networks through mechanical interlocking and covalent grafting (e.g., acylation), significantly improving tensile strength and toughness. Among inorganic nanomaterials, montmorillonite (MMT) and SiO_2 are the most commonly used. MMT's layered structure enhances interfacial area via intercalation, while cation exchange

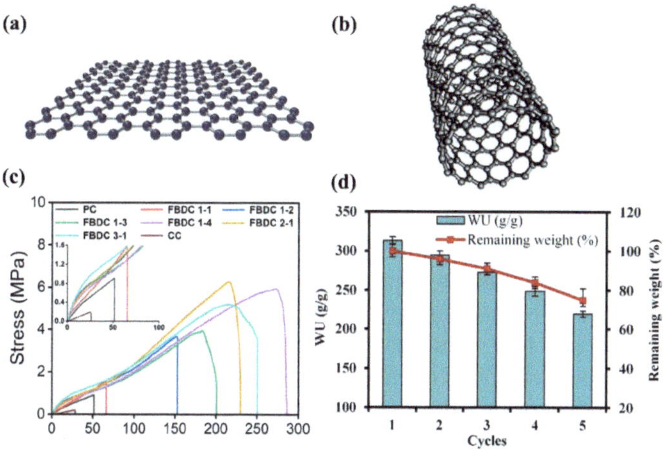

Fig. 1.4 Mechanical properties and water absorption and retention of CBHs: **a** 2D structure of GO. **b** 1D structure of CNTs. Reproduced with the permission from Ref. [73]. **c** Comparison of tensile strain of three types of hydrogels. Reproduced with the permission from Ref. [66]. **d** Water absorption of CMC-g-(PAM-co-PAMPS) hydrogel after cyclic drying. Reproduced with the permission from Ref. [72]

(e.g., Na^+ to Ca^{2+}) increases cross-linking density, significantly improving compressive strength [65]. SiO_2 nanoparticles fill pores and inhibit crack propagation, and their silanol groups form hydrogen bonds with cellulose, thereby enhancing modulus. Both materials optimize mechanical properties through physical reinforcement and stress dispersion.

Cross-linking density is another critical factor influencing mechanical performance. Higher crosslinking densities typically increase rigidity and strength but may compromise flexibility and elongation. For instance, Pan et al. developed highly mechanically robust cellulose hydrogels using long-chain epoxidized vegetable oils as crosslinkers through a dual cross-linking strategy [66]. As shown in Fig. 1.4c, the hydrogels achieved maximum tensile and compressive strains of 290% and 58%, respectively. Compared with conventional cellulose hydrogels, these materials exhibited superior toughness and Young's modulus, highlighting the role of crosslinking strategies in mechanical enhancement.

Additionally, the source and type of cellulose significantly affect mechanical properties. Baraka et al. compared the mechanical behavior of hydrogels fabricated from pure cellulose nanofibers (HCNF) and lignin-containing cellulose nanofibers (HLCNF) [67]. The results showed that HCNF hydrogels exhibited a Young's modulus of 6.53 kPa, significantly higher than the 2.83 kPa of HLCNF hydrogels, underscoring the impact of cellulose type on mechanical behavior.

In summary, the mechanical properties of CBHs are influenced by multiple factors. Future research should continue exploring the relationship between structure and performance to develop novel hydrogels with enhanced and tailored mechanical properties.

1.4.2 Swelling and Water Retention Properties

The swelling and water retention capacities of CBHs stem from their three-dimensional hydrophilic network structure, a key feature that determines their applicability in fields such as agriculture, environmental engineering, biomedical dressings, and ionic electronic devices. The primary factors influencing their swelling and water retention behavior include chemical composition, physical network structure, and crosslinking strategy.

Firstly, the chemical structure dictates hydrophilicity: the more functional groups such as –OH, –COOH, and amine groups on the cellulose backbone, the more hydrogen bonds can form with water molecules, enhancing water uptake [68]. However, excessive hydrophilicity may lead to over-swelling, resulting in a loose or degraded network structure and compromised mechanical stability. The crosslinking density and method are crucial to maintaining the integrity of the network. A high crosslinking density can restrict water diffusion and desorption, thereby reducing swelling, while a low density may result in premature saturation, weak structural integrity, and poor release control [69]. The pore size and interconnectivity of the network affect the swelling rate and capacity, uniformly distributed mesopores (tens to hundreds of nm) facilitate rapid and stable water uptake [70]. Conversely, excessively large pores may cause overly rapid swelling and structural instability, whereas excessively small pores can slow the swelling rate.

Additionally, variations in environmental conditions and preparation techniques can further influence the structure, hydrophilicity, and stability of the hydrogel. Therefore, several strategies have been developed to improve the swelling and water retention properties. One approach involves incorporating nanofillers such as CNCs, CNFs, nanosilica, or metal oxides via covalent or non-covalent interactions to enhance mechanical strength and network stability, thereby improving shape recovery and long-term water retention. For instance, Nyamayaro et al. reported that incorporating CNCs into the hydrogel matrix formed hydrogen bonding networks and ionic complexes, significantly enhancing resistance to over-swelling and improving water retention performance [71]. Another strategy involves designing dual or multi-network structures, such as hydrophilic-hydrophobic networks or secondary crosslinked systems, to balance rapid swelling with slow diffusion while retaining sufficient elasticity. Additionally, chemical modifications, such as grafting hydrophilic groups or introducing ionic cross-linkable monomers, can be employed to enhance swelling capacity and responsiveness. Abdelgelil et al. demonstrated that grafting CMC with PAM and PAMPS yielded a polyelectrolyte hydrogel capable of rapid swelling and retained 70% of its absorption capacity even after five cycles of reuse (Fig. 1.4d), validating the effectiveness of graft copolymerization for durable water retention [72].

The optimization of chemical structure, incorporation of nanofillers, and strategic network design offer promising routes to overcoming traditional limitations in swelling and water retention, thereby enhancing the versatility and competitiveness of CBHs in agriculture, biomedicine, and electronics. Future research should continue to focus on the correlation between structural mechanisms and specific application scenarios, aiming to translate high-performance laboratory outcomes into practical engineering materials.

1.4.3 Porosity and Network Structure Analysis

The porosity and network architecture of CBHs play a pivotal role in determining their ultimate application performance, particularly in aspects such as cell infiltration, substance transport, drug loading capacity, and mechanical strength. The regulation of pore structures governs not only the specific surface area and permeability of the hydrogel, but also correlates closely with internal crosslinking density, mechanical robustness, and biodegradability. Key factors affecting the porosity and network architecture of cellulose hydrogels include cellulose concentration and molecular weight, type and density of crosslinkers, fabrication process, incorporation of nanofillers, and regeneration conditions [74]. In practical applications, precise tuning of pore size, uniformity of distribution, and network compactness is essential to meet specific functional requirements.

First, cellulose concentration and crosslinking density serve as fundamental parameters in controlling pore structures. Systems with lower cellulose concentrations tend to form more porous structures with larger pore sizes but weaker mechanical integrity. Conversely, high cellulose concentrations may lead to overly dense networks that hinder diffusion efficiency. According to Nicu et al., variations in crosslinker type and concentration can effectively modulate pore structures. Their study demonstrated that introducing epoxy cross-linkers with varying functionalities enabled tunable pore sizes ranging from 7 to 46 μm (Fig. 1.5a), while also enhancing pore uniformity and network stability [75]. Similarly, Shang et al. developed a thermosensitive hydrogel using CMC and N-vinylcaprolactam (NVCL) as base materials, and acrylamide (AM) as a cross-linker [76]. SEM characterization (Fig. 1.5b) showed that as the AM content increased, the number of pores in the hydrogel also increased. These results indicate that rational design of the cross-linking system is effective in optimizing the network architecture. Secondly, the fabrication method plays a critical role in determining pore morphology. Among these, freeze-drying is a widely used technique that generates multiscale, continuous porous networks through ice crystal formation and sublimation. Freezing temperature, freezing rate, and solution composition significantly affect the final pore size distribution. For instance, Chen et al. utilized multiple characterizations to show that unidirectional freezing yielded ordered pores ranging from tens to hundreds of nm, whereas non-directional freezing results in more random pore distributions [77].

The incorporation of nanofillers can also significantly enhance pore network structures and related properties. CNFs and CNCs not only improve skeletal rigidity but also promote multiscale pore interconnectivity via hydrogen bonding and electrostatic interactions [78]. In the PVA/CNC hydrogel system developed by Hang et al., the addition of CNCs facilitated the formation of a nanoscale three-dimensional

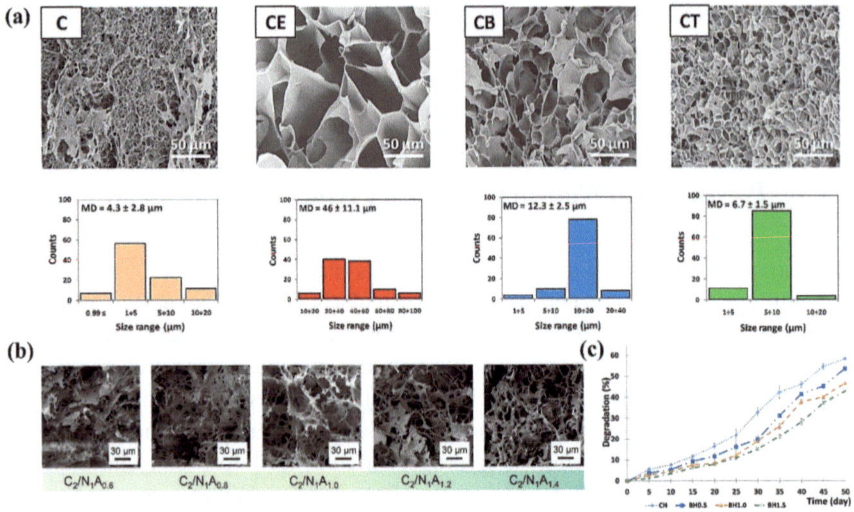

Fig. 1.5 Pore structure and biodegradability of CBHs: **a** SEM images and pore size distribution of hydrogels with different crosslinkers. Reproduced with the permission from Ref. [75]. **b** SEM images of hydrogels with different AM addition amounts. Reproduced with the permission from Ref. [76]. **c** Degradation rate of hydrogel in soil. Reproduced with the permission from Ref. [81]

porous network [79]. This porosity not only substantially increased the specific surface area but also reinforced the network's framework through strong hydrogen bonding between CNC and PVA hydroxyl groups, resulting in more stable pore walls and reduced risk of collapse or closure.

Pore structure is also influenced by the solvent and regeneration system. Common solvent systems such as alkali/urea, LiCl/DMAc, BmimCl ionic liquids, and hydrothermal systems affect cellulose solubility and molecular weight retention, and also govern gel nucleation and phase separation behaviors, ultimately influencing pore size, connectivity, and network compactness [80]. Zhang et al. utilized a deep eutectic solvent composed of ChCl and DL-lactic acid to dissolve ball-milled microcrystalline cellulose and fabricated hydrogels via dropwise coagulation [45]. The resulting regenerated cellulose exhibited sheet- or spherical-like morphology and formed a multiscale porous structure. Compared with hydrogels fabricated directly from ball-milled cellulose, this composite system featured a broader pore size distribution, improved structural stability, and approximately 37% enhancement in compressive mechanical strength. These findings highlight that both solvent system and regeneration method can simultaneously improve pore architecture and overall performance.

1.4.4 Biodegradability and Stability

The biodegradability and environmental stability of CBHs are critical indicators for evaluating their long-term applicability across various fields. These properties are jointly influenced by multiple factors, including chemical composition, cross-linking strategy, nanofiller incorporation, and pore architecture. On the other hand, the cross-linking method and density directly influence the enzymatic permeability of the hydrogel network. Non-degradable epoxy-based or highly dense chemical cross-linking may enhance structural stability but significantly slow down degradation.

In contrast, using enzyme-recognizable crosslinkers such as citric acid or gluconic acid, or employing physical methods like freeze–thaw cycling and ionic crosslinking, can yield structures that balance stability with controllable degradability. For example, Ketheeswaran et al. incorporated biochar as a nano-reinforcement into cellulose hydrogels and conducted a 50-day soil burial degradation test [81]. As shown in Fig. 1.5c, compared with the control group without biochar, the composite CBH exhibited approximately 20% less mass loss, maintaining degradation levels between 42.9 and 58.3%. It also showed improved water retention, pore wall integrity, and mechanical durability, indicating that nanofillers not only strengthen the structure but also significantly slow degradation.

Additionally, pore connectivity and macro-distribution influence degradation pathways: highly interconnected pore networks facilitate enzyme and microbial infiltration, thereby accelerating degradation. Therefore, constructing gradient pore structures via freeze-drying or phase separation techniques allows the inner regions to degrade first while maintaining the integrity of the highly crosslinked outer layers, thereby extending service life and reducing environmental accumulation. Through such design strategies, cellulose hydrogels become highly suitable for applications such as water-retaining and controlled-release agents in agriculture, biomedical scaffolds, wound dressings, and environmental adsorbents.

1.5 Conclusion and Outlooks

This chapter reviewed the sources of raw materials, structural characteristics, and synthesis strategies of CBHs, laying a theoretical foundation for the in-depth discussions in subsequent chapters. As a naturally abundant, structurally diverse, and highly functionalized biopolymer, cellulose plays a central role in hydrogel construction. Its significance lies not only in its excellent biocompatibility and environmental degradability but also in its capacity for structural modulation and functional integration through various physical and chemical approaches. Despite the considerable potential of CBHs in biomedical materials, environmental remediation, and agricultural carriers, their broader development and practical application are still hindered by limitations such as insufficient stability, poor mechanical performance, and immature scalable fabrication techniques. In particular, the choice of synthesis method remains a challenge: although different crosslinking approaches offer distinct advantages in terms of structural robustness, tunability, and sustainability, each comes with inherent trade-offs. Therefore, rational selection and design optimization tailored to specific application scenarios are essential.

Despite the numerous advantages of CBHs, several challenges hinder their practical applications, limiting their industrial scalability and functional expansion. First, the poor solubility of cellulose poses a major technical barrier to CBH fabrication. Extensive hydrogen bonding between cellulose chains makes it insoluble in conventional solvents. Although various solvent systems have been developed, many are costly and environmentally hazardous, restricting large-scale applications. Therefore, the development of green, cost-effective, and efficient dissolution and modification strategies remains a key research priority. Second, CBHs often exhibit insufficient mechanical strength and stability, particularly under high water content, where swelling, rupture, or failure may occur, compromising their long-term performance in complex environments. While the incorporation of physical/chemical cross-linkers or double-network architectures can significantly enhance their mechanical properties, further research is needed to optimize cross-linking strategies that balance mechanical robustness and biocompatibility.

References

1. K. El Bourakadi, F.Z. Semlali, M. Hammi, M. El Achaby, A review on natural cellulose fiber applications: empowering industry with sustainable solutions. Int. J. Biol. Macromol. **281**, 135773 (2024)
2. Y.X. Wang, J.J. Qi, M. Zhang, T. Xu, C.Y. Zheng, Z.H. Yuan, C.L. Si, Cellulose-based aerogels, films, and fibers for advanced biomedical applications. Chem. Eng. J. **497**, 154434 (2024)
3. T. Aziz, W.L. Li, J.G. Zhu, B.B. Chen, Developing multifunctional cellulose derivatives for environmental and biomedical applications: insights into modification processes and advanced material properties. Int. J. Biol. Macromol. **278**, 134695 (2024)
4. Y.H. Ye, L. Yu, E. Lizundia, Y.L. Zhu, C.J. Chen, F. Jiang, Cellulose-based ionic conductor: an emerging material toward sustainable devices. Chem. Rev. **123**, 9204–9264 (2023)
5. W. Zhang, X.X. Ji, M.G. Ma, Emerging MXene/cellulose composites: design strategies and diverse applications. Chem. Eng. J. **458**, 141402 (2023)
6. H. Nasution, H. Harahap, N. F. Dalimunthe, M. H. S. Ginting, M. Jaafar, O. O. H. Tan, H. K. Aruan, A. L. Herfananda, Hydrogel and effects of crosslinking agent on cellulose-based hydrogels: a review. Gels **8** (2022)
7. B.M. Tofanica, A. Mikhailidi, C. Samuil, O.C. Ungureanu, M.E. Fortuna, E. Ungureanu, Advances in cellulose-based hydrogels: current trends and challenges. Gels **10**, 842 (2024)
8. Y. Liu, J.P. Wang, H.Y. Chen, D.D. Cheng, Environmentally friendly hydrogel: a review of classification, preparation and application in agriculture. Sci. Total. Environ. **846**, 157303 (2022)
9. H. Zhu, Z.R. Han, J.H. Cheng, D.W. Sun, Modification of cellulose from sugarcane (Saccharum officinarum) bagasse pulp by cold plasma: dissolution, structure and surface chemistry analysis. Food Chem. **374**, 131675 (2022)
10. J. Wei, H.S. Gao, Y. Li, Y. Nie, Research on the degradation behaviors of wood pulp cellulose in ionic liquids. J. Mol. Liq. **356**, 119071 (2022)

11. C. Verma, V. Singh, A. Alfantazi, Cellulose, cellulose derivatives and cellulose composites in sustainable corrosion protection: challenges and opportunities. Phys. Chem. Chem. Phys. **26**, 11217–11242 (2024)
12. H.Y. Liu, Y. Hu, Y.Y. Liu, R. Hu, X.P. Wu, B. Li, A review of recent advances in biomedical applications of smart cellulose-based hydrogels. Int. J. Biol. Macromol. **253**, 127149 (2023)
13. B.B. Mondal, R. Banik, S. Ghosh, Detailed physicochemical study and thermodynamic aspects of the interaction between nonionic cellulose derivative hydroxyethyl cellulose and anionic surfactant sodium N-dodecanoyl sarcosinate in aqueous media. J. Taiwan Inst. Chem. Eng. **149**, 104982 (2023)
14. A. Amanzholkyzy, S. Zhumagaliyeva, N. Sultanova, Z. Abilov, D. Ongalbek, E. Donbayeva, A. Niyazbekova, Z. Mukazhanova, Hydrogel delivery systems for biological active substances: properties and the role of HPMC as a carrier. Molecules **30**, 1354 (2025)
15. M. He, Y. M. Lin, Y. J. Huang, Y. H. Fang, X. P. Xiong, Research progress of the preparation of cellulose ethers and their applications: a short review. Molecules **30** (2025)
16. C. Chen, Y.W. Xi, Y.X. Weng, Recent advances in cellulose-based hydrogels for tissue engineering applications. Polymers **14**, 3335 (2022)
17. I.G.R. da Silva, B.T.D. Pantoja, G.H.D.R. Almeida, A.C.O. Carreira, M.A. Miglino, Bacterial cellulose and ECM hydrogels: an innovative approach for cardiovascular regenerative medicine. Int. J. Mol. Sci. **23**, 3955 (2022)
18. M. Janmohammadi, Z. Nazemi, A.O.M. Salehi, A. Seyfoori, J. John, M.S. Nourbakhsh, M. Akbari, Cellulose-based composite scaffolds for bone tissue engineering and localized drug delivery. Bioact. Mater. **20**, 137–163 (2023)
19. A. A. Aslam, S. Ul Hassan, M. H. Saeed, O. Kokab, Z. Ali, M. S. Nazir, W. Siddiqi, A. A. Aslam, Cellulose-based adsorbent materials for water remediation: Harnessing their potential in heavy metals and dyes removal. J. Cleaner Prod. **421**, 138555 (2023)
20. L. Yang, H. Wang, Y.N. Yang, Y.P. Li, Self-healing cellulose-based hydrogels: from molecular design to multifarious applications. Carbohydr. Polym. **347**, 122738 (2025)
21. J.O. Cifuente, C. Colleoni, R. Kalscheuer, M.E. Guerin, Architecture, function, regulation, and evolution of α-glucans metabolic enzymes in prokaryotes. Chem. Rev. **124**, 4863–4934 (2024)
22. I. Alvarez-Martínez, F. Pfrengle, On the structure, conformation and reactivity of β-1,4-linked plant cell wall glycans: why are xylan polysaccharides or furanosyl substituents easier to hydrolyze than cellulose? Cellulose **32**, 2145–2165 (2025)
23. Z. Zhang, N. Abidi, L. Lucia, S. Chabi, C.T. Denny, P. Parajuli, S.S. Rumi, Cellulose/nanocellulose superabsorbent hydrogels as a sustainable platform for materials applications: a mini-review and perspective. Carbohydr. Polym. **299**, 120140 (2023)
24. M.D. Li, F.J. Wang, S.Q. Ouyang, Y.C. Liu, Z.H. Hu, Y.T. Wu, J. Qian, Z.H. Li, L.Q. Wang, S.F. Ma, A comprehensive review on preparation and functional application of the wood aerogel with natural cellulose framework. Int. J. Biol. Macromol. **275**, 133340 (2024)
25. A. Sadeghianmaryan, N. Ahmadian, S. Wheatley, H.A. Sardroud, S.A.S. Nasrollah, E. Naseri, A. Ahmadi, Advancements in 3D-printable polysaccharides, proteins, and synthetic polymers for wound dressing and skin scaffolding—a review. Int. J. Biol. Macromol. **266**, 131207 (2024)
26. M.N. Norizan, S.S. Shazleen, A.H. Alias, F.A. Sabaruddin, M.R.M. Asyraf, E.S. Zainudin, N. Abdullah, M.S. Samsudin, S.H. Kamarudin, M.N.F. Norrrahim, Nanocellulose-based nanocomposites for sustainable applications: a review. Nanomaterials **12**, 3483 (2022)
27. K. Yu, L. Yang, N. Zhang, S. Wang, H. Liu, Development of nanocellulose hydrogels for application in the food and biomedical industries: a review. Int. J. Biol. Macromol. **272**, 132668 (2024)
28. Y.E. Qi, Y.Z. Guo, A.A. Liza, G.H. Yang, M.H. Sipponen, J.Q. Guo, H.M. Li, Nanocellulose: a review on preparation routes and applications in functional materials. Cellulose **30**, 4115–4147 (2023)
29. S.P. Bangar, M.M. Harussani, R.A. Ilyas, A.O. Ashogbon, A. Singh, M. Trif, S.M. Jafari, Surface modifications of cellulose nanocrystals: processes, properties, and applications. Food Hydrocolloids **130**, 107689 (2022)

30. Y. Deng, J. Xi, L. Meng, Y. Lou, F. Seidi, W. Wu, H. Xiao, Stimuli-responsive nanocellulose hydrogels: an overview. Eur. Polym. J. **180**, 111591 (2022)
31. Z. Wang, M. Zhu, J. Li, C. Hu, J. Li, R. Xiong, C. Huang, Nanocellulose based hydrogel for flexible sensors: current progress and future perspective. Nano Energy **129**, 109974 (2024)
32. P. Rahmani, A. Shojaei, T. Sakorikar, M. X. Wang, Y. Mendoza Apodaca, M. D. Dickey, Liquid metal nanoparticles physically hybridized with cellulose nanocrystals initiate and toughen hydrogels with piezoionic properties. ACS Nano **18**, 8038–8050 (2024)
33. Y. Li, P. Ren, Z. Sun, R. Xue, D. Ding, W. Tian, F. Ren, Y. Jin, Z. Chen, G. Zhu, High-strength, anti-fatigue, cellulose nanofiber reinforced polyvinyl alcohol based ionic conductive hydrogels for flexible strain/pressure sensors and triboelectric nanogenerators. J. Colloid Interface Sci. **669**, 248–257 (2024)
34. B. Zhang, W. Duan, Y. Wang, L. Dai, B. Cai, L. Kong, J. Fan, G. Zhang, L. Wang, W. Wu, Recent advances of cellulose nanofiber-based materials in cell culture: from population to single-cell, TrAC. Trends Anal. Chem. **166**, 117159 (2023)
35. J.K. Muiruri, J.C.C. Yeo, Q. Zhu, E. Ye, X.J. Loh, Z. Li, Bacterial cellulose: recent advances in biosynthesis, functionalization strategies and emerging applications. Eur. Polym. J. **199**, 112446 (2023)
36. J. Jankau, A. Błażyńska Spychalska, K. Kubiak, M. Jędrzejczak Krzepkowska, T. Pankiewicz, K. Ludwicka, A. Dettlaff, R. Pęksa, Bacterial cellulose properties fulfilling requirements for a biomaterial of choice in reconstructive surgery and wound healing. Front. Bioeng. Biotechnol **9**, 805053 (2022)
37. T.B. Liu, Q.Q. Ren, S. Wang, J.N. Gao, C.C. Shen, S.Y. Zhang, Y.H. Wang, F. Guan, Chemical modification of polysaccharides: a review of synthetic approaches, biological activity and the structure-activity relationship. Molecules **28**, 6073 (2023)
38. A.Z. Najihah, M.Z. Hassan, Z. Ismail, Current trend on preparation, characterization and biomedical applications of natural polysaccharide-based nanomaterial reinforcement hydrogels: a review. Int. J. Biol. Macromol. **271**, 132411 (2024)
39. Z. Ahmadian, F. Kazeminava, M. Afrouz, M. Abbaszadeh, N.T. Mehr, J.A. Shiran, C. Gouda, M. Adeli, H.S. Kafil, A review on the impacts of metal/metal nanoparticles on characteristics of hydrogels: special focus on carbohydrate polymers. Int. J. Biol. Macromol. **253**, 126535 (2023)
40. Y.J. Deng, T.X. Zhu, Y. Cheng, K. Zhao, Z.Y. Meng, J.Y. Huang, W.L. Cai, Y.K. Lai, Recent advances in functional cellulose-based materials: classification, properties, and applications. Adv. Fiber Mater. **6**, 1343–1368 (2024)
41. V. Kokol, Influence of hydroxyethyl and carboxymethyl celluloses on the rheology, water retention and surface tension of water-suspended microfibrillated cellulose. Cellulose **29**, 7063–7081 (2022)
42. S.M. Wang, Y.X. Xie, H.H. Su, Y.R. Luo, M.T. Wang, T. Li, Y.Y. Fu, Delivery of curcumin in a carboxymethyl cellulose and hydroxypropyl methyl cellulose carrier: physicochemical properties and biological activity. Int. J. Biol. Macromol. **239**, 124203 (2023)
43. J.D.D. Carvalho, R.S. Rabelo, M.D. Hubinger, Thermo-rheological properties of chitosan hydrogels with hydroxypropyl methylcellulose and methylcellulose. Int. J. Biol. Macromol. **209**, 367–375 (2022)
44. L.H. Chen, P.S. Doyle, Thermogelling hydroxypropyl methylcellulose nanoemulsions as templates to formulate poorly water-soluble drugs into oral thin films containing drug nanoparticles. Chem. Mater. **34**, 5194–5205 (2022)
45. Y. Zhang, Y.M. Zhu, K.J. Wu, Y.Y. Liu, H.F. Lu, B. Liang, Preparation of hydrogels with self-reinforced mechanical properties using ball-milled microcrystalline cellulose and regenerated cellulose from deep eutectic solvent. New J. Chem. **48**, 7405–7412 (2024)
46. S. Swingler, A. Gupta, H. Gibson, M. Kowalczuk, W. Heaselgrave, I. Radecka, Recent advances and applications of bacterial cellulose in biomedicine. Polymers **13**, 412 (2021)
47. X. Li, C. Wan, T. Tao, H. Chai, Q. Huang, Y. Chai, Y. Wu, An overview of the development status and applications of cellulose-based functional materials. Cellulose **31**, 61–99 (2024)

48. P. Bishnoi, S.S. Siwal, V. Kumar, V.K. Thakur, Cellulose-based smart materials: novel synthesis techniques, properties, and applications in energy storage and conversion devices. Electron **2**, e42 (2024)
49. J.Y. Yang, Y. Chen, L. Zhao, J.H. Zhang, H. Luo, Constructions and properties of physically cross-linked hydrogels based on natural polymers. Polym. Rev. **63**, 574–612 (2023)
50. H.J. Chen, P. Wei, Y. Qi, Y.H. Xie, X.L. Huang, Water-induced cellulose nanofibers/poly(vinyl alcohol) hydrogels regulated by hydrogen bonding for In situ water shutoff. ACS Appl. Mater. Interfaces **15**, 39883–39895 (2023)
51. H.X. Wan, Y. Chen, Y.Z. Tao, P. Chen, S. Wang, X.Y. Jiang, A. Lu, MXene-mediated cellulose conductive hydrogel with ultrastretchability and self-healing ability. ACS Nano **17**, 20699–20710 (2023)
52. T. Benselfelt, G.C. Ciftci, L. Wågberg, J. Wohlert, M.M. Hamedi, Entropy drives interpolymer association in water: Insights into molecular mechanisms. Langmuir **40**, 6718–6729 (2024)
53. H. Hu, F.J. Xu, Rational design and latest advances of polysaccharide-based hydrogels for wound healing. Biomater. Sci. **8**, 2084–2101 (2020)
54. K. Ghosal, S.K. Bhattacharyya, V. Mishra, H. Zuilhof, Click chemistry for biofunctional polymers: from observing to steering cell behaviour. Chem. Rev. **124**, 13216–13300 (2024)
55. T. Luu, K. Gristwood, J.C. Knight, M. Jörg, Click chemistry: reaction rates and their suitability for biomedical applications. Bioconjugate Chem. **35**, 715–731 (2024)
56. S. Luleburgaz, E. Cakmakci, H. Durmaz, U. Tunca, Sustainable polymers from renewable resources through click and multicomponent reactions. Eur. Polym. J. **209**, 112897 (2024)
57. L.M. Grilo, S. Faoro, R. Folkersma, T.M. Lacerda, L. Mazzocchetti, K. Loos, D. Maniar, Biobased networks from Llignin/cellulose via diels-alder click chemistry. ACS Appl. Polym. Mater. **6**, 13723–13734 (2024)
58. M.M. El Sayed, Production of polymer hydrogel composites and their applications. J. Polym. Environ. **31**, 2855–2879 (2023)
59. K.M. Salleh, S. Zakaria, N.A.Z. Armir, M.K. Khairunnisa-Atiqah, B.B. Wang, Electrovalent effects of sodium carboxymethyl cellulose and hydroxyethyl cellulose on regeneration of empty fruit bunch cellulose to a superabsorbent hydrogel. Int. J. Biol. Macromol. **278**, 134816 (2024)
60. R. Blazic, K. Marusic, E. Vidovic, Swelling and viscoelastic properties of cellulose-based hydrogels prepared by free radical polymerization of dimethylaminoethyl methacrylate in cellulose solution. Gels **9**, 94 (2023)
61. N. S. V. Capanema, A. A. P. Mansur, I. C. Carvalho, S. M. Carvalho, H. S. Mansur, Bioengineered water-responsive carboxymethyl cellulose/poly(vinyl alcohol) hydrogel hybrids for wound dressing and skin tissue engineering applications. Gels **9** (2023)
62. B.A. Masry, H.M. Gayed, J.A. Daoud, Gamma radiation synthesis of hydroxyethyl cellulose/acrylic acid/CYANEX 471X hydrogel for silver ions capture from acidic nitrate medium. Cellulose **31**, 4329–4346 (2024)
63. J.W. Gong, L.L. Hou, Y.C. Ching, K.Y. Ching, N.D. Hai, C.H. Chuah, A review of recent advances of cellulose-based intelligent-responsive hydrogels as vehicles for controllable drug delivery system. Int. J. Biol. Macromol. **264**, 130525 (2024)
64. S. Grabska Zielinska, Cross-linking agents in three-component materials dedicated to biomedical applications: a review. Polymers **16**, 2679 (2024)
65. J. Wang, T.L. Song, H.X. Chen, W. Ming, Z.M. Cheng, J.W. Liu, B.L. Liang, Y.T. Wang, G.S. Wang, Bioinspired high-strength montmorillonite-alginate hybrid film: the effect of different divalent metal cation crosslinking. Polymers **14**, 2433 (2022)
66. H.Y. Pan, M.D. Tong, X.M. Wang, B.X. Huang, X.Y. Yu, C.Q. Zhang, Y. Wang, Fully biobased high-strength and high-toughness double cross-linked cellulose hydrogel for flexible electrolytes. ACS Sustain Chem. Eng. **12**, 18231–18244 (2024)
67. F. Baraka, A. Morales, I. Velazco-Cabral, I. Rivilla, J. Labidi, Physically cross-linked cellulose nanofiber (LCNF/CNF) hydrogels: impact of the composition on mechanical and swelling properties. Adv. Compos. Hybrid Mater. **8**, 1–11 (2025)
68. A. Etale, A.J. Onyianta, S.R. Turner, S.J. Eichhorn, Cellulose: a review of water interactions, applications in composites, and water treatment. Chem. Rev. **123**, 2016–2048 (2023)

69. D.K. Li, W. Zhan, W. Zuo, L.P. Li, J. Zhang, G.Y. Cai, Y. Tian, Elastic, tough and switchable swelling hydrogels with high entanglements and low crosslinks for water remediation. Chem. Eng. J. **450**, 138417 (2022)
70. R. Foudazi, R. Zowada, I. Manas Zloczower, D. L. Feke, Porous hydrogels: present challenges and future opportunities. Langmuir **39**, 2092–2111 (2023)
71. K. Nyamayaro, T. Iwase, S.G. Hatzikiriakos, P. Mehrkhodavandi, Cellulose nanocrystal-mediated enhancement of hydrogel anti-swelling and water retention. RSC Sustain **2**, 1543–1550 (2024)
72. A. A. Abdelgelil, A. M. Omer, A. F. Hassan, A. A. Moustafa, M. S. M. Eldin, Enhancement of sandy soil water retention using superabsorbent carboxymethyl cellulose grafted with polyacrylamide and polyacrylamidomethyl propanesulfonic acid copolymer. Sci. Rep. **15** (2025)
73. L. Li, H. Wei, Y.Z. Hao, Y.Z. Li, W. Cheng, Y.A. Ismail, Z.Z. Liu, Carbon nanotube (CNT) reinforced cementitious composites for structural self-sensing purpose: a review. Constr. Build. Mater. **392**, 131384 (2023)
74. Q. Mo, L.J. Huang, Y. Sheng, Z.H. Wei, S.Y. Zhang, Y.S. Li, X.Y. Wang, Y.N. Wang, X.Y. Lu, C.X. Huang, Q.S. Duan, M.G. Xue, Crosslinking strategy and promotion role of cellulose as a composite hydrogel component for three-dimensional printing—a review. Food Hydrocoll. **154**, 110079 (2024)
75. R. Nicu, G. Lisa, R.N. Darie-Nita, M.I. Avadanei, A. Bargan, D. Rusu, D.E. Ciolacu, Tailoring the structure and physico-chemical features of cellulose-based hydrogels using multi-epoxy crosslinking agents. Gels **10**, 523 (2024)
76. H.Y. Shang, X.X. Yang, H. Liu, Temperature-responsive hydrogel prepared from carboxymethyl cellulose-stabilized N-vinylcaprolactam with potential for fertilizer delivery. Carbohydr. Polym. **313**, 120875 (2023)
77. J. Grenier, H. Duval, P. Lv, F. Barou, C. Le Guilcher, R. Aid, B. David, D. Letourneur, Interplay between crosslinking and ice nucleation controls the porous structure of freeze-dried hydrogel scaffolds. Biomater. Adv. **139**, 212973 (2022)
78. Y. Altynov, K. Bexeitova, M. Nazhipkyzy, S. Azat, A. Konarov, D. Rakhman, N. Sahiner, K. Kudaibergenov, Nanocellulose hydrogels from agricultural wastes: methods, properties, and application prospects. Nanoscale **17**, 12580–12619 (2025)
79. X.M. Huang, X. Ao, L.H. Yang, J. Ye, C.W. Wang, Preparation and properties of cellulose nanocrystal-based ion-conductive hydrogels. RSC Adv. **13**, 527–533 (2022)
80. H. Nawaz, A.Y. He, Z. Wu, X.Y. Wang, Y.T. Jiang, A. Ullah, F. Xu, F.W. Xie, Revisiting various mechanistic approaches for cellulose dissolution in different solvent systems: a comprehensive review. Int. J. Biol. Macromol. **273**, 133012 (2024)
81. K. Ketheeswaran, S. Shetranjiwalla, M. Krishnapillai, L. Galagedara, Incorporating biochar to make hydrogel composites with improved structural properties, valorized from waste-paper mill sludge and forestry residues using energy efficient protocols. RSC Sustain. **2**, 3478–3489 (2024)

Chapter 2
Functional Regulation of Cellulose-Based Hydrogels by Organic Materials

Haonan Dai and Gang Wei

Abstract Organic materials play a pivotal role in modulating the structure and properties of cellulose-based hydrogels (CBHs), emerging as a key strategy to enhance their functional diversity and adaptability across a wide range of applications. The incorporation of polysaccharides, synthetic polymers, and functional organic small molecules facilitates the formation of multi-network structures and interfacial synergies, significantly improving the mechanical strength, environmental stability, and stimuli-responsive behavior of hydrogels. These organic components interact with CBH matrix through hydrogen bonding, covalent bonding, hydrophilic/hydrophobic modulation, and active molecule encapsulation, thereby endowing the CBH system with multifunctional properties such as controlled release, improved biocompatibility, and environmental responsiveness. A deeper understanding of these functional regulation mechanisms not only drives practical advancements in biomedicine, environmental remediation, and intelligent sensing, but also provides essential theoretical insights for green synthesis and multiscale structural design of CBHs. Consequently, organic materials have become indispensable functional elements in the development of high-performance CBH systems.

Keywords Cellulose · Hydrogels · Organic materials · Functional regulation · Composite materials

2.1 Introduction

Hydrogels, known for their high water-content, excellent biocompatibility, and tunable three-dimensional (3D) structures, hold significant promise for applications in biomedicine, environmental remediation, sensing technologies, and

H. Dai
College of Chemistry and Chemical Engineering, Qingdao University, Qingdao, China

G. Wei (✉)
School of Polymer Science and Engineering, Qingdao University of Science and Technology, Qingdao, China
e-mail: wei-lab@qust.edu.cn

G. Wei and J. Zhang (eds.), *Cellulose-Based Hydrogels and Aerogels*, Lecture Notes in Nanoscale Science and Technology 41,
https://doi.org/10.1007/978-3-032-03008-5_2

smart responsive systems. In recent years, cellulose-based hydrogels (CBHs) have garnered increasing attention as a key research focus owing to their natural abundance, renewability, and low toxicity [1]. CBHs can be formed through physical or chemical cross-linking of natural cellulose or its derivatives, such as carboxymethyl cellulose, hydroxypropyl methylcellulose, and nanocellulose [2]. They not only exhibit excellent gelation capacity and mechanical flexibility, but also possess inherent biodegradability and hydrophilicity, rendering them highly suitable for applications such as tissue engineering, drug delivery, and wound dressings. Compared to synthetic polymer hydrogels, CBHs offer notable advantages in terms of environmental friendliness and sustainability [3]. However, their intrinsic structural characteristics also impose certain limitations, including relatively low mechanical strength, limited responsiveness, and insufficient sensitivity to external stimuli such as temperature, pH, and light [4]. To overcome these challenges, researchers have explored various strategies, including physical and chemical modifications and composite technologies, to introduce additional functionalities. Furthermore, the microscopic structural differences arising from diverse cellulose sources, such as microbial cellulose, plant-derived cellulose, and cellulose extracted from agricultural waste, offer expanded possibilities for tailoring material performance.

Functional regulation has become a central focus of current hydrogel research, particularly in the development of high-performance biomaterials and smart materials, where functional requirements are increasingly critical. Owing to the abundance of reactive groups, such as hydroxyl and carboxyl groups, in its molecular structure, the CBH matrix provides an excellent platform for chemical modification and functionalization. Sumini et al. proposed a strategy in which hyaluronic acid and silver nanoparticles were introduced into cellulose hydrogels to enhance antibacterial performance and drug release behavior, achieving significant inhibition of *Pseudomonas aeruginosa* [5]. Therefore, functional regulation not only serves as an effective approach to overcoming performance limitations but also contributes to improving structural integrity and biocompatibility, underscoring the potential of CBH composites in achieving multifunctionality.

Among various strategies for functional regulation, the incorporation of organic functional materials is regarded as both efficient and versatile. These materials include small molecule modifiers, functional polymers, biomacromolecules, and natural organic substances. They interact with the cellulose backbone through covalent bonding, non-covalent interactions, or coordination mechanisms to form functional networks [6]. On one hand, organic materials can impart novel functionalities such as antibacterial activity, conductivity, fluorescence, and pH responsiveness to hydrogels. On the other hand, their involvement in directing network architecture can improve structural stability and smart responsiveness. For example, Shi et al. developed a cellulose-based dual-network conductive hydrogel by introducing hydrophobic modifications and forming Schiff base cross-linking with gelatin. The resulting hydrogel exhibited excellent anti-swelling behavior, permeability, freeze–thaw stability, and Na^+-induced conductivity and sensitivity, demonstrating promising applications in flexible sensors [7]. Nevertheless, the incorporation of organic materials may also introduce challenges, including interfacial incompatibility, harsh reaction conditions, and reduced biocompatibility, which must be overcome through rational design and modulation strategies. Overall, organic materials not only broaden the functional scope of CBHs but also enhance their adaptability in complex environments [8].

This chapter presents the advance in the functional regulation of CBHs with various organic components, and summarizes the corresponding interactions and action mechanisms between cellulose hydrogels and organic molecules.

2.2 Types of Organic Materials for Functional Regulation of CBHs

CBHs have been extensively applied in drug delivery, wound dressings, and tissue engineering owing to their excellent biocompatibility, biodegradability, and mechanical properties. However, single-component cellulose hydrogels often exhibit limited functionality and suboptimal performance. To address these limitations, researchers have investigated the incorporation of various organic materials to enhance multifunctionality and expand application potential. This section introduces different categories of organic materials and their applications in CBHs, including natural polymers, synthetic polymers, small organic molecules, conductive organic materials, as well as biomimetic and peptide-based modifiers.

2.2.1 Natural Polymers

Natural polymers are highly regarded for their excellent biocompatibility, biodegradability, and non-toxicity, playing a pivotal role in the functional regulation of CBHs. Natural polysaccharides such as chitosan, alginate, gelatin, and silk fibroin can markedly enhance mechanical strength, degradability, and bioactivity when composited with cellulose, rendering them particularly suitable for applications in biomedical and environmental fields.

Chitosan, a natural polysaccharide derived from crustacean shells, is renowned for its excellent biocompatibility, antibacterial properties, and capacity to promote cell proliferation. Liu et al. developed a multifunctional hydrogel based on quaternized chitosan (QCS) and tunicate cellulose nanocrystals (TCNC), forming a network structure (Fig. 2.1a) that enabled pH-responsive drug release and enhanced wound healing in full-thickness skin infection models [9]. Deng et al. designed a bio-based hydrogel composed of QCS and dialdehyde bacterial cellulose (BC) for wound healing. Owing to dynamic Schiff base interactions, the hydrogel exhibited rapid self-healing and injectability, coupled with excellent antibacterial activity against *E. coli* and *S. aureus* [10]. These chitosan-based hydrogels demonstrated self-healing, antibacterial efficacy, and cost-effectiveness, making them promising candidates for wound repair applications.

Alginate is a naturally derived polysaccharide obtained from seaweed, notable for its excellent gelation ability and biocompatibility. When combined with cellulose, alginate enhances the plasticity and hydration capacity. Georgouvelas et al. prepared cellulose nanofiber (CNF)/alginate composite hydrogel beads with porosity reaching up to 99% [11]. Through in-situ TEMPO-mediated oxidation, hydroxyl groups on cellulose were converted into carboxyl groups, thereby improving the beads' capacity for heavy metal ion removal.

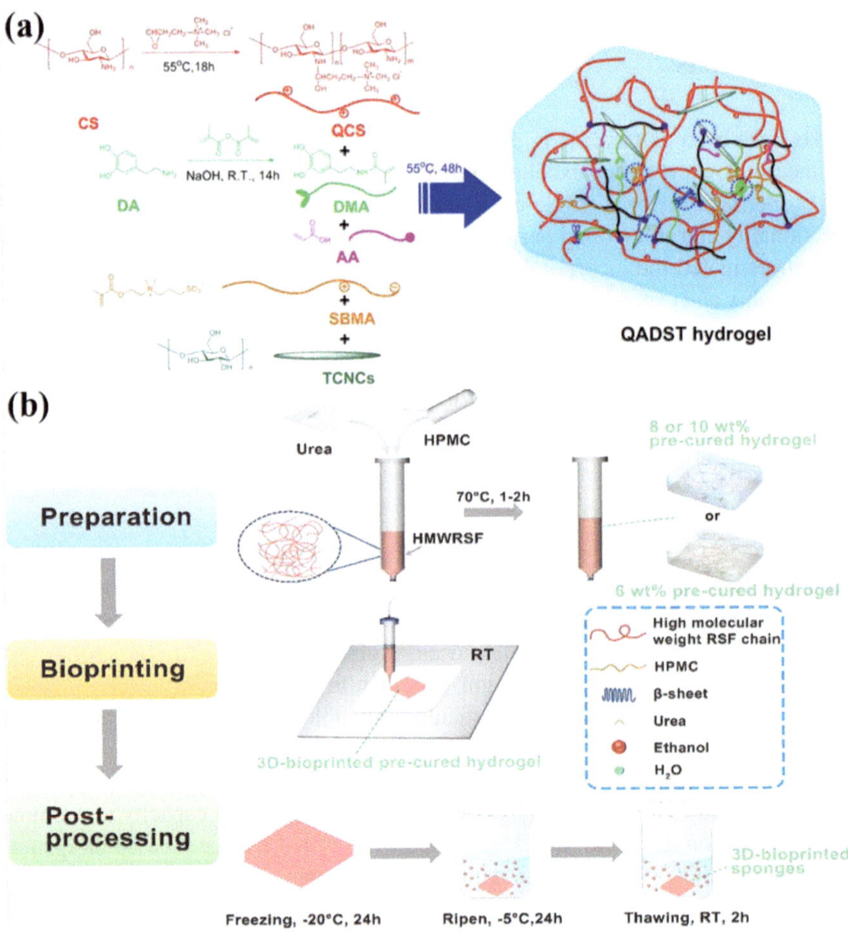

Fig. 2.1 Natural polymers for the regulation of CBHs: **a** Schematic synthesis route of QCS/TCNC hydrogel. Reprinted with the permission from Ref. [9], Copyright 2024, Elsevier Ltd. **b** Schematic illustration of pre-crosslinked HMWRSF-based hydrogels for bioprinting. Reprinted with the permission from Ref. [13], Copyright 2022, Royal Society of Chemistry

Gelatin, a natural protein derived from animal collagen, possesses excellent biocompatibility and biodegradability. When combined with cellulose, it forms hydrogels with enhanced mechanical strength and improved cellular compatibility. Fu et al. developed degradable and recyclable multifunctional strain sensor hydrogels composed of gelatin, CNFs, and Fe^{3+}. The formed CBHs demonstrated outstanding mechanical properties and self-healing capabilities [12].

Silk fibroin, derived from silk, exhibits excellent mechanical properties, biocompatibility, and adjustable degradability. Chen et al. developed a novel regenerated silk fibroin-based printable hydrogel (Fig. 2.1b) by blending high-molecular-weight regenerated silk fibroin (HMWRSF) with a small amount of hydroxypropyl methylcellulose (HPMC) in a urea-containing aqueous solution [13]. The resulting thixotropic hydrogel showed promising potential for bioprinting applications, particularly in tuning mechanical and printing properties.

Natural polymers markedly enhance the performance of CBHs, including mechanical strength, antibacterial activity, and cellular compatibility, and thus hold significant promise for biomedical applications.

2.2.2 *Synthetic Polymers*

Synthetic polymers, characterized by controllable properties and stability, are widely utilized for the functional regulation of CBHs. Common synthetic polymers include polyvinyl alcohol (PVA), polyvinylpyrrolidone (PVP), polyethylene glycol (PEG), and polylactic acid (PLA), which can enhance the properties of CBHs, including mechanical strength, lubricity, degradability, and biocompatibility.

PVA is a water-soluble synthetic polymer renowned for its excellent film-forming ability and biocompatibility. When blended with cellulose, PVA significantly improves mechanical strength and stability. Elsherbiny et al. prepared a self-crosslinked cryogel by combining hydrolyzed PEG and cellulose [14]. The resulting hydrogel demonstrated notable antibacterial and mechanical properties, making it suitable for wound dressing applications.

PVP is a water-soluble synthetic polymer known for its good biocompatibility. PVP-cellulose composite hydrogels have been widely applied in drug delivery systems. Wu et al. developed an injectable, self-healing hydrogel based on PVP and carboxymethyl cellulose (CMC) for drug release [15]. This hydrogel exhibited excellent injectability, biocompatibility, and sustained drug release performance.

PEG is a water-soluble synthetic polymer characterized by low toxicity and good biocompatibility, commonly used in drug delivery and tissue engineering. Getya et al. synthesized PEG-cellulose composite hydrogels using chemically modified cellobiose or cellulose combined with PEG chains of various molecular weights [16]. The resulting hydrogels exhibited distinct morphologies, swelling behaviors, viscoelastic properties, and thermal stability, demonstrating excellent potential for the removal of organic pollutants from water.

PLA is a biodegradable synthetic polymer with favorable biocompatibility and mechanical properties. When combined with cellulose, PLA forms hydrogels with enhanced mechanical strength, suitable for applications in tissue engineering and drug delivery. Yuan et al. synthesized a novel biodegradable composite hydrogel composed of gelatin, chitosan, and PLA for slow-release nitrogen fertilizer [17]. The hydrogel exhibited good mechanical strength and environmental friendliness, highlighting its promise for agricultural applications.

By tailoring molecular structures and functional groups, synthetic polymers enable precise control over the properties of CBHs, offering great potential in biomedicine, particularly in drug delivery and tissue engineering.

2.2.3 Small Organic Molecules

Small organic molecules, including amino acids, vitamins, and other bioactive compounds, are primarily employed in CBHs to modulate their bioactivity and functional properties. These molecules interact with cellulose chains to enhance the hydrogel's bioactivity, promote cell adhesion, and improve tissue repair capabilities.

Amino acids, as fundamental building blocks of peptides and proteins, exhibit diverse biological functions. Studies have demonstrated that amino acids can act as biological gelators, forming thermo-reversible self-assembled hydrogels in combination with natural polysaccharides such as gellan gum. For instance, hydrogels composed of gellan gum and amino acids, including L-lysine, L-arginine, L-aspartic acid, and L-glutamic acid, exhibited approximately 99% antibacterial activity, significantly inhibiting multidrug-resistant bacteria [18]. Hu et al. developed a novel antibacterial hydrogel by combining amino cellulose, dialdehyde xylan, and silver nanoparticles via a green reduction method [19]. The created CBH demonstrated excellent antibacterial performance, particularly against multidrug-resistant bacteria.

Vitamins are organic compounds essential for sustaining life processes, exhibiting antioxidant, anti-inflammatory, and cell growth-promoting properties. Incorporating vitamins into CBHs can enhance their bioactivity and broaden their biomedical applications. Vitamin C, in particular, can serve as a building block for self-assembled nanofiber hydrogels that improve the efficacy of cancer immunotherapy. Zhang et al. conjugated vitamin C with hydrophobic alkyl chains to form amphiphilic molecules capable of self-assembling into injectable nanofiber hydrogels. These hydrogels not only demonstrated self-healing properties but also functioned as efficient delivery platforms for stimulator of interferon genes (STING) agonists, significantly enhancing cancer immunotherapy outcomes [20].

Bioactive compounds, including natural or synthetic antibacterial and anti-inflammatory agents, can be incorporated with cellulose to fabricate bioactive hydrogels. As illustrated in Fig. 2.2a, Yang et al. synthesized a novel pH-responsive hydrogel (RPC/PB) by cross-linking poly(vinyl alcohol)-borax (PB) with resveratrol-grafted cellulose nanofibrils (RPC) [21]. The grafting of bioactive resveratrol onto

CNFs endowed the hydrogel with potent antibacterial and antioxidant activities, significantly enhancing wound healing.

By compounding with cellulose, small molecule organic compounds can modulate the bioactivity and functionality of hydrogels, offering innovative solutions for

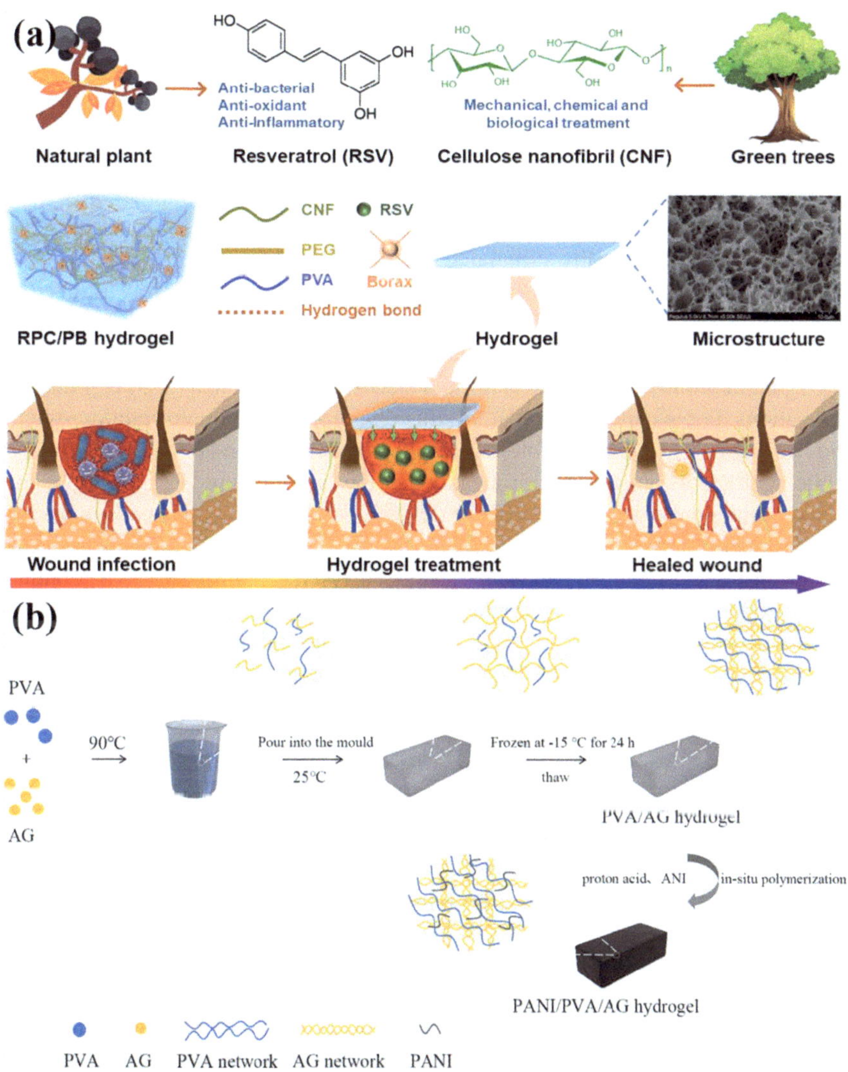

Fig. 2.2 Organic materials for the regulation of CBHs: **a** Synthesis of the RPC/PB hydrogel for wound healing. Reprinted with the permission from Ref. [21], Copyright 2022, Springer. **b** Synthesis of PANI/PVA/AG conductive hydrogel. Reprinted with the permission from Ref. [23], Copyright 2024, Royal Society of Chemistry

biomedical and drug delivery applications. The incorporation of amino acids, vitamins, and other bioactive molecules not only improves hydrogel biocompatibility and antibacterial efficacy but also broadens its potential applications in tissue engineering, wound repair, and cancer therapy.

2.2.4 Conductive Organic Polymers

The application of conductive organic materials in CBHs has seen significant growth, particularly in smart materials, electronic devices, and bioelectronics. By incorporating conductive polymers, the constructed CBHs retain their biocompatibility while acquiring electrical conductivity, thereby broadening their utility in sensors, nerve regeneration, and electrically stimulated tissue engineering.

Polyaniline (PANI), a conductive organic polymer, is extensively utilized in sensors and electronic devices. Recent studies have combined PANI with cellulose to fabricate hydrogels that demonstrate outstanding conductivity and biocompatibility, making them promising for biosensing and electrically stimulated tissue engineering applications. For instance, Wan et al. in situ polymerized PANI onto cellulose/phytic acid hydrogels, producing flexible, stretchable, and highly ion-conductive cellulose/phytic acid/PANI hydrogels with a sandwich-like architecture [22]. These hydrogels were integrated into supercapacitor devices, exhibiting superior performance in flexible supercapacitors and strain sensors, thus highlighting their potential in wearable electronics. Similarly, Fan et al. developed PANI/PVA/agarose (AG) conductive hydrogels by in-situ polymerizing PANI nanofibers within a PVA/AG double-network hydrogel matrix (Fig. 2.2b) [23]. The resulting PANI nanofiber-based hydrogels displayed excellent electrochemical energy storage capacity, photosensitivity, and strain responsiveness, demonstrating their suitability for supercapacitors, flexible sensors, and wearable electronic devices.

PEDOT: PSS is a widely utilized conductive polymer composite renowned for its excellent conductivity and biocompatibility. It has been incorporated with cellulose matrices to fabricate CBHs with enhanced conductivity and biocompatibility, rendering them suitable for applications such as nerve repair and electrically stimulated cellular growth. Zhang et al. developed a novel antifreeze composite hydrogel system based on polyacrylamide (PAM), LiCl, and PEDOT: PSS-coated CNFs (PAM/PEDOT: PSS/CNF) [24]. The incorporation of CNFs enhanced hydrogen bonding interactions within the polymer chains, thereby significantly improving the mechanical strength and conductivity of the hydrogel at the molecular level.

Conductive organic materials endow CBHs with electrical conductivity, thus expanding their applications in bioelectronics and nerve repair, demonstrating considerable scientific and practical significance. Through incorporation with cellulose, conductive polymers such as PANI and PEDOT: PSS not only augment hydrogel conductivity but also improve mechanical properties and biocompatibility, highlighting promising applications in flexible electronics, sensors, and tissue engineering.

2.2.5 *Biomimetic and Peptide-Based Modifiers*

Biomimetic materials and peptide-based molecular modifications have emerged as prominent research focuses in the development of CBHs. The primary advantage of these modifiers lies in their capacity to emulate the structure and function of the natural extracellular matrix (ECM), thereby enhancing the bio-recognition, cell adhesion, and tissue integration properties of hydrogels. Compared to natural and synthetic polymers, peptide-based molecules offer highly tunable functionalities and superior biocompatibility, enabling hydrogels to elicit specific biological responses such as angiogenesis promotion, immunomodulation, and targeted cell homing [25]. For instance, He et al. engineered a double-network CBH by combining oxidized cellulose with alginate functionalized with mesenchymal stem cell (MSC) homing peptide (SKPPGTSS) and cell adhesion peptide (GRGDSP) [26]. This hydrogel exhibited a compressive shape memory effect and effectively enhanced stem cell recruitment and adhesion through the incorporation of specific peptide sequences, thereby facilitating intervertebral disc regeneration.

2.3 Methods of Functional Regulation in CBHs

2.3.1 *Hydrogen Bonding and Electrostatic Interactions*

Functional regulation of CBHs can be effectively achieved through hydrogen bonding and electrostatic interactions. Hydrogen bonding, one of the most prevalent interactions among cellulose molecules, plays a critical role in maintaining the network structure of cellulose hydrogels. By modulating the strength and spatial distribution of hydrogen bonds, key properties such as mechanical strength, swelling behavior, and biocompatibility can be tailored. For example, Zhou et al. incorporated gelatin and glycerol into BC matrices to induce hydrogen bond rearrangement during hydrogel fabrication (Fig. 2.3a). The resulting hydrogel demonstrated a unique combination of high transparency, mechanical robustness, tissue adhesiveness, anti-dehydration capability, biocompatibility, and biodegradability [27].

Electrostatic interactions also significantly contribute to the functional enhancement of hydrogels, particularly when charge modifications are introduced. The incorporation of charged functional groups or the use of oppositely charged components can effectively improve hydrogel biocompatibility and structural stability [28]. For instance, Mihoub et al. fabricated a physically cross-linked CBH by combining cellulose with epichlorohydrin through electrostatic interactions [29]. This CBH was subsequently swollen in a solution containing monomers 2-hydroxyethyl methacrylate and 1,6-hexanediol diacrylate, forming an interpenetrating polymer network utilized for dye removal from aqueous solutions.

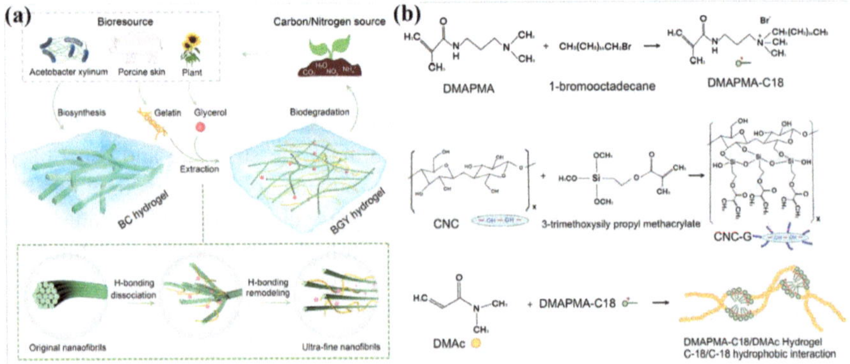

Fig. 2.3 Functional regulation methods of CBHs: **a** Hydrogen bonding-induced synthesis of gelatin and glycerol-modified BC hydrogel. Reprinted with the permission from Ref. [27], Copyright 2023, Elsevier Ltd. **b** Hydrophobic binding for the synthesis of DMAPMA-C18/DMAc hydrogel. Reprinted with the permission from Ref. [39], Copyright 2023, Wiley VCH

However, hydrogen bonding and electrostatic interactions exhibit inherent limitations. The relatively weak nature of hydrogen bonds renders them sensitive to environmental factors such as temperature and pH, which can compromise the stability of the hydrogel's structure and functional properties [30]. Moreover, electrostatic interactions are influenced by ionic strength and solvent conditions, and an excessive charge density or uneven charge distribution may lead to overly strong interactions that hinder molecular release rates [31], thereby diminishing the effectiveness of functional regulation under complex physiological environments.

2.3.2 Covalent Cross-Linking and Graft Modification

Covalent cross-linking and graft modification are pivotal strategies for enhancing the structural stability and functional tunability of CBHs. Compared to hydrogen bonding or electrostatic interactions, covalent bonds confer greater stability, markedly improving mechanical strength and anti-swelling performance under complex environmental conditions. Covalent cross-linking commonly employs cross-linking agents such as glutaraldehyde, EDC/NHS, epichlorohydrin [32], or citric acid [33] to establish permanent 3D networks among cellulose molecules.

Mpouma et al. systematically investigated the influence of glutaraldehyde- and epichlorohydrin-mediated chemical cross-linking of cellulose nanocrystals (CNCs), successfully fabricating multifunctional nanostructured hydrogels with tunable physicochemical properties [34]. Graft modification involves the introduction of functional polymer segments onto the cellulose backbone via chemical reactions including free radical polymerization [35], ring-opening polymerization, or esterification/amidation, thereby imparting stimuli-responsiveness, enhanced bioactivity, or adjustable wettability to the hydrogels. Zhang et al. explored chemical grafting of polylactic acid chains onto CNCs through esterification and ring-opening polymerization, analyzing grafting kinetics under varying temperature, reaction time, and reactant ratio conditions [36].

2.3.3 Hydrophobic/Hydrophilic Interactions

The water absorption and swelling behavior of CBHs are strongly governed by the balance between hydrophobic and hydrophilic groups. Fine-tuning this balance enables synergistic optimization of swelling capacity and mechanical performance. Cellulose inherently possesses abundant hydrophilic hydroxyl groups, which confer excellent water retention and anti-swelling properties to cellulose hydrogels [37].

For instance, Das et al. synthesized nanocomposite hydrogels based on sodium carboxymethyl cellulose (NaCMC) and hydroxyethyl cellulose (HEC), employing citric acid as a cross-linker to enhance mechanical strength and durability for agricultural water management. These hydrogels functioned as water reservoirs, capable of controlled absorption, retention, and release to plants [38]. Conversely, the introduction of hydrophobic moieties promoted interchain hydrophobic interactions that densify the polymer network, thereby improving mechanical toughness and adhesion.

Chen et al. developed physically cross-linked hydrogels by combining a hydrophobic monomer (DMAPMA-C18) with hydrophobic CNCs (CNC-G) via hydrophobic interactions (Fig. 2.3b), resulting in CBHs with notable self-healing and adhesive capabilities [39]. Furthermore, hydrophobic modifications endowed hydrogels with responsiveness to environmental stimuli such as temperature and solvent polarity, facilitating intelligent, controllable swelling and deswelling behavior. Nevertheless, excessive hydrophobization detrimentally affected the hydrophilicity, swelling capacity, and flexibility, which may impair biocompatibility.

Additionally, chemical incompatibility between hydrophobic groups and the cellulose backbone could lead to phase separation and structural heterogeneity. To address this issue, Tohamy et al. grafted both hydrophilic and hydrophobic functional groups to create superabsorbent hydrogels with temperature- and pH-responsive swelling properties, demonstrating excellent water uptake and efficient removal of heavy metal ions from aqueous solutions [40].

2.4 Functions and Properties of Organic Materials-Modified CBHs

2.4.1 *Encapsulation and Release of Bioactive Agents*

Encapsulation and controlled release of bioactive substances constitute critical aspects of functional regulation in CBHs, with broad applications in drug delivery, tissue engineering, and biosensing [41]. Encapsulation methods are primarily categorized into physical encapsulation and chemical bonding. Physical encapsulation depends on the 3D network structure of hydrogels to adsorb and immobilize bioactive molecules without altering their chemical structure. In contrast, chemical bonding involves stable covalent interactions between functional groups of bioactive molecules and reactive groups within the hydrogel network, such as esterification, acylation, and Schiff base reactions. Efficient drug encapsulation is a fundamental requirement for drug carrier systems [42].

Liu et al. encapsulated curcumin within emulsions formed by gelatin and CNCs, subsequently integrating these emulsions with sodium alginate to fabricate stable hydrogel beads, which improved emulsion dispersion and reduced droplet size [43]. Controlled release was achieved by designing hydrogel architectures that enabled sustained release at predetermined rates or respond to specific stimuli, thereby facilitating prolonged therapeutic effects. Naeem et al. grafted 2-acrylamido-2-methylpropane sulfonic acid (AMPS) onto hydroxyethyl cellulose (HEC) to develop an oral controlled-release hydrogel carrier system for delivering Radix Paeoniae Alba (RPA) in vivo [44], demonstrating high potential for treating human diseases, as shown in Fig. 2.4a.

Furthermore, encapsulation systems face several challenges, including maintaining the stability of bioactive molecules, achieving precise control over release kinetics, and synchronizing the biodegradation rate of the carrier with therapeutic needs. Bioactive molecules may lose efficacy due to environmental fluctuations, such as pH variations and enzymatic degradation, during the encapsulation and release processes, thereby diminishing therapeutic outcomes [1]. Concurrently, the degradation rate of CBHs must be carefully matched to release profiles to avoid premature drug release or prolonged retention, both of which could lead to adverse effects [45]. To overcome these limitations, smart responsive mechanisms, such as pH-, temperature-, and enzyme-responsive hydrogels, have been developed to enhance the specificity and timing of drug delivery. For instance, Zhou et al. engineered a pH-responsive, injectable, and self-healing hydrogel via a Schiff base reaction between oxidized hydroxypropyl cellulose (Ox-HPC) and carboxymethyl chitosan (CMCS) (Fig. 2.4b). This hydrogel demonstrated accelerated release of phenylalanine under mildly acidic conditions (pH 6.8) compared to physiological pH (7.4), enabling selective targeting of acidic microenvironments [46]. Overall, the encapsulation and controlled release of bioactive substances present significant potential for advancing drug delivery systems and expanding biomedical applications of CBHs.

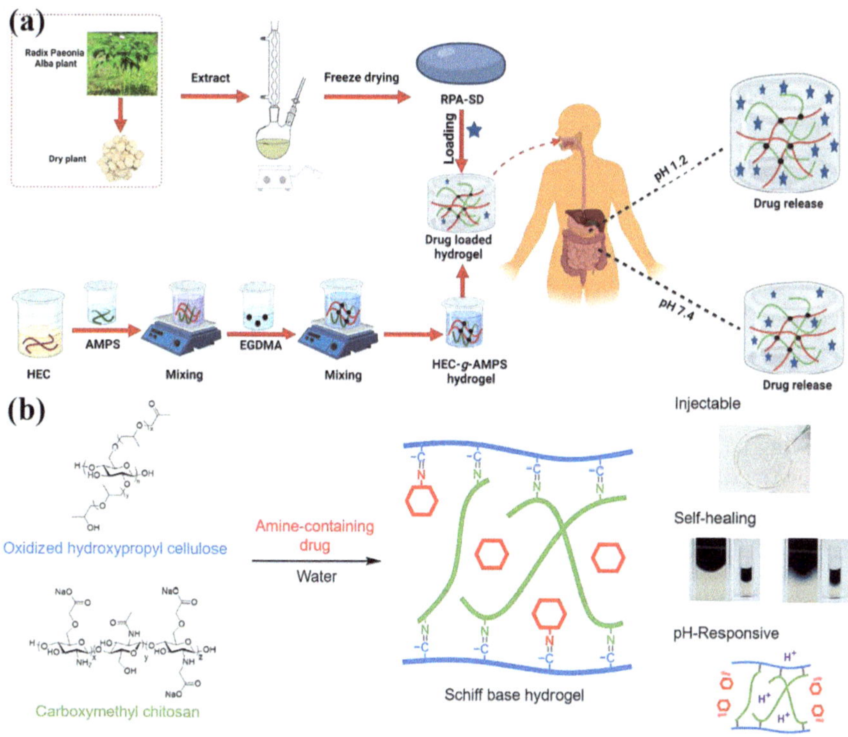

Fig. 2.4 Encapsulation and release of drugs with CBHs: **a** Schematic illustration of RPA encapsulation within the HEC-g-AMPS hydrogel for in vivo drug delivery. Reprinted with the permission from Ref. [44], Copyright 2023, MDPI. **b** Preparation route and pH-responsive characteristics of CBHs formed via reaction between Ox-HPC and CMCS. Reprinted with the permission from Ref. [46], Copyright 2023, Elsevier Ltd

2.4.2 *Charge-Transfer Complexes and Conductivity Enhancement*

CBHs possess inherent biocompatibility and flexibility, making them promising substrates for various biomedical and flexible electronic applications. However, their intrinsically low electrical conductivity restricts their utility in flexible sensors, bioelectronics, and energy storage devices. To overcome this limitation, conductive polymers or charge-transfer complexes are frequently incorporated to construct conductive cellulose hydrogel systems [24].

Among these, PEDOT: PSS has garnered extensive attention in both academic research and industrial fields due to its high electrical conductivity, excellent optical and electrical properties, and facile processability [47]. Yang et al. introduced positively charged conductive polymer cross-linkers into the cellulose hydrogel network, enabling in situ cross-linking of PEDOT: PSS hydrogels to establish continuous conductive networks. This mild and non-toxic preparation approach achieved conductivities up to 3265 $S \cdot m^{-1}$ while preserving excellent biocompatibility, facilitating in-situ detection of cellular electrical signals [48]. Similarly, Patel et al. incorporated spherical nanocellulose into CMCS-based injectable and adhesive hydrogels, resulting in enhanced conductivity and superior adhesion compared to hydrogels composed solely of polymers [49].

PANI, a conjugated conductive polymer, exhibits tunable conductivity through doping, relatively low synthesis cost, and good compatibility with cellulose [50]. Gao et al. fabricated conductive hydrogels composed of CMC-PANI particles, PVA, and phytic acid (PA) [51]. The incorporation of CMC-PANI significantly enhanced the electrical conductivity and mechanical strength of the hydrogels, while also imparting excellent antifreeze, anti-swelling, and reusability properties, rendering them suitable for flexible electronic applications. Additionally, charge-transfer complexes provide innovative strategies to improve conductivity in cellulose hydrogels; nevertheless, achieving an optimal balance between conductivity and other functional properties remains a critical challenge in this field.

2.4.3 Mechanical Strength and Elasticity Enhancement

The mechanical strength and elasticity of CBHs are vital for their applications in soft tissue engineering, wound dressings, and flexible sensors. Although cellulose inherently forms interchain hydrogen bonds, its mechanical properties are often relatively weak. The incorporation of organic materials into CBHs can markedly enhance hydrogel toughness and elasticity via hydrogen bonding, electrostatic interactions, hydrophobic interactions, and covalent crosslinking.

Zheng et al. developed a robust chitosan/templated cellulose nanocrystal (TCNC) double-network hydrogel using synergistic chemical and physical cross-linking combined with chain orientation strategies, achieving excellent anisotropic tensile properties, including improved mechanical strength and elastic modulus [52]. Similarly, Takawa et al. combined bacterial cellulose (BC) with ethylene–vinyl alcohol copolymer (EVOH), yielding hydrogels with enhanced interfacial interactions [53]. As shown in Fig. 2.5a, the increasing of isopropanol (IPA) content promoted polymer precipitation to cover BC fibers, thereby significantly enhancing anisotropy. The compressive strength and storage modulus of BC-EVOH hydrogels (Fig. 2.5b) substantially exceeded those of the individual components and further improved with higher IPA concentrations. These improvements derive from synergistic toughening mechanisms related to cellulose nanostructures, including strengthened cross-linked network integrity, enhanced fiber interface load transfer efficiency, and network densification. Nonetheless, there exists a trade-off between mechanical properties and swelling/water retention capacity. Excessive cross-linking may impede cell migration and biological functions, thus requiring mild synthesis conditions to preserve hydrogel bioactivity and functional components.

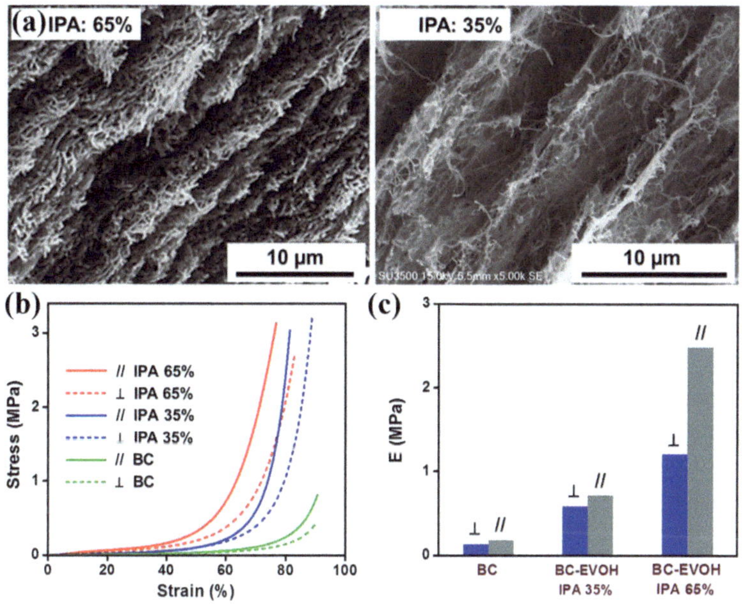

Fig. 2.5 Schematic diagrams and mechanical characterization of cellulose hydrogels: **a** SEM images of BC-EVOH composite hydrogels containing 65% and 35% isopropanol (IPA). **b** Stress–strain curves of BC hydrogels and BC-EVOH composite hydrogels. **c** Tangent elastic modulus and storage modulus of hydrogels. Reprinted with the permission from Ref. [53], Copyright 2022, Royal Society of Chemistry

2.4.4 Swelling, Porosity, and Water Retention Modulation

The water absorption capacity, pore structure regulation, and water retention ability of CBHs are critical determinants of their efficacy as carrier materials. These parameters can be precisely tailored through the design of the hydrogel network structure, such as adjusting crosslinking density and constructing heterogeneous domains, and by incorporating hydrophilic or hydrophobic components like nanocellulose and biopolymers.

For instance, Nyamayaro et al. embedded nanocellulose into polyacrylamide (PAM)-based ionically conductive hydrogels, resulting in a dense network formed via multiple cross-linking mechanisms [37]. This network effectively suppressed excessive swelling and significantly enhanced water retention under dry conditions. The improved retention performance was attributed to strengthened physical cross-linking and electrostatic interactions between nanocellulose and polymer chains, where nanocellulose created hydrogen-bonded and ionic bridge structures that precisely controlled swelling.

In terms of pore structure, alginate/nanocellulose composite hydrogels demonstrated the ability to form microporous spherical architectures with controllable morphology. Georgouvelas et al. introduced TEMPO-oxidized functional groups into alginate/nanocellulose composite microspheres, producing hydrogel beads with porosity reaching 99% [54]. These microspheres exhibited high adsorption efficiency for dyes and heavy metals, achieving dye removal rates up to 91% under a high water-flux of 17,000 L·h^{-}1·m^{-}2·bar^{-}1, indicating strong potential for industrial wastewater treatment applications. Through bidirectional regulation of both network pore size and chemical composition, CBHs achieved high swelling capacity, excellent water retention, and structural stability.

2.4.5 Biodegradability and Environmental Stability

CBHs inherently exhibit good biodegradability due to their natural origin; however, their degradation rate and environmental stability are often modulated through the introduction of organic materials such as crosslinkers, composite polysaccharides, or functional additives [55]. Chemically cross-linked systems with high crosslinking densities, for example, those treated with citric acid or epichlorohydrin, can effectively prolong degradation times in aqueous or soil environments and enhance long-term structural stability, making them suitable for applications in agricultural substrates or environmental filtration.

Organic cross-linkers like citric acid (CA) and epichlorohydrin (ECH) significantly improve the structural integrity and environmental tolerance of cellulose networks while allowing controlled degradation mediated by microbial or enzymatic processes. Yu et al. prepared CA-cross-linked regenerated BC (RC-BC) films, which demonstrated excellent mechanical performance, with fracture strength reaching 93.4 MPa and Young's modulus of 4.2 GPa after 2% CA cross-linking. These films fully degraded within two weeks in soil, indicating favorable ecological compatibility and promising potential as biodegradable packaging materials [56]. To balance environmental stability with bioactivity, multi-network systems have been developed by compositing natural polysaccharides such as gelatin and chitosan, nanocellulose, or plant-derived extracts.

Sannino et al. grafted gelatin and citric acid onto BC, yielding composite hydrogels that exhibit high tensile strength and anti-swelling properties while degrading slowly in pure water, thereby demonstrating potential for tissue engineering or environmentally responsive packaging applications [57]. Despite the improved stability imparted by cross-linking, irreversible residues may form. For instance, excessive cross-linking enhanced mechanical robustness but compromises degradability, whereas lower cross-linking levels favored biodegradation at the expense of structural stability. Additionally, factors such as additive sourcing, cost, and potential toxicity required careful consideration in the design of biodegradable CBHs.

2.4.6 Stimuli-Responsive and Smart Behavior

Stimuli-responsive CBHs exhibit reversible or irreversible changes in their structure, volume, or function in response to external or internal stimuli such as pH, temperature, mechanical force, or enzymes [58]. These responsive mechanisms endow the hydrogels with intelligent behaviors including controlled drug release, self-healing, and shape morphing, rendering them highly attractive for applications in targeted drug delivery, tissue engineering, and flexible devices [59]. By chemically grafting cellulose with pH- or thermo-responsive monomers, environmentally triggered responses can be precisely controlled.

For instance, Li et al. grafted nanocellulose with the pH-responsive monomer acrylic acid (AA) and the thermo-sensitive monomer N-isopropylacrylamide (NIPAM) to fabricate a dual-responsive composite hydrogel [60]. This system could undergo swelling or shrinking upon variations in pH and temperature, enabling effective drug loading and controlled release with excellent cycling stability.

The incorporation of enzyme-cleavable sites or immobilized enzymes such as lysozyme and laccase into the cellulose network facilitates enzyme-driven structural and functional transformations at targeted tissues or environmental sites. Yang et al. immobilized the polymerizable monomer laccase (Lac) into a hybrid hydrogel network composed of aldehyde-modified CNCs (CNC-CHO) and 2-aminoethyl methacrylate hydrochloride (AMH) (Fig. 2.6a), generating a heterogeneous biocatalytic system. Upon laccase immobilization within the CNC network, the hydrogel maintained 71% enzyme activity after 10 successive catalytic cycles and effectively catalyzed phenol degradation [61].

In terms of photothermal responsiveness, cellulose hydrogels can incorporate photothermal agents with strong light absorption and efficient non-radiative heat conversion capabilities, such as gold nanorods, black phosphorus, MXene nanosheets, and polydopamine, thus exhibiting near-infrared (NIR) light sensitivity. Under NIR irradiation, these agents rapidly convert light into heat, inducing hydrogel volume contraction, accelerated drug release, or antibacterial effects [62]. Zhang et al. constructed a thermo-sensitive injectable hydrogel by copolymerizing oxidized CMC (CMC-CHO) with NIPAM and embedding gold nanorods as photothermal agents. Upon NIR excitation, the surface temperature of the hydrogel rapidly increased to 50–60 °C, significantly enhancing doxorubicin (DOX) release rates and demonstrating strong antitumor efficacy both in vitro and in vivo [63]. Similarly, Zhang et al. designed oxidized CNF/PVA composite hydrogels coated with polydopamine, which exhibited photothermal-triggered DOX release under NIR irradiation, maintaining controlled release efficiency and reusability [64]. Despite these advances, challenges persist in balancing response speed and intensity, as highly sensitive responses often compromise structural stability, while more stable structures tend to exhibit weaker responsiveness.

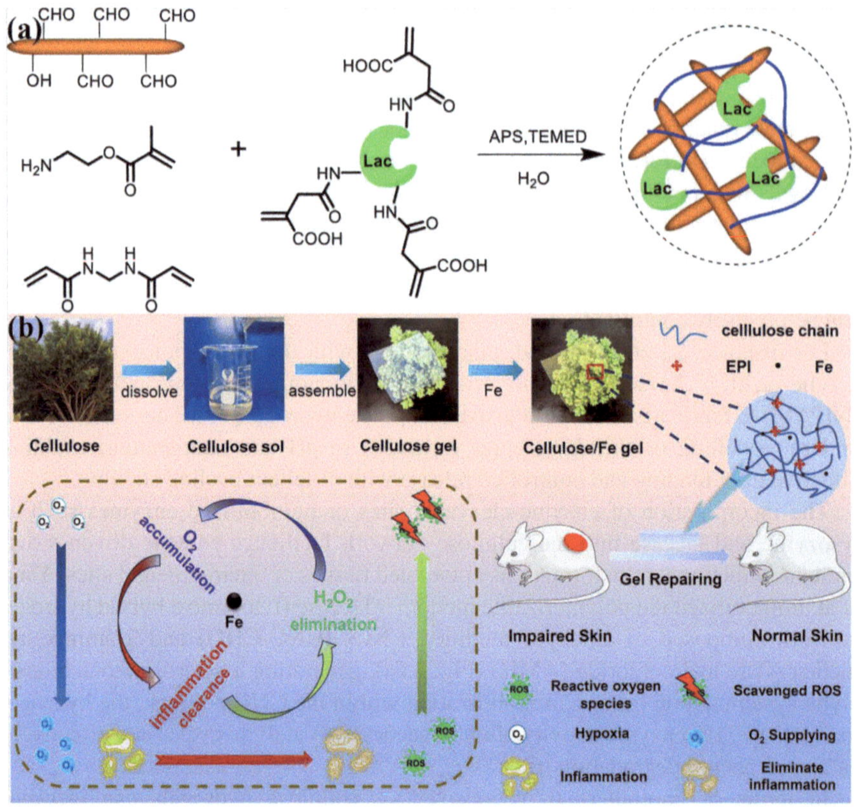

Fig. 2.6 Enhanced functions and properties of CBHs: **a** laccase (Lac)-modified CNC composite hydrogel with catalysis property. Reprinted with the permission from Ref. [61], Copyright 2023, Springer. **b** Cellulose/iron (Fe) composite hydrogel for wound dressing. Reprinted with the permission from Ref. [68], Copyright 2025, Royal Society of Chemistry

2.4.7 *Biocompatibility and Bioactivity*

Cellulose hydrogels inherently exhibit excellent biocompatibility owing to their abundant hydroxyl groups, non-toxicity, and biodegradability [65]. However, for practical biomedical applications including tissue engineering [8], wound dressings, drug delivery, and implantable sensors [66], further incorporation of bioactive organic materials, such as proteins, growth factors, polysaccharides, or cell-adhesive peptides, is essential to promote cell adhesion, proliferation, differentiation, enhance functional performance, and mitigate immune responses [67].

Xue et al. reported a physically and chemically dual-cross-linked 3D porous cellulose/iron composite hydrogel that demonstrated excellent water retention, mechanical stability, and antioxidant properties, thereby significantly accelerating wound healing (Fig. 2.6b) [68]. The incorporation of natural proteins (e.g., gelatin, collagen [8]), polysaccharides (chitosan, alginate), or platelet-rich plasma (PRP) could markedly enhance the bioactivity of cellulose hydrogels. For example, Min et al. developed a sulfonated CNC injectable hydrogel loaded with vascular endothelial growth factor (VEGF), which promoted efficient angiogenesis and substantially improved tissue regeneration [65]. Furthermore, grafting cell-adhesive peptides such as RGD onto the hydrogel matrix further facilitates cellular recognition and adhesion, thereby enhancing biocompatibility [69].

Challenges in enhancing the bioactivity of CBHs include difficulties in maintaining the stability and uniform dispersion of bioactive molecules, which may lose activity or aggregate during preparation and storage [70]. Additionally, bioactive components can trigger immune responses or exhibit mismatched biodegradation rates, thereby compromising long-term material stability and safety. Multi-layer responsive systems further require extensive in vivo evaluation to prevent issues such as residue detachment or immune stimulation [71]. Moreover, achieving synergistic multifunctional regulation and developing intelligent hydrogels capable of adapting to complex physiological environments remain crucial research directions [66]. Yang et al. investigated a composite hydrogel composed of CNFs, PVA, and curcumin-modified silver nanoparticles, which exhibited excellent mechanical strength, self-healing ability, antimicrobial and antioxidant activities, as well as up to 98% cell viability, along with enhanced angiogenesis and tissue repair in both in vitro and in vivo models [72]. Despite these challenges, the rational design and regulation of bioactive organic components offer broad application prospects and significant developmental potential.

2.5 Conclusions

In summary, the integration of organic materials not only enhances the overall performance of CBHs, but also opens new avenues for synergistic functionality and system-level integration. Through molecular design, functional group incorporation, and interfacial engineering, organic materials can optimize intermolecular interactions within the cellulose network, modulate microstructural morphology, and impart novel physicochemical properties, thereby enabling multi-functional performance enhancements. Due to their structural diversity, chemical versatility, and broad reactivity, organic materials serve multiple roles within CBH systems, including mechanical reinforcement, cross-linking, interfacial bridging, and functional modulation. As such, these strategies represent key enablers in the development of intelligent, multifunctional, and high-performance CBHs for a wide array of advanced applications.

References

1. J.W. Gong, L.L. Hou, Y.C. Ching, K.Y. Ching, N.D. Hai, C.H. Chuah, A review of recent advances of cellulose-based intelligent-responsive hydrogels as vehicles for controllable drug delivery system. Int. J. Biol. Macromol. **264**, 130525 (2024)
2. G. Mogli, A. Chiappone, A. Sacco, C.F. Pirri, S. Stassi, Ultrasensitive piezoresistive and Piezo-capacitive cellulose-based Ionic hydrogels for wearable multifunctional sensing. ACS Appl. Electron. Mater. **5**, 205–215 (2022)
3. Y. Altynov, K. Bexeitova, M. Nazhipkyzy, S. Azat, A. Konarov, D. Rakhman, N. Sahiner, K. Kudaibergenov, Nanocellulose hydrogels from agricultural wastes: methods, properties, and application prospects. Nanoscale **17**, 12580–12619 (2025)
4. M.P.H. Pedige, T.A. Asoh, Y. Hsu, H. Uyama, Stimuli-responsive composite hydrogels with three-dimensional stability prepared using oxidized cellulose nanofibers and chitosan. Carbohydr. Polym. **278** (2022)
5. M. Sumini, C.R. de Souza, G.J.S. Andrade, I.R.C. Oliveira, S. Scandorieiro, C.A. Tischer, R.K.T. Kobayashi, G. Nakazato, Cellulose hydrogel with hyaluronic acid and silver nanoparticles: sustained-release formulation with antibacterial properties against. ANTIBIOTICS-BASEL. **12**, 873 (2023)
6. G.Y. Jiang, G.W. Xu, Q.Q. Xia, H.P. Yu, D.W. Zhao, Cellulose gels: functional design and promising smart applications. Small **4**, 2412538 (2025)
7. H.R. Shi, H.X. Huo, H.X. Yang, H.S. Li, J.J. Shen, J.Y. Wan, G.B. Du, L. Yang, Cellulose-based dual-network conductive hydrogel with exceptional adhesion. Adv. Funct. Mater. **34**, 2408560 (2024)
8. C. Chen, Y.W. Xi, Y.X. Weng, Recent advances in cellulose-based hydrogels for tissue engineering applications. Polymers **14**, 3335 (2022)
9. Y.J. Liu, Y.J. Zhang, Q. Yang, Z.R. Yu, M.T. He, Y.F. Zhu, X. Fu, F.J. Meng, Q.L. Ma, L.M. Kong, S.H. Pan, Y.J. Che, Tunicate cellulose nanocrystal reinforced multifunctional hydrogel with super flexible, fatigue resistant, antifouling and self-adhesive capability for effective wound healing. Int. J. Biol. Macromol. **277**, 134337 (2024)
10. L.L. Deng, B.X. Wang, W.Y. Li, Z.L. Han, S.Y. Chen, H.P. Wang, Bacterial cellulose reinforced chitosan-based hydrogel with highly efficient self-healing and enhanced antibacterial activity for wound healing. Int. J. Biol. Macromol. **217**, 77–87 (2022)
11. D. Georgouvelas, H.N. Abdelhamid, U. Edlund, A.P. Mathew, Modified nanocellulose/alginate hydrogel composite beads for purifying mining effluents. Nanoscale Adv. **5**, 5892–5899 (2023)
12. H.C. Fu, B. Wang, J.P. Li, J. Xu, J. Li, J.S. Zeng, W.H. Gao, K.F. Chen, A self-healing, recyclable and conductive gelatin/nanofibrillated cellulose/Fe^{3+} hydrogel based on multi-dynamic interactions for a multifunctional strain sensor. Mater. Horiz. **9**, 1421–1421 (2022)
13. N. Chen, X.B. Zhang, J.Y. Lyu, G.L. Zhao, K. Gu, J. Xia, Z.C. Chen, Z.Z. Shao, Preparation of a novel regenerated silk fibroin-based hydrogel for extrusion bioprinting. Soft Matter **18**, 7360–7368 (2022)
14. D.A. Elsherbiny, A.M. Abdelgawad, B.A. Hemdan, A.S. Montaser, I.E. El-Sayed, S. Jockenhoevel, S. Ghazanfari, Self-crosslinked polyvinyl alcohol/cellulose nanofibril cryogels loaded with synthesized aminophosphonates as antimicrobial wound dressings. J Mater Chem B **11**, 7144–7159 (2023)
15. Z.W. Tang, Y.Q. Yang, Y.W. Pan, M.Q. Yu, X.X. Lin, A.K. Mondal, Biocompatible, Injectable, and self-healing poly(-vinylpyrrolidone)/carboxymethyl cellulose hydrogel for drug release. ACS Omega **9**, 5854–5861 (2024)
16. D. Getya, A. Lucas, I. Gitsov, Composite hydrogels based on poly(ethylene glycol) and cellulose macromonomers as fortified materials for environmental cleanup and clean water safeguarding. Int. J. Mol. Sci. **24**, 7558 (2023)
17. W.J. Yuan, S.H. Li, H.H. Guan, S. Zhang, Y.X. Zhang, M. Zhang, Y. Yu, X.Y. Chen, Preparation and properties of a novel biodegradable composite hydrogel derived from gelatin/chitosan and polylactic acid as slow-release n fertilizer. Polymers **15**, 997 (2023)

18. N. Srivastava, A.R. Choudhury, Thermo-reversible self-assembled novel gellan gum hydrogels containing amino acid biogelators with antibacterial activity. Carbohydr. Polym. **324**, 121462 (2024)
19. Y.J. Hu, N. Li, P.P. Yue, G.G. Chen, X. Hao, J. Bian, F. Peng, Highly antibacterial hydrogels prepared from amino cellulose, dialdehyde xylan, and Ag nanoparticles by a green reduction method. Cellulose **29**, 1055–1067 (2022)
20. H. Zhang, K. Liu, Y.M. Gong, W.J. Zhu, J.F. Zhu, F. Pan, Y. Chao, Z.S. Xiao, Y.B. Liu, X.W. Wang, Z. Liu, Y. Yang, Q. Chen, Vitamin C supramolecular hydrogel for enhanced cancer immunotherapy. Biomaterials **287**, 121673 (2022)
21. G.H. Yang, Z.K. Zhang, K.F. Liu, X.X. Ji, P. Fatehi, J.C. Chen, A cellulose nanofibril-reinforced hydrogel with robust mechanical, self-healing, pH-responsive and antibacterial characteristics for wound dressing applications. J. Nanobiotechnol. **20**, 312 (2022)
22. H.X. Wan, C.R. Qin, A. Lu, A flexible, robust cellulose/phytic acid/polyaniline hydrogel for all-in-one supercapacitors and strain sensors. J Mater Chem A **10**, 17279–17287 (2022)
23. X.W. Fan, Y.L. Tao, F.C. Liu, S. Li, M.A. Liu, Multifunctional polyaniline nanofiber conductive hydrogel for supercapacitors and flexible sensors. New J. Chem. **48**, 18943–18956 (2024)
24. M. Zhang, Y.X. Wang, K. Liu, Y. Liu, T. Xu, H.S. Du, C.L. Si, Strong, conductive, and freezing-tolerant polyacrylamide/PEDOT:PSS/cellulose nanofibrils hydrogels for wearable strain sensors. Carbohydr. Polym. **305**, 120567 (2023)
25. I.G.R. da Silva, B.T.D. Pantoja, G.H.D.R. Almeida, A.C.O. Carreira, M.A. Miglino, Bacterial cellulose and eCM hydrogels: an Innovative approach for cardiovascular regenerative medicine. Int. J. Mol. Sci. **23**, 3955 (2022)
26. C.Y. Ho, C.C. Wang, T.C. Wu, C.H. Kuan, Y.C. Liu, T.W. Wang, Peptide-functionalized double network hydrogel with compressible shape memory effect for intervertebral disc regeneration. Bioeng. Transl. Med. **8**, e10447 (2023)
27. T. Zhou, Z. Qiao, M. Yang, K. Wu, N.N. Xin, J.M. Xiao, X.Y. Liu, C.H. Wu, D. Wei, J. Sun, H.S. Fan, Hydrogen-bonding topological remodeling modulated ultra-fine bacterial cellulose nanofibril-reinforced hydrogels for sustainable bioelectronics. Biosens. Bioelectron. **231**, 115288 (2023)
28. M. Zhang, T. Xu, K. Liu, L.Y. Zhu, C.Y. Miao, T. Chen, M.G. Gao, J.F. Wang, C.L. Si, Modulation and mechanisms of cellulose-based hydrogels for flexible sensors. Susmat. **5**, e255 (2025)
29. M. Mihoub, S. Hamri, T. Bouchaour, M. Popa, D.M. Popa, L.B. Alachaher, M. Hamcerencu, An interpenetrating polymer network hydrogel based on cellulose, applied to remove colorant traces from the water medium: electrostatic interactions analysis. Polymers **14**, 5090 (2022)
30. M. Wohlert, T. Benselfelt, L. Wågberg, I. Furó, L.A. Berglund, J. Wohlert, Cellulose and the role of hydrogen bonds: not in charge of everything. Cellulose **29**, 1–23 (2022)
31. P.P. Sun, M. Wang, T.T. Wu, L.S. Guo, W.J. Han, Covalent crosslinking cellulose/graphene aerogels with high elasticity and adsorbability for heavy metal Ions adsorption. Polymers **15**, 2434 (2023)
32. R.J. Aguado, A. Mazega, Q. Tarrés, M. Delgado-Aguilar, The role of electrostatic interactions of anionic and cationic cellulose derivatives for industrial applications: a critical review. Ind. Crop Prod. **201**, 116898 (2023)
33. N. Thomas, S. Moussaoui, B. Reyes-Suárez, O. Lafon, G.M. Reddy, Dual cross-linked cellulose based hydrogel films. Mater. Adv. **5**, 9210–9219 (2024)
34. J.W. Kim, J. Batta-Mpouma, G. Kandhola, J. Sakon, Covalent crosslinking of colloidal cellulose nanocrystals for multifunctional nanostructured hydrogels with tunable physicochemical properties. Biomacromol **23**, 4085–4096 (2022)
35. J.H. Hu, X.Q. Wang, Modification mechanisms and properties of poplar wood via grafting with 2-hydroxyethyl methacrylate/N, N′-methylenebis(acrylamide) onto cell walls. Polymers **15**, 1861 (2023)
36. Y. Zhang, P. Lan, Z.C. Yu, T. Xia, N. Lin, Grafting kinetics and compatibility simulation in surface modification of cellulose nanocrystals with poly(lactic acid) for composites. Macromolecules **57**, 7689–7700 (2024)

37. K. Nyamayaro, T. Iwase, S.G. Hatzikiriakos, P. Mehrkhodavandi, Cellulose nanocrystal-mediated enhancement of hydrogel anti-swelling and water retention. RSC Sustainability. **2**, 1543–1550 (2024)
38. D. Das, S. Bhattacharjee, S. Bhaladhare, Preparation of cellulose hydrogels and hydrogel nanocomposites reinforced by crystalline cellulose nanofibers (CNFs) as a water reservoir for agriculture use. ACS Appl. Polym. Mater. **5**, 2895–2904 (2023)
39. N. Chen, X. He, Q. Lu, Highly stretchable, repairable, and tough nanocomposite hydrogel physically cross-linked by hydrophobic interactions and reinforced by surface-grafted hydrophobized cellulose nanocrystals. Macromol. Rapid Commun. **44**, 2300053 (2023)
40. H.-A.S. Tohamy, M. El-Sakhawy, B. Strachota, A. Strachota, E. Pavlova, S. Mares Barbosa, S. Kamel, Temperature-and pH-responsive super-absorbent hydrogel based on grafted cellulose and capable of heavy metal removal from aqueous solutions. Gels **9**, 296 (2023)
41. P. He, L. Dai, J. Wei, X. Zhu, J. Li, Z. Chen, Y. Ni, Nanocellulose-based hydrogels as versatile drug delivery vehicles: a review. Int. J. Biol. Macromol. **222**, 830–843 (2022)
42. Y. Du, G. Feng, When nanocellulose meets hydrogels: the exciting story of nanocellulose hydrogels taking flight. Green Chem. **25**, 8349–8384 (2023)
43. S. Liu, Y. Wang, Y. Huang, M. Hu, X. Lv, Y. Zhang, H. Dai, Gelatin-nanocellulose stabilized emulsion-filled hydrogel beads loaded with curcumin: preparation, encapsulation and release behavior. Int. J. Biol. Macromol. **275**, 133551 (2024)
44. A. Naeem, C. Yu, X. Wang, M. Peng, Y. Liu, Y. Liu, Hydroxyethyl cellulose-based hydrogels as controlled release carriers for amorphous solid dispersion of bioactive components of radix paeonia alba. Molecules **28**, 7320 (2023)
45. A.P.A.d. Carvalho, R. Értola, C.A. Conte-Junior, Nanocellulose-based platforms as a multipurpose carrier for drug and bioactive compounds: from active packaging to transdermal and anticancer applications. Int. J. Pharm. **652**, 123851 (2024)
46. Y. Zhou, Z. Zhai, Y. Yao, J.C. Stant, S.L. Landrum, M.J. Bortner, C.E. Frazier, K.J. Edgar, Oxidized hydroxypropyl cellulose/carboxymethyl chitosan hydrogels permit pH-responsive, targeted drug release. Carbohydr. Polym. **300**, 120213 (2023)
47. H. Du, M. Zhang, K. Liu, M. Parit, Z. Jiang, X. Zhang, B. Li, C. Si, Conductive PEDOT:PSS/cellulose nanofibril paper electrodes for flexible supercapacitors with superior areal capacitance and cycling stability. Chem. Eng. J. **428**, 131994 (2022)
48. T. Yang, M. Yang, C. Xu, K. Yang, Y. Su, Y. Ye, L. Dou, Q. Yang, W. Ke, B. Wang, PEDOT: PSS hydrogels with high conductivity and biocompatibility for in situ cell sensing. J Mater Chem B. **11**, 3226–3235 (2023)
49. D.K. Patel, K. Ganguly, J. Hexiu, S.D. Dutta, T.V. Patil, K.-T. Lim, Functionalized chitosan/spherical nanocellulose-based hydrogel with superior antibacterial efficiency for wound healing. Carbohydr. Polym. **284**, 119202 (2022)
50. A.K. Rana, F. Scarpa, V.K. Thakur, Cellulose/polyaniline hybrid nanocomposites: design, fabrication, and emerging multidimensional applications. Ind. Crop Prod. **187**, 115356 (2022)
51. J. Gao, X. Li, L. Xu, M. Yan, Q. Wang, Dual design strategy for carboxymethyl cellulose-polyaniline composite hydrogels as super-sensitive amphibious sensors. Int. J. Biol. Macromol. **280**, 135630 (2024)
52. Y. Zheng, L. Zhang, B. Duan, Anisotropic chitosan/tunicate cellulose nanocrystals hydrogel with tunable interference color and acid-responsiveness. Carbohydr. Polym. **295**, 119866 (2022)
53. S. Takawa, A. Sugawara, T.-A. Asoh, M. Nandi, H. Uyama, Composite hydrogels of bacterial cellulose and an ethylene-vinyl alcohol copolymer with tunable morphological anisotropy and mechanical properties. Mater. Adv. **3**, 5138–5150 (2022)
54. D. Georgouvelas, H.N. Abdelhamid, U. Edlund, A.P. Mathew, In situ modified nanocellulose/alginate hydrogel composite beads for purifying mining effluents. Nanoscale Adv. **5**, 5892–5899 (2023)
55. V.T. Le, S.-W. Joo, M. Berkani, T. Mashifana, H. Kamyab, C. Wang, Y. Vasseghian, Sustainable cellulose-based hydrogels for water treatment and purification. Ind. Crop Prod. **205**, 117525 (2023)

56. Q. Yu, L. Yang, S. Wang, L. Zhang, D. Sun, Citric acid cross-linked regenerated bacterial cellulose as biodegradable film for food packaging. Cellulose **30**, 10273–10284 (2023)
57. A. Sannino, C. Demitri, M. Madaghiele, Biodegradable cellulose-based hydrogels: design and applications. Materials **2**, 353–373 (2009)
58. J. Yuan, T. Curk, Collapse/expansion dynamics and actuation of pH-responsive nanogels. ArXiv preprint. **9**, 2405 (2024)
59. S. Chang, Z. Weng, C. Zhang, S. Jiang, G. Duan, Cellulose-based intelligent responsive materials: a review. Polymers-Basel. **15**, 3905 (2023)
60. Y. Li, L. Zhang, Z. Song, F. Li, D. Xie, Intelligent temperature-pH dual responsive nanocellulose hydrogels and the application of drug release towards 5-fluorouracil. Int. J. Biol. Macromol. **223**, 11–16 (2022)
61. L. Yang, X. Peng, S. Zhang, G. Xie, X. Zhang, Cellulose nanocrystal hybrid hydrogel with immobilized enzyme as heterogeneous biocatalyst for phenolic molecules oxidation. J. Mater. Sci. **58**, 6839–6849 (2023)
62. C. Sang, S. Wang, X. Jin, X. Cheng, H. Xiao, Y. Yue, J. Han, Nanocellulose-mediated conductive hydrogels with NIR photoresponse and fatigue resistance for multifunctional wearable sensors. Carbohydr. Polym. **333**, 121947 (2024)
63. L. Chang, X. Liu, J. Zhu, Y. Rao, D. Chen, Y. Wang, Y. Zhao, J. Qin, Cellulose-based thermo-responsive hydrogel with NIR photothermal enhanced DOX released property for anti-tumor chemotherapy. Colloids Surf. B **218**, 112747 (2022)
64. Z. Zhang, H. Cui, X. Wang, J. Liu, G. Liu, X. Meng, S. Lin, Oxidized cellulose-filled double thermo/pH-sensitive hydrogel for local chemo-photothermal therapy in breast cancer. Carbohydr. Polym. **332**, 121931 (2024)
65. K. Min, G. Tae, Cellular infiltration in an injectable sulfated cellulose nanocrystal hydrogel and efficient angiogenesis by VEGF loading. Biomater. Res. **27**, 28 (2023)
66. K.Y. Tang, J.Z.X. Heng, C.H.T. Chai, C.Y. Chan, B.Q.L. Low, S.M.E. Chong, H.Y. Loh, Z. Li, E. Ye, X.J. Loh, Modified bacterial cellulose for biomedical applications. Chem. Asian J. **17**, e202200598 (2022)
67. S.H. Aswathy, U. NarendraKumar, I. Manjubala, Physicochemical properties of cellulose-based hydrogel for biomedical applications. Polymers **14**, 4669 (2022)
68. M. Xue, X. Pei, J. Zhang, C. Niu, H. Wang, L. Nie, J. Ding, Double cross-linked cellulose hydrogel-supported Fe species for efficient wound healing. RSC Adv. **15**, 7885–7896 (2025)
69. S. Liu, R. Li, K. Dou, K. Li, Q. Zhou, Q. Fu, Injectable thermo-sensitive hydrogel containing ADSC-derived exosomes for the treatment of cavernous nerve injury. Carbohydr. Polym. **300**, 120226 (2023)
70. M.M.H. Rumon, A.A. Akib, S.D. Sarkar, M.A.R. Khan, M.M. Uddin, D. Nasrin, C.K. Roy, Polysaccharide-based hydrogels for advanced biomedical engineering applications. ACS Polym. Au. **4**, 463–486 (2024)
71. W. Cai, S. Mao, Y. Wang, B. Gao, J. Zhao, Y. Li, Y. Chen, D. Zhang, J. Yang, G. Yang, An engineered hierarchical hydrogel with immune responsiveness and targeted mitochondrial transfer to augmented bone regeneration. Adv. Sci. **11**, 2406287 (2024)
72. S. Zhang, B. Gatsi, X. Yao, Y. Jin, H. Amhal, Cellulose nanofiber-reinforced antimicrobial and antioxidant multifunctional hydrogel with self-healing, adhesion for enhanced wound healing. Carbohydr. Polym. **352**, 123189 (2025)

Chapter 3
Functional Regulation of Cellulose-Based Hydrogels by Inorganic Materials

Youyin Xu, Ranran Wang, Xinyue Wang, and Gang Wei

Abstract Cellulose hydrogels are a class of sustainable natural polymeric materials with good biocompatibility, high water retention and degradability. By loading inorganic nanostructures such as metals, metal oxides, carbon-based materials, ceramics and clays as well as inorganic salt ions onto the cellulose gel matrix, precise modulation of functional properties can be achieved. Specific strategies include *in-situ* synthesis and deposition, physical mixing and chemical cross-linking to homogeneously disperse the nanostructures and firmly fix them in a three-dimensional network. These composite systems present a rich variety of functions, such as electrical conductivity, magnetism, photocatalysis and antimicrobial properties. The research hotspots focus on the preparation and property modulation of composites such as cellulose/inorganic nanomaterials, and show promising applications in the fields of flexible electronic sensing, biomedical dressings, environmental pollution control and sustainable energy. This chapter summarizes the functional modulation strategies, research progress and challenges of various types of inorganic nanomaterials modifying cellulose-based hydrogels, and looks forward to their future development in areas such as wearable devices, smart biomaterials and environmental governance.

Keywords Cellulose · Hydrogels · Inorganic nanoparticles · Hybrid materials · Functional regulation

Y. Xu · R. Wang · X. Wang
College of Chemistry and Chemical Engineering, Qingdao University, Qingdao, P. R. China

Y. Xu · R. Wang · X. Wang · G. Wei (✉)
School of Polymer Science and Engineering, Qingdao University of Science and Technology, Qingdao, P. R. China
e-mail: wei-lab@qust.edu.cn

G. Wei and J. Zhang (eds.), *Cellulose-Based Hydrogels and Aerogels*, Lecture Notes in Nanoscale Science and Technology 41,
https://doi.org/10.1007/978-3-032-03008-5_3

3.1 Introduction

Cellulose is one of the richest renewable natural polymer resources on earth, and it is highly valued in green materials research for its good biocompatibility, environmental friendliness and mechanical strength [1]. Cellulose-based hydrogel (CBH), on the other hand, is a 3D hydrophilic network structure constructed using cellulose or its derivatives, which combines the advantages of both natural polymers and gel materials, and exhibits high water retention, flexibility and degradability [2].

With the concept of sustainable development, cellulose-based functional materials research has become a hot spot [3]. Although simple cellulose hydrogel is natural and environmentally friendly, its own function is relatively limited, and it is difficult to satisfy the demand for multifunctionality in complex application scenarios [4]. Therefore, introducing inorganic nanomaterials into cellulose hydrogel systems to achieve functional regulation through interfacial synergy has become an important strategy [5]. Cellulose possesses a high specific surface area and abundant hydroxyl functional groups, which enable it to be used as a template or carrier to effectively load various nanostructured materials [6]. Composite structures prepared using cellulose carriers usually combine high porosity, flexibility and environmental properties with multiple functions such as catalytic, conductive, magnetic, optical and antibacterial [7–9]. These research backgrounds and needs highlight the research significance and value of cellulose-based inorganic composite hydrogels in functional modulation.

Numerous research hotspots focus on different types of inorganic nanomaterials to empower cellulose hydrogels. Metal nanoparticles are commonly used to confer antibacterial or sensing functions to hydrogels [10]. Metal oxide nanomaterials can bring catalytic, optical and magnetic functions [11]. Carbon-based nanomaterials, on the other hand, are mainly used for electrical conductivity and mechanical enhancement [12]. These different types of inorganic components and their composite forms constitute the main direction of research on functional CBHs [13].

In terms of applications, functionalized CBHs show a wide range of prospects [14]. In the field of biomedicine, materials such as cellulose nanofibers or carboxymethyl cellulose are inherently biocompatible, and inorganic antibacterial agents embedded in hydrogels can be used for trauma dressings, wound healing and so on [15, 16]. In the field of flexible electronics and sensors, cellulose hydrogels doped with electrically conductive or ionic conductor materials are important candidates [17]. For environmental protection and energy, CBHs can be applied in water treatment and pollutant removal due to their porous structure and rich functional groups [18]. In addition, CBH electrolytes and super capacitor materials show good potential for application in the field of energy storage and flexible power supply [19]. Therefore, functionalized CBHs have demonstrated increasingly important applications in the fields of biomedicine, environmental treatment, flexible electronics and energy [20].

This chapter comprehensively explores the functional regulation of CBHs through the incorporation of diverse inorganic materials. It begins by categorizing key inorganic components, including metal nanoparticles, metal oxides, carbon nanomaterials, clays, and inorganic ions, and highlights their specific roles in enhancing hydrogel performance for various applications. The core discussion delves into the underlying mechanisms by which these materials interact with the cellulose matrix, encompassing physical interactions, covalent/coordination bonding, nanostructuring effects, and the introduction of inherent properties like photothermal activity or conductivity. Consequently, these inorganic additives significantly modulate critical functional properties: they markedly improve mechanical strength and toughness, precisely control swelling behavior and water retention, enhance environmental stability or tailor biodegradability, impart potent antibacterial/antifouling capabilities, and introduce novel functionalities such as electrical conductivity, magnetism, or unique optical responses.

3.2 Types of Inorganic Materials Used for Functional Regulation

CBHs are degradable but low strength and monofunctional. The introduction of inorganic nanomaterials can significantly expand their functionality. This section focuses on the various inorganic materials used for functional modulation of CBHs.

3.2.1 Metal Nanoparticles

Metal nanoparticles possess immense potential in nanotechnology. Commonly used metal nanoparticles include gold nanoparticles (AuNPs), silver nanoparticles (AgNPs), and copper nanoparticles (CuNPs) [21–23]. Variations in structure, morphology, and size impart distinct physicochemical properties and functionalities to these particles, which are crucial for the functional regulation of CBHs.

AuNPs, as representative noble metal nanomaterials, typically exhibit sizes ranging from 1 to 100 nm and possess unique physicochemical properties distinctly different from their bulk counterparts. These include localized surface plasmon resonance (LSPR), catalytic activity, and excellent biocompatibility. For example, Hoven et al. prepared BC/CMC hydrogels with excellent swelling properties by immersing bacterial cellulose (BC) membranes in carboxymethyl cellulose (CMC) solutions [24]. As shown in Fig. 3.1a, BC/CMC/AuNPs composite hydrogels were obtained by embedding AuNPs into BC/CMC hydrogels by in situ chemical reduction and used for the detection of overexpressed glutathione (GSH) in Alzheimer's disease. Owing to the functional regulation imparted by AuNPs, the BC/CMC/AuNP hydrogel enabled a linear detection range of 50–10,000 nM for GSH with a limit of detection (LOD) of 54.1 nM. In addition to conventional spherical AuNPs, gold nanorods (AuNRs), gold nanostars (AuNSs), and gold nanobipyramids (AuNBPs) have attracted significant attention due to their superior LSPR properties and tunable morphologies. For example, Liz-Marzán et al. synthesized AuNRs and AuNBPs

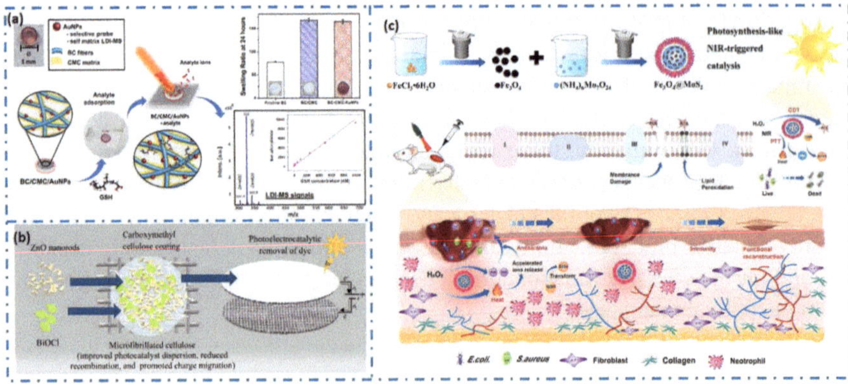

Fig. 3.1 **a** BC/CMC/AuNPs hydrogels for the detection of GSH. Reprinted from Ref. [24], Copyright 2023, Elsevier. **b** ZnO NRs and BiOCl doped CMC hydrogels for photocatalytic removal of dye. Reprinted from Ref. [37], Copyright 2025, Elsevier. **c** Synthesis of $Fe_3O_4@MoS_2$ for antimicrobial application. Reprinted from Ref. [40], Copyright 2025, Wiley

using thermally treated seed solutions, achieving higher yields and tunable aspect ratios of these nanostructures by adjusting the seed amount, thus controlling the near-infrared (NIR) absorption peak positions [25]. Their study provides valuable insights into the controllable synthesis of nanoparticles.

Besides AuNPs, AgNPs are another class of noble metal nanoparticles renowned for their broad-spectrum antimicrobial activity. The antibacterial mechanism primarily involves the release of Ag^+ ions, inducing reactive oxygen species (ROS) generation, disrupting bacterial membranes, and ultimately leading to cell lysis [22]. Compared with conventional antibiotics, AgNPs exhibit remarkable efficacy against resistant strains. Cellulose hydrogels incorporated with AgNPs have demonstrated potent antibacterial properties, making them widely applicable in wound dressings. Wang et al. synthesized CS/AgNPs composites via *in-situ* reduction and incorporated them into oxidized regenerated cellulose (ORC), forming ORC/CS/Lys@AgNP hydrogels through Schiff base reactions with lysine [26]. The resulting hydrogels exhibited excellent wound healing and antibacterial effects, achieving a wound closure rate of 78.5% within five days, significantly outperforming natural healing. Additionally, AgNPs possessed catalytic properties capable of degrading dye pollutants.

CuNPs offer cost-effective advantages and have attracted attention due to their excellent selectivity and catalytic efficiency in various processes. Although AgNPs dominate antibacterial applications, CuNPs have shown promising potential in recent years due to their abundant raw material sources and outstanding electrical conductivity. In a recent study, Talukdar et al. developed copper-based bio-nanocomposites (CBNCS) by loading CuNPs onto nanocellulose, achieving a minimum inhibitory concentration (MIC) of 30 μg/mL and excellent conductivity [27]. These features underscore the potential biomedical applications of CuNPs.

Furthermore, other transition metal nanoparticles such as cobalt (CoNPs), platinum (PtNPs), and palladium (PdNPs), as well as bimetallic or multimetallic composites, have garnered significant interest in catalytic engineering due to their superior catalytic activities [28–30]. In addition, gallium-based liquid metal nanoparticles (LMNPs) offer remarkable conductivity, thermal properties, low vapor pressure, and reduced biotoxicity compared to mercury, making them promising candidates for applications in soft electronics, catalysis, and biomedicine [31]. For instance, Liu et al. constructed EGaIn-PPD@Au nanoparticles by modifying eutectic gallium–indium (EGaIn) liquid metal with para-phenylenediamine (PPD) and grafting AuNPs [32]. These nanoparticles exhibited high electrical and thermal conductivities along with excellent photothermal properties, suitable for wearable sensors, and demonstrated a detection limit of 0.5 mg/L for C-reactive protein (CRP). Moreover, EGaIn-PPD@Au-incorporated PDMS photothermal patches exhibited anti-inflammatory and wound healing capabilities.

3.2.2 Metal Oxides Nanoparticles

Metal oxide nanoparticles are a class of inorganic materials with dimensions ranging from 1 to 100 nm, consisting of metal and oxygen elements. Typical examples include TiO_2, ZnO, Fe_2O_3, and Al_2O_3 nanoparticles [33, 34]. These nanostructured metal oxides exhibit physicochemical properties distinctly different from their bulk counterparts, primarily due to their high specific surface areas and quantum size effects.

TiO_2 nanoparticles (TiO_2 NPs) are among the most extensively studied metal oxide nanomaterials and are categorized into anatase, rutile, and brookite phases based on their crystal structures. Owing to their strong UV absorption, high photocatalytic activity, and excellent chemical stability, TiO_2 NPs are widely utilized as photocatalysts for degrading organic pollutants. For instance, Ezzine et al. anchored TiO_2 NPs onto cellulose-based biochar adsorbents via a simple hydrothermal method, controlling the particle size through the use of surfactants [35]. The resultant composite material exhibited outstanding catalytic performance in the degradation of methyl orange (MO), achieving a decolorization efficiency of 98.87% and demonstrating excellent recyclability.

Similar to TiO_2 NPs, ZnO NPs possess strong UV absorption, photocatalytic activity, and antimicrobial properties, making them highly desirable for applications in semiconductors, coatings, catalysts, batteries, and antibacterial agents [36]. Ang et al. fabricated a photoelectrochemical (PEC) system for methylene blue removal by immobilizing ZnO nanorods (ZnO NRs) and bismuth oxychloride (BiOCl) onto a hydrogel coating composed of CMC and microfibrillated cellulose (MFC) [37]. As shown in Fig. 3.1b, the incorporation of ZnO NRs into cellulose hydrogels significantly enhanced the PEC system's stability, reusability, and photocatalytic efficiency in dye degradation. Notably, the broad-spectrum antimicrobial activity of ZnO NPs is among their most attractive features, enabling effective inhibition of a wide range of bacteria.

Iron oxide nanoparticles are predominantly available in hematite (α-Fe_2O_3), magnetite (Fe_3O_4), and maghemite (γ- Fe_2O_3) forms, among which Fe_3O_4 is particularly notable for its superparamagnetic behavior [38]. Superparamagnetic iron oxide nanoparticles (SPIONs) possess multifunctional theranostic capabilities. Their magnetic properties allow for rapid separation and enrichment in biomarker isolation, localized hyperthermia in cancer therapy, and magnetic field-guided targeted drug delivery systems [39]. For example, inspired by artificial photosynthesis, Wang et al. designed core–shell structured Fe_3O_4@MoS_2 metal oxide nanoparticles for synergistic antibacterial therapy [40]. As depicted in Fig. 3.1c, Fe_3O_4@MoS_2 nanoparticles catalyze the generation of hydroxyl radicals (·OH), achieving chemodynamic therapy (CDT). Additionally, under NIR irradiation, these nanoparticles exhibit excellent photothermal conversion efficiency, enhancing bacterial eradication efficacy through combined CDT and photothermal therapy (PTT).

Al_2O_3 nanoparticles (Al_2O_3 NPs), another crucial metal oxide nanomaterial, are highly valued in industrial applications due to their superior mechanical strength and thermal stability [41]. Al_2O_3 NPs possess high hardness and heat resistance, and their surface acidic sites render them excellent catalyst supports, with potential applications in biomedical material development [42]. For instance, Deng et al. employed atomic layer deposition (ALD) to coat carbon fibers with Al_2O_3 films, creating biomimetic cotton-like hollow Al_2O_3 fibers (HFs) for advanced hemostatic material development [43]. The results demonstrated that Al_2O_3 NPs contributed synergistically through their structural and surface chemical properties, promoting blood concentration, activating coagulation pathways, and facilitating artificial clot formation.

Beyond these conventional oxides, various transition metal oxide nanoparticles-including V_2O_5, Cr_2O_3, MnO_2, CoO, and NiO-exhibit diverse structures and compositions and are widely utilized in catalysis, electrochemistry, optoelectronics, and magnetic materials [44, 45]. Functional regulation of CBHs using metal oxides leverages the unique properties of these nanoparticles and the exceptional matrix characteristics of CBHs, offering expansive potential for applications across multiple fields.

3.2.3 Carbon Nanomaterials

Carbon-based nanomaterials, composed entirely of carbon atoms, feature nanoscale structures with unique atomic arrangements that endow them with exceptional physicochemical properties [46, 47]. The incorporation of carbon-based nanomaterials not only improves the mechanical properties of hydrogels but also imparts electrical or thermal functionalities.

As the most common carbon-based nanomaterial, graphene oxide (GO) is rich in surface functional groups that can interact with cellulose chains through hydrogen bonding or π-π stacking, significantly improving the tensile strength, fracture toughness, and modulus of hydrogels. [48] For example, Huang et al. constructed a ternary composite hydrogel comprising GO, activated carbon (AC), and cellulose nanofibers (CNF), yielding an electrode material with high specific capacitance, flexibility, and processability [49]. In this system, GO serves as the conductive framework, interfacial optimization carrier, and structural stabilizer, providing a new pathway for the development of flexible energy storage devices. Additionally, GO features a high specific surface area and abundant surface active sites, making it an ideal carrier for AC and further enhancing the overall performance of CBHs.

Carbon nanotubes (CNTs) consist of single or multiple graphene sheets rolled into hollow tubular structures. CNTs are categorized into single-walled carbon nanotubes (SWCNTs) and multi-walled carbon nanotubes (MWCNTs) based on the number of layers [50]. In a recent study, Lim et al. fabricated a multifunctional smart hydrogel (PVG/NC) by combining MWCNTs with polyvinyl alcohol (PVA), gelatin, and cellulose nanocrystals (CNC), which was subsequently applied as a flexible electronic skin (E-skin) patch for sensing human motion, temperature, and humidity [51]. As shown in Fig. 3.2a, the incorporation of MWCNTs significantly enhanced the conductivity, adhesion, flexibility, and mechanical properties of the PVG/NC patch while endowing it with photothermal conversion capabilities, offering an advanced material solution for flexible bioelectronics and digital health monitoring.

Distinct from GO and CNTs, carbon quantum dots (CQDs) have demonstrated transformative potential in optoelectronic devices, biomedicine, environmental catalysis, and modern agriculture [52]. Owing to quantum size effects and surface energy trap distributions, the most remarkable feature of CQDs lies in their tunable fluorescence properties. Furthermore, the fluorescence of CQDs is sensitive to environmental pH changes, enabling the development of functional CBHs capable of

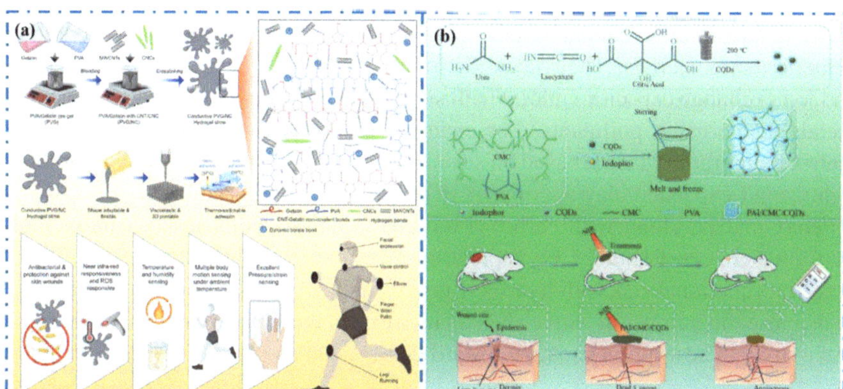

Fig. 3.2 **a** Schematic representation of a multifunctional E-skin hydrogel patch inspired by slime molds for human motion sensing and biological applications. Reprinted from Ref. [51], Copyright 2025, Wiley. **b** Schematic of the preparation of PAI/CMC/CQDs dressing for pH real-time monitoring and accelerated wound healing. Reprinted from Ref. [53], Copyright 2024, Elsevier

monitoring pH variations. As shown in Fig. 3.2b, Chen and his team employed CMC as the framework for a hybrid hydrogel network, crosslinked via hydrogen bonding with polyvinyl alcohol-iodine (PAI) and amino-functionalized CQDs, to construct an RGB monitoring system for quantifying the pH of wounds [53]. The fluorescence response of CQDs could be captured using a smartphone camera and converted into RGB values, facilitating convenient, non-invasive, real-time pH monitoring.

3.2.4 Ceramic and Clay Nanomaterials

With the rapid advancement of nanotechnology, conventional materials are being progressively miniaturized and functionalized, resulting in significantly enhanced properties. Ceramic and clay nanomaterials, owing to their unique physical and chemical characteristics, exhibit tremendous application potential in materials science, energy, environmental engineering, and biomedicine [54, 55].

Ceramic nanomaterials exhibit superior properties compared to their bulk ceramic counterparts, including high strength, hardness, thermal stability, chemical resistance, and electrical/optical features [56]. When interacting with cellulose chains via hydrogen bonding or electrostatic interactions, ceramic materials can significantly improve the compressive and tensile strength of hydrogels, indicating their potential in developing bone repair materials. In a recent study, Lin et al. fabricated multifunctional hydrogel scaffolds (ALG/MCC/HAP/OMMT) using bioceramic hydroxyapatite (HAP), alginate (ALG), microcrystalline cellulose (MCC), and organically modified montmorillonite (OMMT) [57]. Among these, the ceramic component HAP acted as a key mechanical reinforcement agent, enhancing the strength and rigidity of the scaffold. Additionally, HAP functioned as a crosslinker and structural stabilizer, participating in and stabilizing the formation of the gel network.

In contrast, clay nanomaterials exhibit distinct characteristics compared to ceramic nanomaterials. Their layered structures afford high specific surface areas and ion exchange capacities, along with significant adsorption capabilities for heavy metals and organic pollutants. For example, Norouzi et al. developed a mixed-matrix membrane (MMM) based on sustainable materials by incorporating modified lightweight expanded clay aggregates (LECA) doped with iron oxide (Fe_2O_3) into a Pebax matrix, along with carboxyl-modified natural cellulose, to achieve high-performance, eco-friendly CO_2 separation membranes [58]. In this design, LECA was first functionalized with hydroxyl groups and subsequently doped with Fe_2O_3 to increase the number of active sites on its surface, thereby enhancing its reversible reactivity and adsorption capacity toward CO_2.

3.2.5 Inorganic Salts Ions

Inorganic salt ions play a crucial role in the functional regulation of CBHs through multiple mechanisms, including coordination crosslinking, ionic channel construction, and freezing point suppression [59]. These mechanisms endow CBHs with revolutionary properties such as high toughness, superior conductivity, and excellent frost resistance. In CBHs, inorganic salts and ions act as key agents at both the molecular interaction level and in macroscopic performance modulation [60].

Achieving a balance between high conductivity and excellent mechanical performance remains a major challenge in hydrogel design. To address this issue, Dai et al. recently developed a covalently crosslinked composite hydrogel (PGC) composed of cellulose, gelatin, and polyacrylamide (PAM), where effective integration of multiple components was achieved through hydrogen bonding [61]. Subsequently, the PGC hydrogel was immersed in a mixed solution of betaine and $ZnSO_4$, where ion infiltration and interactions further tuned the material properties. The incorporation of Zn^{2+} facilitated coordination complexation with the abundant ligands in PGC, thereby enhancing the mechanical strength and stability of the hydrogel while imparting antibacterial properties. This strategy can be extended to other types of salts, zwitterions, and polymer systems, highlighting its broad adaptability and potential for future development.

Despite these advances, challenges remain in the functional regulation of CBHs via salt ions. Future studies must address issues such as biocompatibility, environmental responsiveness, and the integration of biodegradable devices, while also leveraging computational modeling and green manufacturing technologies to achieve precise material design and sustainable production processes.

3.3 Mechanisms of Functional Regulation in Cellulose Hydrogels

Inorganic nanomaterials systematically modulate the functional output of CBHs through physical adsorption, chemical bonding, structural design, and optical/electrical property transfer [62]. This subsection focuses on the mechanism of functional modulation of inorganic materials in CBHs.

3.3.1 Physical Interactions

Mechanical interactions between inorganic nanomaterials and CBHs are mainly realized through non-covalent forces, consisting of hydrogen bonding interactions, van der Waals effects, electrostatic interactions, and physical embedding/intercalation.

Compared to covalent and anionic bonds, hydrogen bonds have less bonding capacity and are weaker non-covalent interactions. Nonetheless, hydrogen bonding is widely practiced in cellulose hydrogels and functions synergistically at the multi-points in the aggregate polymer hydrogel network, hence significantly improving the whole stability of the hydrogel. For example, Sun et al. used sodium carboxymethyl cellulose (CMC-Na) as a modifier to significantly enhance the dispersion stability of OMMT in the hydrogel system [63]. As indicated in Fig. 3.3a, the hydroxyl and carboxyl groups on CMC-Na can form a network of strong hydrogen bonds with the carboxyl groups or residual hydroxyl groups of poly(acrylic acid) (PAA) chain segments. These hydrogen bonds both enhance the cohesion of the gel structure and provide additional binding sites during adsorption. In addition, the hydroxyl groups on the surface of OMMT can participate in hydrogen bonding or van der Waals forces to further cross-link with the polymer chains. The material combines excellent porous structure, high swelling rate and outstanding adsorption ability, which provides a new feasible solution for the efficient purification of cationic dye wastewater.

Physical interactions play a more important role in cellulose hydrogels. When nanomaterials are used as reinforcing fillers in cellulose hydrogels, the mechanical strength/toughness of the hydrogel can be well modulated, as well as physical entanglement and adsorption can stabilize the inorganic functional materials, thus enhancing stability. In addition, some physical interactions can be altered under the stimulation of external conditions, making cellulose hydrogels physically or chemically responsive, which is potentially instructive for the development of stimuli-responsive hydrogels.

3.3.2 Covalent and Coordination Bonding with Cellulose Matrix

Covalent bonding in CBHs is mainly achieved through the formation of strongly orientated chemical covalent bonds. Common strategies include silane coupling, click chemistry, and so on. Coordination interactions in CBHs are mainly formed through the formation of chemical coordination bonds between reactive groups on cellulose and metal ions [64].

In covalent interactions, strong chemical bonds greatly reduce the leaching rate of nanomaterials, endowing hydrogels with excellent resistance to disintegration or over-swelling under different conditions. For example, Lim et al. designed and constructed a novel soft hydrogel system (CMC-Nb), based on carboxymethyl cellulose functionalized with norbornene (Nb) [65]. As shown in Fig. 3.3b, efficient cross-linking of CMC-Nb molecules was achieved by introducing a cross-linking agent, which led to the rapid construction of the hydrogel network by employing the inverse electron demand Diels–Alder (IEDDA) click chemistry reaction. IEDDA successfully controlled the cross-linking density and gelation rate of the hydrogels and obtained soft hydrogels with tunable mechanical properties and high swelling.

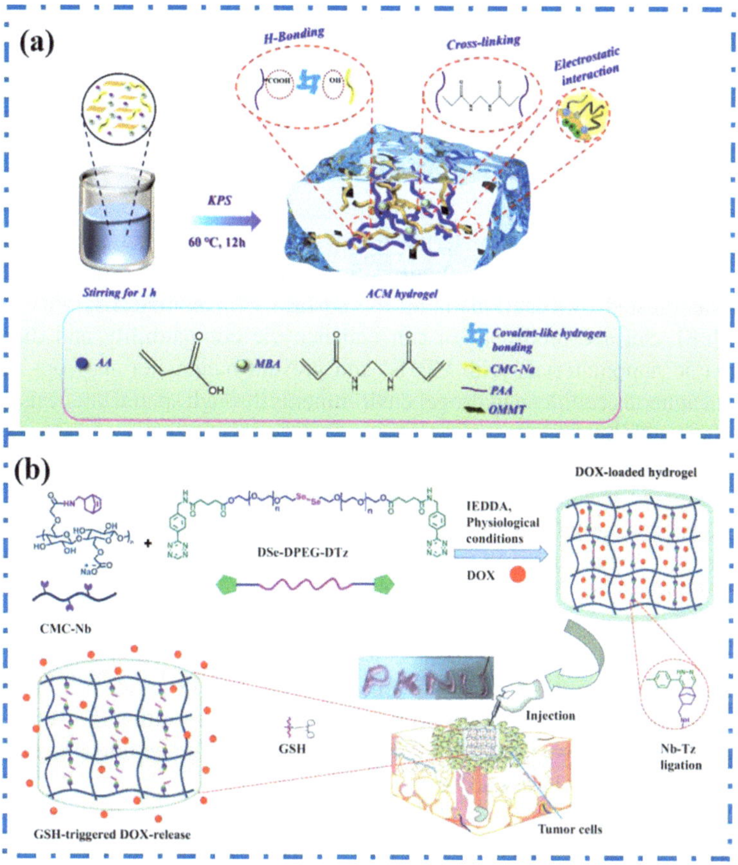

Fig. 3.3 **a** Schematic illustration of the development and drug release mechanism of reduction responsive CMC-based hydrogels. Reprinted from Ref. [63], Copyright 2025, Elsevier. **b** Schematic diagram of the synthesis process of ACM composite hydrogels. Reprinted from Ref. [65], Copyright 2022, Elsevier

These properties are well suited to the needs of injectable hydrogels in the biomedical field.

Jiao and colleagues constructed a p(AM-co-AMPS)/CNF/$ZnSO_4$ hydrogel by introducing CNF and $ZnSO_4$ into covalently crosslinked poly (acrylamide-co-2-acrylamido-2-methylpropanesulfonic) (p(AM-co-AMPS)) network sensor [66]. Among them, covalent cross-linking provided the basic skeleton of the hydrogel to ensure its structural stability and overall mechanical strength, and a tight network was formed between CNF and p(AM-co-AMPS) through a large number of hydrogen bonds. Meanwhile, Zn^{2+} ions can form metal coordination with sulfonate in AMPS and carboxylate in CNF, which further improves the mechanical properties of the hydrogel. It is this synergy of covalent and non-covalent interactions that provides the hydrogel with up to 1092% tensile properties, 54.66 kPa tensile strength, and 244 kJ m^{-3} toughness, combined with skin-like softness.

3.3.3 Nanostructuring and Surface Modification Effects

Nanostructuring of inorganic materials focuses mainly on the size and shape of the nanomaterials [67]. The surface modification effect of inorganic nanomaterials is mainly manifested by functionalizing the surface of nanomaterials with specific groups [68]. Surface modification can enhance the compatibility and dispersion of inorganic nanomaterials with CBHs, and prevent nanomaterials from agglomerating in aqueous cellulose hydrogel environments through spatial site resistance or electrostatic stabilization.

Surface modification of nanomaterials can also introduce functional handles to provide targeted reaction sites to enhance interactions with cellulose or to confer functions such as cellulose hydrogel biorecognition and drug coupling attachment. For example, Li et al. chose SO_4^{2-}/TiO_2 as a catalyst and modified its surface with vinyltrimethoxysilane (VTMS) to improve its compatibility with low-substituted cellulose during acetylation [69]. VTMS-modified SO_4^{2-}/TiO_2 exhibited higher degree of substitution and yield in cellulose acetation reaction compared with the unmodified catalyst. Organic side-chains of the VTMS molecule significantly enhance the SO_4^{2-}/TiO_2 catalyst's affinity water balance, resulting in stronger interfacial interactions between the low-substituted, hydrophilic cellulose generated in the early stage of acetylation and the catalyst.

3.4 Effects of Inorganic Materials on the Functional Properties of CBHs

Based on the above discussion, it can be concluded that CBHs show great potential in the fields of biomedicine, environmental remediation, and flexible electronics due to their excellent biocompatibility, degradability, and hydrophilicity. However, its inherent weaknesses limit the application expansion [70, 71]. The introduction of inorganic materials is an effective strategy to enhance its comprehensive performance. This subsection focuses on the influence of inorganic materials on the functional properties of CBHs, including mechanical strength and toughness, swelling behavior and water retention, biodegradability and environmental stability, antibacterial and antifouling properties, as well as electrical conductivity, magnetism and optical properties.

3.4.1 Mechanical Strength and Toughness Enhancement

Pure cellulose hydrogels usually have weak mechanical properties, including high brittleness, low strength and poor toughness, which make it difficult to withstand complex stresses or long-term use [72]. These shortcomings can be effectively improved by introducing inorganic nanomaterials. Rigid inorganic nanoparticles act as physical cross-linking points or reinforcing fillers, which are uniformly dispersed in the hydrogel network, and can effectively share the effect of external stress on the hydrogel.

The abundant hydroxyl groups in the cellulose structure can form physical/chemical interactions with specific inorganic materials. For example, Chen et al. used the natural anisotropic fibrous skeleton of wood as a high-strength support for in situ graft polymerization of acrylamide (AM) and acrylic acid (AA) monomers on lignocellulosic surfaces under UV light irradiation [73]. As shown in Fig. 3.4a, the cellulose hydrogel with dual network structure was achieved by using iron ions (Fe^{3+}) to form ionic coordination with carboxyl groups on poly(acrylic acid) (PAA), which further constructed physical cross-linking points in the flexible polymer network. The coordination bonds formed by Fe^{3+}-carboxylate ionic coordination can absorb and disperse the stresses during the loading process, further enhancing the toughness. This mechanism not only enhances the mechanical properties of the material, but also opens up new design ideas for the development of multifunctional and high-strength soft hydrogel materials.

In another report, Zhao et al. developed a high adhesive AgCu nanoparticle glue (AgCuNP glue), which was prepared by the simultaneous in situ reduction of Ag^{+} and Cu^{2+} in waterborne polyurethane (WPU) system [74]. The AgCuNP glue entered the yarn structure through the penetration of WPU, which enhanced the bonding between fibers and improved the overall strength and structural stability of the fabric. The method is compatible with conventional textile dyeing and finishing equipment, and has the potential to prepare antimicrobial functional fabrics on a large scale.

3.4.2 Swelling Behavior and Water Retention Control

CBHs are usually highly soluble, which may lead to structural instability, further degradation of mechanical properties, and limitations in applications requiring controlled release or specific shape retention, a challenge that can be well addressed by the introduction of inorganic nanomaterials [75]. In addition, inorganic nanoparticles can occupy the network space and may form microporous structures, resulting in spatial site-blocking or capillarity, which in turn affects water permeation pathways and storage.

For example, Wang et al. found that Al^{3+} and Zn^{2+} have a synergistic effect during cellulose depolymerization by means of multidimensional characterization and theoretical simulations [76]. As shown in Fig. 3.4b, Al^{3+} preferentially interacts strongly

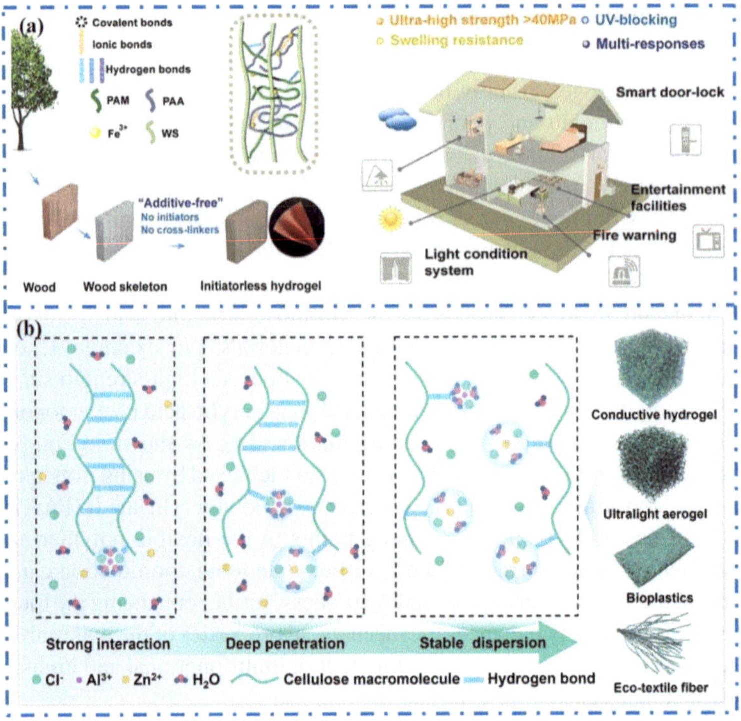

Fig. 3.4 **a** Schematic representation of the preparation of PAA_n-Fe_m/PAM/WS hydrogels and their application in multifunctional sensors. Reprinted from Ref. [73], Copyright 2025, Elsevier. **b** Schematic representation of the synergistic effect of Al^{3+} and Zn^{2+} in cellulose depolymerization. Reprinted from Ref. [76], Copyright 2025, American Chemical Society

electrostatically with the hydrogen bonds between cellulose chains by virtue of its high charge density and its very small hydration radius allows it to rapidly penetrate into the cellulose microfibril structure. In contrast, Zn^{2+} produces an effective solvation shielding effect through its larger hydration radius, stabilizing the dissolved chain segments and preventing them from re-aggregating. The study also systematically evaluated the potential application of this inorganic salt system in the preparation of functional materials, especially in the direction of hydrogels.

The metal–ligand interactions formed between inorganic ions and cellulose chains, as well as the optimization of the network structure brought about by the modulation of the hydration state, significantly improved the swelling capacity and water retention properties of the hydrogels.

3.4.3 Biodegradability and Environmental Stability

Cellulose hydrogels may degrade too rapidly under wet, microbial-rich environmental conditions, affecting longevity, while their stability needs to be a concern under dry, sterile environmental conditions or when long-term implantation is required [77]. The introduction of specific inorganic materials can regulate the biodegradability and environmental stability of CBHs.

For example, Chen et al. prepared a green nanocomposite printing ink (TOBC/SA/Xls) by using TEMPO-oxidized bacterial cellulose (TOBC) and sodium alginate (SA) as the matrix, and introduced slab silicate nanoclay (Xls) as an inorganic enhancer [78]. As shown in Fig. 3.5a, TOBC/SA/Xls hydrogel scaffolds with fine structure and excellent conformability were successfully constructed by a simple material extrusion 3D printing process without temperature control. Notably, these inorganic nanosheets acted as pillars between the network pores and effectively resisted hydration swelling and external stresses, which significantly enhanced the morphological stability of the hydrogel. The introduction of Xls not only improved pore connectivity and network pore homogeneity, but also provided a mild protective and slow-release environment for protein factors through weak electrostatic and hydrogen bonding interactions. This long-lasting drug release behavior, combined with the good support ability of the material itself for cell proliferation, suggests that the hydrogel has a broad application prospect in the field of drug-carrying and drug-releasing, tissue repair and functional bio-scaffolds.

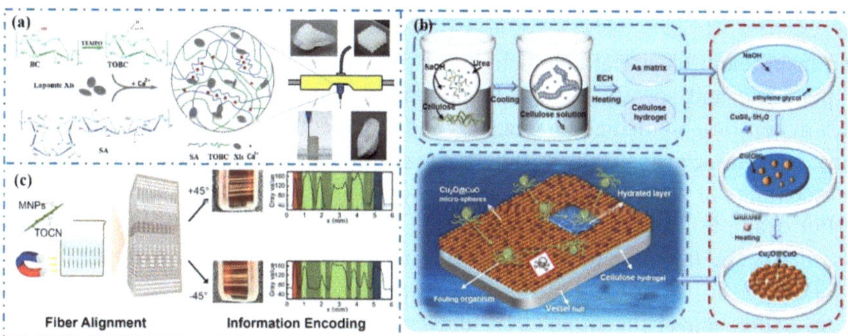

Fig. 3.5 **a** Schematic diagram of TOBC/SA/Xls hydrogel scaffold manufacturing process. Reprinted from Ref. [78], Copyright 2020, Elsevier. **b** Schematic diagram of the preparation process and antifouling mechanism of Cu_2O@CuO doped cellulose hydrogel. Reprinted from Ref. [81], Copyright 2022, Elsevier. **c** Schematic representation of CNFs functionalized with magnetic iron oxide nanoparticles for application in information coding. Reprinted from Ref. [83], Copyright 2025, American Chemical Society

3.4.4 Antibacterial and Antifouling Properties

As a biobased material, cellulose itself is prone to be a breeding ground for microorganisms, leading to bio-contamination, infection risk and material property degradation [79]. By introducing specific inorganic nanomaterials, cellulose hydrogels are endowed with antimicrobial or antifouling properties, which greatly improve their service life and performance [80]. The introduction of inorganic nanomaterials can not only endow CBHs with bacteria-killing properties, but also reduce the initial adhesion of biomolecules and microorganisms by improving the properties of the hydrogel surfaces to achieve resistance to bacterial contamination.

In a related report, Yang and colleagues loaded Cu_2O@CuO nanospheres in situ onto the surface of cellulose hydrogel by liquid phase reduction method, as shown in Fig. 3.5b [81]. The results showed that the antimicrobial activity and mechanical properties of the materials were significantly enhanced with the increase in the content of Cu_2O@CuO nanospheres. Among them, the hydrogel containing 37.5% precursors exhibited 100% killing rate against *E. coli*. In addition, the hydration layer formed on the surface of the hydrogel effectively inhibited the attachment of green algae, with a 40% reduction in the number of attachments. The adsorption of proteins was also reduced by 53.13%, demonstrating good resistance to protein contamination.

3.4.5 Conductivity, Magnetic, and Optical Properties

The insufficient electrical conductivity of pure cellulose hydrogels and their lack of magnetic and specific optical functionalities limit their applications in fields such as sensors, biomedical diagnostics and therapeutics. The introduction of inorganic materials can enable more functionalization of hydrogels, thus overcoming the shortcomings of pure hydrogels in applications [82].

For example, Rojas et al. used TEMPO oxidized cellulose nanofibers (TOCN) as a matrix and introduced magnetic ferrite nanoparticles to obtain magnetic TOCN (M-TOCN) [83]. As shown in Fig. 3.5c, the M-TOCN enable the fibrillar network to reversibly adjust the orientation under an external field, thereby dynamically adjusting the angle between the incident light and the fiber array to achieve controllable and programmable light reflection and transmission effects. Based on the combination of magnetic field guidance and gas phase gelation, this strategy not only breaks through the limitations of static optical materials, but also provides a completely new technological path for generating, writing and hiding information on demand.

Looking ahead, the composite of inorganic materials with cellulose hydrogels thus further realizes a high degree of integration and synergy of electric/magnetic/optical and other multifunctionalities, and develops flexible biosensors and wearable electronic devices with high sensitivity and stability. Meanwhile, exploring the modulation of cell behavior by electrical/magnetic stimulation in neural interfaces and tissue engineering, as well as optimizing the efficiency and controllability of therapies deserve greater attention in the development of biomaterials for biomedical applications.

3.5 Conclusion

In summary, functional regulation by introducing inorganic materials, including metal nanoparticles, metal oxides, carbon-based materials, clay minerals and inorganic ions, can effectively break through the above limitations. At the mechanism level, the inorganic components form a multilevel synergistic network with the cellulose matrix through physical interactions, chemical bonding and unique nanostructured effects. At the performance level, inorganic materials not only significantly improve the mechanical strength, controlled swelling and environmental stability of CBHs, but also give them multiple intelligent response characteristics and functional scalability, thus upgrading traditional hydrogels from passive carriers to active responsive intelligent platforms.

Nonetheless, the field is still facing problems such as the lack of dispersion uniformity of inorganic nanoparticles, doubtful long-term biosafety, structural biomimetic design challenges in complex application scenarios, and the green economy of large-scale production. Future research should be devoted to developing in-situ modification and precise assembly techniques to optimize dispersion, establishing a system to evaluate the slow release and toxicity of inorganic components, and exploring multi-scale ordered composite structures. Promote the development of low-energy synthesis and renewable recycling process, and then achieve the breakthrough of cellulose-inorganic composite hydrogel in the scale of tissue engineering, soft robotics and intelligent sensing and other cutting-edge fields.

References

1. B.Z. Liu, F.F. Sun, P.Y. Zhu, K. Wang, L.C. Peng, Y.L. Zhuang, H. Li, Preparation of multi-barrier and multi-functional paper-based materials by chitosan, ethyl cellulose and green walnut husk biorefinery products for sustainable food packaging. Int. J. Biol. Macromol. **278**, 134557 (2024)
2. K. Liu, Y.X. Wang, W. Liu, C.Y. Zheng, T. Xu, H.S. Du, Z.H. Yuan, C.L. Si, Bacterial cellulose/chitosan composite materials for biomedical applications. Chem. Eng. J. **494**, 153014 (2024)
3. H.D. Zhang, X.T. Gan, Y.Y. Yan, J.P. Zhou, A sustainable dual cross-linked cellulose hydrogel electrolyte for high-performance zinc-metal batteries. Nano-Micro Lett. **16**, 106 (2024)
4. Z. Zhang, N. Abidi, L. Lucia, S. Chabi, C.T. Denny, P. Parajuli, S.S. Rumi, Cellulose/nanocellulose superabsorbent hydrogels as a sustainable platform for materials applications: a mini-review and perspective. Carbohyd. Polym. **299**, 120140 (2023)
5. Z. Chen, T. Si, X. Gao, K.L. Chen, Z. Yu, Y.T. Zhang, L.C. Peng, J. Shang, H. Zhang, Biomimetic transformation of inferior parenchyma cellulose into a sustainable, functional medical hydrogel inspired by the dermis. Adv. Healthc. Mater. **14**, 2500698 (2025)

6. V. Sethi, M. Kaur, A. Thakur, P. Rishi, A. Kaushik, Unravelling the role of hemp straw derived cellulose in CMC/PVA hydrogel for sustained release of fluoroquinolone antibiotic. Int. J. Biol. Macromol. **222**, 844 (2022)
7. A. Bora, D. Sarmah, M.A. Rather, M. Mandal, N. Karak, Nanocomposite of starch, gelatin and itaconic acid-based biodegradable hydrogel and ZnO/cellulose nanofiber: a pH-sensitive sustained drug delivery vehicle. Int. J. Biol. Macromol. **256**, 128253 (2024)
8. K. Enoch, A.A. Somasundaram, Development of mn_3o_4 nanoparticles and justicia adhatoda extract incorporated carboxymethyl cellulose/carbopol hydrogels: rheological and in vitro bioactivity for sustained drug delivery system. Int. J. Biol. Macromol. **307**, 141713 (2025)
9. Z. Waheed, F. Ahmad, B. Mushtaq, S. Ahmad, S.R. Habib, A. Rasheed, M.S. Zafar, F. Sefat, M. Saeinasab, F. Azam, Biowaste rice husk derived cellulosic hydrogel incorporating industrial cotton waste nonwoven for wound dressing. Int. J. Biol. Macromol. **281**, 136412 (2024)
10. S.A. Ovalle-Serrano, L.A. Díaz-Serrano, C. Hong, J.P. Hinestroza, C. Blanco-Tirado, M.Y. Combariza, Synthesis of cellulose nanofiber hydrogels from fique tow and ag nanoparticles. Cellulose **27**, 9947 (2020)
11. T. Su, X.F. Mao, Z.R. Wang, Y.H. Pan, B. Xu, W.Z. Yang, H. Xu, Cellulose nanocrystal-infused polymer hydrogel imbued with ferric-manganese oxide nanoparticles for efficient antinomy removal. J. Hazard. Mater. **476**, 135097 (2024)
12. M.M. Chen, M. Zhou, Y.Y. Wang, C.H. Mao, S.J. Pang, T. Meng, X.D. Yang, Carboxymethyl cellulose and sodium alginate-enhanced hydrogel for carbon dots loading: a novel platform for pH sensing and sensitive detection of Al^{3+} and Ag^{+}. Int. J. Biol. Macromol. **307**, 141955 (2025)
13. H.S. Du, W.M. Liu, M.L. Zhang, C.L. Si, X.Y. Zhang, B. Li, Cellulose nanocrystals and cellulose nanofibrils based hydrogels for biomedical applications. Carbohyd. Polym. **209**, 130 (2019)
14. X.S. Pan, J. Li, N. Ma, X.J. Ma, M. Gao, Bacterial cellulose hydrogel for sensors. Chem. Eng. J. **461**, 142062 (2023)
15. S. Bhaladhare, D. Das, Cellulose: a fascinating biopolymer for hydrogel synthesis. J. Mater. Chem. B **10**, 1923 (2022)
16. H.Q. Xue, C. Zhu, Y.F. Wang, Q.C. Gu, Y.Y. Shao, A.Q. Jin, X.F. Zhang, L.J. Lei, Y.L. Li, Stimulus-responsive cellulose hydrogels in biomedical applications and challenges. Mater. Today Bio **32**, 101814 (2025)
17. Z.Q. Li, M.L. Wang, Y. Li, J.M. Ren, C.H. Pei, Effect of cellulose nanocrystals on bacterial cellulose hydrogel for oil-water separation. Sep. Purif. Technol. **304**, 122349 (2023)
18. K. Enoch, A.A. Somasundaram, Rheological insights on carboxymethyl cellulose hydrogels. Int. J. Biol. Macromol. **253**, 127481 (2023)
19. D.H. Zhang, J.Y. Jian, Y.T. Xie, S.S. Gao, Z. Ling, C.H. Lai, J.F. Wang, C.P. Wang, F.X. Chu, M.J. Dumont, Mimicking skin cellulose hydrogels for sensor applications. Chem. Eng. J. **427**, 130921 (2022)
20. Y.J. Tang, Z.X. Fang, H.J. Lee, Exploring applications and preparation techniques for cellulose hydrogels: a comprehensive review. Gels **10**, 365 (2024)
21. B.H. Zhang, J.L. Yang, Q.J. Jiang, R.P. Zheng, L.T. Qiu, J.X. Chen, R.Y. You, Z.X. Yan, Y.D. Lu, AuNPs/c-CNF/PVA hydrogel sers sensor for comprehensive detection of antitumor drug separation and enrichment. Chem. Eng. J. **516**, 164282 (2025)
22. M. Sumini, C.R. de Souza, G.J.S. Andrade, I.R.C. Oliveira, S. Scandorieiro, C.A. Tischer, R.K.T. Kobayashi, G. Nakazato, Cellulose hydrogel with hyaluronic acid and silver nanoparticles: sustained-release formulation with antibacterial properties against. Antibiotics-Basel **12**, 873 (2023)
23. S.L. Nejad, H. Shekarchizadeh, Visual tracking of real-time freshness of fish using an agar hydrogel colorimetric indicator containing CuNPs/NCQDs. Carbohyd. Polym. **343**, 122477 (2024)
24. P. Lormaneenopparat, J. Yukird, N. Rodthongkum, V.P. Hoven, Bacterial cellulose composite hydrogel for pre-concentration and mass spectrometric detection of thiol-containing biomarker. Int. J. Biol. Macromol. **253**, 126855 (2023)

25. A. Sánchez-Iglesias, N. Winckelmans, T. Altantzis, S. Bals, M. Grzelczak, L.M. Liz-Marzán, High-yield seeded growth of monodisperse pentatwinned gold nanoparticles through thermally induced seed twinning. J. Am. Chem. Soc. **139**, 107 (2017)
26. Q. Sun, X.L. Dong, J. Xu, T. Wang, Silver-infused lysine crosslinked hydrogel with oxidized regenerated cellulose for prospective advanced wound dressings. Int. J. Biol. Macromol. **264**, 130675 (2024)
27. R. Baruah, M.P. Hazarika, A.M. Das, G.N. Sastry, D. Nath, K. Talukdar, Green synthesis of nanocellulose supported cu-bionanocomposites and their profound applicability in the synthesis of amide derivatives and controlling of food-borne pathogens. Carbohyd. Polym. **330**, 121786 (2024)
28. S. Ashiri, E. Mehdipour, Preparation of a novel palladium catalytic hydrogel based on graphene oxide/chitosan NPs and cellulose nanowhiskers. Rsc Adv. **8**, 32877 (2018)
29. X.M. Liu, X. Zhou, C.Y. Yang, W.L. Yang, G. Liu, Y.H. Li, G.Y. Zhang, X. Zhao, Surfactant-free synthesis of cubr nps decorated by Pt for glucose and nitrite sensors. J. Ind. Eng. Chem. **124**, 323 (2023)
30. Y. Anwar, H.S.H.M. Ali, W.U. Rehman, H.A. Hemeg, S.A. Khan, Antibacterial films of alginate CoNi-coated cellulose paper stabilized Co NPs for dyes and nitrophenol degradation. Polymers **13**, 4122 (2021)
31. S.Y. Tang, D.R.G. Mitchell, Q.B. Zhao, D. Yuan, G.L. Yun, Y.X. Zhang, R.R. Qiao, Y.L. Lin, M.D. Dickey, W.H. Li, Phase separation in liquid metal nanoparticles. Matter **1**, 192 (2019)
32. M.Y. Guan, Z.Y. Huang, Z.T. Bao, Y.R. Ou, S.Y. Zou, G.Z. Liu, Gold nanoparticles incorporated liquid metal for wearable sensors and wound healing. Chem. Eng. J. **508**, 161120 (2025)
33. A.A. Oun, S. Shankar, J.W. Rhim, Multifunctional nanocellulose/metal and metal oxide nanoparticle hybrid nanomaterials. Crit. Rev. Food Sci. **60**, 435 (2020)
34. Z. Ahmadian, F. Kazeminava, M. Afrouz, M. Abbaszadeh, N.T. Mehr, J.A. Shiran, C. Gouda, M. Adeli, H.S. Kafil, A review on the impacts of metal/metal nanoparticles on characteristics of hydrogels: special focus on carbohydrate polymers. Int. J. Biol. Macromol. **253**, 126535 (2023)
35. M. Ezzine, A.S. El-Shafie, K.M. Youssef, M. El-Azazy, Bifunctional tio_2-cellulose based nanocomposites for synergistic adsorptive-photocatalytic removal of methyl orange: response modelling and optimization. Int. J. Biol. Macromol. **306**, 141753 (2025)
36. K.Y. Huang, H. Zeng, Q.X. Zhou, Heatwave enhance the adaptability of chlorella pyrenoidosa to zinc oxide nanoparticles: regulation of interfacial interactions and metabolic mechanisms. Water Res. **279**, 123466 (2025)
37. N.Y.Y. Lim, S.L. Chiam, C.P. Leo, S.Y. Pung, C.K. Chang, W.L. Ang, Suppression of charge recombination using microfibrillated cellulose in carboxymethyl cellulose coatings containing ZnO nanorods and biocl during photoelectrocatalysis. Int. J. Biol. Macromol. **308**, 142421 (2025)
38. T. Su, X.X. Zhang, Z.R. Wang, Y. Guo, X.Y. Wei, B. Xu, H.T. Xia, W.Z. Yang, H. Xu, Cellulose nanocrystal-based polymer hydrogel embedded with iron oxide nanorods for efficient arsenic removal. Carbohyd. Polym. **331**, 121855 (2024)
39. Y.F. Wang, Y.J. Zhang, Y.B. Zhao, Z. Zhao, J. Yao, L. Zou, Y. Zhang, Y. Guan, Y.J. Zhang, Magnetite nanoparticles with high affinity toward target protein for efficient and facile bio-separation. Adv. Sci. **12** (2025)
40. J.M. Xiao, X.P. An, F. Tang, X. Dai, S. Zhang, X.L. Zhu, J. Shen, J. Yuan, D.L. Gan, M.Q. Wang, Photosynthesis-inspired nir-triggered Fe_3O_4@MoS_2 core-shell nanozyme for promoting mrsa-infected diabetic wound healing. Adv. Healthc. Mater. **14**, 2404525 (2025)
41. T. Tang, W. Ding, W. Fu, S. Tang, X. Zhang, Scale-dependent anomalous behavior of confined water between Al_2O_3 layers. Nano Res. **18**, 94907417 (2025)
42. M.Q. Zhong, Y.Z. Li, J.C. Wu, Q.Q. Hu, Y.Q. Hu, Q. Du, C. Ji, H.M. Cao, L. Ji, Light promotion and different reaction pathways enhance photothermocatalytic syngas production for cellulose steam reforming over Ni/θ-Al_2O_3. Adv. Funct. Mater. **34**, 2405453 (2024)
43. C.L. Li, F. Li, J.Y. Chen, H.M. Wu, Y. Lin, C.Y. Chen, P. Zhang, Q.G. Wang, J. Liu, G.Y. Deng, Flexible biomimetic hollow Al_2O_3 fibers for safe and effective hemostasis. Mater. Design **213**, 110365 (2022)

44. M.R. Tariq, M. Ahmad, M.U.D. Naik, I. Khan, B.L. Zhang, A comprehensive review of the advancement of transition metal oxide nanocomposites for microwave absorption. Coordin. Chem. Rev. **533**, 216535 (2025)
45. A.B. Sengul, E. Asmatulu, Toxicity of metal and metal oxide nanoparticles: a review. Environ. Chem. Lett. **18**, 1659 (2020)
46. B.W. Sui, Y.P. Zhang, L. Huang, Y.X. Chen, D.W. Li, Y.F. Li, B. Yang, Fluorescent nanofibrillar hydrogels of carbon dots and cellulose nanocrystals and their biocompatibility. Acs Sustain. Chem. Eng. **8**, 18492 (2020)
47. H. Nawaz, X. Zhang, S. Chen, T.T. You, F. Xu, Recent studies on cellulose-based fluorescent smart materials and their applications: a comprehensive review. Carbohyd. Polym. **267**, 118135 (2021)
48. J.R. Liu, S.H. Lv, Y.L. Mu, J.H. Tong, L.P. Liu, T.X. He, Q. Zeng, D.Q. Wei, Applied research and recent advances in the development of flexible sensing hydrogels from cellulose: a review. Int. J. Biol. Macromol. **281**, 136100 (2024)
49. Q. Wu, C. Jiang, Y. Zhao, Y.K. Li, S.T. Yu, L. Huang, Cellulose nanofiber-based hybrid hydrogel electrode with superhydrophilicity enabling flexible high energy density supercapacitor and multifunctional sensors. Int. J. Biol. Macromol. **276**, 134003 (2024)
50. L. Wu, L.W. Li, M.S. Fan, P. Tang, S.T. Yang, L.J. Pan, H. Wang, Y.Z. Bin, Strong and tough PVA/PAA hydrogel fiber with highly strain sensitivity enabled by coating mwcnts. Compos. Part a-Appl. S. **138**, 106050 (2020)
51. H. Kim, S.D. Dutta, M.J. Jeon, J. Lee, H. Park, Y. Seol, K.T. Lim, Bioinspired shape reconfigurable, printable, and conductive "e-skin" patch with robust antibacterial properties for human health sensing. Adv. Funct. Mater. **35**, 2504088 (2025)
52. F. Montazeri, A. Ghasedi, B. Mahdavi, E. Koushki, The effect of i*n-situ* cellulosic matrix on the photophysical properties of white emissive cqds. Carbon. Lett. **34**, 399 (2024)
53. Y.Y. Yang, C. Zhang, M. Gong, Y. Zhan, Z.K. Yu, C. Shen, Y.H. Zhang, L. Yu, Z.X. Chen, Integrated photo-inspired antibacterial polyvinyl alcohol/carboxymethyl cellulose hydrogel dressings for ph real-time monitoring and accelerated wound healing. Int. J. Biol. Macromol. **238**, 124123 (2023)
54. Q. Mo, L.J. Huang, Y. Sheng, Z.H. Wei, S.Y. Zhang, Y.S. Li, X.Y. Wang, Y.N. Wang, X.Y. Lu, C.X. Huang, Q.S. Duan, M.G. Xue, Crosslinking strategy and promotion role of cellulose as a composite hydrogel component for three-dimensional printing—a review. Food Hydrocolloid. **154**, 110079 (2024)
55. L. Alves, E. Ferraz, J.A.F. Gamelas, Composites of nanofibrillated cellulose with clay minerals: a review. Adv. Colloid. Interfac. **272**, 101994 (2019)
56. C. Hotton, J. Sirieix-Plénet, G. Ducouret, T. Bizien, A. Chennevière, L. Porcar, L. Michot, N. Malikova, Organisation of clay nanoplatelets in a polyelectrolyte-based hydrogel. J. Colloid Interf. Sci. **604**, 358 (2021)
57. X.Q. Chen, J.Y. Liu, Y.N. Bu, T. Wu, J.J. Fan, H.Q. Yan, Q. Lin, Dodecyl glycoside intercalated organo-montmorillonite promoted biomimetic alginate/microcrystalline cellulose/nanohydroxyapatite composite hydrogels for bone tissue engineering. Int. J. Biol. Macromol. **310**, 143304 (2025)
58. M.E. Kojabad, E. Bekhradinassab, A. Norouzi, M. Moradian, New pebax mixed-matrix membrane incorporating natural cellulose and lightweight expanded clay aggregate-iron oxide for enhanced co_2 separation. J. Environ. Chem. Eng. **13**, 117352 (2025)
59. Z. Abbasi, Z. Esfandiari, H. Rostamabadi, Postbiotic-loaded κ-carrageenan hydrogels double cross-linked with carboxymethyl cellulose and calcium ions. Food Hydrocolloid. **168**, 111551 (2025)
60. S.Y. Lu, X.N. Hu, B. Xu, C.Y. Bai, T.H. Wang, T. Ma, S. Yi, Effects of different salt ions on the structure and rheological behavior of sulfated cellulose nanocrystal hydrogel. Food Hydrocolloid. **151**, 109799 (2024)
61. Q.Y. Luo, S.Y. Yang, Z.W. Wu, J.G. Dai, M. Wang, Y.T. Xu, L.Z. Dai, Construction of multifunctional composite hydrogels *via* zwitterionic osmosis, the hofmeister effect, and metal complexation for flexible sensors. Compos. Sci. Technol. **265**, 111138 (2025)

62. B. Lu, H.H. Cheng, L.T. Qu, Inorganic hydrogel based on low-dimensional nanomaterials. ACS Nano **18**, 2730 (2024)
63. Y. Zhao, S. Sun, Carboxymethyl cellulose sodium modified poly(acrylic acid) and organic montmorillonite nanocomposite hydrogels with excellent purification ability for cationic dye wastewater. Int. J. Biol. Macromol. **308**, 142521 (2025)
64. H. Gao, Z. Wang, T. Lin, G. Zhang, X. Han, C. Cheng, Dynamic covalent bonding based polysaccharide hydrogels and their applications in wound dressings: a review. Colloids Surf. B **254**, 114878 (2025)
65. I. Ali, M. Gulfam, S.H. Jo, J.W. Seo, A. Rizwan, S.H. Park, K.T. Lim, Reduction-responsive and bioorthogonal carboxymethyl cellulose based soft hydrogels cross-linked via iedda click chemistry for cancer therapy application. Int. J. Biol. Macromol. **219**, 109 (2022)
66. M.Q. Shi, Y. Liang, C.Y. Zhang, N. Li, Y.F. Li, X.J. Shi, Z.H. Qin, T.F. Jiao, Cellulose nanofibril enhanced ionic conductive hydrogels with high stretchability, high toughness and self-adhesive ability for flexible strain sensors. Int. J. Biol. Macromol. **293**, 139297 (2025)
67. X. He, Q.Y. Lu, Design and fabrication strategies of cellulose nanocrystal-based hydrogel and its highlighted application using 3D printing: a review. Carbohyd. Polym. **301**, 120351 (2023)
68. A. Shavandi, S. Hosseini, O.V. Okoro, L. Nie, F.E. Babadi, F. Melchels, 3D bioprinting of lignocellulosic biomaterials. Adv. Healthc. Mater. **9**, 2001472 (2020)
69. Q.Q. Li, X.J. Zhao, K.Z. Shi, C.H. Zhang, N. Suo, J. She, Q. Qian, R. An, L.C. Li, Hydrophobic surface modification of SO_4^{2-}/TiO_2 solid acid for enhanced catalytic cellulose acetylation. Acs Sustain. Chem. Eng. **13**, 4302 (2025)
70. A.K. Rana, V.K. Gupta, P. Hart, V.K. Thakur, Cellulose-alginate hydrogels and their nanocomposites for water remediation and biomedical applications. Environ. Res. **243**, 117889 (2024)
71. C. Zollfrank, C. Selhuber-Unkel, Recent advances in the field of bioinspired materials and systems. Adv. Funct. Mater. **34**, 2408945 (2024)
72. S. Butylina, S.Y. Geng, K. Laatikainen, K. Oksman, Cellulose nanocomposite hydrogels: from formulation to material properties. Front. Chem. **8**, 655 (2020)
73. L.Z. Wang, J. Wei, M.Q. You, Y.C. Jin, D.G. Li, Z.Y. Xu, A.P. Yu, J.S. Li, C.C. Chen, Initiatorless polymerization of mechanically robust hydrogels reinforced by cellulose of wood skeleton as multifunctional sensors. Carbohyd. Polym. **354**, 123345 (2025)
74. T.T. Cao, Y. Zhou, L.L. Wang, J. Zhang, B. Zhao, Washable antibacterial cellulose fabric with improved mechanical strength through constructing an adhesive polymer coating by nano glue. Int. J. Biol. Macromol. **315**, 144414 (2025)
75. Z.D. Zhu, S.Y. Fu, N. Lavoine, L.A. Lucia, Structural reconstruction strategies for the design of cellulose nanomaterials and aligned wood cellulose-based functional materials—a review. Carbohyd. Polym. **247**, 116722 (2020)
76. X. Li, Z.H. Chen, X. Guan, H.C. Jiang, M. Yan, L.L. Zhang, J.X. Ma, L. Wang, Z.G. Wang, Decoding dual-ion synergy in $AlCl_3/ZnCl_2$ hydrates: an atomic "interaction-penetration-dispersion" mechanism for ambient cellulose valorization. ACS Nano **19**, 20991 (2025)
77. D.M. Nascimento, Y.L. Nunes, M.C.B. Figueirêdo, H.M.C. de Azeredo, F.A. Aouada, J.P.A. Feitosa, M.F. Rosa, A. Dufresne, Nanocellulose nanocomposite hydrogels: technological and environmental issues. Green Chem. **20**, 2428 (2018)
78. J.X. Wei, B.X. Wang, Z. Li, Z.T. Wu, M.H. Zhang, N. Sheng, Q.Q. Liang, H.P. Wang, S.Y. Chen, A 3D-printable tempo-oxidized bacterial cellulose/alginate hydrogel with enhanced stability *via* nanoclay incorporation. Carbohyd. Polym. **238**, 116207 (2020)
79. X. Lan, T. Du, J.C. Zhuo, T.Y. Wang, R. Shu, Y.C. Li, W.T. Zhang, Y.W. Ji, Y.R. Wang, X.Y. Yue, J.L. Wang, Advances of biomacromolecule-based antibacterial hydrogels and their performance evaluation for wound healing: a review. Int. J. Biol. Macromol. **279**, 135577 (2024)
80. T. Zivari-Ghader, M.R. Rashidi, M. Mehrali, Biological macromolecule-based hydrogels with antibacterial and antioxidant activities for wound dressing: a review. Int. J. Biol. Macromol. **279**, 134578 (2024)

81. X. Zhang, Y.C. Feng, D.C. Gao, W. Ma, C.Z. Jin, X.M. Jiang, J.G. Lin, F. Yang, Functionalization of cellulosic hydrogels with Cu_2O@CuO nanospheres: toward antifouling applications. Carbohyd. Polym. **282**, 119136 (2022)
82. G. Sharma, A. Verma, A. García-Peñas, A. Kumar, P. Dhiman, T.T. Wang, J. Amirian, Polysaccharide-based biopolymeric magnetic hydrogels for remediation of antibiotics from aqueous solution. Int. J. Biol. Macromol. **283**, 137555 (2024)
83. J.H. Xu, Y.J. Zou, H.J.Y. Chen, Z.M. Wan, A. Takagi, Z.G. Wang, J. Yu, L. Liu, Y. Lu, Y.M. Fan, O.J. Rojas, Magnetoresponsive cellulose nanofiber hydrogels: dynamic structuring, selective light transmission, and information encoding. ACS Nano **19**, 14063 (2025)

Chapter 4
Cellulose-Based Hydrogels for Woung Dressing

Minghao Zhang, Shiyan Chen, and Huaping Wang

Abstract Cellulose-based hydrogels (CBHs) have emerged as highly promising materials for advanced wound dressings, offering biocompatibility and unique physicochemical features. In this chapter, we focus on the major categories of cellulose used in hydrogels dressings, including bacterial cellulose, plant-derived cellulose, regenerated cellulose, and cellulose derivatives, and discuss their fabrication strategies, functional enhancements, and therapeutic effects. The integration of new functionalities, including antimicrobial activity, stimuli-responsiveness, mechanical enhancement, self-healing capability, hemostatic ability, and injectability, enables these systems to adapt to dynamic wound environments and address complex clinical needs. Despite substantial progress, challenges such as limited scalability and reproducibility of manufacturing, bioactive stability under sterilization, and the economic feasibility of highly functionalized systems continue to impede clinical translation. Overall, CBHs offer a versatile and evolving platform with the potential to redefine wound care through their adaptability, safety, and integration with advanced regenerative technologies.

Keywords Cellulose · Hydrogel · Wound healing · Dressing · Biomaterials

4.1 Introduction

As the primary protective barrier against the external environment, human skin performs pivotal functions including sensory perception, thermoregulation, inhibition of transepidermal water loss, mechanical/chemical protection of underlying

M. Zhang (✉)
School of Materials Science and Engineering, Shanghai Institute of Technology, Shanghai, PR China
e-mail: mhzhang@sit.edu.cn

S. Chen (✉) · H. Wang
State Key Laboratory of Advanced Fiber Materials, College of Materials Science and Engineering, Donghua University, Shanghai, PR China
e-mail: chensy@dhu.edu.cn

G. Wei and J. Zhang (eds.), *Cellulose-Based Hydrogels and Aerogels*, Lecture Notes in Nanoscale Science and Technology 41,
https://doi.org/10.1007/978-3-032-03008-5_4

tissues, and defense against pathogens [1]. Nevertheless, dermatological pathologies or traumatic insults such as burns, physical injuries, and decubitus ulcers frequently compromise cutaneous integrity. Such profound loss of skin function and structure causes severe homeostatic disruption, ultimately leading to debilitating disability or mortality [2]. Wound dressings play a pivotal role in wound care, with advanced types providing a moist environment, protecting the wound from external contaminants, and promoting tissue regeneration [3].

Traditional wound dressings, such as bandages and cotton wool made from natural cellulose fibers, provide only passive protection. They are often ineffective in managing exudate, preventing infection, or promoting active tissue regeneration. Additionally, frequent dressing changes, required to prevent infection, and the risk of adherence to the wound due to absorbed exudate remain major drawbacks of these conventional wound care methods [4]. Hydrogels are cross-linked hydrophilic polymer networks renowned for their exceptional ability to absorb and retain large volumes of aqueous solutions, including biological fluids. These materials have become a key focus in wound management research due to their ability to maintain an optimal moist microenvironment, promote cellular migration and proliferation, thereby enhancing wound healing [5, 6].

As a primary component of traditional dressings, cellulose has attracted considerable attention for its unique combination of biocompatibility, tunability, mechanical strength, and sustainability. The hydroxyl-rich structure of cellulose facilitates chemical modifications, enabling the design of hydrogels with tailored properties such as pH responsiveness, self-healing capability, and antimicrobial activity [7]. From a sustainability standpoint, cellulose-based hydrogels (CBHs) are particularly advantageous as they derive from renewable sources, aligning with eco-friendly design principles in biomedical material development. Furthermore, they allows integration with bioactive factors [8], nanoparticles [9–11], or smart sensors [4], significantly expanding their therapeutic potential for both acute and chronic wound management.

This chapter systematically reviews and highlights key advances in CBH wound dressings, encompassing raw material types, design strategies, functionalization and performance enhancement, as well as remaining challenges. Special emphasis is placed on emerging features, including antimicrobial activity, stimuli-responsiveness, mechanical enhancement, self-healing capability, hemostatic ability, and injectability, which contribute to the development of next-generation wound care technologies.

4.2 Fundamentals and Evolution of Wound Dressings

4.2.1 Essential Requirements for Ideal Wound Dressings

Wound healing is a complex biological process involving various phases, including hemostasis, inflammation, proliferation, and remodeling (Fig. 4.1) [12, 13]. Effective wound management is crucial for preventing infections, accelerating the healing process, and minimizing scarring. An ideal wound dressing should maintain a moist environment conducive to healing, permit gas exchange, absorb excess exudate, and protect against microbial invasion [14, 15]. Compared with dry dressings, wet dressings accelerate healing, as skin regeneration occurs optimally in a moist environment without inflammation or eschar formation [16, 17]. The dressing must allow oxygen diffusion through its matrix, as it supports cellular metabolism and angiogenesis [18]. Biocompatibility and non-toxicity are equally important to prevent inflammation and delayed healing, especially in chronic wounds, where impaired vascularization and persistent inflammation hinder recovery [4]. Additionally, the material should conform closely to the wound bed to eliminate dead space, minimize trauma during dressing changes, and enhance comfort and adherence. These principles have driven the shift from passive gauze to biointeractive materials, with hydrogels emerging as advanced systems that meet these physiological requirements [19, 20].

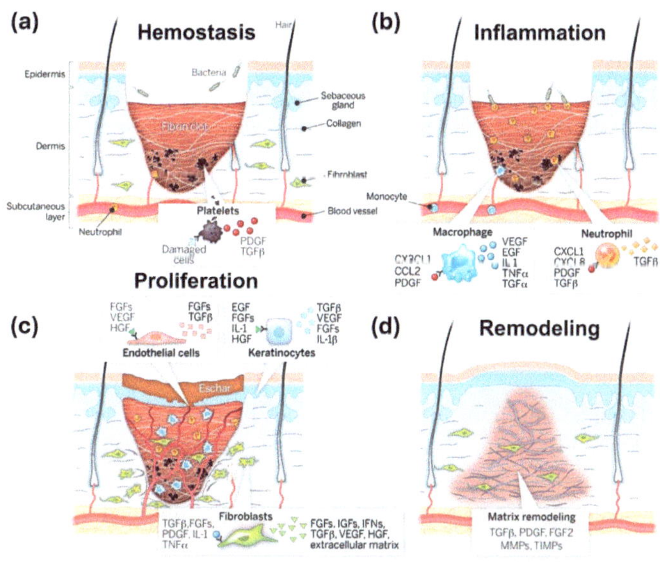

Fig. 4.1 The four stages of wound healing. **a** Hemostasis, **b** inflammation, **c** proliferation, and **d** remodeling. Reproduced with the permission from Ref. [13]. Copyright 2014, American Association for the Advancement of Science

4.2.2 Advantages and Key Characteristics of Hydrogel Dressings

Historically, gauze and bandages were widely used for their low cost and simplicity. However, they are dry, non-occlusive, and require frequent changes, which can damage new tissue and cause pain. Gauze also lacks moisture retention and fails to regulate the wound microenvironment. While foams offer better absorption, they are often bulky and lack adhesion. Hydrogels, composed of 3D networks of hydrophilic polymers, can absorb significant amounts of water or exudate without losing structural integrity. Their high water content and 3D porous structure closely mimic the natural extracellular matrix [21]. Hydrogels can also be molded into various shapes and tailored for specific softness or stiffness in accordance with the distinct requirements of different wound types [22]. For example, chronic wounds generally necessitate softer hydrogels to minimize mechanical stress on the tissue, whereas acute wounds may benefit from stiffer hydrogels that offer greater structural support [23]. Many hydrophilic polymers have been used to construct hydrogels through various chemical or physical cross-linking methods, including natural polymers such as chitosan (CS), alginate, hyaluronic acid, and gelatin, as well as synthetic polymers such as polyvinyl alcohol (PVA), polyethylene glycol (PEG), and polyacrylamide (PAM) [24]. The choice of polymer or polymer combination is determined by the desired properties and the specific requirements of the wound [5]. CBHs further advance the field due to their excellent biocompatibility, biodegradability, and non-immunogenic nature. Their degradation products, notably glucose, are non-toxic and can be metabolized by cells [25]. In addition, cellulose offers exceptional mechanical properties, enabling the production of hydrogels that retain structural integrity. Its hydrophilic nature allows for high water-holding capacity, creating a moist wound environment that supports tissue regeneration. Together, these advantages make CBHs highly adaptable and effective for addressing the demands of modern wound care.

4.3 CBHs for Wound Dressings: Types and Fabrication

CBHs are classified into four primary categories based on the source and chemical structure of cellulose: bacterial cellulose (BC), plant-derived cellulose, regenerated cellulose, and cellulose derivatives. Each type presents distinct fabrication strategies and biomedical functionalities suitable for wound dressing applications. This section systematically examines these classes, highlighting their physicochemical characteristics, fabrication methods, and wound healing potential.

4.3.1 Plant-Derived CBH Dressings

Plant-based cellulose, sourced from wood pulp, cotton, or agricultural residues, is abundantly available and cost-effective [26]. Native plant cellulose can be chemically processed to produce hydrogels with suitable swelling capacity, porosity, and moisture retention, supporting wound healing [27]. In parallel, cellulose can also be converted into nanoscale forms such as cellulose nanocrystals (CNCs) and cellulose nanofibers (CNFs). These nanocellulose materials offer increased surface area, improved mechanical properties, and greater potential for chemical modification [28].

Nanoellulose can be extracted from various types of plants for use in developing CBH wound dressings. Yin et al. developed a Schiff-base crosslinked hydrogel using microcrystalline cellulose from pineapple peel and CS, achieving a swelling ratio exceeding 4000% and hemostatic potential (Fig. 4.2 a). The use of agricultural by-products for hydrogel fabrication represents a resource-saving and eco-friendly approach, and the resulting composite hydrogel shows potential for application in wound dressing development [29]. Tang et al. prepared oxidized soybean CNC/PAM hydrogels and demonstrated that CNCs modulate rheology, enhance stiffness, and improve pH responsiveness and drug delivery, thereby effectively accelerating wound healing in a wound model [30]. In another study, Bajpai et al. developed composite hydrogel films composed of cellulose nanowhiskers derived from paper and polysodium acrylate, loaded with minocycline. The films exhibited sustained drug release, non-thrombogenic and non-hemolytic behavior. Their demonstrated antibacterial and antifungal activity further supports their potential as safe and effective hydrogel dressings for wound care applications [31]. In addition to CNCs, CNFs have also been utilized in hydrogel dressing systems. Jafri et al. developed a PVA-PEG/glycerin hydrogel incorporating CNFs and lemongrass essential oil via freeze–thaw processing. The optimized formulation showed high swelling, porosity, and gel fraction, alongside strong antibacterial activity against S. aureus and B. subtilis. These properties suggest its potential as a multifunctional wound dressing with enhanced moisture retention and antimicrobial efficacy [32].

These CBH dressings also demonstrate excellent mechanical properties. Ahmad et al. reinforced a sol–gel synthesized cellulose hydrogel with nonwoven cotton, enhancing its mechanical stability and exudate absorption (Fig. 4.2b). The addition of titania particles endowed the composite with antimicrobial properties while retaining air permeability [33]. Yin et al. incorporated CNCs extracted from plant cellulose into a gelatin-hyaluronic acid hydrogel, which supported NIH-3T3 fibroblast proliferation. The CNCs significantly enhanced the property of the hydrogels and play a vital role according to the rheology and swelling results [34]. In another study, Deng et al. combined fenugreek gum with cellulose to form a hydrogel with synergistic hemostatic and healing activity, while its hydrogen-bonded network enhanced mechanical properties. Moreover, the resulting hydrogel featured a porous architecture with high water absorption capacity and good biocompatibility [35]. Collectively, these studies demonstrate that plant-derived CBH dressing can be tailored

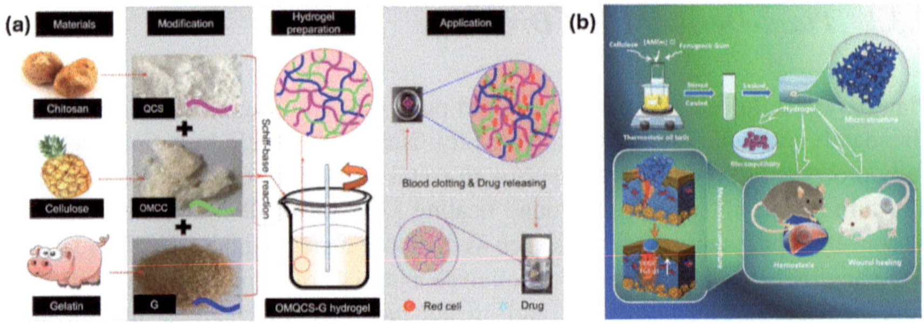

Fig. 4.2 Design and fabrication of hydrogel dressings based on plant-derived cellulose. **a** The OMQCS-G hydrogel, prepared through a brief fabrication process using cellulose derived from pineapple, is employed as a wound dressing. Reproduced with the permission from Ref. [29]. Copyright 2025, Springer Nature. **b** Development of the fenugreek gum-cellulose composite hydrogel for wound dressing applications. Reproduced with the permission from Ref. [35]. Copyright 2020, Elsevier

through chemical modification and composite formulation to improve hydrogel performance in terms of mechanical strength, swelling behavior, biocompatibility, and wound-healing efficacy.

4.3.2 BC Hydrogel Dressings

BC is a high-purity, nanofibrillar form of cellulose produced by certain bacterial strains such as *Acetobacter xylinum* [36]. Under steady-state culture conditions, *Acetobacter xylinum* converts glucose into cellulose at the gas/liquid interface to form a regular hydrogel film, while dynamic culture conditions yield flocculent or spherical cellulose [37]. The microstructure of BC hydrogel differs significantly from plant cellulose, featuring a unique 3D network of nanofibers (Fig. 4.3a) [38, 39]. This structure formed through *Acetobacter xylinum's* biosynthetic process where 1.5 nm microfibers aggregate into 3–4 nm fiber bundles, eventually assembling into ribbon-like structures via hydrogen bonding [40]. BC exhibits exceptional properties compared to plant-derived cellulose: its 3D nanofiber network and surface-enriched hydroxyl groups enable a water-holding capacity up to 90 times its weight. The biosynthesis pathway ensures high purity ($\leq$99.9%) without lignin or hemicellulose contamination, requiring minimal purification. BC also demonstrates high crystallinity (70–80%) due to the ordered crystallization of fiber bundles within cell wall pores, along with remarkable mechanical strength (Young's modulus of 0.1–10 GPa for dry films). These attributes–including nanofibrillar morphology, superior water retention, mechanical robustness, and biocompatibility–collectively position BC as an ideal wound dressing material.

Produced via microbial fermentation, BC hydrogels are typically harvested directly in the form of hydrogel films for wound dressing applications. These films

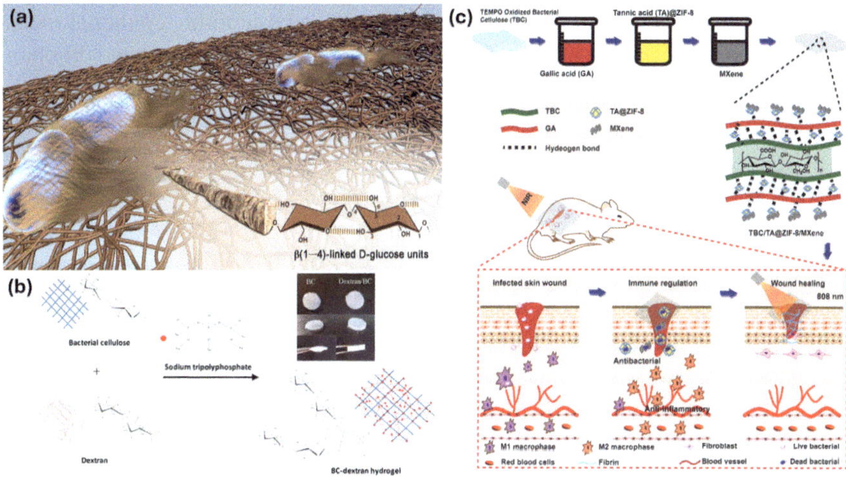

Fig. 4.3 Design and fabrication of hydrogel dressings based on BC. **a** Representative scheme of the 3D nanofibrous network produced by *Acetobacter xylinum* bacteria. Reproduced with the permission from Ref. [39]. Copyright 2016, Elsevier. **b** Schematic illustration depicting the chemical synthesis of the dextran modified BC hydrogel. Reproduced with the permission from Ref. [42]. Copyright 2017, Springer Nature. **c** Schematic illustration of the hydrogel dressing fabrication from BC, TA@ZIF-8 and MXene, along with the proposed mechanism for infected wound healing. Reproduced with the permission from Ref. [43]. Copyright 2023, Cell Press

undergo purification and sterilization processes prior to use. To enhance their functionality, BC hydrogels, whether in film or bulk form, can be chemically modified, physically combined with other materials, or crosslinked with bioactive agents to improve their structural, biological, and therapeutic performance. For example, Suneetha et al. impregnated fungal-derived carboxymethyl CS (CMCS) into BC via solution immersion (Fig. 4.3b). The resulting CMCS-BC hydrogel demonstrated enhanced improving hydrophilicity, cell adhesion, and antibacterial properties against *E. coli* and *S. aureus* [41]. Lin et al. fabricated a dextran-modified BC hydrogel with enhanced elasticity, reduced cytotoxicity, and significantly accelerated wound closure in animal models [42]. Cheng et al. prepared a BC hydrogel decorated with TA@ZIF-8 and MXene nanosheets via solution immersion (Fig. 4.3c). Under near infrared light irradiation, this hydrogel eradicated bacterial infection while promoting neovascularization and wound closure [43]. BC can also be processed into nanofiber suspensions through homogenization for subsequent modification or cross-linking to reconstruct hydrogels. Mo et al. developed a BC/PVA hydrogel crosslinked with citric acid, which exhibited superior swelling behavior (3485.3%) and no cytotoxicity to fibroblast cells, has certain potential for highly absorbent dressing [44]. Liu et al. fabricated a BC hydrogel grafted with quaternary ammonium functional and crosslinked with the gelatin-heparin system, showing antibacterial inhibition zones of 3 cm implemented with *Staphylococcus aureus*, and promoting epithelial reconstruction and angiogenesis in vivo, indicating its potential for use as wound dressings [45].

4.3.3 Regenerated CBH Dressings

Common regenerated cellulose products include viscose, rayon, and Lyocell, which are already well established in industrial production. For wound dressing applications, regenerated cellulose can be processed into hydrogels by combining it with polymers or crosslinking agents, offering a uniform structure and good biocompatibility. The preparation of such hydrogels typically requires a chemical pretreatment and dissolution step. Dissolution is commonly achieved using ionic liquids, alkali/urea systems, or solvents such as DMAc/LiCl. Subsequent regeneration is carried out by coagulating the cellulose in anti-solvents like ethanol or water, resulting in the formation of CBH dressing.

For instance, Kiselioviene et al. fabricated transparent regenerated cellulose hydrogel sheets by hardening hydrolyzed cellulose acetate in acetone and incorporating glycerol to enhance softness and flexibility. The resulting hydrogels exhibited strong moisture retention, good transparency, satisfactory mechanical strength, and effective antimicrobial activity when loaded with neomycin. These combined properties contributed to their potential in supporting wound healing [46]. Similarly, Mao et al. developed a regenerated hydrogel by dissolving BC in a low-temperature NaOH-urea-H_2O system. The hydrogel was formed by crosslinking with epichlorohydrin and incorporated with MXene nanosheets. The resulting electroactive BC-based hydrogel promoted accelerated wound healing under electrical stimulation [47]. Sun et al. synthesized oxidized regenerated cellulose (ORC) and incorporated CS-silver nanoparticles crosslinked with lysine. The hydrogel sustained moist conditions for 86 h, showed mechanical integrity, and promoted almost full recovery in murine models within 13 days [48]. The key advantage of regenerated CBHs lies in their uniform structure, tunable properties, and compatibility with various chemical modifications, enabling the design of dressings with desirable mechanical strength, moisture retention, and biocompatibility for effective wound care.

4.3.4 Cellulose Derivative Hydrogel Dressings

Cellulose derivatives, which are chemically modified forms of native cellulose, have emerged as key materials in the development of hydrogel wound dressings, offering versatile platforms for hydrogel fabrication due to their structural adaptability, chemical functionality, and adjustable compatibility with bioactive compounds, enabling precise tailoring to meet diverse requirements in wound management.

Among these, carboxymethyl cellulose (CMC) is predominantly employed in hydrogel dressings. Sadeghi et al. crosslinked CMC with human hair keratin and incorporated clindamycin-loaded halloysite nanotubes. This composite hydrogel wound dressing exhibited slow drug release, potent antibacterial activity, and promoted fibroblast proliferation [49]. Javanbakht et al. developed a CMC hydrogel incorporating tetracycline-loaded UiO-66 MOF particles. The hydrogel film demonstrated controlled antibiotic release and inhibition zones of 1.3 and 1.7 cm against *E. coli* and *S. aureus*, respectively [50]. Pagano et al. designed CMC-based microparticles that formed an in situ hydrogel upon contact with wound exudate, enabling bioadhesion and keratinocyte proliferation via controlled release of glycyrrhetinic acid [51]. In another study, Soleimani et al. created CMC-gelatin hydrogels loaded with cefdinir, achieving significant bacterial inhibition and accelerated wound healing in animal model [52]. Jirawitchalert et al. utilized cotton waste to fabricate both hydrogel and electrospun fiber wound dressings. CMC derived from cotton demonstrated swelling ratios of 1718% and good water vapor permeability suitable for wound healing [53].

In addition, other types of cellulose derivatives are widely used in the development of hydrogel dressings. Hydroxyethyl cellulose (HEC) is another common derivative. Jamlang et al. demonstrated that HEC hydrogels exhibited inherent antimicrobial activity and exudate absorption, with maximum water uptake of 173.8% [54]. El Fawal et al. supplemented HEC with tungsten oxide to create hydrogel membranes exhibiting anti-inflammatory and antibacterial properties, while HEC improved the safety of tungsten oxide toward normal human cells [55]. Jiang et al. prepared a multifunctional CBH incorporating ZnO nanoparticles and menthol using HEC, cationic β-cyclodextrin, and epichlorohydrin. The hydrogel demonstrated strong antibacterial activity, slow menthol release, and favorable mechanical and moisture retention properties. Its non-toxic profile and synergistic therapeutic effects highlight its promise as an effective wound dressing for pain relief and infection control [56].

Hydroxypropyl Methylcellulose (HPMC) is also known for good chemical stability, film-forming capability, and high hydrophilicity. Some commercial wound gels use HPMC blended with alginate or collagen. Prabu et al. prepared gatifloxacin-loaded sodium alginate hydrogel membranes enhanced with HPMC or HPC, glycerol, and glutaraldehyde. The hydrogels showed improved tensile strength, moisture control, and sustained drug release over 48 h. Their antimicrobial protection and controlled release behavior indicated strong potential as safe and effective wound dressings for prolonged therapeutic applications [57].

Oxidized cellulose also shows significant promise, typically synthesized through periodate oxidation or TEMPO-mediated processes. Wang et al. developed a sprayable hydrogel composed solely of TEMPO-oxidized cellulose nanofibers (TOCN), forming a 3D network in water without additional crosslinkers. The hydrogel exhibited rapid hemostatic capacity, excellent biocompatibility, and promoted full skin regeneration–including sebaceous glands and hair follicles–in both rat and porcine wound models, confirming its promise for advanced wound healing applications [58].

In summary, the diversity of cellulose sources and modification routes allows the fabrication of hydrogels tailored for specific clinical demands. Through variation in functional groups, crosslinking methods, and additive combinations, their mechanical, biological, and chemical properties can be finely tuned for specific wound environments.

4.4 Functionalization and Performance Enhancement of CBH Dressings

To comprehensively optimize therapeutic outcomes in wound management, CBHs are increasingly functionalized through advanced strategies that impart multifaceted properties including self-healing capability, stimuli-responsiveness, enhanced mechanical strength, antimicrobial activity, and injectability. These enhancements enable hydrogel dressings to adapt to dynamic wound environments and provide more precise and effective therapy, thereby facilitating accelerated tissue regeneration, improved infection control, and overall enhanced wound healing efficacy.

4.4.1 Integration of Antimicrobial Functionality

Bacterial infection is a major barrier to effective wound healing, often resulting in delayed tissue regeneration and increased clinical complications. To address this, the integration of antimicrobial functionality into CBH dressings has become a central strategy in their development. By incorporating antibacterial agents, such as antibiotics, metal nanoparticles, and plant-derived extracts, CBHs can provide localized infection control while maintaining key wound dressing properties such as hydration and biocompatibility.

For instance, Karimi et al. developed a composite hydrogel based on CMCS, sodium CMC, and agarose, incorporating ciprofloxacin-loaded silk fibroin/polydopamine nanoparticles. The hydrogel exhibited controlled drug release under alkaline conditions and retained over 80% of antibiotic bioactivity, effectively inhibiting the growth of Pseudomonas aeruginosa. These findings highlight its potential as a multifunctional wound dressing for infection control and tissue repair [59]. Metal particles also can be incorporated into CBHs to provide antimicrobial and antioxidant functions. Sachin et al. developed a nanocomposite hydrogel film by embedding Ag-doped ZnO nanoparticles into a PVA-CMC hydrogel matrix (Fig. 4.4a). In vitro release studies confirmed a sustained and pH-responsive release of Zn^{2+} and Ag^{+}, supporting their potential for targeted antimicrobial activity. The film's superior antimicrobial properties, structural durability, and bioactivity demonstrate its promise as an advanced multifunctional dressing for infection control and wound repair [60]. Halogen-containing compounds such as iodine also can be used in

CBH films to enhance antibacterial performance. Jantrawut et al. fabricated hydrogel films using low methoxyl pectin, gelatin, and CMC, crosslinked with glutaraldehyde and calcium chloride. The optimized formulation exhibited excellent fluid uptake, water retention, and mechanical integrity. When loaded with povidone iodine, the film showed good antibacterial activity property against *Staphylococcus aureus*, indicating its potential for use in absorbent antimicrobial wound dressings while its clinical efficacy remains to be further investigated [61].

Natural plant extracts can be added to CBH dressings to create bioactive systems with controlled antimicrobial function. Wannaboworn et al. developed a superabsorbent CMC–PEG hydrogel incorporating Quercus infectoria nutgall extract. The hydrogel demonstrated exhibited strong antimicrobial activity against pathogens of clinical importance associated with wound infections, superior antioxidant, and showed excellent fluid uptake (~2300%). Its biological performance suggest it as a promising candidate for managing chronically infected wounds [62]. Furthermore, there are also cases of antibacterial effects of combining CBH dressings with enzymes. Huber et al. developed a bioactive hydrogel composed of CMC and succinyl CS, incorporating a dual-enzyme system that continuously generates hydrogen peroxide (Fig. 4.4b). The hydrogel exhibited antimicrobial activity, biocompatibility with fibroblasts, and rapid gelation. This enzyme-triggered system

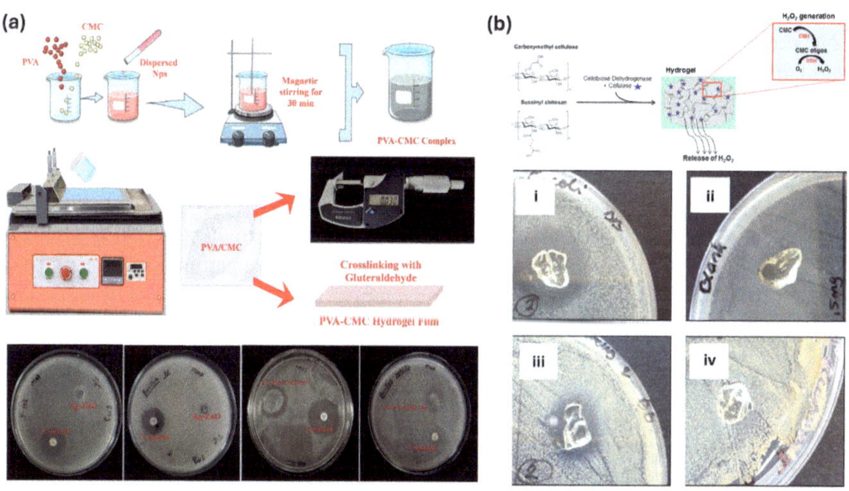

Fig. 4.4 Fabrication and evaluation of CBH dressings with antimicrobial functionality integration. **a** Preparation of PVA-CMC hydrogel film and evaluation of antimicrobial activity using the agar well diffusion method. Reproduced with the permission from Ref. [60]. Copyright 2025, Elsevier. **b** Preparation of succinyl CS/CMC hydrogels incorporating a dual-enzyme hydrogen peroxide generation system. Antimicrobial activity of the succinyl CS/CMC hydrogel was evaluated using a zone of inhibition (ZOI) assay after 24 h. A hydrogel lacking TrlCel served as the negative control. (i) ZOI against *E. coli* (23 mm). (ii) Negative control against *E. coli*. (iii) ZOI against *S. aureus* (14 mm). (iv) Negative control against *S. aureus*. Reproduced with the permission from Ref. [63]. Copyright 2017, American Chemical Society

provided a controlled antimicrobial environment and addressed the dosing issues of conventional topical wound treatments, offering a promising strategy for treating chronic wounds [63]. These diverse antimicrobial integration strategies enable CBHs to actively combat infection while supporting tissue regeneration, paving the way for the development of next generation wound dressings.

4.4.2 Development of Stimuli-Responsive Properties

Smart hydrogel systems respond to specific stimuli in the wound microenvironment, triggering controlled release of therapeutic agents. pH-responsiveness is particularly important in infected or diabetic wounds where the local environment becomes acidic. Fan et al. designed a pH-responsive hydrogel using dialdehyde CMC crosslinked by tobramycin, with borneol-loaded β-cyclodextrin as an auxiliary agent. The hydrogel released drugs under mildly acidic wound conditions. Animal studies confirmed its effectiveness in accelerating healing, highlighting its potential as a dual drug-loaded cellulose-based wound dressing platform [64].

To further improve the stimuli-responsive behavior of hydrogels, Hua et al. developed a triple-responsive system that reacts to glucose, pH, and temperature variations. The hydrogel was constructed using polydopamine-coated tunicate CNCs and phenylboronic acid-modified quaternized chitosan (QCS). This design enabled controlled insulin release under diabetic wound conditions by incorporating insulin (INS) into the system. In vivo studies demonstrated significantly improved wound healing compared to commercial dressings, underscoring its potential for diabetic wound care [65]. More visible stimulus-responsive strategies have also been explored. Arafa et al. developed a pH-sensing hydrogel using HEC grafted with itaconic acid and loaded with Curcuma longa extract. The hydrogel visibly changed color in response to different pH values, providing real-time wound monitoring [66]. These CBH systems allow real-time responsiveness to wound condition changes and offer personalized therapeutic release based on physiological cues.

4.4.3 Enhancement of Mechanical Performance

Hydrogels must maintain structural integrity and flexibility to ensure effective wound coverage. However, they are often too brittle or lack sufficient strength for use on high-motion areas. To address this limitation, composite formulations and crosslinking strategies are commonly employed to enhance their mechanical performance. Notably, CBHs demonstrate superior mechanical performance compared to many traditional hydrogel systems, making them particularly suitable for applications requiring both durability and flexibility. Moreover, their mechanical behavior can be further optimized through tailored structural design.

Tan et al. designed a supramolecular hydrogel membrane integrating oxidized cellulose, QCS, and long-chain polyurethane via acylhydrazone bonds. The resulting structure exhibited programmable strain-adaptive microdomain entanglement, in which interpenetrating entangled microdomains within a dynamically crosslinked hydrogel matrix enhanced tear and fatigue resistance under extreme mechanical stress (Fig. 4.5a). Its dynamic crosslinking and in situ stiffening mechanisms make it particularly suited for treating wounds in mechanically active tissues [67]. In addition, biomimetic composite designs have also been explored to reinforce the mechanical performance of CBHs. Zhong et al. developed a dual-reinforced hydrogel wound dressing using a PVA-borax (PB) matrix enhanced with dopamine (DA)-grafted oxidized carboxymethyl cellulose (OCMC) and cellulose nanofibers. Neomycin served as both an antibacterial agent and dynamic cross-linker. The hydrogel exhibits ultrastretchability exceeding 3000% without fracture, accompanied by rapid self-healing, improved biocompatibility, and strong antimicrobial activity, showing great potential for advanced wound healing applications, especially in joint areas (Fig. 4.5b) [68]. These advancements illustrate the critical role of mechanical optimization in expanding the clinical utility of CBH dressings.

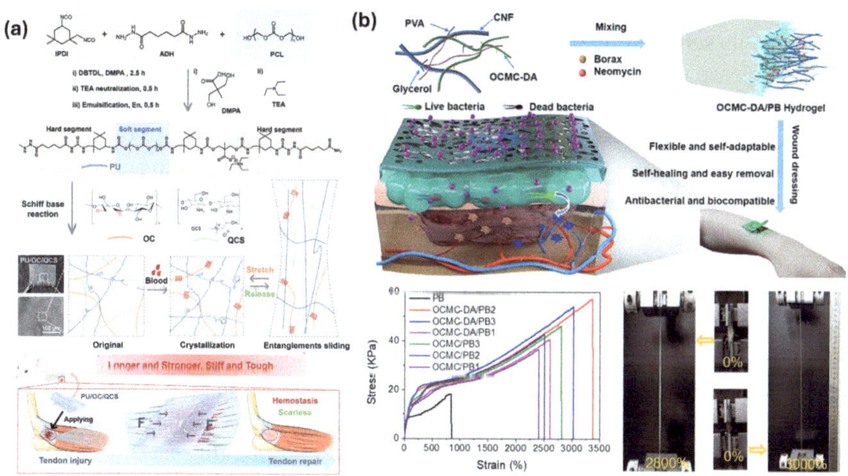

Fig. 4.5 Fabrication and structure of CBH dressings with improved mechanical properties. **a** Schematic illustration of the fabrication and wound healing application of supramolecular hydrogel-based dressings. Reproduced with the permission from Ref. [67]. Copyright 2024, American Chemical Society. **b** Schematic illustration, tensile stress–strain curves, and visual evidence of OCMC-DA/PB hydrogel dressings demonstrating their excellent stretchability. Reproduced with the permission from Ref. [68]. Copyright 2021, American Chemical Society

4.4.4 Implementation of Self-Healing Mechanisms

Advancements in self-healing hydrogels represent a significant breakthrough in wound dressing development, particularly for applications requiring extended wear and adaptability to mechanically dynamic environments. In CBHs, self-healing functionality is typically achieved through the incorporation of dynamic covalent bonds, such as Schiff base linkages, or non-covalent interactions, including hydrogen bonding and host–guest complexes. These reversible interactions allow the hydrogel matrix to autonomously repair physical damage, restoring both structure and function.

Schiff base dynamic bond hydrogels can be formed through interactions between functional groups on cellulose derivatives, including oxidized cellulose, aminated cellulose, or other polymers containing amino or aldehyde groups. Xu et al. developed a covalently crosslinked, biodegradable hydrogel composed of CMCS and oxidized CMC (OCMC), incorporating tetracycline-loaded microspheres. The hydrogel exhibited self-healing capability; when cut into halves and placed in contact at 37 °C, the separated pieces rejoined into a single continuous structure within 2 h, forming a strong junction that remained intact under gravity. This dynamic gelation was attributed to the Schiff base reaction between CMCS and oxidized CMC. In addition to its self-healing property, the hydrogel demonstrated sustained antibacterial drug release without burst effects, supported stem cell proliferation, and showed enhanced antibacterial activity. These combined features confirm its potential as a multifunctional platform for both wound healing and drug delivery [69].

In addition, dynamic boronate ester bonds have been employed to construct self-healing hydrogels containing cellulose components. Threepopnatkul et al. formulated a self-healing hydrogel composed of polyvinyl alcohol, corn starch, boric acid, and CNCs. The optimized 3:1 PVA-to-starch ratio offered balanced mechanical and swelling properties. Incorporation of CNCs enhanced strength, while all formulations demonstrated self-healing in wet conditions [70]. These hydrogels not only heal structurally but also restore appropriate mechanical properties, making them ideal for dynamic or irregular wound sites. Overall, these strategies demonstrate the potential to achieve self-healing in CBHs through dynamic bonding mechanisms. These self-healing capabilities contribute to enhanced durability and functionality of hydrogel dressings.

4.4.5 Optimization of Hemostatic Functionality

Effective hemostasis is a critical function of wound dressings, particularly in acute or trauma-induced injuries where rapid blood loss must be controlled. CBHs have gained considerable attention in hemostatic applications due to their intrinsic biocompatibility, hydrophilicity, and capacity for physical and chemical modification.

By combining cellulose or its derivatives with functional polymers and bioactive compounds, various hydrogel systems have been designed to accelerate clot formation, promote cell aggregation, and adhere to bleeding surfaces.

One strategy involves enhancing water absorption and clotting through structural design. Pan et al. prepared porous hydrogel dressings using PVA, human-like collagen (HLC), and CMCS via freeze–thaw cycling, with Tween 80 (T80) serving as a pore-forming agent. The resulting hydrogels exhibited high swelling capacity, moisture permeability, hemostatic efficiency, and bacterial barrier properties. The hemostatic performance was attributed to the hydrophilicity of the hydrogels, with PVA-HLC-T80 showing the highest water absorption. The porous structure further improved the water absorption capacity of the hydrogel. Consequently, both PVA-HLC-T80 and PVA-HLC-CS-T80 hydrogels demonstrated effective control of heavy bleeding. In addition, In vivo studies using full-thickness wound models revealed accelerated healing, increased collagen deposition, and enhanced re-epithelialization, indicating their promise as multifunctional wound dressings [71].

Another approach utilizes adhesiveness and protein-binding mechanisms to achieve rapid clotting. Zhou et al. developed a hydrogel composed of CMC-graft-tyramine and MXene, further functionalized by immersion in tannic acid (TA). TA acted as a multifunctional hydrogen bond donor, enhancing both the adhesiveness and hemostatic performance of the hydrogel. The strong adhesive strength of the TA-treated hydrogel contributed significantly to its in vivo hemostatic efficacy by allowing firm adherence to wound surfaces, thus forming a physical barrier that rapidly halted bleeding. In addition, TA formed a 3D network through strong hydrogen bonding with plasma proteins, further inhibiting plasma flow. These features highlight the potential of the hydrogel as an effective hemostatic dressing [72].

Hemostatic performance can also be enhanced through interactions between bioactive functional groups and cell surface. Liu et al. fabricated a nanocomposite hydrogel comprising carboxylated CNFs, gelatin, and aminated silver nanoparticles ($Ag–NH_2$ NPs). The CNF/G/Ag hydrogel demonstrated enhanced mechanical strength, self-recovery, antibacterial activity, and hemostatic function. The uncrosslinked and protonated amine groups in $Ag–NH_2$ NPs and gelatin interacted electrostatically with negatively charged residues on red blood cell membranes, leading to strong hemagglutination. The hydrogel also adhered effectively to wound surfaces and promoted platelet aggregation. Additionally, both in vitro and in vivo wound models confirmed its biocompatibility and wound-healing potential, supporting its suitability as a CBH wound dressing with multifunctional properties [73].

Collectively, these studies demonstrate that CBHs can serve as effective platforms for hemostatic management and wound recovery. Continued optimization of these systems may contribute to improved clinical outcomes in both emergency and surgical wound care settings.

4.4.6 Advancement of Injectable CBH Systems

Injectable hydrogels facilitate minimally invasive application in irregular or deep wounds, offering enhanced patient comfort and ease of administration. These systems rapidly gel upon application, allowing effective adaptation to complex wound geometries. Zhang et al. developed a directly injectable hydrogel with high self-adaptability based on a CMC/poly(γ-glutamic acid) system. The hydrogel promoted fibroblast proliferation, enhanced collagen deposition, restored skin function, and promoted VEGF expression in dermal wound models. Its favorable adaptability and injectability highlight its potential as a promising candidate for skin regeneration during dermal wound healing [74].

Injectable CBHs can also be integrated with advanced fabrication techniques to support on-site customization of wound dressings. Teoh et al. explored 3D-printed methyl cellulose hydrogels enhanced with salts to achieve gelation at skin temperature. Loaded with burn-treatment drugs, the hydrogels demonstrated good printability, swelling, and controlled release within 6 h. These findings underscore the potential of directly printing methylcellulose-based wound dressings onto patient wounds for precise and personalized treatment [75].

To meet more complex therapeutic demands, multifunctional injectable systems have also been engineered. Sun et al. developed a tunable hydrogel platform using PNIPAM matrices reinforced with nanocellulose, offering injectable and solid-like formulations. The hydrogels demonstrated temperature-responsive viscosity, mechanical adaptability, and drug release control. Solid-like variants showed antimicrobial activity, while injectable hydrogel maintained high drug bioactivity, highlighting their versatility for tailored therapeutic delivery and wound healing applications [76]. These advancements demonstrate the versatility of injectable CBHs in wound care, highlighting their conformability to complex wound sites and tunable properties, which make them valuable platforms for next-generation personalized wound management.

4.5 Challenges

Despite the remarkable progress in the development of CBHs for wound dressings, several challenges hinder their clinical translation and commercial scalability. These span material processing, performance stability, and economic feasibility. Understanding these limitations is critical to guiding future development and promoting clinical translation.

Scalability and Reproducibility in Manufacturing: Producing high-quality CBHs at industrial scale while maintaining uniformity and functional consistency remains a primary challenge. The variability of raw cellulose from plant sources, including crystallinity and purity, leads to batch-to-batch inconsistencies. Landin et al. noted batch-to-batch variation in microcrystalline cellulose performance [77], which complicated standardization. For BC, the reproducibility of produced nanofibers over sub-cultured generations is also a critical factor [78]. Moreover, hydrogels with complex structures, such as those incorporating nanocellulose, dual-network polymers, or responsive polymer, may rely on multi-step synthesis protocols involving toxic solvents or energy-intensive processing. Translating these laboratory procedures to continuous manufacturing processes remains an unresolved bottleneck.

Sterilization Methods and Stability of Incorporated Bioactivities: Sterilization is a critical step before clinical use but can adversely affect hydrogel structure and the stability of incorporated therapeutics. Autoclaving may cause hydrogel shrinkage or breakdown and alter the efficacy of embedded bioactivities. Maintaining functionality without compromising sterility requires exploring milder sterilization alternatives, which in turn increases product and process complexity.

Balancing Complex Multifunctionality with Cost-Effectiveness: Highly functionalized hydrogels, those with self-healing, antimicrobial, and stimuli-responsive behaviors, often involve expensive raw materials (e.g., MXene, silver nanoparticles) or elaborate synthetic procedures. While these features improve healing outcomes, they raise concerns about production cost and market competitiveness. Clinical and commercial viability requires balance, maximizing efficacy without introducing excessive complexity that impedes affordability and scalability.

4.6 Conclusion

In summary, we focused on CBHs as a promising platform for advanced wound care, owing to their inherent biocompatibility and unique physicochemical properties. Varied forms such as BC, plant-derived nanocellulose, regenerated cellulose, and cellulose derivatives exhibit specific advantages and tunable properties, enabling tailoring to diverse requirements. Emerging innovations in CBH dressing include smart responsive platforms for detecting infection or pH changes, along with injectable and self-healing formats that expand applicability in complex conditions and dynamic wound environments. However, challenges such as manufacturing reproducibility and the scalability of complex functionalities continue to hinder widespread adoption. The development of CBH applications in wound healing continues to advance, driven by their strong potential to revolutionize wound management. As research and development efforts intensify, CBHs are poised to play an increasingly central role in the evolution of next-generation wound care technologies.

References

1. T.A. Harris-Tryon, E.A. Grice, Microbiota and maintenance of skin barrier function. Science **376**, 940–945 (2022)

2. A.W.C. Chua, Y.C. Khoo, B.K. Tan, K.C. Tan, C.L. Foo, S.J. Chong, Skin tissue engineering advances in severe burns: review and therapeutic applications. Burns Trauma **4**, s41038–41016–40027-y (2016)
3. E. Rezvani Ghomi, M. Niazi, S. Ramakrishna, The evolution of wound dressings: From traditional to smart dressings. Polym. Adv. Technol. **34**, 520–530 (2023)
4. M. Farahani, A. Shafiee, Wound healing: from passive to smart dressings. Adv. Healthc. Mater. **10**, 2100477 (2021)
5. N.S. Bora, K.N. Dutta, A. Talukder, B. Gogoi, S. Gam, K. Deka, G. Deka, B.J. Sahariah, Chapter 11—cellulose-based hydrogels for wound dressing and wound healing applications, in *Cellulose-based hydrogells*, ed. by K. Deshmukh, S. Barua, S. Baruah, C. Mustansar Hussain (Elsevier, 2025), pp 229–257
6. Q. Chai, Y. Jiao, X. Yu, Hydrogels for biomedical applications: their characteristics and the mechanisms behind them. Gels **3**, 6 (2017)
7. B. Darbasizadeh, Y. Fatahi, B. Feyzi-barnaji, M. Arabi, H. Motasadizadeh, H. Farhadnejad, F. Moraffah, N. Rabiee, Crosslinked-polyvinyl alcohol-carboxymethyl cellulose/ZnO nanocomposite fibrous mats containing erythromycin (PVA-CMC/ZnO-EM): fabrication, characterization and in-vitro release and anti-bacterial properties. Int. J. Biol. Macromol. **141**, 1137–1146 (2019)
8. T. Kaur, A. Joshi, N. Singh, Natural cocktail of bioactive factors conjugated on nanofibrous dressing for improved wound healing. Biomater. Adv. **143**, 13 (2022)
9. L. Mao, L. Wang, M.Y. Zhang, M.W. Ullah, L. Liu, W.W. Zhao, Y. Li, A.A.Q. Ahmed, H.Y. Cheng, Z.J. Shi, G. Yang, In situ synthesized selenium nanoparticles-decorated bacterial cellulose/gelatin hydrogel with enhanced antibacterial, antioxidant, and anti-inflammatory capabilities for facilitating skin wound healing. Adv. Healthc. Mater. **10**, 16 (2021)
10. D.C. Contreras, M.A. Cisternas, R.C. Congreve, K. Varaprasad, K. Chandrasekaran, E.R. Sadiku, Antimicrobial hydrogel scaffolds from barba de viejo microfibers, alginate and ago nanoparticles via green synthesis. Int. J. Biol. Macromol. **284**, 11 (2025)
11. N.S.V. Capanema, A.A.P. Mansur, S.M. Carvalho, L.L. Mansur, C.P. Ramos, A.P. Lage, H.S. Mansur, Physicochemical properties and antimicrobial activity of biocompatible carboxymethylcellulose-silver nanoparticle hybrids for wound dressing and epidermal repair. J. Appl. Polym. Sci. **135**, 18 (2018)
12. S. Singh, A. Young, C.-E. McNaught, The physiology of wound healing. Surgery (Oxford) **35**, 473-477 (2017)
13. B.K. Sun, Z. Siprashvili, P.A. Khavari, Advances in skin grafting and treatment of cutaneous wounds. Science **346**, 941–945 (2014)
14. S.D. Bhoyar, K. Malhotra, B. Madke, Dressing materials: a comprehensive review. J Cutan Aesthet Surg **16**, 81–89 (2023)
15. R. Laurano, M. Boffito, G. Ciardelli, V. Chiono, Wound dressing products: a translational investigation from the bench to the market. Eng. Regen. **3**, 182–200 (2022)
16. Z. Fan, B. Liu, J. Wang, S. Zhang, Q. Lin, P. Gong, L. Ma, S. Yang, A novel wound dressing based on Ag/graphene polymer hydrogel: Effectively kill bacteria and accelerate wound healing. Adv. Funct. Mater. **24**, 3933–3943 (2014)
17. E. Rezvani Ghomi, S. Khalili, S. Nouri Khorasani, R. Esmaeely Neisiany, S. Ramakrishna, Wound dressings: Current advances and future directions. J. Appl. Polym. Sci. **136**, 47738 (2019)
18. H.M. Nguyen, T.T. Ngoc Le, A.T. Nguyen, H.N. Thien Le, T.T. Pham, Biomedical materials for wound dressing: Recent advances and applications. RSC Adv. **13**, 5509–5528 (2023)
19. M.H. Norahan, S.C. Pedroza-González, M.G. Sánchez-Salazar, M.M. Álvarez, G. Trujillo de Santiago, Structural and biological engineering of 3D hydrogels for wound healing. Bioact. Mater. **24**, 197–235 (2023)
20. I. Gardikiotis, F.-D. Cojocaru, C.-T. Mihai, V. Balan, G. Dodi, Borrowing the features of biopolymers for emerging wound healing dressings: a review. Int. J. Mol. Sci. **23**, 8778 (2022)
21. E.M. Ahmed, Hydrogel: preparation, characterization, and applications: a review. J. Adv. Res. **6**, 105–121 (2015)

22. A.M.S. Costa, J.F. Mano, Extremely strong and tough hydrogels as prospective candidates for tissue repair—a review. Eur. Polym. J. **72**, 344–364 (2015)
23. T. Xia, W. Liu, L. Yang, A review of gradient stiffness hydrogels used in tissue engineering and regenerative medicine. J. Biomed. Mater. Res. Part A **105**, 1799–1812 (2017)
24. Y. Liang, J. He, B. Guo, Functional hydrogels as wound dressing to enhance wound healing. ACS Nano **15**, 12687–12722 (2021)
25. G. Kaur, G. Narayanan, D. Garg, A. Sachdev, I. Matai, Biomaterials-based regenerative strategies for skin tissue wound healing. ACS Appl. Bio Mater. **5**, 2069–2106 (2022)
26. J. Pennells, I.D. Godwin, N. Amiralian, D.J. Martin, Trends in the production of cellulose nanofibers from non-wood sources. Cellulose **27**, 575–593 (2020)
27. S.M.F. Kabir, P.P. Sikdar, B. Haque, M.A.R. Bhuiyan, A. Ali, M.N. Islam, Cellulose-based hydrogel materials: chemistry, properties and their prospective applications. Prog. Biomater. **7**, 153–174 (2018)
28. B. Thomas, M.C. Raj, A.K.B, R.M.H, J. Joy, A. Moores, G.L. Drisko, C. Sanchez, Nanocellulose, a versatile green platform: From biosources to materials and their applications. Chem. Rev. **118**, 11575–11625 (2018)
29. H.S. Yin, N. Li, P.Q. Song, J. Liao, M.X. Xiao, C.H. Zhou, H.H. Huang, Eco-friendly oxidized microcrystalline cellulose/quaternized chitosan/gelatin hydrogels based on schiff-base reaction as wound dressing and naringin release kinetic. Cellulose **32**, 1709–1727 (2025)
30. L. Tang, B. Wang, S.R. Bai, B. Fan, Z.X. Zhang, L. Zhang, F.Z. Wang, Structure regulation of oxidized soybean cellulose nanocrystals/poly-acrylamide hydrogel and its application potential in wound dressing. Int. J. Biol. Macromol. **281**, 14 (2024)
31. S.K. Bajpai, V. Pathak, B. Soni, Minocycline-loaded cellulose nano whiskers/poly(sodium acrylate) composite hydrogel films as wound dressing. Int. J. Biol. Macromol. **79**, 76–85 (2015)
32. N.S. Jafri, D.N. Jimat, W.M.F.W. Nawawi, Y.A. Nor, A. Amid, Effect of incorporating cellulose nanofibers and lemongrass essential oil in polyvinyl alcohol-polyethylene glycol/glycerin hydrogel for wound dressing. Iium. Eng. J. **25**, 99–115 (2024)
33. F. Ahmad, B. Mushtaq, F.A. Butt, M.S. Zafar, S. Ahmad, A. Afzal, Y. Nawab, A. Rasheed, Z. Ulker, Synthesis and characterization of nonwoven cotton-reinforced cellulose hydrogel for wound dressings. Polymers **13**, 4098 (2021)
34. F. Yin, L. Lin, S. Zhan, Preparation and properties of cellulose nanocrystals, gelatin, hyaluronic acid composite hydrogel as wound dressing. J. Biomat. Sci-Polym. E. **30**, 190–201 (2019)
35. Y.D. Deng, X.S. Yang, X.D. Zhang, H.H. Cao, L.Z. Mao, M.M. Yuan, W.Z. Liao, Novel fenugreek gum-cellulose composite hydrogel with wound healing synergism: facile preparation, characterization and wound healing activity evaluation. Int. J. Biol. Macromol. **160**, 1242–1251 (2020)
36. A.J. Brown, Xliii-on an acetic ferment which forms cellulose. J. Chem. Soc. T. **49**, 432–439
37. R.E. Cannon, S.M. Anderson, Biogenesis of bacterial cellulose. CRC Crit. Rev. Microbiol. **17**, 435–447 (1991)
38. S. Li, D. Huang, J. Yang, B. Zhang, X. Zhang, G. Yang, M. Wang, Y. Shen, Freestanding bacterial cellulose-polypyrrole nanofibres paper electrodes for advanced energy storage devices. Nano Energy **9**, 309–317 (2014)
39. H.G. de Oliveira Barud, R.R. da Silva, H. da Silva Barud, A. Tercjak, J. Gutierrez, W.R. Lustri, O.B. de Oliveira, S.J.L. Ribeiro, A multipurpose natural and renewable polymer in medical applications: bacterial cellulose. Carbohydr. Polym. **153**, 406–420 (2016)
40. C.H. Haigler, M. Benziman, *Cellulose and other natural polymer systems: biogenesis, structure, and degradation* (Springer, US, Boston, MA, 1982)
41. M. Suneetha, S.-Y. Won, S.M. Zo, S.S. Han, Fungal carboxymethyl chitosan-impregnated bacterial cellulose hydrogel as wound-dressing agent. Gels **9**, 184 (2023)
42. S.P. Lin, H.N. Kung, Y.S. Tsai, T.N. Tseng, K.D. Hsu, K.C. Cheng, Novel dextran modified bacterial cellulose hydrogel accelerating cutaneous wound healing. Cellulose **24**, 4927–4937 (2017)
43. F. Cheng, X. Yi, J. Dai, Z. Fan, J. He, Y. Huang, H. Li, Photothermal MXene@zn-mof-decorated bacterial cellulose-based hydrogel wound dressing for infectious wound healing. Cell Rep. Phys. Sci. **4**, 101619 (2023)

44. M. Mo, C. Wu, Y. Chen, Bacterial cellulose-based superabsorbent hydrogel for wet wound dressing. Molecules **30**, 737 (2025)
45. R. Liu, J. Cai, N. Qin, K. Zhang, T. Li, H. Luo, D. Guo, Nontoxic chemical crosslinked bacterial cellulose-heparin-gelatin composite hydrogel as antibacterial dressing. J. Mar. Sci. Technol. **178**, 29–38 (2024)
46. S. Kiselioviene, O. Baniukaitiene, V. Harkavenko, N.A. Babenko, J. Liesiene, Cellulose hydrogel sheets for wound dressings. Cellul. Chem. Technol. **50**, 915–923 (2016)
47. L. Mao, S.M. Hu, Y.H. Gao, L. Wang, W.W. Zhao, L.N. Fu, H.Y. Cheng, L. Xia, S.X. Xie, W.L. Ye, Z.J. Shi, G. Yang, Biodegradable and electroactive regenerated bacterial cellulose/MXene (Ti3C2Tx) composite hydrogel as wound dressing for accelerating skin wound healing under electrical stimulation. Adv. Healthc. Mater. **9**, 13 (2020)
48. Q. Sun, X. Dong, J. Xu, T. Wang, Silver-infused lysine crosslinked hydrogel with oxidized regenerated cellulose for prospective advanced wound dressings. Int. J. Biol. Macromol. **264**, 130675 (2024)
49. S. Sadeghi, J. Nourmohammadi, A. Ghaee, N. Soleimani, Carboxymethyl cellulose-human hair keratin hydrogel with controlled clindamycin release as antibacterial wound dressing. Int. J. Biol. Macromol. **147**, 1239–1247 (2020)
50. S. Javanbakht, M. Nabi, M. Shadi, M.M. Amini, A. Shaabani, Carboxymethyl cellulose/tetracycline@uio-66 nanocomposite hydrogel films as a potential antibacterial wound dressing. Int. J. Biol. Macromol. **188**, 811–819 (2021)
51. C. Pagano, P. Calarco, A. Di Michele, M.R. Ceccarini, T. Beccari, S. Primavilla, S. Scuota, F. Marmottini, D. Ramella, M. Ricci, L. Perioli, Development of sodium carboxymethyl cellulose based polymeric microparticles for in situ hydrogel wound dressing formation. Int. J. Pharmaceut. **602**, 120606 (2021)
52. Z. Soleimani, H. Baharifar, N. Najmoddin, K. Khoshnevisan, Evaluation of carboxymethyl cellulose/gelatin hydrogel-based dressing containing cefdinir for wound healing promotion in animal model. Gels **11**, 38 (2025)
53. S. Jirawitchalert, S. Mitaim, C.-Y. Chen, N. Patikarnmonthon, Cotton cellulose-derived hydrogel and electrospun fiber as alternative material for wound dressing application. Int. J. Biomater. **2022**, 2502658 (2022)
54. J. Jamlang, M.T. Maranon, G.J. Rigor, T. Tumolva, Developing a physically cross-linked hydroxyethyl cellulose hydrogel for wound dressing applications. Mater. Sci. Forum **947**, 3–12 (2019)
55. G.F. El Fawal, M.M. Abu-Serie, M.A. Hassan, M.S. Elnouby, Hydroxyethyl cellulose hydrogel for wound dressing: fabrication, characterization and in vitro evaluation. Int. J. Biol. Macromol. **111**, 649–659 (2018)
56. L. Jiang, Y. Han, J. Xu, T. Wang, Preparation and study of cellulose-based ZnO NPs@hec/c-β-cd/menthol hydrogel as wound dressing. Biochem. Eng. J. **184**, 108488 (2022)
57. D. Prabu, A.F. Majdalawieh, I.A. Abu-Yousef, K. Inbasekaran, T. Balasubramaniam, N. Nallaperumal, C.J. Gunasekar, Preparation and characterization of gatifloxacin-loaded sodium alginate hydrogel membranes supplemented with hydroxypropyl methylcellulose and hydroxypropyl cellulose polymers for wound dressing. Int. J Pharm. Investig. **6**, 86–95 (2016)
58. L. Wang, Y. Liu, W.T. Duan, F. Cao, Q.X. Lv, S.J. Zhang, J.Z. She, L.Y. Yang, B.L. He, Y. Hou, L. Kong, T.Q. Dai, R.Z. Ning, B.L. Cai, Sprayable oxidized cellulose nanofiber hydrogel with rapid hemostatic ability for skin wound healing. Int. J. Biol. Macromol. **310**, 14 (2025)
59. T. Karimi, F. Mottaghitalab, H. Keshvari, M. Farokhi, Carboxymethyl chitosan/sodium carboxymethyl cellulose/agarose hydrogel dressings containing silk fibroin/polydopamine nanoparticles for antibiotic delivery. J. Drug Deliv. Sci. Technol. **80**, 104134 (2023)
60. K.S. Sachin, J.T. Koshy, D. Sangeetha, Development of Ag doped ZnO nanoparticle loaded carboxymethyl cellulose/polyvinyl alcohol hydrogel film for enhanced wound dressing. Inorg. Chem. Commun. **178**, 114606 (2025)
61. P. Jantrawut, J. Bunrueangtha, J. Suerthong, N. Kantrong, Fabrication and characterization of low methoxyl pectin/gelatin/carboxymethyl cellulose absorbent hydrogel film for wound dressing applications. Materials **12**, 1628 (2019)

62. S. Wannaboworn, S. Lethongkam, S. Bilhman, T. Waen-ngoen, M. Yingkajorn, M. Phengmak, S.P. Voravuthikunchai, S. Paosen, Superabsorbent crosslinked carboxymethyl cellulose-PEG hydrogel containing a nutgall extract as a broad antimicrobial agent for wound dressing applications. J. Drug Deliv. Sci. Technol. **99**, 106001 (2024)
63. D. Huber, G. Tegl, A. Mensah, B. Beer, M. Baumann, N. Borth, C. Sygmund, R. Ludwig, G.M. Guebitz, A dual-enzyme hydrogen peroxide generation machinery in hydrogels supports antimicrobial wound treatment. ACS Appl. Mater. Inter. **9**, 15307–15316 (2017)
64. X. Fan, L. Yang, T. Wang, T. Sun, S. Lu, PH-responsive cellulose-based dual drug-loaded hydrogel for wound dressing. Eur. Polym. J. **121**, 109290 (2019)
65. S. Hua, Y. Zhang, Y. Zhu, X. Fu, L. Meng, L. Zhao, L. Kong, S. Pan, Y. Che, Tunicate cellulose nanocrystals strengthened injectable stretchable hydrogel as multi-responsive enhanced antibacterial wound dressing for promoting diabetic wound healing. Carbohydr. Polym. **343**, 122426 (2024)
66. A.A. Arafa, A.A. Nada, A.Y. Ibrahim, M.K. Zahran, O.A. Hakeim, Greener therapeutic pH-sensing wound dressing based on curcuma longa and cellulose hydrogel. Eur. Polym. J. **159**, 110744 (2021)
67. L. Tan, C. Huyan, Y.Q. Wang, M.H. Li, D. Liu, M.H. Liu, Z. Luo, K.Y. Cai, Y. Hu, Mechanically robust hemostatic hydrogel membranes with programmable strain-adaptive microdomain entanglement for wound treatment in dynamic tissues. ACS Nano **18**, 8360–8382 (2024)
68. Y. Zhong, F. Seidi, C. Li, Z. Wan, Y. Jin, J. Song, H. Xiao, Antimicrobial/biocompatible hydrogels dual-reinforced by cellulose as ultrastretchable and rapid self-healing wound dressing. Biomacromol **22**, 1654–1663 (2021)
69. W. Xu, X. Gao, H. Tan, S. Li, T. Zhou, J. Li, Y. Chen, Covalent and biodegradable chitosan-cellulose hydrogel dressing containing microspheres for drug delivery and wound healing. Mater. Today Commun. **33**, 104163 (2022)
70. P. Threepopnatkul, J. Sokjorhor, P. Homchan, W. Khammee, Preparation of hydrogel with self-healing properties based on poly(vinyl alcohol)/corn starch/zeolite/cellulose nanocrystals for wound dressing applications. Mater. Today: Proc. **52**, 2467–2473 (2022)
71. H. Pan, D.D. Fan, W. Cao, C.H. Zhu, Z.G. Duan, R.Z. Fu, X. Li, X.X. Ma, Preparation and characterization of breathable hemostatic hydrogel dressings and determination of their effects on full-thickness defects. Polymers **9**, 28 (2017)
72. X.Y. Zhou, W.D. Ma, J.H. Jiang, J.B. Dang, R.F. Lv, H.B. Wang, M.N. Ma, D.H. Sun, M. Zhang, Non-antibiotic dependent photothermal antibacterial hemostatic MXene hydrogel for infectious wounds healing. Biomater. Adv. **169**, 16 (2025)
73. R. Liu, L. Dai, C.L. Si, Z.G. Zeng, Antibacterial and hemostatic hydrogel via nanocomposite from cellulose nanofibers. Carbohydr. Polym. **195**, 63–70 (2018)
74. W.J. Zhang, X.X. Wang, J.P. Ma, R. Yang, Y. Hu, X.Y. Tan, B. Chi, Adaptive injectable carboxymethyl cellulose/poly (γ-glutamic acid) hydrogels promote wound healing. Biomater. Adv. **136**, 12 (2022)
75. J.H. Teoh, F.T.A. Shakoor, C.-H. Wang, 3D printing methyl cellulose hydrogel wound dressings with parameter exploration via computational fluid dynamics simulation. Pharm. Res. **39**, 281–294 (2022)
76. Q.Y. Sun, S.Y. Tao, G. Bovone, G. Han, D. Deshmukh, M.W. Tibbitt, Q. Ren, P. Bertsch, G. Siqueira, P. Fischer, Versatile mechanically tunable hydrogels for therapeutic delivery applications. Adv. Healthc. Mater. **13**, 10 (2024)
77. M. Landín, R. Martínez-Pacheco, J.L. Gómez-Amoza, C. Souto, A. Concheiro, R.C. Rowe, Effect of batch variation and source of pulp on the properties of microcrystalline cellulose. Int. J. Pharm. **91**, 133–141 (1993)
78. J. Wood, C. van der Gast, D. Rivett, J. Verran, J. Redfern, Reproducibility of bacterial cellulose nanofibers over sub-cultured generations for the development of novel textiles. Front. Bioeng. Biotechnol. **10**, 876822 (2022)

Chapter 5
Cellulose-Based Hydrogels for Controlled Drug Delivery

Raluca Nicu and Diana E. Ciolacu

Abstract The use of biopolymers in controlled drug delivery has attracted significant attention due to their properties such as, biocompatibility, biodegradability and non-toxicity. Among these, cellulose is one of the most attractive natural polymers, which can be found either in plants (cell wall or various organs) or can be produced by bacteria or animals. This chapter provides a comprehensive overview of cellulose-based hydrogels and their potential in controlled drug delivery. From the beginning, the role of hydrogels in improving the efficacy and safety of drug delivery is briefly presented. Their unique properties that make them ideal carriers for controlled drug release, like porosity, swelling behavior, mechanical properties, and responsiveness to environmental stimuli, are highlighted. The mechanisms underlying the drug release processes are also discussed. The latest and most important developments in cellulose-based hydrogels, the integration of nanocellulose into the hydrogels' matrix, as well as the preparation of stimuli-responsive hydrogels and composite/hybrid hydrogels are discussed in detail. All these systems are presented in close connection with their successful application in different drug delivery routes (oral, ocular, nasal, dermal and transdermal). The chapter concludes by addressing some limitations (i.e., the limited loading capacity of hydrophobic drugs and the need for increased control over release kinetics) and mentioning some future research directions (i.e., more eco-friendly synthesis methods and advanced formulations).

Keywords Cellulose-based hydrogels · Nanocellulose · Stimuli sensitivity · Controlled drug delivery

R. Nicu · D. E. Ciolacu (✉)
Department of Natural Polymers, Bioactive and Biocompatible Materials, "Petru Poni" Institute of Macromolecular Chemistry, Iaşi, Romania
e-mail: dciolacu@icmpp.ro

G. Wei and J. Zhang (eds.), *Cellulose-Based Hydrogels and Aerogels*, Lecture Notes in Nanoscale Science and Technology 41,
https://doi.org/10.1007/978-3-032-03008-5_5

5.1 Overview of Hydrogels in Drug Delivery

Over the years, the researchers have actively explored suitable and promising alternatives for drug administration in both humans and animals, in order to enhance the biodistribution and bioavailability of therapeutic agents and, consequently, to improve the dose–response relationship. In this context, drug delivery systems (DDS) have emerged as a strategic solution to overcome biological barriers and enable efficient, targeted delivery of drugs [1].

An ideal DDS should provide controlled delivery and release the drug with effective rate, specifically at the target site [2]. Among various candidates for DDS, the hydrogels have attracted significant attention due to their unique physicochemical and biological properties. Hydrogels are hydrophilic, three-dimensional viscoelastic polymer networks with high water content (typically 70–99%), enabling diffusion and attachment of molecules and cells [3]. Their soft, flexible, and stimuli-responsive nature, in combination with properties such as biocompatibility, biodegradability, minimal immunogenicity, and non-toxicity, make hydrogels suitable for a wide range of biomedical applications, including drug delivery. Additional advantages include their ability to encapsulate poorly soluble drugs, exhibit bioadhesive properties, maintain mechanical strength and elasticity, and swell repeatedly under physiological conditions without dissolving [1, 2, 4]. Moreover, the porous structure of hydrogels facilitates sustained drug release, allowing for the gradual elution of therapeutics at the targeted site, maintaining high local drug concentrations over extended periods [5]. These characteristics make hydrogels highly versatile and efficient in supporting drug retention and controlled release, reinforcing their role in advanced drug delivery technologies [2, 6].

5.1.1 Hydrogels' Features and Their Role in Drug Delivery

The effectiveness of hydrogels as matrices in controlled drug delivery is decisively influenced by its three-dimensional (3D) structure and implicitly by their features, such as porosity and mesh size, swelling behavior, mechanical strength, biological properties, degradability, etc. [7]. The hydrogels 3D-structure enables the effective immobilization and protection of bioactive compounds and also facilitates their release under specific biological conditions, tailored to the distinct environments of different organs and tissues [8]. Their chemical and physical similarity to the extracellular matrix, high water content, and bioadhesive properties further enhance their adaptability and integration at application sites. This makes hydrogels the ideal carriers for localized and systemic drug delivery [9]. Depending on the design and crosslinking chemistry, the hydrogels can form stable or biodegradable structures that can dissolve after delivering the therapeutic load [1].

Porosity is a fundamental feature of hydrogels as matrices for controlled drug delivery, as it dictates the internal voids and structural arrangement, and facilitates the drug release. This property is influenced by polymerization techniques and crosslinking density, so it can be tailored and optimized so that the designed hydrogels will allow precise and controlled drug release [7]. The mesh size of the network determines the interaction between the drug and hydrogel network, thus controlling its diffusion. Depending on the size of the drug molecules, three cases can be encountered, namely: (a) when the drug molecules are smaller, the release is governed mainly by diffusion; (b) when the mesh size of the network and the drug are similar, the diffusion process is slowed down, allowing a prolonged release of the drug; (c) a small ratio between the drug and mesh size leads to its capture, and network degradation or its expansion is necessary for drug release [7]

Swelling behavior is also a crucial factor in hydrogels application as matrices for controlled drug delivery because it significantly impacts the drug release and diffusion. The swelling behavior is influenced by numerous factors, such as hydrogel's chemical compositions, cross-linking density, and solvent characteristics (i.e., pH, temperature). The influence of chemical composition manifested through the presence of hydrophilic groups, for example, amide, carboxyl or hydroxyl groups [10]. The hydrogels with increased crosslinking density have a reduced swelling ratio because they have fewer spaces available for water molecules to enter and swell the hydrogels and also the network is more rigid compared to non-crosslinked ones, depending on the density and spatial distribution of the crosslink points [11]. Modifications in ionic strength and pH in the surrounding environment may affect the swelling characteristics of hydrogels. For instance, the acidic environments had protonated the amine groups, which led to a swelling reduction. On the contrary, the basic environments may cause deprotonation of the carboxylic acid groups that lead towards increased swelling [12]. As for temperature, it affects the swelling behavior of certain hydrogels, especially thermoresponsive ones, which can swell or shrink in response to temperature changes [13].

Mechanical strength defines hydrogels capability of maintaining their structural integrity under certain mechanical forces and plays a pivotal role in drug delivery, especially for the sensitive therapeutic agent delivery (i.e., proteins) that needs protection whilst transit towards the target site [14]. However, challenges such as the high-water content and network heterogeneity of hydrogels often compromise their mechanical stability, posing limitations to their practical utility. To address these issues, the mechanical performance of hydrogels is enhanced through innovative strategies, including the development of double-network hydrogels and the introduction of chemically or ionically crosslinked hydrogels [7]. Due to careful control of such mechanical characteristics, systems for delivering drugs have been optimized to achieve the required kinetics for the release of drugs, enhanced compatibility of tissues, and increased therapeutic efficacy [13].

Biological properties, such as biocompatibility, bioadhesion or biodegradability, are features that enable hydrogels to interact with biological systems and facilitate the drugs release in a controlled way [15]. *Biocompatibility* refers to the ability of materials to coexist with living tissues and minimize inflammation or immune responses after their administration or implantation. *Bioadhesion* is the ability to adhere to the biological surfaces or biological tissues. Such characteristics are particularly relevant to local drug delivery, where the adhesion to targeted sites (i.e., mucosal membranes, wounds) is required in order to increase the penetration of the drug in the tissues, and also to provide a sustained release of the drug at the targeted site [16]. *Degradability* is a key attribute tailored for biomedical applications, allowing these materials to break down into biocompatible by-products that can be safely metabolized or excreted by the human body. This property ensures that drug delivery systems based on hydrogels perform their intended therapeutic functions and then degrade at a controlled rate to minimize long-term adverse effects. The degradability rate and mechanism are often achieved through hydrolysis or enzymatic degradation and they are influenced by the chemical composition of hydrogel, including the nature of the polymeric backbone and the presence of degradable linkages [7].

5.1.2 Hydrogels' Morphology

The hydrogels differ in size, architecture and function, which dictate the routes of delivery into human body. As shown in Fig. 5.1, there are features in hydrogels with length scales ranging from cm to nm, such as (i) macroscopic hydrogels with pores of 10 to 500 μm; (ii) the mesh size (the distance between polymer molecules in the network) from 5 to 100 nm; (iii) the atomic (molecular) scale, where the drugs can interact with polymer chains of the subnamometer order [17]. The multiscale features of hydrogels can lead to large differences regarding their application and even to completely different functional characteristics [18]. Moreover, different sizes and structures of the hydrogels determine the manner of their transport and adherence, if they are introduced into blood vessels, airways or the gastrointestinal tract [9]. Currently, hydrogels are classified according to their particle size as *macrogels*, *microgels* and *nanogels*, respectively (Fig. 5.2) [17].

Macrogels are defined as particle sizes > 100 μm and are generally utilized for transdermal administration or direct injection/implantation around the defected wound or tumor tissue. Most macrogels used to treat diseases are delivered locally [9]. Macrogels are hydrogels with dimensions larger than millimetres. Macrogels have multiple advantages, such as high drug loading capacity, high stability, and responsiveness to sensitive stimuli. However, they are difficult to administer due to their huge size. Therefore, they can be used for in situ delivery, including transdermal administration, direct injection, or direct application to the surface of the surgical cavity [14].

Microgels refer to hydrogels with dimensions in the colloidal regime (<100 μm), particularly with size of 0.5–10 μm [9]. They have unique properties of both hydrogels and colloidal particles, such as structural integrity, compartmentalization, orthogonal functionalization, softness, deformability, permeability, stimuli-responsiveness, reversible swelling, and adaptability, and have been found to be promising for a wide

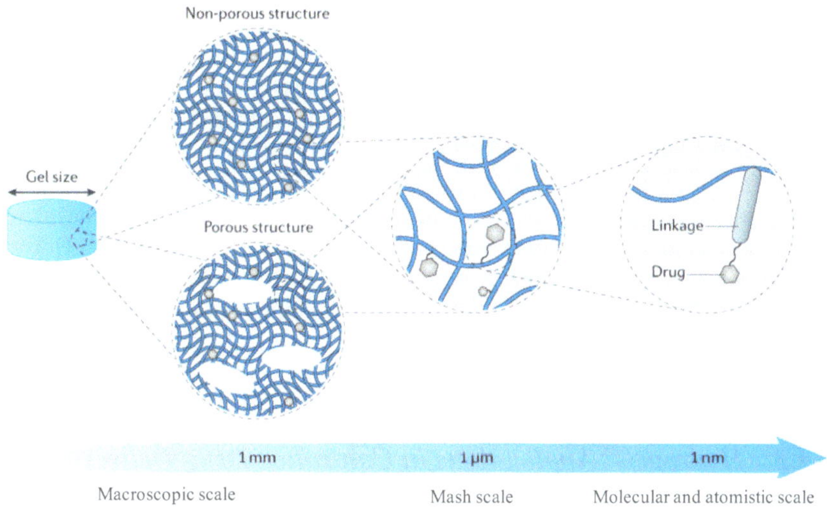

Fig. 5.1 Multiscale features of hydrogels

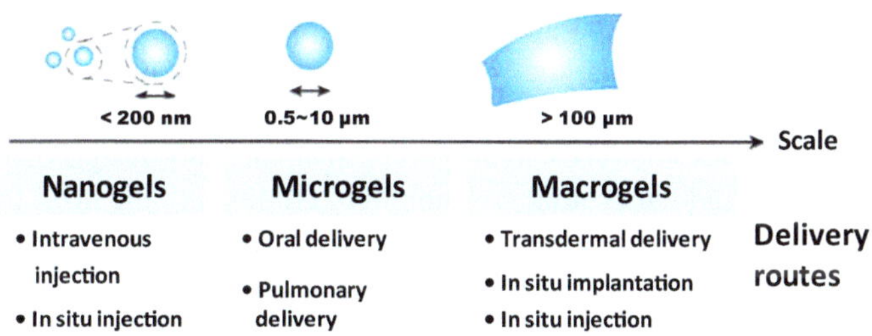

Fig. 5.2 Hydrogels' particle size and their delivery routes. Reproduced with the permission from Ref. [9]. Copyright 2024, Elsevier

range of applications in various fields including controlled drug release [18]. Microgels smaller than 5 μm are commonly used for oral or pulmonary drug delivery but are generally considered unsuitable for intravascular injection due to their rapid clearance from circulation and vulnerability to phagocytosis by macrophages within the bloodstream. For gastrointestinal or lung cancer, the microgels administered orally or via pulmonary routes can effectively deliver antitumor drugs. Following oral administration, the drug-encapsulated microgels enable the active substance to rapidly reach the target site, such as gastrointestinal tumors, and exert its therapeutic effect [14].

Nanogels have dimensions smaller than 200 nm, typically between 10 and 100 nm, and are also known as so-called hydrogel nanoparticles. They have gained tremendous interest as nanoparticulate systems for controlled drug delivery, since they combine the unique hydrogel characteristics with submicron particle size [19]. In essence, nanogels have properties similar to those of their macrogels/microgels counterparts, but stands out through a series of advantages caused by the size effects, such as enhanced permeability; targeted and controlled release with diminishing side effects; reaching areas of the body that are not easily accessed by other hydrogels); high drug loading due to the large surface area; high stability in blood circulation, biodegradability, and biocompatibility [9, 18, 20–23]. However, the nanogels have some disadvantages: are expensive; some fractions of the particles are in the micrometer range; scaling up is not easy due to the mean size and weight [20]. However, these disadvantages are minimal compared to the multitude of advantages presented above, so the nanogels have attracted great attention for nanomedicine applications, for instance, as pharmaceutical drug carriers.

5.1.3 Hydrogels' Applications in Controlled Drug Delivery

Hydrogels are suitable carriers for multiple delivery routes, as presented in Fig. 5.3 such as oral [24, 25], ocular [26–28], nasal [29, 30], vaginal [31], rectal [32], and dermal or transdermal [33, 34]. The choice of drug delivery method is guided by the goal of maximizing its therapeutic efficacy, but also the patient compliance. Equally important is how the drug is released, whether the treatment requires short- or long-term availability, and whether the release should be continuous or pulsatile. To support these therapeutic needs, hydrogels must be carefully designed: they should either safely biodegrade after drug consumption, eliminating the need for surgical removal, or allow for reuse by refilling the drug [14, 17].

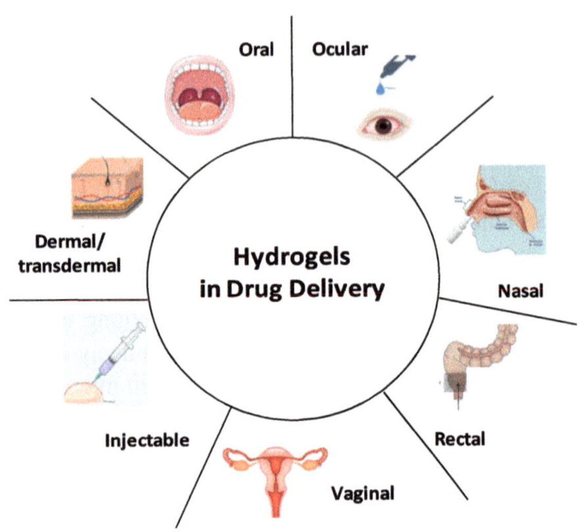

Fig. 5.3 Routes of administration of hydrogel in drug delivery

Oral drug administration is one of the most widely used methods of drug delivery, since it is comfortable to patients and requires low costs [35]. In addition, the gastrointestinal tract (GIT) is characterized by a large area of systemic absorption that extends from the mouth to the anus. There is a progressive increase in pH from the stomach (pH 1.3–3.5) and small intestine (pH 5.5–6.8) to the colon (pH 6.4–7.0). This range of pH acts as an environmental stimulus for drug release and pH-sensitive polymers are used as the device for drug release because their swelling and deswelling are dependent on the change in pH [36]. However, there are some disadvantages regarding this type of administration, mainly related to the pH of the environment through which the drug transits the GIT. For example, the stomach is not a very hospitable environment for many drugs, so they must survive at a low pH between 1.5 and 5.0, but also in the presence of digestive enzymes that can destroy many therapeutic molecules [37]. Consequently, localized drug delivery to the lower GIT (e.g., colon) is only possible when the drug is protected from the hostile environment of the upper GIT, minimizing side effects that occur due to unwanted drug release or unnecessary systemic absorption [38].

Ocular route of drug delivery using hydrogels is considered an attractive alternative to conventional formulations, such as standard eye drops, due to the ability of hydrogels to resist natural eye defense mechanisms like tear drainage and blinking, as well as their effectiveness in overcoming low corneal permeability [26, 27]. Studies have shown that up to 95% of a drug delivered via eye drops is lost through tear drainage, with only about 5% reaching intraocular tissues. To address this limitation, increasing the contact time between the drug and the cornea is essential for improving the treatment of ocular pathologies. In this context, various ophthalmic formulations have been developed to prolong ocular surface contact and enable sustained drug release [39]. Hydrogels, in particular, show strong potential as topical treatments for eye diseases, as they can adhere to the ocular surface and modulate drug release over time. However, physiological challenges such as low corneal permeability, rapid tear turnover, and frequent blinking continue to hinder effective ocular drug delivery, making hydrogel-based systems a promising focus for ongoing research and innovation [7].

Nasal administration is an alternative to parenteral or oral administration, which is characterized by low bioavailability [40]. This nasal route of administration is supported by several features, such as the large surface area and high permeability of the nasal epithelium, rich vascularization of the submucosa, avoidance of first-pass hepatic metabolism, and relatively high blood flow, which promotes rapid absorption [41]. Conventionally, the nasal route of administration is used to treat local diseases, such as rhinitis and nasal congestion. However, in more recent studies, much more attention has been paid to nasal administration of drugs as a promising route of drug delivery for systemic therapy [40, 42, 43].

Vaginal and rectal drug administrations are effective for localized treatments and drugs undergoing extensive first-pass metabolism [44]. The vaginal route suits women's health drugs like microbicides and hormones due to its large surface area, rich blood vessels, and avoidance of hepatic metabolism [45]. The rectal route is used for local conditions such as hemorrhoids and systemic delivery by bypassing the liver

[46]. Challenges include patient discomfort and drug leakage causing variable drug availability. Thermoreversible in situ mucoadhesive gels can increase residence time, reduce leakage, and improve drug efficacy [47].

Dermal and transdermal drug delivery offer a painless, accessible alternative to oral or parenteral routes due to the skin's large surface area and ease of access, improving patient compliance [48]. Dermal delivery targets the skin's outer layers, while transdermal delivery transports drugs into systemic circulation via the dermis [49]. Hydrogels enable prolonged drug administration, easy removal, and bypass first-pass metabolism. These formulations are often used for treating skin conditions like inflammation, eczema, and psoriasis [50, 51].

Injectable hydrogels are advanced drug delivery systems administered via minimally invasive injection, reducing patient discomfort and costs, even in hard-to-reach areas [52]. They require shear-thinning behavior to ease injection and rapidly regain viscosity for stability at the target site [53]. Optimal viscosity, elasticity, gelation time, and temperature sensitivity ensure smooth flow during injection and solidification at body temperature. These hydrogels enable precise control of drug localization and release kinetics, suitable for targeted treatments like cancer, diabetes, and gene therapy [7, 54].

5.2 Hydrogels' Mechanisms in Controlled Drug Delivery

The therapeutic benefit is not totally decided by drug effectiveness. Instead, it is also affected by the techniques involved in the drug formulation and delivery within the body. Specifically, the delivery mode largely affects the therapeutic efficacy since it changes numerous factors, including pharmacological activity, pharmacokinetics, drug distribution, cellular uptake and metabolism, excretion and clearance, and toxicity [55]. Over the years, the researchers have been looking for suitable and promising alternatives for drug administration, in order to improve the biodistribution and bioavailability of drugs in vivo and, consequently, to improve the dose–response relationship. In this sense, drug delivery systems (DDS) arise as a promising strategy to overcome the biological barriers and to perform an efficient drug delivery, in which hydrogels have gained a lot of interest in the pharmaceutical field [1]. Basically, DDS is a formulation to introduce a therapeutic substance in our body that can control the rate, amount, time, and place of release to improve the efficiency and safety [56].

An ideal DDS should: (i) deliver an optimal drug dose to the target site; (ii) control release kinetics to enhance therapy and minimize side effects; and (iii) ensure drug availability at specific times and durations [8]. Controlled or sustained DDS release drugs at a predetermined rate, improving compliance and reducing side effects [57, 59]. Hydrogels are promising carriers due to tunable swelling, mechanical properties, and ease of modification, enabling controlled release [5, 59]. Their drug release typically has an initial burst followed by sustained diffusion-controlled release, governed by swelling and drug–matrix interactions [55]. This allows hydrogels to maintain therapeutic drug levels locally over extended periods. Overall, hydrogels maintain

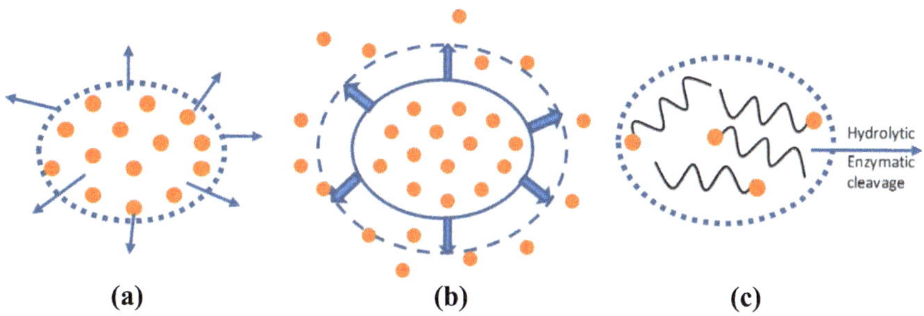

Fig. 5.4 Schematic representation of drug release mechanisms from hydrogels: (**a**) diffusion-controlled; (**b**) swelling-controlled; (**c**) chemically-controlled (● – drug molecule) [60]

a high local concentration of the therapeutic agent over an extended period through controlled release mechanisms influenced by diffusion, swelling, chemical structure, or environmental stimuli.

Diffusion-controlled release is the most common mechanism for drug release from hydrogels. These systems can function as either reservoirs or matrices, allowing drug diffusion through their water-filled mesh or pores (Fig. 5.4a) [5]. In reservoir-type drug delivery systems, drug molecules are encapsulated at high concentration in the hydrogel core, which is surrounded by a membrane. Here, drug release is time-independent and generally follows Fickian first law of diffusion. In contrast, matrix-type systems have drug molecules homogeneously dispersed throughout the hydrogel, resulting in time-dependent release that typically follows Fickian second law of diffusion [60].

Swelling-controlled release mechanism involves the swelling and expansion of the hydrogel beyond its boundary, in contact with a bio-fluid, facilitating the drug release along with the polymer chain relaxation, as schematically presented in Fig. 5.4b. The process otherwise referred to as Case II transport, supports time-independent, constant drug release kinetics and mostly fit to zero-order model [5]. The higher the swelling rate of the hydrogel, the higher the drug release rate. Consequently, in this case, at least two factors become important in controlling the swelling process, namely the rate and ability of the hydrogels to absorb liquids from the environment, as well as the thickness of the hydrogels [60].

Chemically-controlled release mechanism (Fig. 5.4c) involves the existence of chemical reactions within the hydrogel matrix that are responsible for drug release. Typically, these reactions involve enzymatic or hydrolytic cleavage of the polymer network, either by bulk or surface erosion. The cleavage of the polymer chains becomes, in these chemically controlled release systems, the rate-determining step, so that the diffusion process becomes less significant in drug release [60].

5.3 Cellulose-Based Hydrogels as Matrices in Controlled Drug Delivery

In recent years, the biomass-based materials have become key elements in obtaining hydrogels for targeted, controlled, and smart delivery systems, due to their unique properties primarily related to their easy availability, relatively low cost, biodegradability, biocompatibility, and non-toxicity [61]. This category also includes cellulose, the best-known natural renewable polymer in the world, existing in plants, animals and some bacteria. Cellulose is a linear polymer composed of D-glucose units connected by (1–4) glycosidic bonds and contains three free hydroxyl groups on each glucose unit, allowing for the functional modification of cellulose [4, 9].

Cellulose and its derivatives are highly versatile in biomedical applications, allowing obtaining hydrogels with different shapes and features. The cellulose-based hydrogels can be produced either from native, non-derivatized cellulose [62] or from cellulose derivatives, such as methylcellulose (MC) [63], carboxymethyl cellulose (CMC) [64], hydroxypropyl methylcellulose (HPMC) [65] or hydroxyethyl cellulose (HEC) [66]. They can also contain cellulose in the form of nanocrystals (CNC) [67, 68], nanofibrills (NFC) [50, 69] and bacterial nature (BC) [70–72] or they can even be composite/hybrid hydrogels [73, 74]. All these types of cellulose-based hydrogels have gained significant attention since unlike synthetic polymers based-hydrogels, it has been shown not to trigger the immune responses or cause toxicity, being well-tolerated by biological systems [4, 75].

In drug delivery systems, the cellulose-based hydrogels have been extensively studied, being recommended by a number of suitable intrinsic properties, such as biocompatibility, biodegradability, hydrophilic nature, excellent mechanical properties, and versatile functionality [57, 76]. The biocompatibility of cellulose-based materials ensures the safety of their long-term, and their biodegradability minimizes the risk of long-term accumulation in the body. Also, being derived from renewable sources, cellulose hydrogels offer a sustainable alternative to synthetic polymers-based hydrogels, reducing environmental impact [77]. The hydrophilic nature allows hydrogels to retain large amounts of water, creating an optimal environment for drug encapsulation and sustained release. The cellulose-based hydrogels features can be easily adjusted through chemical or physical crosslinking to achieve desired swelling behavior and mechanical strength in order to best serve the intended purpose [57]. Due to the versatile functionality, the cellulose hydrogels can be engineered to respond to external stimuli, such as pH, temperature, enzymes, etc. enabling targeted and controlled drug release. This tunability is essential for personalized drug delivery systems, allowing for customized release rates based on the drug's properties and patient needs. Therefore, more advanced and smarter hydrogel-based DDS can be developed that can bind the drug in their network by covalent or physical linkages to further control its release profile, as well as taking advantage at the body temperature and the variable pH ranges in the different body parts in order to do a space-controlled drug release [1]. Moreover, the combination of cellulose-based materials with other natural polymers [62] or nanoparticles [69, 78] can enhance its

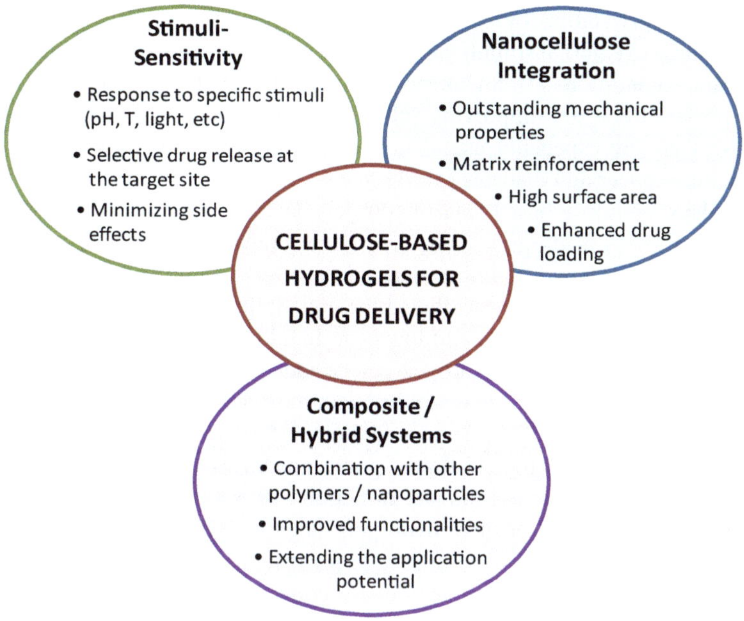

Fig. 5.5 Recent trends in cellulose-based hydrogels for controlled drug delivery

properties (i.e., mechanical strength) or introduce new functionalities (i.e., antibacterial properties). At the same time, the multitude and diversity of these combinations can expand the use of cellulose-based hydrogels, offering sustainable and effective alternatives to traditional materials [6].

The research in cellulose-based hydrogels for controlled drug delivery has focused on several innovative trends (Fig. 5.5) including the hydrogels with stimuli-responsive behavior for controlled drug release, the nanocellulose based-materials use as individual component or as reinforcing agent to enhance drug loading due to their high surface area, or the combination of cellulose with other polymers or nanoparticles to add new functionalities and features. These delivery systems based on cellulosic materials exist both individually or in combination with others, but they are essentially aimed to create more efficient and accurate drug administration systems.

5.3.1 Stimuli-Responsive Hydrogels for Controlled Drug Delivery

Designing efficient delivery matrices that enable targeted drug release and dose reduction while maintaining therapeutic effects is a major goal in drug delivery. This can be achieved by creating systems sensitive to physical (e.g., temperature, light), chemical (e.g., pH, redox), or biological (e.g., enzymes) stimuli [79, 80]. Hydrogels' responsiveness to external stimuli has paved the way for smart materials that adapt to physiological conditions [7]. Such responsive systems offer controlled drug release, reducing side effects and improving selectivity compared to conventional delivery methods. They are especially useful in treating diseases with local physiological changes, like cancer and diabetes [5, 61].

Nowadays, cellulose-based smart hydrogels have gained much attention for the delivery of a series of drugs in vitro or in vivo. Preparation of stimuli-responsive DDS using cellulose and cellulose derivatives offer a significant advantage over conventional materials, owing to their biocompatibility, as they are nontoxic and nonallergenic, biodegradability, and low cost. In addition, some specific properties of cellulose derivatives-based hydrogels can be added, such as high transparency and swelling degree (over 80%); compatibility with most active pharmaceutical components; similarity with living tissues due to their soft consistency and low interfacial tension with water or biological fluids; good bioadhesiveness, providing intimate contact between the drug and the site of action [48, 61, 79]. Figure 5.6 illustrate different types of stimuli-sensitive hydrogels, beginning with pH and temperature-sensitive hydrogels that are most attractive for drug delivery, since the biological mechanisms of the human body are largely connected to these stimuli [81], and continuing with magnetic-, ionic or electric-responsive hydrogels.

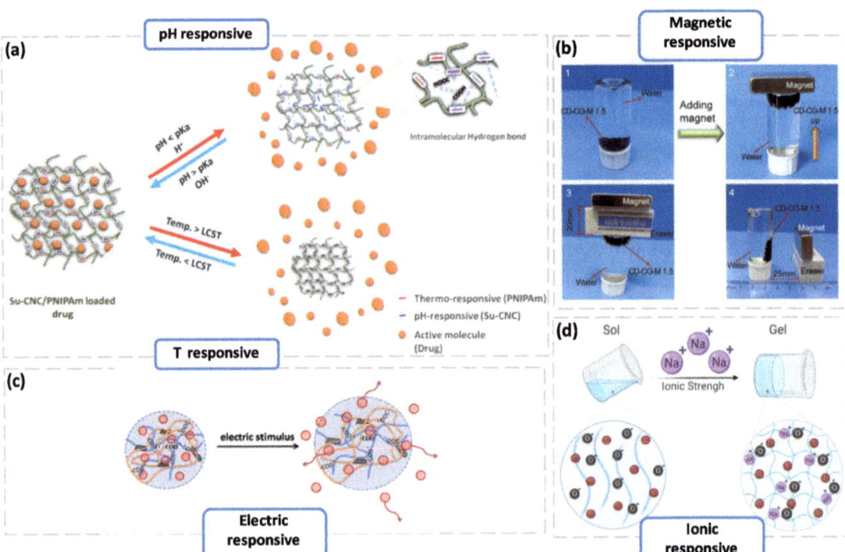

Fig. 5.6 Illustrations of different types of stimuli-sensitive hydrogels: (**a**) pH and thermo-responsive. Reproduced with the permission from Ref. [82]; (**b**) magnetic responsive. Reproduced with the permission from Ref. [118]; (**c**) photo-thermal responsive. Reproduced with the permission from Ref. [85]; and (**d**) ionic strength responsive hydrogels. Reproduced with the permission from Ref. [85]

pH-responsive *cellulose-based hydrogels.*

The most attractive targeted delivery route is controlled by the pH gradient, considering that the human gastrointestinal tract (GIT) has the advantage of having a wide pH range, which varies from 1 to 7.5, namely the stomach has a pH between 1 and 3, the saliva between 5 and 6, the small intestine 6.6 - 7.5 and colon 6.4 - 7.0. The main characteristic of pH-responsive hydrogels is the presence of ionizable functional groups within the hydrogel network, which by protonation or deprotonation modify the electrostatic interactions and subsequently affect the swelling behavior of the hydrogel and implicitly the drug release rate. They can be anionic (e.g., containing carboxylic acid groups) or cationic (e.g., containing amine groups), depending on the type of ionizable groups. They can deliver drugs specifically to the desired location in the body based on pH gradients, minimizing side effects and maximizing efficacy. For example, the hydrogel systems exhibiting minimal swelling at acidic pH and maximal swelling around neutral pH can be used as potential oral drug carriers for intestine or colon-specific delivery [81, 86] (Fig. 5.6).

Cellulose derivatives (i.e., carboxymethyl cellulose, hydroxypropyl cellulose, cellulose acetate) have been studied in pH-responsive drug delivery applications for a long time and have proven to be very effective [61]. For instance, carboxymethyl cellulose (CMC) is highly popular in pH-responsive hydrogels fabrication due to the pendant carboxyl group on CMC side chains with protonation - deprotonation behavior. At low pH values, the interactions between polymer chains become dominant over solvent-polymer interactions due to the protonation of carboxyl groups and strong hydrogen bonding interaction between them. In contrast, at alkaline pH values, these carboxyl groups are ionized, and their electrostatic repulsions cause extended hydrogel network structure leading to increased swelling properties of the hydrogel. In addition, the CMC-based hydrogels are non-toxic and can be obtained at relatively low cost [87–89]. However, their applications as DDS are quite limited due to their low mechanical strength. To overcome these shortcomings, usually CMC is modified or combined with various polymers, natural or synthetic, or diverse nanoparticles to produce matrices with suitable features for DDS use.

For example, pH-responsive hydrogels were prepared from carboxymethyl cellulose (CMC) and 2-hydroxyethylacrylate (2-HEA), a grafting agent used to enhance CMC's adhesion, flexibility, and mechanical properties. These hydrogels were synthesized via radical polymerization as transdermal delivery systems for naringenin (NRG), a compound with antioxidant, anti-inflammatory, and anticancer effects used in treating atopic dermatitis. The hydrogels exhibited a stable gel network with good resilience and adhesion, swelling more at pH 7.5 and 8.5 than at pH 5.5 due to electrostatic repulsion from ionized carboxyl groups. In vitro skin permeation studies demonstrated that the CMC/2-HEA hydrogel improved NRG penetration, making it an effective system for treating skin lesions related to pH imbalance [89]. Similarly, Jeong et al. [90] enhanced the mechanical strength and swelling capacity of CMC hydrogels by synthesizing a dual-component system combining CMC with carboxymethyl β-cyclodextrin (cmβCD). The storage modulus of the cmβCD/CMC gel was nearly eight times higher than that of pure CMC gel, and compressive strength was also improved, while maintaining high swelling ability. Notably, the swelling ratio at pH 7.4 was 1.7 times greater than at pH 1.2. The hydrogels were tested for drug-loading and release using tetracycline (TC) as a model drug, showing efficient loading capacity and strong antibacterial activity against *Staphylococcus aureus*. Additionally, they exhibited no cytotoxicity toward human dermal fibroblasts, supporting their potential as safe, non-toxic, and site-specific drug delivery systems.

Hui Li Tan et al. utilized the pH-sensitive carboxyl groups of CMC and pectin, both non-toxic and cost-effective polymers, to create CMC/pectin hydrogel beads via ionotropic gelation and electron beam crosslinking for targeted colon delivery of diclofenac sodium. This drug, with a short half-life, requires controlled release to minimize stomach irritation and enable sustained release at colonic pH 7.4. The 15% CMC/5% pectin formulation formed rigid beads that remained intact and reached swelling equilibrium within 24 h in colonic conditions [36]. Similarly, Yin et al. developed self-healing hydrogels cross-linked via hydrazone bonds between dialdehyde CMC and hydrazide-modified pectin for localized silibinin delivery in lung cancer. These hydrogels exhibited pH-sensitive sustained release, biodegradability, cell compatibility, enhanced antitumor efficacy, and reduced silibinin toxicity [91].

Chen et al. developed a pH-sensitive injectable hydrogel for controlled delivery of hydrophobic drugs using hydrazide- and aldehyde-modified CMC as precursors. These were mixed with drug-loaded micelles of PEO-b-PDPA, which enhance hydrophobic drug loading and enable pH-triggered release via a hydrophobic-to-hydrophilic transition. The cellulose-based hydrogel encapsulates the micelles, providing a second tunable barrier for drug release. Gelation time, mechanical strength, release rate, and degradation were controlled by adjusting precursor concentration and oxidation degree. Faster gelation occurred in low pH, beneficial for acidic tissue environments. Cell viability tests confirmed minimal cytotoxicity of the hydrogels and precursors [92]. A novel hydrogel was prepared from sodium CMC and hydroxyethyl cellulose (HEC) using ammonium persulfate as initiator and methylenebisacrylamide (MBA) as crosslinker, showing sensitivity to pH and ionic strength. Bovine serum albumin (BSA) release studies revealed low release at pH 1.2 (17.8%) and high release at pH 7.4 (85.2%), indicating potential for controlled oral drug delivery [66]. Similarly, MBA-crosslinked polyampholytic hydrogels based on CMC and chitosan exhibited enhanced stability and swelling over a wide pH range. Aspirin release was higher in acidic (pH 2.1) than alkaline (pH 7.4) conditions, with swelling increasing alongside CMC concentration [93]. Hydroxypropyl cellulose (HPC), hydroxyethyl cellulose (HEC), and cellulose acetoacetate (CAA) are cellulose derivatives studied for pH-sensitive drug release matrices. Zhou et al. prepared a pH-responsive hydrogel by mixing oxidized HPC (Ox-HPC) with carboxymethyl chitosan (CMCS), linking phenylalanine as a model drug via imine bonds. The hydrogel showed faster phenylalanine release at pH 6.8 (tumor-like environment, ~ 60% in 8 h) compared to pH 7.4 (normal tissue, ~ 12 h), demonstrating its potential for targeted drug delivery in tumor microenvironments [94].

Kwon et al. synthesized pH-sensitive hydrogels from hydroxyethyl cellulose (HEC) and hyaluronic acid (HA) cross-linked with divinyl sulfone and loaded them with isoliquiritigenin (ILTG), a compound with antimicrobial and anticancer properties. These hydrogels showed over 70% drug release at pH 7, enhancing antimicrobial activity against *Propionibacterium acnes*, and exhibited excellent skin permeability, primarily via hair follicles, making them promising for treating pH-related skin lesions [95]. Liu et al. developed a dual-responsive cellulose-based hydrogel from CAA and cystamine dihydrochloride (CYS) containing pH-sensitive enamine and redox-active disulfide bonds. This hydrogel showed reversible sol–gel transitions triggered by pH and redox changes. Rhodamine B release was slow at neutral and intestinal pH (~63% in 25 h), but rapid at acidic stomach pH (88% in 3 h) due to proton-catalyzed enamine bond cleavage. Reducing agents like Dithiothreitol (DTT) further enhanced release to nearly 100% after 10 h, by disrupting disulfide linkages [95, 96] (Table 5.1).

Temperature-responsive *cellulose-based hydrogels.*

Generally, the thermosensitive hydrogels are smart hydrogels that respond to temperature changes, altering its shape from sol to gel phase at a specific range of temperature [111]. They contain thermosensitive groups that cause the material to respond to temperature (i.e., change in solubility or phase) and are classified into two types: (i) upper critical solution temperature hydrogels (UCST), those that respond positively to the temperature, and lower critical solution temperature (LCST), those that react negatively to temperature [112]. Temperature-sensitive hydrogels contain both hydrophobic and hydrophilic component in structures and the thermal response is a consequence of the delicate balance between the hydrophobic and hydrophilic portions of the polymer. Temperature changes the interaction between hydrophilic and hydrophobic segments in the polymer with water molecules, thus can induce a change in the solubility of the cross-linked network, causing the sol–gel phase transition [113]. Common thermosensitive materials include poly(N-isopropyl acrylamide) (PNIPAAm), PEG-PLGA block copolymers and pluronics (also known as poloxamers) [114]. PNIPAAm is known for its LCST behavior, typically around 32 °C in aqueous solutions, close to body temperature, which makes it very suitable for drug delivery [115].

Grafting PNIPAM onto cellulose-based materials is a common strategy for creating thermosensitive hydrogels with drug delivery applications. For example, methyl cellulose (MC), which has an LCST between 60–80 °C, can be grafted with PNIPAM to lower its LCST closer to body temperature [97]. Other polymers and salts have also been used for this purpose. Itoh et al. incorporated pectin into MC to reduce gelation temperature below physiological levels, creating sustained-release systems for paracetamol, particularly beneficial for dysphagic or elderly patients. These formulations demonstrated suitable rheological properties and remained in the rat stomach for over 3 h, allowing prolonged drug release [98]. Similarly, Bhowmik et al. used i-carrageenan and KCl to lower MC's sol–gel transition temperature from 60 °C to 33.5 °C in ophthalmic formulations containing pilocarpine hydrochloride. The results showed improved ocular bioavailability, extended precorneal residence time, and reduced dosing frequency, enhancing patient compliance [99].

Methylcellulose (MC) has also been combined with Pluronic F127 (Poloxamer 407), a thermoresponsive copolymer composed of polyethylene oxide (PEO) and

Table 5.1 Cellulose and cellulose derivative-based smart hydrogels for controlled drug delivery

Material	Stimulus	Drug	Delivery route	Ref
CMC/2-HEA	pH	Naringenin	Dermal	[89]
cmβCD/CMC		Tetracycline	Dermal	[90]
CMC/pectin		Diclofenac	Colon	[36]
DACMC/ Hydrazide-pectin		Silibinin	Lungs	[91]
CMC/HEC		BSA	Oral	[66]
CMC/Chitosan		Aspirin	Oral	[93]
Ox-HPC/CMCS		Phenylalanine	Tumors	[94]
HEC/HA		Isoliquiritigenin	Tumors	[95]
CAA/CYS	pH/ redox	Rhodamine B	-	[96]
MC/PNIPAM	Temperature	-	-	[97]
MC/pectin		Paracetamol	Oral	[98]
MC/ i-carrageenan/ KCl		Pilocarpine hydrochloride	Oftalmic	[99]
MC/ Pluronic F127		Etidronat	Oral	[100]
CMC/PNIPAAm/AM	pH/ Temperature	Tetracycline	Intestinal	[87]
CMC/ Fe_3O_4/QCMC	Magnetic field	Aciclovir	Herpes	[101]
CMC/$CoFe_2O_3$/Fe_3O_4		Doxorubicin	Tumor	[102]
CMC/MNPs		Dopamine hydrochloride	Neurological	[103]
CMC/Fe_2O_3		Doxorubicin	Tumor	[104]
HEC/Fe_2O_3		Doxorubicin	Tumor	[105]
C/β-CD/Fe_3O_4		5-Fluorouracil	Tumor	[83]
CMC/rNc/PVA/Fe_2O_3	pH/ magnetic	Naringin	Antioxidant	[106]
CMC/HEC	Ionic strength	BSA	Oral	[66]
CMC/Gel		5-fluorouracil	Tumor	[107]
HPMC		Hydrochlorothiazide	Oral	[108]
HPMC/GG		Salbutamol sulphate	Nasal	[109]
CMC-CHO/GNRs/ P(NIPAM-co-AH)	NIR/ Temperature	Doxorubicin	Tumor	[110]

polypropylene oxide (PPO) blocks. At low temperatures, Pluronic F127 is water-soluble, but forms micelles and gels at body temperature due to its hydrophobic PPO core and hydrophilic PEO shell. Its bioadhesiveness, safety, and ability to gel at low concentrations make it ideal for drug delivery [116]. Rangabhatla et al. used an MC/Pluronic F127 formulation for targeted etidronate delivery in osteogenesis therapy, achieving sustained in vitro release for over 28 days and improved mechanical strength and degradation resistance [100]. Kim et al. developed thermoresponsive injectable hydrogels with MC, Pluronic F127, and ammonium sulfate for docetaxel delivery. The gelation temperature ranged from 15 to 40 °C, and drug release lasted over 35 days, maintaining stability and showing complete tumor eradication at high doses, without adverse effects [117].

N-isopropyl acrylamide (PNIPAAm) has been used to impart temperature sensitivity to CMC hydrogels, complementing their inherent pH responsiveness due to carboxyl groups. Song et al. [87] developed a dual-network hydrogel composed of CMC and a copolymer of PNIPAAm and acrylamide (AM), loaded with TC for potential treatment of intestinal inflammation. This structure enhanced the hydrogel's mechanical strength and enabled dual responses to both pH and temperature. The copolymerization of NIPAM and AM adjusted the LCST close to physiological temperature. The hydrogel exhibited excellent biocompatibility and responsiveness: TC release increased from 20 to 73% as pH shifted from 4.6 to 6.8, and from 39 to 84% as temperature rose from 36 °C to 39.6 °C, reaching 94% above 39.6 °C due to gel collapse. This demonstrates the hydrogel's suitability for sustained, targeted drug delivery while minimizing toxicity to healthy tissue.

Magnetic field-*responsive cellulose-based hydrogels.*

Magnetically responsive hydrogels refer to hydrogels with special responses to magnetic fields. This type of hydrogels is extremely attractive for controlled drug delivery because the drug release behavior can be easily tuned by an external magnetic field through noncontact external stimuli, along with noninvasiveness, and fewer side effects on healthy tissues [61, 83]. Unlike other stimuli-responsive hydrogels, whose stimuli-responsive properties originate from the hydrogel networks themselves, the magnetically -responsive hydrogels are always made by combining hydrogel networks with magnetic nanoparticles (MNPs) [118]. MNPs including iron oxides (e.g., Fe_2O_3, Fe_3O_4) [119] or transition metal ferrites ($CoFe_2O_4$, $MnFe_2O_4$) [120] have been widely used as promising materials for magnetically targeted drug delivery, due to their unique magnetic property, high stability in aqueous media and favorable biocompatibility properties [61]. Nevertheless, among these candidate, Fe_3O_4 nanoparticles are the most commonly used in biomedical domain with the advantages of good biocompatibility, high magnetization ability and relative ease of preparation and functionalization. The incorporation of MNPs allows hydrogels, inherently insensitive, to become sensitive to external magnetic stimuli and, therefore, to be magnetically actuated, to deform and transform under the action of magnetic fields, in a remote and tunable manner [118]. Furthermore, magnetic nanoparticles can be easily integrated with other stimuli-responsive hydrogels to achieve sensitivity to multiple stimuli, expanding the applicability of magnetic hydrogels [106, 121].

MNPs can be incorporated to cellulose or derivatized cellulose hydrogels by polymerization [101], precipitation or sorption on the surface of the chains [122] or as cellulose coated magnetic nanoparticles [103]. Cellulose acts as a polymer backbone chain and provides active sites for the binding of MNPs via hydrogen bonding or electrostatic interactions [83]. Various types of magnetically receptive cellulose-based hydrogels have been prepared with various morphologies, with different types of cellulose derivatives, like carboxymethyl cellulose [121], cellulose acetate [123], hydroxyethyl cellulose [105] and hydroxypropylmethyl cellulose [124]. The magnetically receptive CMC-based hydrogels, composed of MNPS modified with CMC, have been prepared by various synthetic methods: physically embedding MNPs in CMC particles [101], using MNPs as crosslinking agents for CMC [102], or coating MNPs with CMC by physical adsorption or conjugation onto the MNPs surface [103]. For instance, the magnetic CMC particles (m-CMC) were synthesized by embedding the magnetic ferrite particles (Fe_3O_4) in CMC and quaternized-CMC (QCMC) particles, obtained separately in a single step, using water-in-oil microemulsion polymerization and divinylsulfone as a crosslinking agent (50–100 mol% relative to the CMC repeating unit). Acyclovir (Ac), used in antiviral therapy for herpes viruses, was chosen as the model drug for the absorption and release studies. The modified forms of CMC particles had similar Ac loading capacities but m-CMC particles had a higher release capacity (i.e., the loading efficiencies were 3.74 mg/g for CMC, 3.67 mg/g for m-CMC and 10.62 mg/g for QCMC, respectively at 24 h). The loading capacity of the particles was increased by chemical modification and prolonged drug release time was accomplished (i.e., 0.8 mg/g Ac release by CMC particles and 2.4 mg/g Ac by QCMC particles) [101].

Simply incorporating MNPs into the hydrogel matrix has several limitations, primarily related to the fact that there may be a continuous release of MNPs into the human environment, thus producing potential effects of accumulation and, possibly, toxicity in tissues. In order to overcome these problems, Barbucci et al. fabricated hybrid magnetic CMC hydrogels using MNPs as crosslinking agent for CMC chains, and creating a three-dimensional structure with covalent bonds [102]. Hybrid CMC-MNPs hydrogels with $CoFe_2O_3$ and Fe_3O_4 nanoparticles were loaded with antitumor drugs (DOX) and the kinetics of drug release was determined under the effect of an alternating magnetic field (AMF). In addition, MNPs were functionalized with (3-aminopropyl) trimethoxysilane (APTMS) in order to introduce amino groups on the surface, allowing the formation of amidic groups with the CMC carboxylic ones. Under the application of AMF, the CMC-MNPs hybrid hydrogels loaded with DOX show a drug release greater than that showed by CMC hydrogels crosslinked with 1,3-diaminopropane. The application of low-frequency (40 kHz) AMF increases the amount of DOX initially released from CMC-MNPs hydrogels, a trend generally observed when a gel containing free or cross-linked MNPs and loaded with a drug is examined under AMF stimulus. Typically, magnetic actuation of iron oxide MNPs triggers drug release from temperature-sensitive materials by modifying their permeability, without having any major effect on the environment [102].

Aguilera et al. applied CMC coating to MNPs to develop Fe_3O_4/CMC systems capable of crossing the blood–brain barrier (BBB) for neurological drug delivery. The CMC coating provided a hydrophilic, anti-agglomeration surface, enhancing nanoparticle dispersion and bioavailability. In vitro assays using dopamine hydrochloride and a BBB model confirmed successful translocation across endothelial cell layers, with no cytotoxicity or morphological alterations observed in HLMVE cells after 24–48 h. These findings highlight the potential of CMC-coated MNPs for targeted CNS therapies with reduced dosage and side effects. Similarly, Guo et al. [104] synthesized CMC-stabilized Fe_2O_3 nanocarriers via a one-step hydrothermal method. These carriers showed high magnetic saturation (54.6 emu/g), excellent biocompatibility (MTT assay with L929 cells), and a doxorubicin (DOX) loading efficiency of 92.5% due to abundant surface carboxyl groups. Drug release was stable at pH 7.4 and significantly accelerated at pH 5.0, driven by protonation of CMC carboxyl groups, making them ideal for pH-responsive drug delivery to acidic tumor environments.

Bekaroglu et al. [105]. also surface-modified superparamagnetic Fe_3O_4 nanoparticles with hydroxyethyl cellulose (HEC). Although the HEC coating slightly reduced magnetization, the nanoparticles retained their superparamagnetic behavior. Importantly, the coating minimized oxidation, a concern during hyperthermia treatments due to free radical formation, and reduced cytotoxicity, as confirmed by in vitro tests at 1 g/L HEC. Doxorubicin (DOX) was successfully loaded onto the HEC-coated Fe_3O_4 nanoparticles, demonstrating comparable antitumor activity to pure DOX in human breast adenocarcinoma cells. Another approach for creating magnetically sensitive cellulose-based hydrogels involves the *in-situ* formation of MNPs by immersing hydrogels in Fe^{2+}/Fe^{3+} solutions, followed by co-precipitation. Using this method, Dai et al. [106] developed dual pH- and magneto-responsive hydrogels from CMC, regenerated nanocellulose (rNc), and PVA. The incorporation of Fe_3O_4 conferred magnetic responsiveness, while 2% rNc addition optimized thermal stability, swelling capacity, and the loading/entrapment efficiency of the antioxidant drug naringin. To enhance drug loading and incorporate Fe_3O_4 nanoparticles, Lin and al. [83] developed a cellulose-based hydrogel composed of cellulose (C) grafted with β-cyclodextrin (β-CD). Fe_3O_4 nanoparticles were introduced via a simple dropwise method by dispersing them in the β-CD/C solution, followed by gelation in a $CaCl_2$ coagulating bath. The resulting magnetic β-CD/C hydrogel beads exhibited rapid, reversible swelling–deswelling behavior under an external magnetic field (EMF). Grafted β-CD doubled the drug loading capacity (from 6.78% to 13.26%), while the embedded MNPs enabled EMF-controlled drug release. The release of 5-fluorouracil could be finely tuned by switching the EMF on or off and adjusting the Fe_3O_4 content.

Other cellulose-*based hydrogels sensitive to stimuli.*

pH, temperature and magnetic field-sensitive hydrogels are the most well-known and used stimuli-responsive cellulose hydrogels for controlled drug delivery. However, there are also other stimuli that cause drug release from cellulose-based hydrogels, namely ionic-strength-, electric- or photo/light-responsive.

The use of *ionic strength-sensitive hydrogels* is another approach for controlled drug delivery, which implies the uses of ionic strength of the surrounding environment as a trigger for drug release. Usually, these types of hydrogels include ionizable groups in their matrix which can dissociate when immersing the hydrogel in solution and undergo conformational changes in response to ions (i.e., K^+, Na^+, Ca^{2+}, etc.) and induce changes in its swelling properties [61, 125]. For example, carboxymethyl cellulose/hydroxyethyl cellulose (CMC/HEC) hydrogels, used in drug delivery studies, exhibit both pH sensitivity, due to carboxyl groups, and ionic strength sensitivity, attributed to the ionic exchange capacity of the same groups. When alternately immersed in ultrapure water and PBS buffer, the hydrogels showed a lower swelling capacity in PBS (60 g/g) than in water (90 g/g), explained by the "charge screening effect" of PBS cations. These cations neutralize charges on the polymer chains, reducing repulsion and hydrogel expansion. This also led to an "on–off" switching behavior in swelling. Furthermore, the hydrogels demonstrated effective BSA release control: only 17.8% released in simulated gastric fluid (pH 1.2), compared to 85.2% in intestinal fluid (pH 7.4), suggesting strong protection in the stomach and targeted intestinal release [66]. Similar behavior was observed in bacterial cellulose/polyacrylamide (BC/PAM) hydrogels synthesized via accelerated microwave irradiation using MBA as a cross-linker. The swelling ratio decreased with increasing NaCl concentration, due to the "charge screening effect" of Na^+ ions, which neutralize the hydrogel's carboxylate groups and reduce electrostatic repulsion. Likewise, the in vitro release of theophylline was influenced by both polymer–drug interactions and the hydrogel's swelling behavior in the dissolution media [126].

Carboxymethyl cellulose (CMC) was combined with gelatin (Gel) and HA to form a polyelectrolyte complex (PEC) hydrogel, with sulfonated reduced graphene oxide (SRGO) added to enhance mechanical and conductive properties and assess ionic strength sensitivity. Swelling tests in NaCl solutions (0–0.2 M) at pH 1.2 showed anti-polyelectrolyte behavior for SRGO-free hydrogels (increased swelling with ionic strength), whereas SRGO-based hydrogels exhibited reduced swelling due to sulfonyl group-induced polyelectrolyte effects. Similarly, 5-fluorouracil (5-FU) release decreased with higher ionic strength in SRGO-based hydrogels, suggesting their potential for targeted, ionic strength-sensitive drug delivery in gastric cancer therapy [107]. Ionic strength significantly influences drug release from hydroxypropyl methylcellulose (HPMC) matrices. Asare-Addo et al. examined how varying NaCl concentrations (0–0.4 M) and pH levels affected hydrochlorothiazide release. Drug release increased with ionic strength, rising from 28% in water to 48% at 0.49 M NaCl [108]. HPMC was also used with gellan gum (GG) by Salunke et al. to develop a mucoadhesive in situ nasal gel, which forms a 3D network through ionic interactions and hydrogen bonding. In simulated nasal fluid, the formulation transitioned rapidly from sol to gel and showed 97.34% drug release in 11 h. Histopathology confirmed no tissue damage, suggesting this HPMC/GG gel is a promising system for nasal delivery of salbutamol sulphate [109].

Light-responsive hydrogels are a type of stimuli-responsive hydrogel that is currently of widespread interest, with minimal side effects because lighting is non-invasive and does not require additional reagents for remote control. By adjusting the irradiation parameters, such as the illumination power density or period, the irradiation dose may be precisely regulated [125]. In virtue of practicality, convenience and suitability for biological applications, near infrared (NIR) laser is treated as the trigger of the photo-responsive hydrogels. In addition, NIR light has good tissue penetration and no phototoxicity to cells) [127]. For instance, a multi-responsive (pH/NIR) hydrogel was developed by $Ca2^{+}$ cross-linking of polydopamine (PDA) and TEMPO-oxidized cellulose nanofibrils (TOCNFs) [128]. PDA served as a photothermal and drug carrier, while TOCNFs provided a 3D structure. The hydrogel enabled on-demand tetracycline hydrochloride (TH) release under low pH or NIR irradiation and showed sustained drug release (77%) over 24 h in PBS (pH = 5.0) without NIR. It also exhibited antibacterial activity against *S. aureus* and *E. coli*. With its pH/NIR responsiveness, extended release, and antibacterial properties, this hydrogel offers a promising option for skin disease drug delivery [128]. A dual photo/thermal-responsive hydrogel was developed by embedding gold nanorods (GNRs) into a matrix of oxidized carboxymethyl cellulose (CMC-CHO) and P(NIPAM-co-AH) copolymer. GNRs enabled NIR-triggered photothermal effects to enhance DOX release for in vivo antitumor therapy in mice. The hydrogel showed good biocompatibility, biodegradability, and a tunable LCST slightly above body temperature. DOX-loaded hydrogels exhibited reduced toxicity and improved therapeutic efficacy, further enhanced by NIR stimulation. With its cellulose-based degradability and responsive drug release, the system is a promising vehicle for localized cancer treatment [110].

Photo-responsive hydrogels with interpenetrating networks were prepared from alginate, TOCNFs, and polyacrylamide, cross-linked by Fe^{3+} and N,N'-methylenebis-acrylamide. Alginate and TOCNFs served as photo-sensitive components, with TOCNFs also enhancing mechanical strength. BSA and dexamethasone were loaded as model drugs, and UV irradiation increased BSA release. The addition of TOCNFs significantly improved Young's modulus and compressive strength, while hydrogels with varying CNF content showed similar release profiles with enhanced UV-triggered drug release [129].

5.3.2 *Nanocellulose-Based Hydrogels for Controlled Drug Delivery*

The cellulose nanomaterials, also known as nanocellulose, have attracted rapidly growing interest in the last few years. Due to the excellent inherent physical and chemical properties, like high specific surface area, remarkable mechanical properties (tensile strength, elastic modulus) together with biocompatibility, biodegradability and renewability, the nanocellulose materials present broad application perspectives [130]. Compared to cellulose, the nanocellulose materials have at least one dimension smaller than 100 nm. Nanocellulose materials are found in three different forms based on its morphology, namely, CNFs, CNCs, and BC [131]. CNFs sometimes referred to as nanofibrous cellulose or cellulose nanofibers, are flexible fibers with a diameter of 5–10 nm and length of 500 nm to several micrometers, and consist of amorphous and crystalline domains. CNCs have rod-like structure and thus are also referred as rod-like cellulose microcrystals, nanocrystalline cellulose and cellulose nanowhiskers, which are cylindrical (5–30 nm in diameter) and elongated (100–500 nm for plant cellulose and 100 nm to several micrometers for tunicate and algae). CNCs have the highest tensile strength among the three types of nanocelluloses due to its crystalline nature, which make it one of the best materials for reinforcement in composites [132]. BC is also known as biocellulose or microbial cellulose and is the purest form of cellulose, being the primary metabolic product in the form of swollen membranes by some bacteria of the genera *Agrobacterium, Rhizobium, Sarcina and Gluconacetobacter*. BC has unique macromolecular structural features such as thin subfibrils, which form microfibrils because of strong inter- and intramolecular hydrogen bonding [72].

Nanocellulose-based hydrogels are considered to be an ideal option as matrices for controlled drug release due to their biocompatibility which means it does not cause any immunological response in the body, the large specific surface area which is favorable for drug loading and also due to their modifiability which results in the possibility of adjusting drug loading and release by introducing special chemical groups [125]. As a result, the nanocellulose hydrogels have been demonstrated to sustainably deliver various kinds of drugs via different routes of administration, in which nanocellulose significantly improves the hydrogel properties and tunes the drug releasing profile [55]. Moreover, the nanocellulose materials they have found application in controlled drug delivery, either as an individual component of hydrogels or as part of a mixture or composite (i.e., as reinforcing agent) [133]. For instance, as an individual component of DDS, the nanocellulose has emerged as an excellent material for drug encapsulation due to its mechanical strength since it can endure physico-chemical barriers in the body, such as the stomach's acidic pH. On the other hand, the incorporation of nanocellulose as reinforcing agent in drug delivery systems, in the form of nanowhiskers or nanofibers, has become a widely used practice due to the numerous benefits provided by the nanoscale [131]. Higher surface area at the nanoscale enhances the drug's solubility and dissolution rate. In addioton, the incorporating nanocellulose, such as cellulose nanocrystals and nanofibers, into hydrogels has enhanced mechanical strength and drug loading efficiency. These nanostructures offer high surface area and reinforce the hydrogel matrix, improving its performance in drug delivery applications [130].

CNFs-based *hydrogels.*

CNF hydrogels have been explored for controlled drug delivery, both alone and in composites with other polymers. To improve drug loading and release, CNF surfaces are often chemically modified. Paukkonen et al. studied TOCNF hydrogels as matrices for releasing metronidazole (MZ), nadolol (NAD), and ketoprofen (KETO), before and after freeze-drying. Results showed that freeze-drying and rehydration had no impact on drug release or rheological properties, suggesting these hydrogels can be stored dry and rehydrated when needed [134]. Hujaya et al. developed CNF-based multi-ion complex hydrogels using oppositely charged cellulose nanofibril derivatives, with anionic (carboxylic groups) and cationic (quaternary ammonium groups), in acetate buffer (pH 4–5) at 0.5% w/w. The resulting hydrogels showed approximately tenfold increase in storage and loss moduli compared to individual dispersions. Doxorubicin-loaded hydrogels exhibited pH-responsive release: approximately 65% drug retention at pH 7.4 after the initial burst, and complete release at pH 4 within 5 days. No cytotoxicity was observed up to 900 mg/mL on NIH3T3 cells, supporting their potential as green biomaterials for biomedical use [135]. CNFs have been used as reinforcing agents in composite hydrogels with both synthetic and natural polymers, often to impart stimuli responsiveness. Orasugh et al. [136] developed HPMC/CNF composite hydrogel films for transdermal delivery of ketorolac tromethamine (KT). CNFs (up to 1.00 wt%) enhanced the storage modulus and tensile strength of HPMC films while reducing their moisture affinity. Drug release studies in PBS (pH 7.4) showed prolonged KT release over 8 h from the nanocomposites, with tunable release profiles depending on CNF content, indicating potential for controlled delivery of hydrophobic drugs (Table 5.2).

Dual responsive hydrogels were developed by combining TOCNFs with thermoresponsive poly(N-isopropylacrylamide) (PNIPAAm). pH-responsiveness was tuned by adjusting the CNFs' carboxyl content during oxidation; higher carboxylation enhanced responsiveness but reduced compressive strength. The swelling behavior varied with temperature: hydrogels were swollen and translucent below LCST, and shrunken and opaque above it. Methylene blue (MB) release followed the Higuchi model best at pH 10, and the Korsmeyer–Peppas model at pH 2, indicating ion exchange–driven interactions between MB and CNF–PNIPAAm hydrogels [74].

CNFs have been used to reinforce polymer matrices for pH-sensitive hydrogels. Azhar et al. developed CNF-reinforced PVA hydrogels to control cisplatin release,

Table 5.2 CNFs-based hydrogels for controlled drug delivery

Material	Stimulus	Drug	Delivery route	Ref
TOCNFs hydrogels	–	Metronidazole Nadolol Ketoprofen	Oral	[134]
CNFs-based multi-ion complex hydrogels	–	Doxorubicin	Tumor	[135]
HPMC/CNFs composite hydrogel	–	Ketorolac tromethamine	Oral	[136]
TEMPO-CNFs/ PNIPAAm	pH/ Temperature	Methylene blue	-	[74]
CNFs/PVA		Cisplatin	Oral	[137]
SA/TOCNFs	pH	Probiotics	Oral	[138]
SA/CNFs	pH	IBuprofen	Oral	[139]
FA-NC/ GMA-HEMA/EGDMA	pH	Curcumin	Tumors	[58]
PDA/TOCNFs hydrogels	pH/ NIR	Tetracycline hydrochloride	Skin	[128]
ALG/TEMPO-CNFs/PAM	Light	BSA; Dexamethasone	Oral	[129]

reducing delivery in the stomach and enhancing it in the small intestine. Drug-loaded CNFs (0.37 wt%) were incorporated into PVA and crosslinked with 1% glutaraldehyde. Swelling and drug release were influenced by both CNF content and pH. At pH 7.4, hydrogels with 1 wt% CNFs sustained drug release, minimizing side effects. Kinetic modeling showed Korsmeyer–Peppas best fit, indicating diffusion-controlled release, suitable for intestinal delivery of cisplatin [137]. Zhang et al. [138] prepared pH-responsive hydrogel macrospheres from sodium alginate (SA) and TOCNFs via extrusion into $CaCl_2$ for crosslinking. These macrospheres were stable in simulated gastric fluid, protecting encapsulated probiotics, and swelled in intestinal fluid to release them. Similarly, Yuhang et al. [139] used SA matrices with NFCs to create pH-responsive hydrogels for sustained ibuprofen (IBU) release. CNFs were well-dispersed, and a 2:1 CNFs:SA ratio yielded optimal morphology. Swelling increased from 7 to 19 times dry weight as pH changed from 1.5 to 11, with faster IBU release at alkaline pH and slower at acidic pH. Release followed Fickian diffusion and erosion per the Korsmeyer-Peppas model.

CNCs-based *hydrogels.*

Most studies on CNCs for drug delivery focus on their role as reinforcing agents in composite hydrogels [125]. Thermo-responsive PNIPAm hydrogels reinforced with CNCs were synthesized via free-radical polymerization. Increasing CNC content decreased thermal stability but enhanced mechanical strength (elastic and viscous moduli) of tested hydrogels. CNCs also raised the LCST from approximately 32 °C to 36–39 °C, closer to body temperature, due to increased hydrophilicity and swelling. These hydrogels showed good metronidazole loading at room temperature, followed by an initial burst and sustained release at 37 °C [67]. The simple thermo-sensitive PNIPAm-CNCs was transformed into dual (temperature and pH) sensitive hydrogels, by Emam et al., through free chain polymerization reaction of the modified CNCs with succinic acid (Su-CNCs) in the presence of NIPAm. Three different DS of Su-CNCs were used to investigate their effects on the swelling rate of the prepared hydrogel. However, the hydrogel Su-CNC/PNIPAm 2 (DS = 1.8) represented the highest swelling as well as a remarkable high response to the pH trigger. The selected hydrogel was used as a template for in vitro Famotidine drug release study by exploitation of their pH and temperature sensitive properties. The results revealed that the drug release % enhances by altering the pH from 10 to 2 at 25 °C over a period more than 6 h [82]. CNCs-reinforced temperature-responsive hydrogels were prepared by incorporation of CNCs in amphiphilic copolymers like poly(ε-caprolactone-*co*-lactide)-b-poly(ethylene–glycol)-b-poly(ε-caprolactone-*co*-lactide) (PCLA) [140]. The presence of CNCs in the porous network and the sulfate groups in the CNCs may control the structural stability of the hydrogel through hydrophobic interactions between CNCs and PCLA copolymers and eventually modify the sol–gel transition temperature. Thus, the hydrogel retains good mobility under 30 °C, but transform into gel at a temperature of 34 °C, close to the physiological ones. In addition, CNCs controlled the release properties with the slightly faster release for hydrophilic cationic protein and sustained the release of the hydrophobic drug DOX (Table 5.3).

Table 5.3 CNCs- and BNC-based hydrogels for controlled drug delivery

Material	Stimulus	Drug	Delivery route	Ref
PNIPAm/Su-CNCs	pH/ Temperature	Famotidine	Oral	[82]
CNCs/ PCLA	Temperature	Doxorubicin	Tumor	[140]
CNCs/ Pluronic F127	Temperature	Pilocarpine hydrochloride	Oftalmic	[68]
FA-NC/GMA-HEMA/ EGDMA	pH	Curcumin	Tumor	[58]
CNCs/Fe_2O_3/SA	pH/ magnetic	Ibuprofen	Oral	[142]
CMC/CNCs/Fe_3O_4/ PMAA	Temperature/Magnetic filed	Ibuprofen	Oral	[143]
BNC matrices	-	Famotidine Tizanidine	Oral	[72]
PMGly/BNC	pH	Diclofenac sodium salt	Oral, Dermal	[80]
BNC/β-CD	pH	Berberine hydrochloride	Oral	[144]
BNC/ Poloxamers 338/ 407	pH	Octenidine	Skin	[145]
BNC/PAM hydrogels	Ionic strength	Theophylline	-	[126]

Moreover, CNCs have been used as nanofillers in thermoresponsive Pluronic F127 (P-127), a nontoxic triblock copolymer of poly(ethylene oxide) – poly(propylene oxide) – poly(ethylene oxide) (PEO100 – PPO65 – PEO100), soluble in water at low temperatures [68, 141]. In both cases, the aim was to obtain thermosensitive hydrogels with special rheological and mechanical properties by incorporating CNCs. For instance, Kushan et al. [141] showed that moderate loading of CNCs up to 3% result in more softened P-127 hydrogels by disrupting closely packed micelles, and enhancing deformability. These hydrogels fluidize under high shear but rapidly recover gel state at body temperature, useful for dual-targeted drug delivery. On the other hand, Orasugh et al. [68] developed more robust P-127 hydrogels with 0.8–1.2% CNCs for ophthalmic drug delivery. CNCs reduced the gelation concentration from 18% to 16.6% (w/v), and prolonged pilocarpine hydrochloride (PL) release over 20 h, with drug release decreasing from 87.26% (pure P-127) to 52.89% at 0.8% CNCs. The 1.2% CNC formulation showed the best sustained release, highlighting CNCs' role in gelation, strength, and drug release control. Anirudhan et al. developed a complex CNC-based drug delivery system for pH-controlled, targeted curcumin delivery. Nanocellulose (NC) was conjugated with folic acid (FA) via EDC–NHS (1-Ethyl-3(3-dimethylaminopropyl) carbodiimide - *N*-hydroxysuccinimide) coupling to target folate receptors overexpressed on cancer cells. FA-NC was polymerized with glycidyl methacrylate (GMA) to enhance chain length, drug encapsulation, and biocompatibility, and hydroxyethyl methacrylate (HEMA) for chemical stability and tissue-like properties. The resulting FA-NC/GMA-HEMA/EGDMA system achieved 89.2% drug loading at pH 8.0, with maximum release at pH 5.5. In vitro studies showed cytotoxicity against cancer cells, with enhanced encapsulation attributed to hydrogen bonding, and MTT/apoptosis assays confirmed targeted delivery via FA receptors [58].

Supramaniam et al. [142] has proven the ability of CNCs to obtain magnetic field-sensitive hydrogels, developing a pH-sensitive based hydrogel beads composed of magnetic nanocellulose (m-CNCs) and SA as a drug carrier for ibuprofen. M-CNCs were prepared by co-precipitation of Fe (II) and Fe (III) ions in aqueous solution containing CNCs with NaOH and were incorporated into SA-based hydrogel beads in order to enhance its mechanical strength and regulating drug release behavior. The results showed that, besides the magnetic property, the presence of m-CNCs improved the physical and mechanical properties of SA-hydrogel beads and its swelling degree, along with controlled and sustained drug release from 30 to 330 min. Also, the CNCs incorporation could lead to targeting, detecting, and potentially treating cancer tissue by magnetic resonance imaging. Later, Markovic et al. developed dual-responsive hybrid hydrogels based on poly(methacrylic acid) (PMAA), CMC, CNCs, and Fe_3O_4 particles for targeted ibuprofen delivery. CNCs enhanced PMAA's mechanical properties and encapsulated poorly water-soluble ibuprofen, while CMC stabilized magnetite particles. Under an external magnetic field, the hydrogels exhibited a magnetocaloric effect, enabling targeted drug delivery with reduced circulation time. Increasing CNC or magnetite content decreased swelling and cumulative ibuprofen release. The system allows tunable release kinetics by adjusting components, offering controlled, prolonged ibuprofen delivery that reduces gastrointestinal irritation [143].

BNC-based *hydrogels.*

Bacterial nanocellulose (BNC) is obtained from bacteria cultivation and possesses a three-dimensional nanostructure with high mechanical performance [55]. Unlike CNCs and CNFs, whose purification requires chemical processes, including harsh alkaline or acidic treatments, BNC is the purest form of cellulose, being produced by several bacteria of the genera *Gluconacetobacter, Agrobacter, Sarcina*, as a metabolic product. BNC has unique properties including high water holding capacity, high chemical purity and crystallinity, hyperfine network structure and impressive tensile strength [72, 84, 144]. These intrinsic properties make BNC a suitable material for countless applications, including the drug delivery.

In recent year, the research in BNC as a drug delivery system has increased significantly due to its ultra-fine nano-dimensional network structure. For instance, Badshah et al. [72] studied the in vitro release of Famotidine (poorly water-soluble drug) and Tizanidine (a highly water-soluble drug) loaded in different concentrations in single-excipient-based BNC matrices, with the objective to overcome the problems associated with the conventional tablet dosage forms. In addition, the BNC matrices were prepared under various conditions, i.e., never drid, partially dried and freeze dried. The release medium was a simulated gastric fluid (0.1N HCl, pH = 1.2) and the temperature was adjusted to 37 °C. Variation in drug-loading capacity of the hydrated and freeze-dried matrices was observed, with the hydrated matrices showing a higher drud loading, probably due to the hydrophilic nature and porous fibrous network. The drug release profile shows that more than 90% of the drug (20 – 40 mg/mL loaded Famotidine and 6 mg/mL loaded Tizanidine) was released within the first 15 min, the hydrophilic nature and porous network of BNC probably resulting in the faster drug release. Moreover, the amount and the water solubility of the drug loaded in BNC matrices did not affect its release behavior, all formulations displayed am immediate drug release pattern, which is achieved in conventional formulations by adding a number of excipients. The results demonstrated the possibility of using BC hydrogels as a single polymer-based oral administration, without adding additional excipients. To improve drug solubility, drugs like berberine hydrochloride (BBH) are complexed with β-cyclodextrin (β-CD) and loaded into BNC hydrogels. Lin Xiao et al. [144] reported that BBH's solubility at 25 °C increased by ~ 102% with 16 mM β-CD. BBH/β-CD-loaded BNC hydrogels prepared by physical absorption showed higher drug loading and sustained release in simulated gastric (pH 1.2) and intestinal fluids (pH 6.8), with faster release at pH 6.8. In vitro antibacterial tests confirmed effective inhibition of *S. aureus*, *P. aeruginosa*, and *E. coli*. Combined with BNC's biocompatibility, these hydrogels are promising for oral treatment of gastrointestinal infections.

BNC-based hydrogels for controlled drug delivery were prepared also to respond to externali stimuli. For instance, in the study of Alkhatib et al., a pH-sensitive BNC-based drug delivery system was developed to extend the retention time of antiseptic Octenidine up to 1 week, by the combination of the advantages of two materials, BNC and Poloxamers, highly attractive due to their biocompatibility and low toxicity. BNC was modified by incorporation of differently concentrated Poloxamers 338 and 407 as micelles and gels. The results show the Octenidine was slowly released up to one week depending on the type and concentration of poloxamer. Moreover, the water binding and high compression stability properties were attained. In addition, the antimicrobial activity against *Pseudomonas aeruginosa* and *Staphylococcus aureus* was reported [145]. Saidi et al. [80] reported a pH-responsive composite material based on BNC and poly(N-methacryloyl glycine) (PMGly), by in situ free radical polymerization of MGly within the swollen BC tridimensional network. The polymers containing amino acid moieties attached to their side chain were already studied as responsive drug-release systems. The obtained materials display good thermal stability (up to 200 °C), good mechanical properties (storage modulus is over 700 MPa and Young's modulus between 3.5 and 4.9 GPa), along with high water uptake capacity. The results of the in vitro MTT assay showed that PMGly/BC nanocomposites are non-cytotoxic to HaCaT cells, for 72 h. The in vitro release profile of diclofenac sodium salt (DCF) from the nanocomposites into simulated body fluids at different pH values demonstrates the pH-responsive behaviour of these materials. Besides, DCF is mainly retained in the nanocomposites at pH 2.1 and released at pH 7.4, revealing their potential for the controlled release of DCF in dermal as well as in oral drug delivery applications.

5.3.3 Composite/hybrid Cellulose-Based Hydrogels for Controlled Drug Delivery

Hybrid and composite hydrogels both incorporate additional components to enhance properties but differ in integration methods. Hybrid hydrogels combine different polymer networks, often natural and synthetic, to create materials with improved mechanical strength, tunable degradation, stimuli-responsiveness, and biocompatibility [21]. Composite hydrogels embed particles like nanoparticles or inorganics into a single polymer network to enhance mechanical strength or conductivity, prepared via dissolution/dispersion and cross-linking [86]. Combining cellulose with biopolymers or nanoparticles yields hybrid/composite hydrogels with improved antibacterial and mechanical properties, expanding their applications in controlled drug delivery [146].

For instance, the poor mechanical stability of alginate hydrogels, limiting their drug carrier efficacy, was improved by adding cellulose fibers, such as cotton linter (CL), microcrystalline cellulose (MCC), and microfibrillated cellulose (MFC). Using metformin hydrochloride (MH), a water-soluble drug with approximately 50–60% bioavailability and 1.5–4.5 h half-life, as a model, results showed that cellulose enhanced the mechanical and thermal stability of alginate. Additionally, micro- and nanoscale cellulose improved drug encapsulation and slowed MH release in simulated intestinal fluid [147]. Cellulose/lignin hydrogel blends were developed for targeted, controlled paracetamol delivery, studying structure–property relationships based on lignin type, such as organosolv hardwood lignin (TCA) and lignosulfonate (TCS) [59]. The molecular interactions between components play an important role in their viscoelastic behavior, namely the organosolv lignin introduced hydrophobicity, reducing swelling, while ionic sulfonate groups in TCS increased it. Mechanical properties depended on composition, with a cellulose/TCA ratio of 95:5 showing the highest compression modulus due to reinforcement via intermolecular interactions. Paracetamol release improved with cellulose/TCA lignin, attributed to reduced paracetamol-cellulose interactions [59]. In another study, a hydrogel biocomposite was formed by combining dialdehyde cellulose (DAC) from sugarcane waste and chitosan (3:1 w/w). DAC, produced via periodate oxidation, contain aldehyde groups used to create cross-linked gels by an amine-containing polymer such as chitosan by a spontaneously occurring Schiff base reaction. The hydrogel showed an encapsulation efficiency of 70.63% in PBS and 55.23% in acidic medium. Drug release was higher in 0.1 M HCl (7.32–90.57%) than in PBS pH 7.4 (8.24–58.79%) over 5 h. DAC/chitosan hydrogels demonstrate strong potential as drug delivery matrices for methylene blue [148] (Table 5.4).

A particular case of composites is represented by nanocomposites obtained incorporating different types of nanocellulose into polymers matrix. The nanocellulose particles can be used as such or sometimes chemical modification of its groups is practiced to improve the interaction between the polymer matrix and nanocellulose. For instance, Ahlen et al. [78] developed polyvinyl alcohol (PVA) contact lenses loaded with chitosan-poly(acrylic acid) nanoparticles and CNCs for controlled ophthalmic

Table 5.4 Cellulose-based composite hydrogels for controlled drug delivery

Material	Stimulus	Drug	Delivery route	Ref
ALG/CL; ALG/MCC ALG/MFC	-	Metformin hydrochloride	Oral	[147]
C/TCA; C/TCS	-	Paracetamol	Oral	[59]
DAC/CS	-	Methylene blue	Oral	[148]
PVA/CS-PAA/CNCs			Ophthalmic	[78]
ALG/CNFs		Ampicilin		[149]
DA-CNCs/CS	pH	Theophylline	Oral	[150]
CNCs-GMA/Starch	-	Vitamin B12	Oral	[151]
ALG/CCNCs-AAG	-	Ceftazidime hydrate	-	[152]
AuNPs/CMC-PMA	pH	Diltiazem hydrochloride; Diclofenac sodium	Transdermal	[153]
CMC/GO hydrogel beads	pH	Doxorubicin	Tumors	[154]
g-CMC/GO	pH	Methotrexate	Tumors	[155]
GNP- CNF/QGG-PVA	-	Diltiazem hydrochloride	Transdermal	[156]

drug delivery. Chitosan nanoparticles degrade in physiological lysozyme (0.2 mM), while CNC integration extended drug release up to 28 h. In another study, ampicillin-loaded alginate/CNFs composites improved mechanical and swelling properties compared to alginate alone. These composites showed immediate ampicillin release followed by sustained release over 500 min, with significant chemical interaction between CNFs and ampicillin enhancing drug loading. Release rates could be tuned by adjusting composite composition [149].

Chemical modifications of nanocellulose include periodate oxidation to aldehyde groups [150], vinyl group introduction via glycidyl methacrylate [151], and cationization with 2,3-epoxypropyl trimethylammonium chloride [152]. For example, a nanocomposite hydrogel of periodate-oxidized dialdehyde CNC (DA-CNC) and chitosan (CS) was developed for controlled theophylline delivery. DA-CNC acted as both matrix and crosslinker, confirmed by FT-IR showing bonding between chitosan amino and DA-CNC aldehyde groups. Increasing chitosan raised the isoelectric point (alkaline shift) due to more free amino groups and increased swelling ratio likely from reduced crosslinking. Drug release was pH-dependent, with ~ 85% cumulative release in gastric fluid (pH 1.5) and 23% in intestinal fluid (pH 7.4) [150]. Mauricio et al. [151] developed covalent microhydrogel composites via ultrasound-assisted radical polymerization from starch and CNCs chemically modified with glycidyl methacrylate (GMA) vinyl groups. CNCs acted as covalent cross-linkers, slowing vitamin B12 release by approximately 2.9 times compared to starch-only microgels, acting as a drug release retardant. Lin et al. [152] created double-membrane hydrogels from cationic CNCs (CCNCs) and anionic alginate (AAG) via electrostatic interactions. CCNCs were cationized with (2,3-epoxypropyl) trimethylammonium chloride. The system enabled co-delivery of two drugs with distinct release profiles:

rapid ceftazidime hydrate release from the alginate outer membrane (first 3 days), and sustained epidermal growth factor release from the CCNC inner membrane (from day 4 to day 12). This design of the double-membrane hydrogels offers synergistic drug release and potential to overcome drug resistance.

Nanocomposite hydrogels can be formed by embedding inorganic nanoparticles into cellulose-based matrices or grafting cellulose onto particle surfaces to add stimuli sensitivity. Mandal et al. [153] developed a biodegradable, non-cytotoxic nanocomposite hydrogel by in-situ forming gold nanoparticles (AuNPs) on CMC hydrogel crosslinked with poly(methacrylic acid). This hydrogel showed sustained release over 3 days of transdermal drugs diltiazem hydrochloride (~85%) and diclofenac sodium (~79%), enhancing skin permeability and drug delivery. By modifying CMC with graphene oxide (GO) can produce pH-sensitive hydrogels. Rasoulzadeh et al. [154] prepared biodegradable CMC/GO hydrogel beads crosslinked with $FeCl_3 \bullet 6H_2O$ for controlled release of doxorubicin (DOX). CMC provides pH-responsive drug release, while GO increases swelling and loading capacity via hydrogen bonding. The release profile of DOX at pH 6.8 and 7.4 indicated its strongly pH dependence. However, the interactions between GO and DOX based on H-bonding could be unstable under acidic conditions, which resulted in faster drug release rate in pH 6.8. Similar, Jiao et al. grafted CMC onto graphene oxide via hydrothermal treatment to create a pH-sensitive drug carrier for methotrexate (MTX). In vitro release tests showed MTX release strongly depended on pH: 4.76% at pH 1.0 (6 h), 10.9% at pH 6.8 (3 h), and sharply increased to 67.4% (6 h) and 82.2% (48 h) at pH 7.4. The system exhibited low cytotoxicity on NIH-3T3 cells and superior in vivo antitumor activity against HT-29 colon cancer cells compared to free MTX, highlighting its promise for colon cancer therapy [155].

Efficient transdermal composites were designed to deliver diltiazem hydrochloride (DH) by Anirudhan et al., combining the CNF and AuNPs in order to increase the drug penetration across the skin. The AuNP- CNF nanofiller was prepared by the reduction of $HAuCl_4 \cdot 3H2O$ with sodium citrate dihydrate (Na_3Cit) in an aqueous suspension of CNF and followed by their incorporation into a polymeric matrix formed by quaternized guar gum (QGG) and PVA. The available hydroxyl groups on the surface of AuNP-CNF provide higher encapsulation efficiency on the basis of hydrogen bonding interaction with hydroxyl group of CNF and carbonyl group of DH. Also, the composites released 70% and 76% of DH drug over 24 h periods when CNF concentrations were 1.0 wt% and 4.0 wt%, respectively [156].

5.4 Conclusions

In the last decades, hydrogels have significantly advanced the field of controlled drug delivery, providing innovative and adaptable polymer matrices for targeted therapy. Evolving from simple and single-component release systems, the modern hydrogels now support complex, multicomponent formulations capable of delivering multiple therapeutics in a spatially and temporally controlled manner. Nanogels, a subclass of hydrogels, offer additional advantages due to their nanoscale size. They can be administered intravenously and are capable of intracellular drug delivery. Their rapid response to environmental stimuli makes them ideal for triggered release applications, particularly when intracellular conditions act as cues.

Cellulose-based hydrogels stand out for their biocompatibility, biodegradability, tunable physicochemical properties, and environmental sustainability. Their structure can be readily modified to accommodate targeted and stimuli-responsive drug release, making them especially valuable in biomedical applications. From the studies briefly presented in this chapter, several conclusions can be drawn regarding the use of cellulose-based hydrogels in the controlled drug delivery, namely: the hydrogels responsive to pH, temperature, light, or magnetic fields provide controlled, site-specific drug release; the reinforcement with nanocellulose improves hydrogel mechanical strength and thermal stability, broadening their applicability in controlled drug delivery under different physiological conditions; the nanocellulose-based hydrogels generally exhibit good biocompatibility and low cytotoxicity; the complex nanocomposites combining cellulose-based materials with other synthetic/natural polyerms, along with anorganic particles (i.e., magnetic) enable targeted and controlled release, improving therapeutic efficacy and minimizing side effects; the developed cellulose based-hydrogels can sustain drug release over hours to days, allowing improved drug bioavailability and reduced dosing frequency.

Despite their benefits, the cellulose hydrogel-based drug delivery systems face several challenges, related to burst drug release, limited hydrophobic drug loading, or slow responsiveness of stimuli-sensitive hydrogels. However, ongoing research has suggested various strategies to overcome these limitations, including chemical functionalization, use of smart cross-linkers, and incorporation of nanostructures like cellulose or nanowhiskers to enhance strength and responsiveness. Future research should prioritize the development of "green" and biocompatible solvents and cross-linkers, injectable and in situ forming hydrogels, and smart systems responsive to different stimuli. Additionally, functionalized hydrogels offer a cost-effective way to enhance efficacy and selectivity. In conclusion, cellulose-based hydrogels and their derivatives, but especially those based on nanocellulose, are about to play an increasingly important role in controlled drug delivery. Due to continuous technological advances, interdisciplinary collaboration and a strong focus on clinical translation, cellulose hydrogel-based systems have immense potential for improving therapeutic outcomes in the coming years.

References

1. A.M. Ribeiro, M. Magalhães, F. Veiga, A. Figueiras, Cellulose-based hydrogels in topical drug delivery: a challenge in medical devices, in *Cellulose-Based Superabsorbent Hydrogels*, ed. by M. Mondal (Springer, Cham, 2018)
2. D.E. Ciolacu, R. Nicu, F. Ciolacu, Cellulose-based hydrogels as sustained drug-delivery systems. Materials **13**, 5270 (2020)

3. S. Cascone, G. Lamberti, Hydrogel-based commercial products for biomedical applications: a review. Int. J. Pharm. **573**, 118803 (2020)
4. L. Liu, D. Wu, H. Tu, M. Cao, M. Li, L. Peng, J. Yang, Applications of hydrogels in drug delivery for oral and maxillofacial diseases. Gels **9**, 146 (2023)
5. R. Narayanaswamy, V.P. Torchilin, Hydrogels and their applications in targeted drug delivery. Molecules **24**, 603 (2019)
6. R. Saadan, A. Ihammi, M. Chigr, A. Fatimi, A short overview of the formulation of cellulose-based hydrogels and their biomedical applications. Eng. Proc. **81**, 3 (2024)
7. B. Liu, K. Chen, Advances in hydrogel-based drug delivery systems. Gels **10**, 262 (2024)
8. E. Hamidi, R. Alhaloush, Ş Şenol, E. Akyol, An overview on current trends and future outlook of hydrogels in drug delivery. Sigma J Eng Nat Sci **41**(5), 1047–1061 (2023)
9. J. Gong, L. Hou, Y.C. Ching, K.Y. Ching, N.D. Hai, C.H. Chuah, A review of recent advances of cellulose-based intelligent-responsive hydrogels as vehicles for controllable drug delivery system. Int. J. Biol. Macromol. **264**, 130525 (2024)
10. M. Dattilo, F. Patitucci, S. Prete, O.I. Parisi, F. Puoci, Polysaccharide-based hydrogels and their application as drug delivery systems in cancer treatment: a review. J. Funct. Biomater. **14**, 55 (2023)
11. R. Nicu, G. Lisa, R.N. Darie-Nita, M.I. Avadanei, A. Bargan, D. Rusu, D.E. Ciolacu, Tailoring the structure and physico-chemical features of cellulose-based hydrogels using multi-epoxy crosslinking agents. Gels **10**, 523 (2024)
12. M.K. Azeem, A. Islam, R.U. Khan, A. Rasool, M.A.R. Qureshi et al., Eco-friendly three-dimensional hydrogels for sustainable agricultural applications: current and future scenarios. Polym. Adv. Technol. **34**, 3046–3062 (2023)
13. H. Hameed, S. Faheem, A.C.P. Santos, H.S. Sarwar, M. Jamshaid, A comprehensive review of hydrogel-based drug delivery systems: classification, properties, recent trends, and applications. AAPS PharmSciTech **25**, 64 (2024)
14. C. Markand, S. Vaishnav, V. Khairnar, S. Kale, Advancements in hydrogel-based drug delivery systems: a review. Int. J. of Pharm. Sci. **2**(4), 586–597 (2024)
15. N.H. Thang, T.B. Chien, D.X. Cuong, Polymer-based hydrogels applied in drug delivery: an overview. Gels **9**, 523 (2023)
16. Z. Fu, S. Xiao, P. Wang, J. Zhao, Z. Ling et al., Injectable, stretchable, toughened, bioadhesive composite hydrogel for bladder injury repair. RSC Adv. **13**, 10903 (2023)
17. J. Li, D. Mooney, Designing hydrogels for controlled drug delivery. Nat. Rev. Mater. **1**, 16071 (2016)
18. J. Li, X. Jia, L. Yin, Hydrogel: diversity of structures and applications in food science. Food Rev. Int. **37**(3), 313–372 (2021)
19. S. Jacob, A.B. Nair, J. Shah, N. Sreeharsha, S. Gupta, P. Shinu, Emerging role of hydrogels in drug delivery systems. Tissue Eng. Wound Management. Pharm. **13**, 357 (2021)
20. P.N. Kendre, T.S. Satav, Current trends and concepts in the design and development of nanogel carrier systems. Polym. Bull. **76**, 1595–1617 (2019)
21. S.J. Buwalda, T. Vermonden, W.E. Hennink, Hydrogels for therapeutic delivery: current developments and future directions. Biomacromol **18**, 316–330 (2017)
22. I. Neamtu, A.G. Rusu, A. Diaconu, L.E. Nita, A.P. Chiriac, Basic concepts and recent advances in nanogels as carriers for medical applications. Drug Delivery **24**(1), 539–557 (2017)
23. L.E. Nita, A.P. Chiriac, A. Diaconu, N. Tudorachi, L. Mititelu-Tartau, Multifunctional nanogels with dual temperature and pH responsiveness. Int. J. Pharm. **515**, 165–175 (2016)
24. Y. Zhao, B. Ran, X. Xie, W. Gu, X. Ye, J. Liao, Developments on the smart hydrogel-based drug delivery system for oral tumor therapy. Gels **8**, 741 (2022)
25. Z.C. Yin, Y.L. Wang, K. Wang, A pH-responsive composite hydrogel beads based on agar and alginate for oral drug delivery. J. Drug Deliv. Sci. Technol. **43**, 12e18 (2018)
26. G. Fang, X. Yang, Q. Wang, A. Zhang, B. Tang, Hydrogels-based ophthalmic drug delivery systems for treatment of ocular diseases. Mat. Sci. Eng. C **127**, 112212 (2021)
27. N. Dubashynskaya, D. Poshina, S. Raik, A. Urtti, Y.A. Skorik, Polysaccharides in ocular drug delivery. Pharmaceutics **12**, 22 (2020)

28. C.R. Lynch, P.P.D. Kondiah, Y.E. Choonara, L.C. du Toit, N. Ally, V. Pillay, Hydrogel biomaterials for application in ocular drug delivery. Front. Bioeng. Biotechnol. **8**, 228 (2020)
29. B. Fan, H. Jia, H. Yu, Y. Xu, R. Li et al., Hydrogel-based intranasal delivery systems for the treatment of allergic rhinitis. Adv. Therap. **6**, 2300035 (2023)
30. A.L. Schilling, Y. Kulahci, J. Moore, E.W. Wang, S.E. Lee, S.R. Little, A thermoresponsive hydrogel system for long-acting corticosteroid delivery into the paranasal sinuses. J. Control. Release **330**, 889–897 (2021)
31. R. AlAnsari, B. Hasan, G.R. Deen, U. Torsten, Hydrogel- and nanocomposite-based drug-delivery strategies in the treatment of vaginal infections. Polymers **16**, 775 (2024)
32. Y. Ye, W. Zhong, R. Luo, H. Wen, Z. Ma et al., Thermosensitive hydrogel with emodin-loaded triple-targeted nanoparticles for a rectal drug delivery system in the treatment of chronic non-bacterial prostatitis. J Nanobiotechnology **22**(1), 33 (2024)
33. H. Labie, M. Blanzat, Hydrogels for dermal and transdermal drug delivery. Biomater. Sci. **11**, 4073 (2023)
34. L.C. Wong, J.H. Poh, W.T. Tan, B.K. Khor, V. Murugaiyah et al., Cellulose hydrogel development from unbleached oil palm biomass pulps for dermal drug delivery. Int. J. Biol. Macromol. **224**, 483–495 (2023)
35. A. Nokhodchi, S. Raja, P. Patel, K. Asare-Addo, The role of oral controlled release matrix tablets in drug delivery systems. BioImpacts **2**(4), 175–187 (2012)
36. H.L. Tan, L.S. Tan, Y.Y. Wong, S. Muniyandy, K. Hashim, J. Pushpamalar, Dual crosslinked carboxymethyl sago pulp/pectin hydrogel beads as potential carrier for colon-targeted drug delivery. J. Appl. Polym. Sci. **43416** (2016)
37. S. Javanbakht, A. Shaabani, Carboxymethyl cellulose-based oral delivery systems. Int. J. Biol. Macromol. **133**, 21–29 (2019)
38. L.E. McCoubrey, A. Favaron, A. Awad, M. Orlu, S. Gaisford, A.W. Basi, Colonic drug delivery: formulating the next generation of colon-targeted therapeutics. J. Control. Release **353**, 1107–1126 (2023)
39. N. Khan, M. Aqil, S.S. Imam, A. Ali, Development and evaluation of a novel *in situ* gel of sparfloxacin for sustained ocular drug delivery: *in vitro* and *ex vivo* characterization. Pharm. Dev. Technol. **20**(6), 662–669 (2015)
40. P. Katare, T.P. Medhe, A. Nadkarni, M. Deshpande, R.K. Tekade et al., Nasal drug delivery system and devices: an overview on health effects. ACS Chem. Health Saf. **31**, 127–143 (2024)
41. D. Kapoor, M. Patel, R.B. Vyas, C. Lad, B. Lal, Site Specific drug delivery through nasal route using bioadhesive polymers. J. Drug Deliv. Ther. **5**, 1–9 (2015)
42. N.N. Gonzalez, G. Rassu, M. Cossu, L. Catenacci, M.L. Sorrenti et al., A thermosensitive chitosan hydrogel: an attempt for the nasal delivery of dimethyl fumarate. Int. J. Biol. Macromol. **278**, 134908 (2024)
43. J.T.W. Wang, A.C. Rodrigo, A.K. Patterson, K. Hawkins, M.M.S. Aly et al., Enhanced delivery of neuroactive drugs via nasal delivery with a self-healing supramolecular gel. Adv. Sci. **8**, 2101058 (2021)
44. A.R. Kumar, A.S.S. Aeila, Sustained release matrix type drug delivery system: an overview. World J. Pharm. Pharm. Sci. **9**, 470–480 (2019)
45. A.M. dos Santos, S.G. Carvalho, V.H.S. Araujo, G.C. Carvalho, MP.D. Gremiao, M. Chorilli, Recent advances in hydrogels as strategy for drug delivery intended to vaginal infections. Int. J. Pharm. **590**, 119867 (2020)
46. R. Rathi, Sanshita; A. Kumar, V. Vishvakarma, K. Huanbutta, I. Singh, T. Sangnim, Advancements in rectal drug delivery systems: clinical trials, and patents perspective. Pharmaceutics **14**, 2210 (2022)
47. A.K. Gupta, Environmental responsive hydrogels: a novel approach in drug delivery system. J. Drug Deliv. Ther. **2**, 1–8 (2012)
48. L. Vlaia, G. Coneac, I. Olariu, V. Vlaia, D. Lupuleasa, Cellulose-derivatives-based hydrogels as vehicles for dermal and transdermal drug delivery, in *Emerging Concepts in Analysis and Applications of Hydrogels* (IntechOpen, Rijeka, 2016)

49. B.J. Kong, A. Kim, S.N. Park, Properties and in vitro drug release of hyaluronic acid-hydroxyethyl cellulose hydrogels for transdermal delivery of isoliquiritigenin. Carbohydr. Polym. **147**, 473–481 (2016)
50. G. Sarkar, J.T. Orasugh, N.R. Saha, I. Roy, A. Bhattacharyya et al., Cellulose nanofibrils/chitosan based transdermal drug delivery vehicle for controlled release of ketorolac tromethamine. New J. Chem. **41**, 15312–15319 (2017)
51. N. Tavakoli, M. Minaiyan, M. Heshmatipour, R. Musavinasab, Transdermal iontophoretic delivery of celecoxib from gel formulation. Res. Pharm. Sci. **10**, 419–428 (2016)
52. A. Raeisi, F. Farjadian, Commercial hydrogel product for drug delivery based on route of administration. Front. Chem. **12**, 1336717 (2024)
53. A.P. Mathew, S. Uthaman, K. Cho, C.S. Cho, I.K. Park, Injectable hydrogels for delivering biotherapeutic molecules. Int. J. Biol. Macromol. **110**, 17–29 (2018)
54. S. Javanbakht, H. Namazi, Doxorubicin loaded carboxymethyl cellulose/graphene quantum dot nanocomposite hydrogel films as a potential anticancer drug delivery system. Mater. Sci. Eng. C **87**, 50–59 (2018)
55. P. He, L. Dai, J. Wei, X. Zhu, J. Li, Z. Chen, Y. Ni, Nanocellulose-based hydrogels as versatile drug delivery vehicles: a review, Int. J. Biol. Macromol. **222A**, 830–843 (2022)
56. K. Sangsuriyonk, N. Paradee, A. Sirivat, Electrically controlled release of anticancer drug 5-fluorouracil from carboxymethyl cellulose hydrogels. Int. J. Biol. Macromol. **165A**, 865–873 (2020)
57. B. Suna, M. Zhang, J. Shenc, Z. He, P. Fatehi, Y. Ni, Applications of cellulose-based materials in sustained drug delivery systems. Curr. Med. Chem. **26**, 2485–2501 (2019)
58. T.S. Anirudhan, V. Manjusha, V. Chithra Sekhar, A new biodegradable nano cellulose-based drug delivery system for pH-controlled delivery of curcumin. Int. J. Biol. Macromol. **183**, 2044–2054 (2021)
59. M. Culebras, A. Barrett, M. Pishnamazi, G.M. Walker, M.N. Collins, Wood-derived hydrogels as a platform for drug-release systems. ACS Sustain. Chem. Eng. **9**, 2515–2522 (2021)
60. P. Ghasemiyeh, S.M. Samani, Hydrogels as drug delivery systems; pros and cons. Trends Pharm. Sci. Tech. **5**(1), 7–24 (2019)
61. T.M.S.U. Gunathilake, Y.C. Ching, C.H. Chuah, N.A. Rahman, N.S. Liou, Recent advances in celluloses and their hybrids for stimuli-responsive drug delivery. Int. J. Biol. Macromol. **158**, 670–688 (2020)
62. R. Nicu, D.E. Ciolacu, A.R. Petrovici, D. Rusu, M. Avadanei et al., 3D Matrices for enhanced encapsulation and controlled release of anti-inflammatory bioactive compounds in wound healing. Int. J. Mol. Sci. **24**, 4213 (2023)
63. M. Dewan, Methyl cellulose based hydrogel for controlled drug delivery and tissue engineering application: a review. Asian J. Chem. **35**(10), 2312–2322 (2023)
64. K. Pratinthong, W. Punyodom, P. Jantrawut, K. Jantanasakulwong, W. Tongdeesoontorn et al., Modification of a carboxymethyl cellulose/Poly(vinyl alcohol) hydrogel film with citric acid and glutaraldehyde crosslink agents to enhance the anti-inflammatory effectiveness of triamcinolone acetonide in wound healing. Polymers **16**, 1798 (2024)
65. M.Y. Jeong, S. Kim, H.R. Kim, J. Jeon, S.S. Won et al., Dexamethasone nanocrystals-embedded hydroxypropyl methylcellulose hydrogel increases cochlear delivery and attenuates hearing loss following intratympanic injection. Carbohydr. Polym. **345**, 122546 (2024)
66. X. Wen, D. Bao, M. Chen, A. Zhang, C. Liu, R. Sun, Preparation of CMC/HEC crosslinked hydrogels for drug delivery. BioRes. **10**(4), 8339–8351 (2015)
67. K. Zubik, P. Singhsa, Y. Wang, H. Manuspiya, R. Narain, Thermo-responsive Poly(N-Isopropylacrylamide)-cellulose nanocrystals hybrid hydrogels for wound dressing. Polymers **9**, 119 (2017)
68. J.T. Orasugh, G. Sarkar, N.R. Saha, B. Das, A. Bhattacharyya et al., Effect of cellulose nanocrystals on the performance of drug loaded in situ gelling thermo-responsive ophthalmic formulations. Int. J. Biol. Macromol. **124**, 235–245 (2019)
69. J. Lu, W. Zhu, L. Dai, C. Si, Y. Ni, Fabrication of thermo- and pH-sensitive cellulose nanofibrils-reinforced hydrogel with biomass nanoparticles. Carbohydr. Polym. **215**, 289–295 (2019)

70. S. Li, A. Jasim, W. Zhao, L. Fu, M. Wajid, et a., Fabrication of pH-electroactive bacterial cellulose/polyaniline hydrogel for the development of a controlled drug release system. ES Mater. Manuf. **1**, 41–49 (2018)
71. V. Sharma, D. Kapil, B. Singh, Fabrication of bacterial derived natural polysaccharide-based copolymer network hydrogels for use in drug delivery applications. Hybrid Advances **6**, 100263 (2024)
72. M. Badshah, H. Ullah, S.A. Khan et al., Preparation, characterization and in-vitro evaluation of bacterial cellulose matrices for oral drug delivery. Cellulose **24**, 5041–5052 (2017)
73. D.E. Ciolacu, R. Nicu, D.M. Suflet, D. Rusu, R.N. Darie-Nita, N. Simionescu, G. Cazacu, F. Ciolacu, Multifunctional hydrogels based on cellulose and modified lignin for advanced wounds management. Pharmaceutics **15**, 2588 (2023)
74. N. Masruchin, B.D. Park, V. Causin, Dual-responsive composite hydrogels based on TEMPO-oxidized cellulose nanofibril and poly(N-isopropylacrylamide) for model drug release. Cellulose **25**, 485–502 (2018)
75. H. Abou-Yousef, S. Dacrory, M. Hasanin, E. Saber, S. Kamel, Biocompatible hydrogel based on aldehyde-functionalized cellulose and chitosan for potential control drug release. Sustain. Chem. Pharm. **21**, 100419 (2021)
76. L.H. Fu, C. Qi, M.G. Ma, P. Wan, Multifunctional cellulose-based hydrogels for biomedical applications. J. Mater. Chem. B **7**, 1541 (2019)
77. D. Rusu, D.E. Ciolacu, B.C. Simionescu, Cellulose-based hydrogels in tissue engineering applications. Cellul. Chem. Technol. **53**(9), 907–923 (2019)
78. M. Åhlén, G.K. Tummala, A. Mihranyan, Nanoparticle-loaded hydrogels as a pathway for enzyme-triggered drug release in ophthalmic applications. Int. J. Pharm. **536**(1), 73–81 (2018)
79. S. Ghorbani, H. Eyni, S.R. Bazaz, H. Nazari, L.S. Asl et al., Hydrogels based on cellulose and its derivatives: applications, synthesis, and characteristics. Polym. Sci. A **60**(6), 707–722 (2018)
80. L. Saïdi, C. Vilela, H. Oliveira, A.J.D. Silvestre, C.S.R. Freire, Poly(N-methacryloyl glycine)/nanocellulose composites as pH-sensitive systems for controlled release of diclofenac. Carbohydr. Polym. **169**, 357–365 (2017)
81. S.M.F. Kabir, P.P. Sikdar, B. Haque, M.A.R. Bhuiyan, A. Ali, M.N. Islam, Cellulose-based hydrogel materials: chemistry, properties and their prospective applications. Prog. Biomater. **7**, 153–174 (2018)
82. H.F. Emam, T.I. Shaheen, Design of a dual pH and temperature responsive hydrogel based on esterified cellulose nanocrystals for potential drug release. Carbohydr. Polym. **278**, 118925 (2022)
83. F. Lin, J. Zheng, W. Guo et al., Smart cellulose-derived magnetic hydrogel with rapid swelling and deswelling properties for remotely controlled drug release. Cellulose **26**, 6861–6877 (2019)
84. X. Shi, Y. Zheng, C. Wang, L. Yue, K. Qiao, G. Wang, L. Wang, H. Quan, Dual stimulus responsive drug release under the interaction of pH value and pulsatile electric field for a bacterial cellulose/sodium alginate/multiwalled carbon nanotube hybrid hydrogel. RSC Adv. **5**, 41820 (2015)
85. F. Andrade, M.M. Roca-Melendres, E.F. Durán-Lara, D. Rafael, S. Schwartz, Stimuli-responsive hydrogels for cancer treatment: the role of pH, light. Ion. Strength Magn. Field. Cancers **13**, 1164 (2021)
86. X. Shen, J.L. Shamshina, P. Berton, G. Gurauc, R.D. Rogers, Hydrogels based on cellulose and chitin: fabrication, properties, and applications. Green Chem. **18**, 53 (2016)
87. F. Song, J. Gong, Y. Tao, Y. Cheng, J. Lu, H. Wang, A robust regenerated cellulose-based dual stimuli-responsive hydrogel as an intelligent switch for controlled drug delivery. Int. J. Biol. Macromol. **176**, 448–458 (2021)
88. K. Sangsuriyonk, N. Paradee, A. Sirivat, Electrically controlled release of anticancer drug 5-fluorouracil from carboxymethyl cellulose hydrogels. Int. J. Biol. Macromol. A **165**, 865–873 (2020)

89. S.H. Park, H.S. Shin, S.N. Park, A novel pH-responsive hydrogel based on carboxymethyl cellulose/2-hydroxyethyl acrylate for transdermal delivery of naringenin. Carbohydr. Polym. **200**, 341–352 (2018)
90. D. Jeong, S.W. Joo, Y. Hu, V.V. Shinde, E. Cho, S. Jung, Carboxymethyl cellulose-based superabsorbent hydrogels containing carboxymehtyl β-cyclodextrin for enhanced mechanical strength and effective drug delivery. Eur. Polym. J. **105**, 17–25 (2018)
91. L. Yin, W. Duan, Y. Chen, D. Chen, Y. Wang, S. Guo, J. Qin, Biodegradable hydrogel from pectin and carboxymethyl cellulose with Silibinin loading for lung tumor therapy. Int. J. Biol. Macromol. **243**, 125128 (2023)
92. N. Chen, H. Wang, C. Ling, W. Vermerris, B. Wang, Z. Tong, Cellulose-based injectable hydrogel composite for pH-responsive and controllable drug delivery. Carbohydr. Polym. **225**, 115207 (2019)
93. N.F. Kassem, A.S. Negm, N. Shukry, S. El-Kalyouby, Synthesis of pH-sensitive CMC / chitosan hydrogels by gamma irradiation. J. Compos. Biodegradable Polym. **2**, 22–30 (2014)
94. Y. Zhou, Z. Zhai, Y. Yao, J.C. Stant, S.L. Landrum et al., Oxidized hydroxypropyl cellulose/ carboxymethyl chitosan hydrogels permit pH-responsive, targeted drug release. Carbohydr. Polym. **300**, 120213 (2023)
95. S.S. Kwon, B.J. Kong, S.N. Park, Physicochemical properties of pH-sensitive hydrogels based on hydroxyethyl cellulose–hyaluronic acid and for applications as transdermal delivery systems for skin lesions. Eur. J. Pharm. Biopharm. **92**, 146–154 (2015)
96. H. Liu, L. Rong, B. Wang, R. Xie, X. Sui et al., Facile fabrication of redox/pH dual stimuli responsive cellulose hydrogel. Carbohydr. Polym. **176**, 299–306 (2017)
97. W. Liu, B. Zhang, W.W. Lu, X. Li, D. Zhu et al., A rapid temperature-responsive sol–gel reversible poly(N-isopropylacrylamide)-g-methylcellulose copolymer hydrogel. Biomaterials **25**, 3005–3012 (2004)
98. K. Itoh, T. Hatakeyama, T. Shimoyama, S. Miyazaki, A. D'Emanuele, D. Attwood, *In situ* gelling formulation based on methylcellulose/pectin system for oral-sustained drug delivery to dysphagic patients. Drug Dev. Ind. Pharm. **37**(7), 790–797 (2011)
99. B. Bhowmick, G. Sarkar, D. Rana, I. Roy, N.R. Saha et al., Effect of carrageenan and potassium chloride on an in situ gelling ophthalmic drug delivery system based on methylcellulose. RSC Adv. **5**, 60386 (2015)
100. A.S.L. Rangabhatla, V. Tantishaiyakul, K. Oungbho, O. Boonrat, Fabrication of pluronic and methylcellulose for etidronate delivery and their application for osteogenesis. Int. J. Pharm. **499**(1–2), 110–118 (2016)
101. S. Butun, F.G. Ince, H. Erdugan, N. Sahiner, One-step fabrication of biocompatible carboxymethyl cellulose polymeric particles for drug delivery systems. Carbohydr. Polym. **86**(2), 636–643 (2011)
102. R. Barbucci, G. Giani, S. Fedi, S. Bottari, M. Casolaro, Biohydrogels with magnetic nanoparticles as crosslinker: characteristics and potential use for controlled antitumor drug-delivery. Acta Biomater. **8**(12), 4244–4252 (2012)
103. G. Aguilera, C.C. Berry, R.M. West, E.G. Monterrubio, A.A. Molina et al., Carboxymethyl cellulose coated magnetic nanoparticles transport across a human lung microvascular endothelial cell model of the blood–brain barrier. Nanoscale Adv. **1**, 671 (2019)
104. X. Guo, L. Xue, W. lv, Q. liu, R. Li, Z. Li, J. Wang, Facile synthesis of magnetic carboxymethylcellulose nanocarriers for pH-responsive delivery of doxorubicin. New J. Chem. **39**, 7340–7347 (2015)
105. M.G. Bekaroğlu, Y. İşçi, S. İşçi, Colloidal properties and in vitro evaluation of Hydroxy ethyl cellulose coated iron oxide particles for targeted drug delivery. Mater. Sci. Eng. C **78**, 847–853 (2017)
106. H. Dai, H. Zhang, L. Ma, H. Zhou, Y. Yu et al., Green pH/magnetic sensitive hydrogels based on pineapple peel cellulose and polyvinyl alcohol: synthesis, characterization and naringin prolonged release. Carbohydr. Polym. **209**, 51–61 (2019)
107. N. Gülçelik, K.N. Çifci, N. Alemdar, The effect of sulfonated reduced graphene oxide on the properties of ionic strength sensitive PEC film comprising protein/polysaccharides combined system. Int. J. Biol. Macromol. **281**(4), 136490 (2024)

108. K. Asare-Addo, B.R. Conway, H. Larhrib, M. Levina, A.R. Rajabi-Siahboomi et al., The effect of pH and ionic strength of dissolution media on in-vitro release of two model drugs of different solubilities from HPMC matrices. Colloids Surf. B Biointerfaces **111**, 384–391 (2013)
109. S.R. Salunke, S.B. Patil, Ion activated in situ gel of gellan gum containing salbutamol sulphate for nasal administration. Int. J. Biol. Macromol. **87**, 41–47 (2016)
110. L. Chang, X. Liu, J. Zhu, Y. Rao, D. Chen et al., Cellulose-based thermo-responsive hydrogel with NIR photothermal enhanced DOX released property for anti-tumor chemotherapy. Colloids Surf. B Biointerfaces **218**, 112747 (2022)
111. B. Khan, A. Arbab, S. Khan, et al., Recent progress in thermosensitive hydrogels and their applications in drug delivery area. MedComm—Biomater Appl. **2**, e55 (2023)
112. R. Fan, Y. Cheng, R. Wang, T. Zhang, H. Zhang et al., Thermosensitive hydrogels and advances in their application in disease therapy. Polymers **14**, 2379 (2022)
113. H. Huang, X. Qi, Y. Chen, Z. Wu, Thermo-sensitive hydrogels for delivering biotherapeutic molecules: a review. Saudi Pharm. J. **27**(7), 990–999 (2019)
114. J.M. Caronia, D.W. Sorensen, H.M. Leslie, J.H. van Berlo, S.M. Azarin, Adhesive thermosensitive gels for local delivery of viral vectors. Biotechnol. Bioeng. **116**(9), 2353–2363 (2019)
115. G. Fundueanu, M. Constantin, P. Ascenzi, Poly(N-isopropylacrylamide-co-acrylamide) cross-linked thermoresponsive microspheres obtained from preformed polymers: Influence of the physico-chemical characteristics of drugs on their release profiles. Acta Biomater. **5**(1), 363-373 (2009)
116. J. García-Couce, M. Tomás, G. Fuentes, I. Que, A. Almirall, L.J. Cruz, Chitosan/Pluronic F127 thermosensitive hydrogel as an injectable dexamethasone delivery carrier. Gels **8**, 44 (2022)
117. J.K. Kim, Y.W. Won, K.S. Lim, Y.H. Kim, Low-molecular-weight methylcellulose-based thermo-reversible gel/pluronic micelle combination system for local and sustained docetaxel delivery. Pharm. Res. **29**, 525–534 (2012)
118. Z. Li, Y. Li, C. Chen, Y. Cheng, Magnetic-responsive hydrogels: From strategic design to biomedical applications. J. Control. Release **335**, 541–556 (2021)
119. D. Li, M. Deng, Z. Yu, W. Liu, G. Zhou et al., Biocompatible and stable GO-coated Fe_3O_4 nanocomposite: a robust drug delivery carrier for simultaneous tumor MR imaging and targeted therapy. ACS Biomater. Sci. Eng. **4**(6), 2143–2154 (2018)
120. B. Hermenegildo, C. Ribeiro, L. Pérez-Álvarez, José L. Vilas, David A. Learmonth, Rui A Sousa, P. Martins, S. Lanceros-Méndez, Hydrogel-based magnetoelectric microenvironments for tissue stimulation. Colloids Surf. B Biointerfaces **181**, 1041–1047 (2019)
121. N. Movagharnezhad, P.N. Moghadam, Hexamethylene diamine/carboxymethyl cellulose grafted on magnetic nanoparticles for controlled drug delivery. Polym. Bull. **74**, 4645–4658 (2017)
122. R.S. Tade, M.P. More, V.K. Chatap, P.O. Patil, P.K. Deshmukh, Fabrication and *In vitro* drug release characteristics of magnetic nanocellulose fiber composites for efficient delivery of nystatin. Mater. Res. Express **5**, 116102 (2018)
123. A. Zaher, S. Li, K.T. Wolf, F.N. Pirmoradi, O. Yassine, L. Lin, N.M. Khashab, J. Kosel, Osmotically driven drug delivery through remote-controlled magnetic nanocomposite membranes. Biomicrofluidics **9**, 054113 (2015)
124. G.R. Mahdavinia, S. Ettehadi, M. Amini, M. Sabzi, Synthesis and characterization of hydroxypropyl methylcellulose-g-poly(acrylamide)/LAPONITE RD nanocomposites as novel magnetic- and pHsensitive carriers for controlled drug release. RSC Adv. **5**, 44516 (2015)
125. Y. Deng, J. Xi, L. Meng, Y. Lou, F. Seidi, W. Wu, H. Xiao, Stimuli-responsive nanocellulose hydrogels: an overview. Eur. Polym. J. **180**, 111591 (2022)
126. M. Pandey, M.C.I.M. Amin, N. Ahmad, M.M. Abeer, Rapid synthesis of superabsorbent smart-swelling bacterial cellulose/acrylamide-based hydrogels for drug delivery. Int. J. Polym. Sci. **5**, 10 (2013)

127. J. Zhao, X. Liang, H. Cao, T. Tan, Preparation of injectable hydrogel with near-infrared light response and photo-controlled drug release. Bioresour. Bioprocess. **7**, 1 (2020)
128. Y. Liu, Y. Sui, C. Liu, C. Liu, M. Wu, B. Li, Y. Li, A physically crosslinked polydopamine/nanocellulose hydrogel as potential versatile vehicles for drug delivery and wound healing. Carbohydr. Polym. **188**, 27–36 (2018)
129. F. Zhou, S. Wu, C. Rader, J. Ma, S. Chen et al., Crosslinked ionic alginate and cellulose-based hydrogels for photoresponsive drug release systems. Fibers Polym. **21**(1), 45–54 (2020)
130. R. Nicu, F. Ciolacu, D.E. Ciolacu, Advanced functional materials based on nanocellulose for pharmaceutical/medical applications. Pharmaceutics **13**(8), 1125 (2021)
131. M. Das, S. Lalsangi, S. Santra, R. Banerjee, Nanocellulose as a carrier for improved drug delivery: progresses and innovation. J. Drug Deliv. Sci. Technol. **97**, 105743 (2024)
132. Y. Ai, Z. Lin, W. Zhao, M. Cui, W. Qi, R. Huang, R. Su, Nanocellulose-based hydrogels for drug delivery. J. Mater. Chem. B **11**, 7004 (2023)
133. K. Dhali, M. Ghasemlou, F. Daver, P. Cass, B. Adhikari, A review of nanocellulose as a new material towards environmental sustainability. Sci. Total. Environ. **775**, 145871 (2021)
134. H. Paukkonen, M. Kunnari, P. Laurén, T. Hakkarainen, V.V. Auvinen et al., Nanofibrillar cellulose hydrogels and reconstructed hydrogels as matrices for controlled drug release. Int. J. Pharm. **532**(1), 269–280 (2017)
135. S.D. Hujaya, G.S. Lorite, S.J. Vainio, H. Liimatainen, Polyion complex hydrogels from chemically modified cellulose nanofibrils: structure-function relationship and potential for controlled and pH-responsive release of doxorubicin. Acta Biomater. **75**, 346–357 (2018)
136. J.T. Orasugh, N.R. Saha, D. Rana, G. Sarkar, Md.M.R. Mollick et al., Jute cellulose nanofibrils/hydroxypropylmethylcellulose nanocomposite: a novel material with potential for application in packaging and transdermal drug delivery system. Ind. Crops Prod. **112**, 633–643 (2018)
137. F.F. Azhar, E. Shahbazpour, A. Olad, PH sensitive and controlled release system based on cellulose nanofibers-poly vinyl alcohol hydrogels for cisplatin delivery. Fibers Polym. **18**(3), 416–423 (2017)
138. H. Zhang, C. Yang, W. Zhou, Q. Luan, W. Li et al., A pH-responsive gel macrosphere based on sodium alginate and cellulose nanofiber for potential intestinal delivery of probiotics. ACS Sustain. Chem. Eng. **6**, 13924–13931 (2018)
139. Y. Li, C. Wang, Y. Luan, W. Liu, T. Chen, P. Liu, Z. Liu, Preparation of pH-responsive cellulose nanofibril/sodium alginate based hydrogels for drug release. J. Appl. Polym. Sci. **139**, e51647 (2022)
140. V.H.G. Phan, M. Murugesan, H. Huong, T.T. Le, T.H. Phan et al., Cellulose nanocrystals-incorporated thermosensitive hydrogel for controlled release, 3D printing, and breast cancer treatment applications. ACS Appl. Mater. Interfaces **14**, 42812–42826 (2022)
141. E. Kushan, E. Senses, Thermoresponsive and injectable composite hydrogels of cellulose nanocrystals and pluronic F127. ACS Appl. Bio Mater. **4**, 3507–3517 (2021)
142. J. Supramaniam, R. Adnan, N.H.M. Kaus, R. Bushra, Magnetic nanocellulose alginate hydrogel beads as potential drug delivery system. Int. J. Biol. Macromol. **118A**, 640–648 (2018)
143. M.D. Markovic, S.I. Seslija, V.Dj. Ugrinovic, M. Kunaver, V.V. Panic, et al., Green pH- and magnetic-responsive hybrid hydrogels based on poly(methacrylic acid) and Eucalyptus wood nanocellulose for controlled release of ibuprofen. Cellulose **28**, 11109–11132 (2021)
144. L. Xiao, A.J. Poudel, L. Huang, Y. Wang, A.M.E. Abdalla, G. Yang, Nanocellulose hyperfine network achieves sustained release of berberine hydrochloride solubilized with β-cyclodextrin for potential anti-infection oral administration. Int. J. Biol. Macromol. **153**, 633–640 (2020)
145. Y. Alkhatib, M. Dewaldt, S. Moritz, R. Nitzsche, D. Kralisch, D. Fischer, Controlled extended octenidine release from a bacterial nanocellulose/Poloxamer hybrid system. Eur. J. Pharm. Biopharm. **112**, 164–176 (2017)
146. Y. Sun, J. Wang, D. Li, F. Cheng, The recent progress of the cellulose-based antibacterial hydrogel. Gels **10**, 109 (2024)

147. T. Guo, Y. Pei, K. Tang, X. He, J. Huang, F. Wang, Mechanical and drug release properties of alginate beads reinforced with cellulose. J. Appl. Polym. Sci. **44495** (2017)
148. D. Sondari, A.S. Hanifah, P. Amanda, S. Hermanto, A.A. Septevani et al., Hydrogel of dialdehyde cellulose/chitosan crosslinked with sodium trimetaphosphate and its application for drug delivery matrices. Waste Biomass Valorization **15**, 3739–3748 (2024)
149. R. Poonguzhali, S. Khaleel Basha, V. Sugantha Kumari, Synthesis of alginate/nanocellulose bionanocomposite for in vitro delivery of ampicillin. Polym. Bull. **75**, 4165–4173 (2018)
150. Q. Xu, Y. Ji, Q. Sun, Y. Fu, Y. Xu, L. Jin, Fabrication of cellulose nanocrystal/chitosan hydrogel for controlled drug release. Nanomaterials **9**, 253 (2019)
151. M.R. Mauricio, P.G. da Costa, S.K. Haraguchi, M.R. Guilherme, E.C. Muniz, A.F. Rubira, Synthesis of a microhydrogel composite from cellulose nanowhiskers and starch for drug delivery. Carbohydr. Polym. **115**, 715–722 (2015)
152. N. Lin, A. Gèze, D. Wouessidjewe, J. Huang, A. Dufresne, Biocompatible double-membrane hydrogels from cationic cellulose nanocrystals and anionic alginate as complexing drugs codelivery. ACS Appl. Mater. Interfaces **8**, 6880–6889 (2016)
153. B. Mandal, A.P. Rameshbabu, S. Dhara, S. Pal, Nanocomposite hydrogel derived from poly (methacrylic acid)/carboxymethyl cellulose/AuNPs: a potential transdermal drugs carrier. Polymer **120**, 9–19 (2017)
154. M. Rasoulzadeh, H. Namazi, Carboxymethyl cellulose/graphene oxide bio-nanocomposite hydrogel beads as anticancer drug carrier agent. Carbohydr. Polym. **168**, 320–326 (2017)
155. Z. Jiaoa, B. Zhanga, C. Li, W. Kuanga, J. Zhang et al., Carboxymethyl cellulose-grafted graphene oxide for efficient antitumor drug delivery. Nanotechnol. Rev. **7**(4), 291–301 (2018)
156. T.S. Anirudhan, Syam S. Nair, Chithra Sekhar. V, Deposition of gold-cellulose hybrid nanofiller on a polyelectrolyte membrane constructed using guar gum and poly(vinyl alcohol) for transdermal drug delivery. J. Membrane Sci. **539**, 344–357 (2017)

Chapter 6
Cellulose-Based Hydrogels for Tissue Engineering

Guanghui Gu, Qi Zhang, and Yongming Xi

Abstract Cellulose-based hydrogels (CBHs) have emerged as promising biomaterials in tissue engineering due to their biocompatibility, biodegradability, mechanical tunability, and abundance as a sustainable resource. They serve as excellent scaffolds for supporting cell proliferation, differentiation, and extracellular matrix deposition across a wide range of biomedical applications. This chapter reviews recent advances in the development and applications of CBHs in bone, cartilage, myocardial tissue engineering, organoid culture, and other regenerative therapies. Through various formulations, such as the incorporation of nanocellulose, surface functionalization with bioactive motifs, and integration with conductive or mineralized components, CBHs demonstrate enhanced printability, mechanical properties, and biological performance. While in vitro and preclinical studies have shown promising results, challenges such as long-term biocompatibility, clinical scalability, and functional integration remain. Continued research is required to validate their performance in vivo and to advance these systems toward clinical translation. Overall, CBHs offer a versatile and cost-effective platform for next-generation tissue engineering strategies.

Keywords Cellulose-based hydrogels · Tissue engineering · Biomaterials · Bone regeneration · Cartilage repair

G. Gu · Y. Xi (✉)
Department of Orthopedic Surgery, The Affiliated Hospital of Qingdao University, Qingdao, China
e-mail: xym700118@163.com

Q. Zhang
Department of Radiology, Shandong Provincial Hospital Affiliated to Shandong First Medical University, Jinan, China

G. Wei and J. Zhang (eds.), *Cellulose-Based Hydrogels and Aerogels*, Lecture Notes in Nanoscale Science and Technology 41,
https://doi.org/10.1007/978-3-032-03008-5_6

6.1 Introduction

Trauma, disease, infection, birth defects, diabetes, and aging can all lead to tissue or organ defects, failure, and potentially life-threatening conditions. Current treatments for tissue or organ failure include autografting and allografting, which aim to replace damaged tissues or organs [1, 2]. However, the shortage of organ donors and the high costs associated with transplantation make it a challenging treatment option [3]. Furthermore, post-transplant care and long-term treatments impose a significant financial burden on both families and society. Tissue engineering techniques offer an alternative approach for clinicians by providing strategies to replace damaged tissues or organs using suitable biomaterials [4]. It is a multidisciplinary field that aims to reconstruct functional tissues and organs by using scaffolds to provide structural support, nutrients, and essential factors for cell growth [5, 6]. In recent years, significant progress and extensive research have been achieved in the field of tissue engineering (TE) [7–9].

Hydrogels are widely recognized as three-dimensional (3D) scaffolds due to their unique properties, including high porosity, excellent water retention, and close resemblance to the natural extracellular matrix (ECM) of tissues [10]. In addition, hydrogels can be readily functionalized to closely mimic the characteristics of native tissues or organs [11]. Furthermore, hydrogels can encapsulate therapeutic agents, enable controlled and targeted release, and provide a fundamental structural framework essential for tissue engineering [12].

As the most abundant natural biopolymer, cellulose has been widely utilized in the development of hydrogels for tissue engineering due to its biodegradability, excellent mechanical stability, non-toxicity, and cost-effectiveness [13]. In addition, the abundance of hydroxyl groups in cellulose facilitates the fabrication of diverse hydrogel structures with customizable properties tailored for advanced tissue engineering applications [6]. This chapter summarizes recent advances in the development and application of cellulose-based hydrogels (CBHs) across various fields, including bone tissue engineering (BTE), cartilage tissue engineering (CTE), myocardial tissue engineering (MTE), organoid cultivation, and other emerging biomedical applications.

6.2 CBHs for BTE Applications

BTE explores methods to restore the structural and functional integrity of bone damaged by trauma or disease [14]. Developing and fabricating suitable materials for BTE is challenging, as the biomaterials must not only exhibit good biocompatibility and biodegradability but also meet the mechanical requirements of bone tissue and possess an appropriately interconnected porous structure [15, 16]. For biomaterials used in BTE, a pore size of 300 μm or larger and compressive strength ranging from 5 to 224 MPa are recommended [17, 18]. While larger pore sizes promote cell infiltration and vascularization, they often compromise mechanical strength. Therefore, it is essential to design biomaterials that achieve an optimal balance between porosity and mechanical properties.

Traditional materials such as metals, ceramics, and glass lack biological recognition, are prone to inducing inflammation, and generally do not possess osteoinductive properties [6, 19]. Hydrogels composed of natural or synthetic polymers can provide a 3D structure to support cell growth and differentiation. However, protein-based polymeric materials often lack sufficient mechanical strength and stability, while polysaccharides generally lack cell-binding motifs necessary for effective cell infiltration. Although synthetic polymers offer adequate mechanical properties and structural stability, they typically lack bioactivity and fail to support cell attachment [20–22]. CBHs can meet the mechanical and biological requirements for BTE when formulated by combining two or more complementary compounds.

Extensive research has been conducted on CBHs for BTE [23–26]. For instance, Li et al. presented an innovative approach to combating bacterial infections in open bone defects by developing a mineralized hydrogel with dual functionality: antimicrobial activity and osteogenic potential [27]. In their study, cellulose nanocrystals (CNCs) were employed as stabilizing agents for the synthesis of silver nanoparticles (AgNPs). The high aspect ratio and abundance of hydroxyl groups in CNCs allowed for effective stabilization and controlled formation and dispersion of AgNPs, thereby preventing their aggregation. These nanoparticles were integrated into a hydrogel composed of polyvinyl alcohol (PVA) and polydopamine-modified hydroxyapatite (PDA@HAP). Additionally, CNCs served as a substrate for tannic acid (TA) coating, introducing numerous phenolic groups that further enhance interfacial adhesion. The constructed CBH not only exhibited excellent antibacterial efficacy but also enabled the sustained release of Ca^{2+} and PO_4^{3-} ions, promoting bone regeneration (Fig. 6.1a). In this investigation, during the green synthesis process, CNCs not only prevented the aggregation of AgNPs but also facilitated TA coating, thereby enhancing the stability and antibacterial properties of the CBH.

3D bioprinting has been at the forefront of tissue engineering due to its ability to fabricate customized, complex structures that more accurately mimic the architecture of internal tissues or organs compared to conventional methods [28]. The incorporation of cellulose nanofibers (CNFs) into bio-ink formulations significantly enhanced the mechanical properties of 3D-printed hydrogels, thereby providing structural support for bone regeneration [29]. Maturavongsadit et al. developed a nanocellulose/chitosan-based bioink optimized for extrusion-based bioprinting, specifically for BTE applications [30], as shown in Fig. 6.1b. The bioink formulation included thermo-responsive chitosan, glycerophosphate, hydroxyethyl cellulose, and CNCs, achieving rapid gelation kinetics (<7 s at 37 °C) while maintaining structural integrity after printing. In their study, CNCs functioned as multifunctional additives, including the functions of (i) reinforcing mechanical strength, (ii) improving printability, (iii) enhancing osteogenic potential, and (iv) ensuring biocompatibility. Their incorporation effectively addressed key challenges in BTE, such as replicating biomimetic mechanical properties and supporting cell differentiation. In another study, researchers developed a novel bioactive hydrogel reinforced

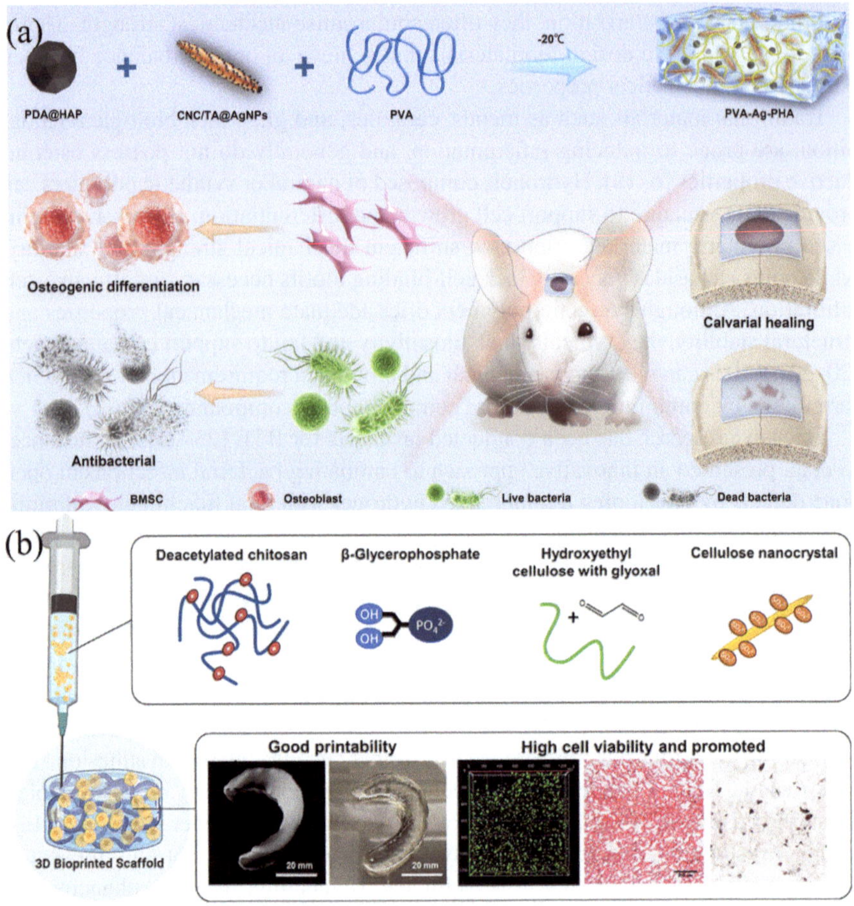

Fig. 6.1 CBHs for BTE applications: **a** Synthesis of PVA-Ag-PHA hydrogel and mechanisms for promoting bone regeneration. Adapted from Li et al. [27], Copyright 2024 Elsevier B.V. **b** 3D-printed CBH exhibiting excellent printability and enhanced bone formation potential. Adapted from Maturavongsadit et al. [30], Copyright 2021 American Chemical Society

with mineralized 2,2,6,6-tetramethylpiperidine-1-oxyl (TEMPO)-oxidized bacterial cellulose (m-TOBC) nanofibers and mesoporous silica nanoparticles (MSNs) loaded with dimethyloxalylglycine (DMOG), to promote bone regeneration [24]. The shear-thinning behavior of bacterial cellulose (BC) nanofibers improved the rheological properties of the hydrogel ink, enabling stable and continuous extrusion during 3D printing. In this system, CNFs, particularly m-TOBC, functioned as a structural backbone, a mineralization template, and a bioactive component, collectively contributing to enhanced osteogenic performance.

6.3 CBHs for CTE Applications

As a non-vascular tissue, cartilage contains a low density of connective tissue cells and is prone to damage due to disease, aging, or sports-related injuries [31]. Its avascular nature and limited cellularity make cartilage regeneration particularly challenging. Surgical procedures such as debridement, chondroplasty, osteochondral transplantation, and microfracture have been employed to treat cartilage defects; however, these approaches often fail to fully restore damaged cartilage and are associated with risks of infection and prolonged recovery times [32, 33].

The ECM of cartilage is primarily composed of chondrocytes, proteoglycans, and collagens [33]. Hydrogels are considered ideal biomaterials for CTE due to their structural and biochemical similarity to native cartilage ECM. However, a major obstacle in CTE is the limited cell migration and proliferation within the scaffold [34]. The mechanical strength and biocompatibility of CBHs make them a promising strategy for CTE.

Numerous CBHs have been reported for applications in CTE [35–38]. Yu et al. developed double-network (DN) microcrystalline cellulose (MCC) hydrogels using a chemical-physical multi-step cross-linking strategy [39]. The formed CBHs could address the limitations of conventional single-network (SN) hydrogels, such as poor mechanical strength and excessive swelling. As shown in Fig. 6.2. MCC was employed as the primary polysaccharide to construct both SN and DN hydrogel systems. The DN hydrogels exhibited smaller pore sizes, which contributed to improved resistance to swelling and mechanical degradation. The DN hydrogels exhibited superior compressive strength (4.53 MPa), excellent compression recovery, and fatigue resistance, making them more suitable for cartilage scaffold applications. Overall, MCC functioned as the structural and functional backbone of the hydrogels, imparting high mechanical strength, biocompatibility, and cartilage-like properties crucial for CTE. Another related study also employed cellulose to construct DN hydrogels for CTE via 3D printing [40]. In the study, CNFs enhanced the rheological properties of the bioink, resulting in improved shape fidelity during the printing process.

6.4 CBHs for MTE Applications

Myocardial infarction (MI) is a leading cause of death among cardiovascular diseases and results in irreversible damage to myocardial tissue [41]. Heart transplantation remains the ultimate solution for MI; however, it is severely limited by the scarcity of suitable donors. Alternative treatments, such as mechanical circulatory support devices and pharmacological therapies, are also associated with significant side effects that cannot be overlooked [42]. Given the shortcomings of traditional therapies for end-stage cardiac failure, hydrogels have emerged as promising materials in MTE to support cardiac function restoration [43]. The goal of MTE is to

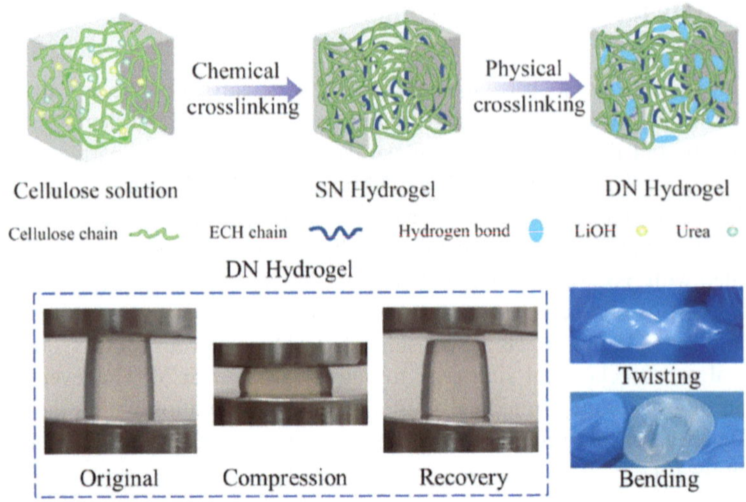

Fig. 6.2 Synthesis of SN and DN hydrogels, along with their excellent shape fidelity. Adapted from Yu et al. [39], Copyright 2023 Elsevier B.V

repair damaged heart tissue through the application of bioengineered constructs [44]. By mimicking the ECM of heart tissue, hydrogels—owing to their biodegradable nature and porous architecture—facilitate cell migration and proliferation. A key requirement in MTE is the restoration of the electrical conductivity of the heart muscle, which cannot be achieved by hydrogels alone. However, hydrogels must also possess appropriate mechanical properties and support the sustained release of bioactive factors to effectively contribute to cardiac repair. Due to their biocompatibility, biodegradability, ease of functionalization, and favorable mechanical properties, CBHs represent a promising class of biomaterials for MTE. Researchers have explored hybrid CBHs incorporating carboxymethyl cellulose (CMC) for MTE [44]. In these systems, CMC functioned as a hydration modulator, mechanical stabilizer, and biocompatibility enhancer, working synergistically with alginate and synthetic polymers to form biomimetic scaffolds that address critical challenges in cardiac tissue regeneration.

Hua et al. developed an innovative silk fibroin/hydroxypropyl cellulose (SF/HPC) composite hydrogel loaded with folic acid-modified extracellular vesicles (EVs) derived from AC16 cells (Fig. 6.3a) [45]. In this study, HPC, a natural thermo-sensitive polymer, contributed to the hydrogel's injectability and thermo-responsive gelation. In combination with silk fibroin, the hydrogel mimicked the softness of native biological tissue, providing mechanical support to the damaged myocardium. Moreover, the hydrogel enabled sustained EV release, which facilitated the effective internalization of EVs by cardiomyocytes.

In another study, researchers developed a nanocellulose/polypyrrole (TOCN-PPy) composite hydrogel for MTE (Fig. 6.3b) [46]. The hydrogel was synthesized by cross-linking TEMPO-oxidized CNF (TOCN) with iron ions, followed by in-situ

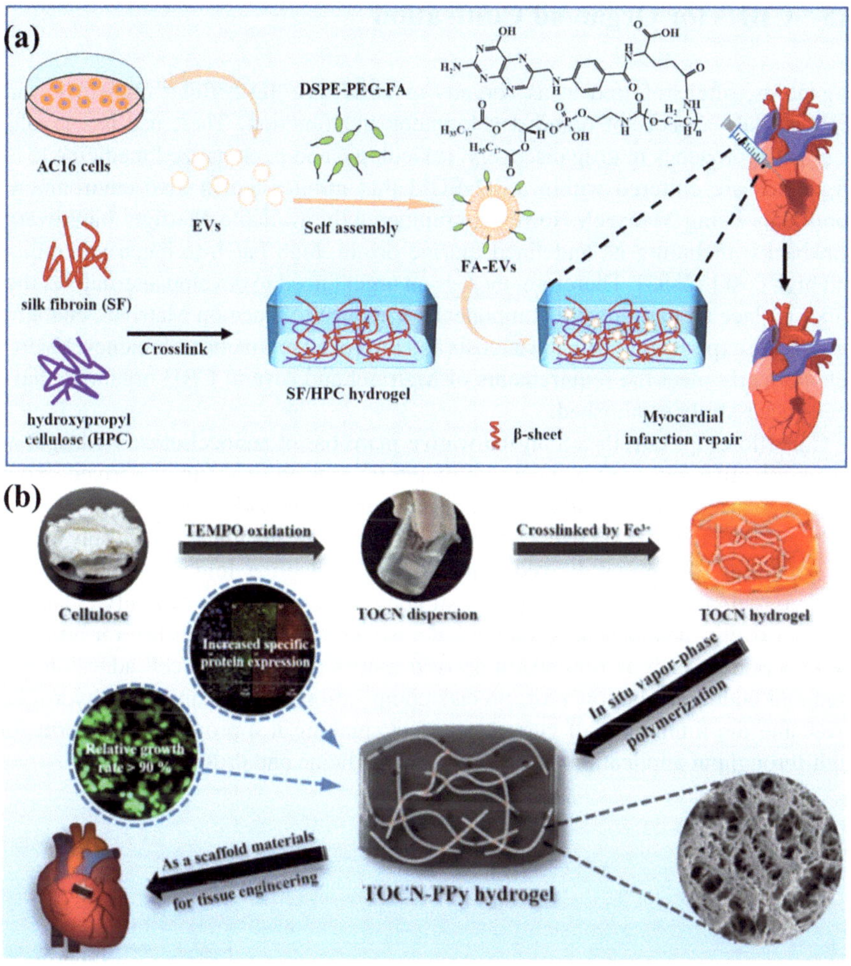

Fig. 6.3 CBHs applied in myocardial tissue engineering (MTE): **a** SF/HPC hydrogel encapsulated with EVs for cardiac repair. Adapted from Hua et al. [45], Copyright 2024 IOP Publishing. **b** Schematic of the preparation of TOCN-PPy hydrogel and its application in MTE. Adapted from Hou et al. [46], Copyright 2024 Springer

polymerization of pyrrole monomers. This fabrication process yielded a hydrogel with mechanical strength (50–75 kPa) and electrical conductivity (5×10^{-5} to 1×10^{-3} $S{\cdot}cm^{-1}$) comparable to native cardiac tissue, along with excellent biocompatibility. Similar CBHs incorporating gold nanoparticles and carbon nanotubes have also been reported to enhance electrical conductivity and mechanical properties [47–49]. Additionally, other studies have explored CBHs for heart valve tissue repair, demonstrating their versatility in cardiovascular applications [50].

6.5 CBHs for Organoid Cultivation

Organoids, often referred to as "organs in vitro" are 3D cellular constructs that replicate key aspects of organ development and function. Their use has revolutionized approaches in drug discovery, toxicology, and personalized medicine [51]. Organoids are cultured within a 3D ECM that mimics the in vivo environment, commonly using Matrigel. However, commercially available Matrigel has several drawbacks, including its undefined murine origin, high batch-to-batch variability, and high cost [52, 53]. Therefore, there is an urgent need to develop alternatives that either replace animal-derived components or reduce reliance on Matrigel, enabling large-scale experiments with lower costs and improved production efficiency. Hydrogels perfectly meet the requirements of Matrigel and several CBH organoids have been successfully established.

Curvello et al. introduced an innovative plant-based nanocellulose hydrogel as a cost-effective and well-defined alternative to conventional tumor-derived extracellular matrices for the culture of small intestinal organoids (Fig. 6.4a) [54]. This hydrogel, containing 0.1% nanocellulose, was mechanically tunable to closely mimic the properties of animal-derived matrices while enabling ionic crosslinking and osmotic regulation through a salt-free strategy. Oxidized CNFs not only enhanced the hydrogel's mechanical strength but also provided active sites for grafting RGD peptides, such as fibronectin-derived motifs that promote cell adhesion and organoid budding. The low cost, precise composition, and scalability of hydrogels overcame the limitations of current matrices, making it a promising platform for high-throughput applications in regenerative medicine and drug screening.

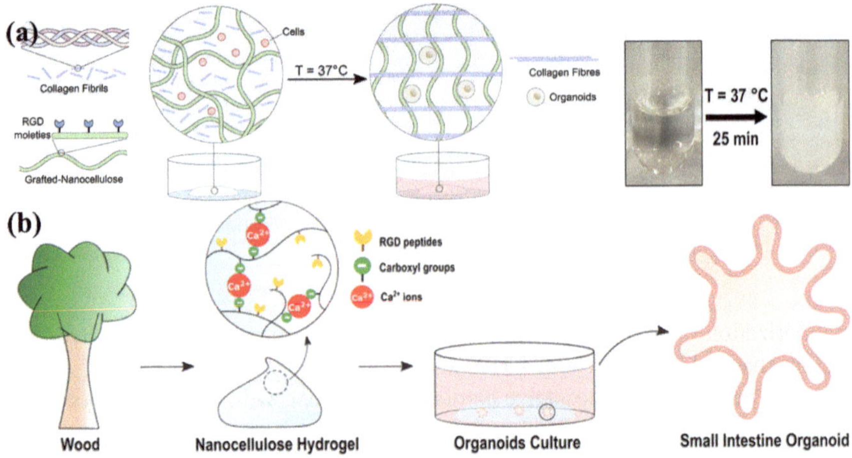

Fig. 6.4 CBHs applied in organoid cultivation: **a** Preparation of thermo-responsive collagen/nanocellulose (COL-NC) hydrogel and its sol–gel transition behavior. Adapted from Curvello et al. [54], Copyright 2021 Elsevier BV. **b** Small intestinal organoids cultured in a wood-derived cellulose hydrogel. Adapted from Curvello et al. [56], Copyright 2020 Wiley–VCH GmbH

Human liver organoids have also been successfully developed using CNF-based hydrogels, demonstrating hepatic differentiation outcomes comparable to, or even exceeding, those achieved with Matrigel, thereby offering a viable alternative for liver research [55]. Additionally, collagen/nanocellulose composite hydrogels have been investigated for culturing intestinal organoids (Fig. 6.4b) [56]. These hydrogels supported organoid growth and epithelial budding, highlighting their potential as an effective scaffold. Notably, matrix stiffness was identified as a critical factor in organoid development. These CBHs offer performance comparable to animal-derived matrices, with the added advantages of sustainability, cost-effectiveness, and batch-to-batch consistency.

Although these studies have demonstrated the successful establishment of organoids, the field remains in its early stages, and further validation in in vivo models, along with scalable approaches for clinical translation, is still required.

6.6 CBHs for Other Tissue Engineering Applications

Meniscus injuries are among the most prevalent knee joint conditions, often resulting from acute sports trauma or age-associated degenerative processes [57]. The limited blood supply to the meniscus hinders its natural healing capacity, and surgical interventions typically offer only temporary pain relief without restoring full function [58]. Meniscus tissue engineering offers a potential solution for the treatment of severe meniscal damage by enabling structural and functional restoration. In this context, 3D bioprinting has gained attention for its capacity to produce customized implants. By utilizing medical imaging data, 3D bioprinting can accurately replicate the meniscus's complex and irregular anatomy as well as adapt to the specific characteristics of individual injury types [59]. For instance, Lan et al. examined the use of TEMPO-oxidized CNF/alginate (TCNF/ALG) composite hydrogels as bioinks for the 3D bioprinting of functional human meniscus tissue (Fig. 6.5a,b) [60]. Their study aimed to optimize the hydrogel formulation for bioprinting performance in one way, and in another way to evaluate the redifferentiation potential of human meniscus fibrochondrocytes (hMFCs) encapsulated within the hydrogel. This strategy could produce ECM components resembling those of native inner meniscus tissue. The incorporation of TCNF endowed the hydrogel with critical shear-thinning properties and high zero-shear viscosity, allowing for precise 3D printing. Additionally, the hydrogel provided both biomechanical support and a favorable microenvironment for biological functionality.

In another study, Sinna et al. developed an injectable hydrogel composed of PVA grafted with glycidyl methacrylate and reinforced with CNFs (PVA-g-GMA/CNF) for meniscus tissue engineering [61]. The incorporation of CNFs significantly enhanced the compressive strength of the hydrogel. Among the tested formulations, the hydrogel containing 10% PVA-g-GMA and 0.7% CNF was identified as optimal, meeting the mechanical and biological criteria required for meniscus repair. Thanks to its ability to form highly ordered structures, cellulose offers excellent mechanical

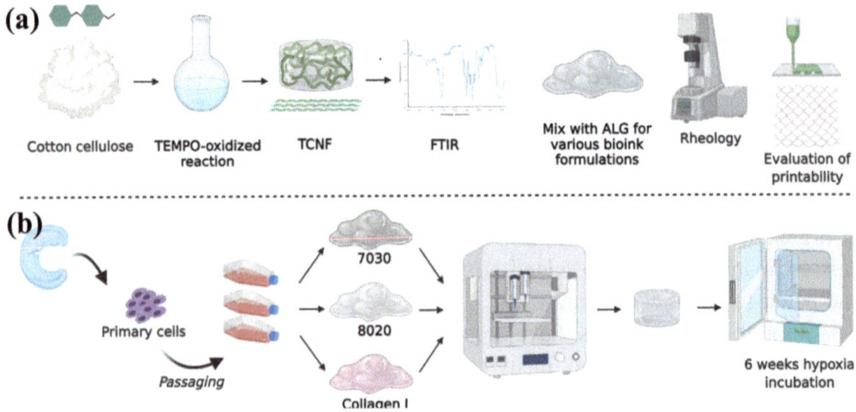

Fig. 6.5 CBHs applied in meniscus TE: **a** Schematic illustration of hydrogel preparation and physicochemical characterization. **b** Cell-laden hydrogel construct designed for meniscus tissue engineering applications. Adapted from Lan et al. [60], Open access

strength and flexibility. The addition of CNFs significantly improved the mechanical integrity of injectable hydrogel systems, making them more suitable for biomedical applications [62].

Other tissue engineering applications have also been explored using CBHs, including intervertebral disc repair [63], vascular tissue engineering [64], breast tissue reconstruction [65], neuro-endovascular repair [66], and gastrointestinal tissue regeneration [67].

6.7 Conclusion

CBHs, derived from an abundant, renewable, and sustainable natural resource, have shown remarkable potential in a wide range of biomedical applications, particularly in tissue engineering. Their intrinsic biocompatibility, biodegradability, and mechanical tunability make them ideal candidates for constructing scaffolds that mimic the ECM and support cell proliferation, differentiation, and tissue regeneration. The incorporation of cellulose into hydrogel systems not only improves mechanical strength and structural integrity but also introduces abundant hydroxyl groups, offering reactive sites for chemical modification, biofunctionalization, and drug loading. These features enable the design of highly customized, application-specific hydrogels.

The ultimate performance of CBHs depends heavily on formulation parameters, especially the type and concentration of cellulose derivatives, the cross-linking method, and the inclusion of bioactive factors such as growth factors, stem cells, or therapeutic agents. These design elements significantly influence the hydrogel's biological activity, degradation rate, and mechanical behavior under physiological conditions. While in vitro and in vivo animal studies have yielded promising results, key challenges remain before clinical translation can be achieved. One of the primary concerns is the long-term biocompatibility and potential toxicity of CBHs, particularly after chemical modification or prolonged implantation. Most existing studies are limited to short-term observations in animal models, with few progressing to rigorous preclinical or clinical trials.

To bridge these gaps, future research on CBHs should focus on several key areas. First, comprehensive long-term biocompatibility studies are essential, particularly in large animal models and eventually in human clinical trials, to fully evaluate their safety and therapeutic potential. Second, standardizing fabrication protocols is crucial to ensure reproducibility, scalability, and minimize batch-to-batch variability, which is a critical requirement for clinical translation. Third, the development of smart CBHs incorporating advanced functionalities—such as stimuli-responsive behavior, self-healing capabilities, or on-demand drug release—can significantly enhance their biomedical performance. Lastly, integrating CBHs with emerging technologies, including 4D bioprinting, gene editing (e.g., CRISPR/Cas9), and biosensing systems, could greatly expand their therapeutic scope and enable precision-engineered solutions for complex tissue regeneration challenges.

In conclusion, although still in the early stages of development, CBHs represent a promising and versatile platform for next-generation tissue engineering. With continued interdisciplinary collaboration and translational research efforts, CBHs hold strong potential to revolutionize regenerative medicine and improve patient outcomes across a spectrum of clinical applications.

Acknowledgements This study was supported by the Taishan Scholar Project of Shandong Province, China (No. tstp20250511 to Yongming Xi).

References

1. G.H. Gu, Z.C. Cui, X.F. Du, P. He, C. Rong, H. Tao, G. Wei, Y.M. Xi, Recent advances in biomacromolecule-reinforced 2d material (2dm) hydrogels: From interactions, synthesis, and functionalization to biomedical applications. Adv. Funct. Mater. **34**, 8367 (2024)
2. H.Z. Luo, R.T. Cha, J.J. Li, W.S. Hao, Y. Zhang, F.S. Zhou, Advances in tissue engineering of nanocellulose-based scaffolds: a review. Carbohydr. Polym. **224**, 115144 (2019)
3. T. Dvir, B.P. Timko, D.S. Kohane, R. Langer, Nanotechnological strategies for engineering complex tissues. Nat. Nanotechnol. **6**, 13–22 (2011)
4. F. Xu, C. Dawson, M. Lamb, E. Mueller, E. Stefanek, M. Akbari, T. Hoare, Hydrogels for tissue engineering: addressing key design needs toward clinical translation. Front. Bioeng. Biotechnol. **10**, 849831 (2022)
5. R. Hama, A. Ulziibayar, J.W. Reinhardt, T. Watanabe, J. Kelly, T. Shinoka, Recent developments in biopolymer-based hydrogels for tissue engineering applications. Biomolecules **13**, 280 (2023)
6. C. Chen, Y.W. Xi, Y.X. Weng, Recent advances in cellulose-based hydrogels for tissue engineering applications. Polymers **14**, 3335 (2022)
7. P. Gupta, S. Sharma, S. Jabin, S. Jadoun, Chitosan nanocomposite for tissue engineering and regenerative medicine: a review. Int. J. Biol. Macromol. **254**, 127660 (2024)

8. Z.K. Wang, Y. Dong, Y.Y. Wang, Collagen-based biomaterials for tissue engineering. ACS Biomater. Sci. Eng. **9**, 1132–1150 (2023)
9. L.Q. Guo, Z.H. Liang, L. Yang, W.Y. Du, T. Yu, H.Y. Tang, C.D. Li, H.B. Qiu, The role of natural polymers in bone tissue engineering. J. Control. Release **338**, 571–582 (2021)
10. B. Maharjan, J. Park, V.K. Kaliannagounder, G.P. Awasthi, M.K. Joshi, C.H. Park, C.S. Kim, Regenerated cellulose nanofiber reinforced chitosan hydrogel scaffolds for bone tissue engineering. Carbohydr. Polym. **251**, 117023 (2021)
11. R.X. Xie, W.C. Zheng, L.D. Guan, Y.J. Ai, Q.L. Liang, Engineering of hydrogel materials with perfusable microchannels for building vascularized tissues. Small **16**, 1902838 (2020)
12. A.K. Tamo, Nanocellulose-based hydrogels as versatile materials with interesting functional properties for tissue engineering applications. J. Mat. Chem. B **12**, 7692–7759 (2024)
13. N. Lin, A. Dufresne, Nanocellulose in biomedicine: current status and future prospect. Eur. Polym. J. **59**, 302–325 (2014)
14. J. Melke, S. Midha, S. Ghosh, K. Ito, S. Hofmann, Silk fibroin as biomaterial for bone tissue engineering. Acta Biomater. **31**, 1–16 (2016)
15. A.R. Short, D. Koralla, A. Deshmukh, B. Wissel, B. Stocker, M. Calhoun, D. Dean, J.O. Winter, Hydrogels that allow and facilitate bone repair, remodeling, and regeneration. J. Mat. Chem. B **3**, 7818–7830 (2015)
16. K.F. Leong, C.M. Cheah, C.K. Chua, Solid freeform fabrication of three-dimensional scaffolds for engineering replacement tissues and organs. Biomaterials **24**, 2363–2378 (2003)
17. G. Turnbull, J. Clarke, F. Picard, P. Riches, L.L. Jia, F.X. Han, B. Li, W.M. Shu, 3d bioactive composite scaffolds for bone tissue engineering. Bioact. Mater. **3**, 278–314 (2018)
18. A.A. Zadpoor, Bone tissue regeneration: the role of scaffold geometry. Biomater. Sci. **3**, 231–245 (2015)
19. Y.B. Zhang, X.C. Liu, L.D. Zeng, J. Zhang, J.L. Zuo, J. Zou, J.X. Ding, X.S. Chen, Polymer fiber scaffolds for bone and cartilage tissue engineering. Adv. Funct. Mater. **29**, 1903279 (2019)
20. M. Janmohammadi, Z. Nazemi, A.O.M. Salehi, A. Seyfoori, J. John, M.S. Nourbakhsh, M. Akbari, Cellulose-based composite scaffolds for bone tissue engineering and localized drug delivery. Bioact. Mater. **20**, 137–163 (2023)
21. M.S.B. Reddy, D. Ponnamma, R. Choudhary, K.K. Sadasivuni, A comparative review of natural and synthetic biopolymer composite scaffolds. Polymers **13**, 1105 (2021)
22. E.T. Moghadam, M. Yazdanian, M. Alam, H. Tebyanian, A. Tafazoli, E. Tahmasebi, R. Ranjbar, A. Yazdanian, A. Seifalian, Current natural bioactive materials in bone and tooth regeneration in dentistry: a comprehensive overview. J. Mater. Res. Technol-JMRT **13**, 2078–2114 (2021)
23. M.H. Kim, B.S. Kim, H. Park, J. Lee, W.H. Park, Injectable methylcellulose hydrogel containing calcium phosphate nanoparticles for bone regeneration. Int. J. Biol. Macromol. **109**, 57–64 (2018)
24. X.K. Liu, H.R. Hu, J.H. Ma, B.X. Wang, Mineralized cellulose nanofibers reinforced bioactive hydrogel remodels the osteogenic and angiogenic microenvironment for enhancing bone regeneration. Carbohydr. Polym. **357**, 123480 (2025)
25. B. Maharjan, J. Park, V.K. Kaliannagounder, G.P. Awasthi, M.K. Joshi, C.H. Park, C.S. Kim, Regenerated cellulose nanofiber reinforced chitosan hydrogel scaffolds for bone tissue engineering, Carbohydr. Polym. **251** (2021)
26. D. Burger, M. Beaumont, T. Rosenau, Y. Tamada, Porous silk fibroin/cellulose hydrogels for bone tissue engineering via a novel combined process based on sequential regeneration and porogen leaching. Molecules **25**, 5097 (2020)
27. X.C. Li, Y.X. Pang, L. Guan, L. Li, Y.L. Zhu, A.K. Whittaker, B. Yang, S. Zhu, Q. Lin, Mussel-inspired antimicrobial hydrogel with cellulose nanocrystals/tannic acid modified silver nanoparticles for enhanced calvarial bone regeneration. Int. J. Biol. Macromol. **270**, 132419 (2024)
28. L. Xu, H. Qin, Z.L. Cheng, W.B. Jiang, J. Tan, X. Luo, W.H. Huang, 3d-printed personalised prostheses for bone defect repair and reconstruction following resection of metacarpal giant cell tumours. Ann. Transl. Med. **9**, 3400 (2021)

29. J.X. Wei, B.X. Wang, Z. Li, Z.T. Wu, M.H. Zhang, N. Sheng, Q.Q. Liang, H.P. Wang, S.Y. Chen, A 3d-printable tempo-oxidized bacterial cellulose/alginate hydrogel with enhanced stability via nanoclay incorporation. Carbohydr. Polym. **238**, 116207 (2020)
30. P. Maturavongsadit, L.K. Narayanan, P. Chansoria, R. Shirwaiker, S.R. Benhabbour, Cell-laden nanocellulose/chitosan-based bioinks for 3d bioprinting and enhanced osteogenic cell differentiation. ACS Appl. Bio Mater. **4**, 2342–2353 (2021)
31. Y. Krishnan, A.J. Grodzinsky, Cartilage diseases. Matrix Biol. **71–72**, 51–69 (2018)
32. X. Xu, Y.J. Liang, X.F. Li, K. Ouyang, M.Y. Wang, T. Cao, W.C. Li, J.Q. Liu, J.Y. Xiong, B.Q. Li, J. Xia, D.P. Wang, L. Duan, Exosome-mediated delivery of kartogenin for chondrogenesis of synovial fluid-derived mesenchymal stem cells and cartilage regeneration. Biomaterials **269**, 120539 (2021)
33. K.D. Ngadimin, A. Stokes, P. Gentile, A.M. Ferreira, Biomimetic hydrogels designed for cartilage tissue engineering. Biomater. Sci. **9**, 4246–4259 (2021)
34. M. Hafezi, S.N. Khorasani, M. Zare, R.E. Neisiany, P. Davoodi, Advanced hydrogels for cartilage tissue engineering: recent progress and future directions. Polymers **13**, 4199 (2021)
35. L.L. Gu, T. Li, X.B. Song, X.T. Yang, S.L. Li, L. Chen, P.J. Liu, X.Y. Gong, C. Chen, L. Sun, Preparation and characterization of methacrylated gelatin/bacterial cellulose composite hydrogels for cartilage tissue engineering. Regen. Biomater. **7**, 195–202 (2020)
36. M. Nouri-Felekori, N. Nezafati, M. Moraveji, S. Hesaraki, T. Ramezani, Bioorthogonal hydroxyethyl cellulose-based scaffold crosslinked via click chemistry for cartilage tissue engineering applications. Int. J. Biol. Macromol. **183**, 2030–2043 (2021)
37. A. Mahajan, Z.S. Zaidi, A. Shukla, R. Saxena, D.S. Katti, Functionally graded hydrogels with opposing biochemical cues for osteochondral tissue engineering. Biofabrication **16**, 035020 (2024)
38. H.B. Zhao, Y.J. Zhang, Y.S. Liu, P.H. Zheng, T. Gao, Y. Cao, X.Z. Liu, J.B. Yin, R.J. Pei, Forming cellulose nanofibril-reinforced hyaluronic acid hydrogel for cartilage regeneration. Biomacromol **22**, 5097–5107 (2021)
39. X.D. Yu, X. Li, L. Kan, P. Pan, X. Wang, W.T. Liu, J.S. Zhang, Double network microcrystalline cellulose hydrogels with high mechanical strength and biocompatibility for cartilage tissue engineering scaffold. Int. J. Biol. Macromol. **238**, 124113 (2023)
40. L. Trachsel, C. Johnbosco, T. Lang, E.M. Benetti, M. Zenobi-Wong, Double-network hydrogels including enzymatically crosslinked poly-(2-alkyl-2-oxazoline)s for 3d bioprinting of cartilage-engineering constructs. Biomacromol **20**, 4502–4511 (2019)
41. Y.Y. Wang, J.J. Wang, C.X. Liu, J.J. Li, K.Y. Lu, Q. Yu, Y.X. Zhang, Z.Y. Shen, Injectable decellularized extracellular matrix hydrogel loaded with exosomes encapsulating curcumin for prevention of cardiac fibrosis after myocardial infarction. J. Mater. Sci. Technol. **167**, 50–58 (2023)
42. M. Vaduganathan, B.L. Claggett, P.S. Jhund, J.W. Cunningham, J.P. Ferreira, F. Zannad, M. Packer, G.C. Fonarow, J.J.V. McMurray, S.D. Solomon, Estimating lifetime benefits of comprehensive disease–modifying pharmacological therapies in patients with heart failure with reduced ejection fraction: a comparative analysis of three randomised controlled trials. Lancet **396**, 121–128 (2020)
43. S.G. Alamdari, A. Alibakhshi, M. de la Guardia, B. Baradaran, R. Mohammadzadeh, M. Amini, P. Kesharwani, A. Mokhtarzadeh, F. Oroojalian, A. Sahebkar, Conductive and semiconductive nanocomposite-based hydrogels for cardiac tissue engineering. Adv. Healthcare Mater. **11**, e2200526 (2022)
44. S. Sedighim, Y.Q. Chen, C.L. Xu, R. Mohindra, H.A. Liu, D.K. Agrawal, F.G. Thankam, Carboxymethyl cellulose-alginate interpenetrating hydroxy ethyl methacrylate crosslinked polyvinyl alcohol reinforced hybrid hydrogel templates with improved biological performance for cardiac tissue engineering. Biotechnol. Bioeng. **120**, 819–835 (2023)
45. Y. Hua, Z. He, Y. Ni, L. Sun, R. Wang, Y. Li, X. Li, G. Jiang, Silk fibroin and hydroxypropyl cellulose composite injectable hydrogel-containing extracellular vesicles for myocardial infarction repair. Biomed. Phys. Eng. Express **10**, 5001 (2024)

46. R.Q. Hou, Y.Y. Xie, R. Song, J.K. Bao, Z.Q. Shi, C.X. Xiong, Q.L. Yang, Nanocellulose/polypyrrole hydrogel scaffolds with mechanical strength and electrical activity matching native cardiac tissue for myocardial tissue engineering. Cellulose **31**, 4523–4532 (2024)
47. C.Y. Sun, Y.Y. Xie, H.F. Zhu, X. Zheng, R.Q. Hou, Z.Q. Shi, J. Li, Q.L. Yang, Highly electroactive tissue engineering scaffolds based on nanocellulose/sulfonated carbon nanotube composite hydrogels for myocardial tissue repair. Biomacromol **24**, 5989–5997 (2023)
48. H. Tohidi, N. Maleki, A. Simchi, Conductive, injectable, and self-healing collagen-hyaluronic acid hydrogels loaded with bacterial cellulose and gold nanoparticles for heart tissue engineering. Int. J. Biol. Macromol. **280**, 135749 (2024)
49. Q.H. Liang, S.H. Chen, S.F. Hua, W.H. Jiang, J.M. Zhan, C.Y. Pu, R.R. Lin, Y.T. He, H.H. Hou, X.Z. Qiu, Biomimetic versatile anisotropic, electroactive cellulose hydrogel scaffolds tailored from fern stem serving as nerve conduit and cardiac patch. Adv. Sci. **12**, 2400002 (2025)
50. N.A.F. Ma, D.Y. Cheung, J.T. Butcher, Incorporating nanocrystalline cellulose into a multifunctional hydrogel for heart valve tissue engineering applications. J. Biomed. Mater. Res. A **110**, 76–91 (2022)
51. H. Clevers, Modeling development and disease with organoids. Cell **165**, 1586–1597 (2016)
52. Y.R. Lou, L. Kanninen, T. Kuisma, J. Niklander, L.A. Noon, D. Burks, A. Urtti, M. Yliperttula, The use of nanofibrillar cellulose hydrogel as a flexible three-dimensional model to culture human pluripotent stem cells. Stem Cells Dev. **23**, 380–392 (2014)
53. C.S. Hughes, L.M. Postovit, G.A. Lajoie, Matrigel: A complex protein mixture required for optimal growth of cell culture. Proteomics **10**, 1886–1890 (2010)
54. R. Curvello, G. Kerr, D.J. Micati, W.H. Chan, V.S. Raghuwanshi, J. Rosenbluh, H.E. Abud, G. Garnier, Engineered plant-based nanocellulose hydrogel for small intestinal organoid growth. Adv. Sci. **8**, 2002135 (2021)
55. M. Krüger, L.A. Oosterhoff, M.E. van Wolferen, S.A. Schiele, A. Walther, N. Geijsen, L. De Laporte, L.J.W. van der Laan, L.M. Kock, B. Spee, Cellulose nanofibril hydrogel promotes hepatic differentiation of human liver organoids. Adv. Healthc. Mater. **9**, 1901658 (2020)
56. R. Curvello, D. Alves, H.E. Abud, G. Garnier, A thermo-responsive collagen-nanocellulose hydrogel for the growth of intestinal organoids. Mater. Sci. Eng. C-Mater. Biol. Appl. **124**, 112051 (2021)
57. H. Kwon, W.E. Brown, C.A. Lee, D.A. Wang, N. Paschos, J.C. Hu, K.A. Athanasiou, Surgical and tissue engineering strategies for articular cartilage and meniscus repair. Nat. Rev. Rheumatol. **15**, 550–570 (2019)
58. S. Bansal, E.R. Floyd, M.A. Kowalski, E. Aikman, P. Elrod, K. Burkey, J. Chahla, R.F. LaPrade, S.A. Maher, J.L. Robinson, J.M. Patel, Meniscal repair: the current state and recent advances in augmentation. J. Orthop. Res. **39**, 1368–1382 (2021)
59. Z.X. Zhou, J.J. Wang, C.Q. Jiang, K.W. Xu, T.J. Xu, X.N. Yu, J.H. Fang, Y.Y. Yang, X.S. Dai, Advances in hydrogels for meniscus tissue engineering: a focus on biomaterials, crosslinking, therapeutic additives. Gels **10**, 114 (2024)
60. X.Y. Lan, Z.Y. Ma, A.R.A. Szojka, M. Kunze, A. Mulet-Sierra, M.J. Vyhlidal, Y. Boluk, A.B. Adesida, Tempo-oxidized cellulose nanofiber-alginate hydrogel as a bioink for human meniscus tissue engineering. Front. Bioeng. Biotechnol. **9**, 766399 (2021)
61. J. Sinna, R. Jeencham, P. Mueangkhot, S. Sophon, P. Noralak, R. Raksapakdee, P.O. Numpaisal, Y. Ruksakulpiwat, Development of poly(vinyl alcohol) grafted glycidyl methacrylate/cellulose nanofiber injectable hydrogels for meniscus tissue engineering. Polymers **15**, 4230 (2023)
62. C. Li, R.H. Deng, M. Yang, F.Z. Yuan, C. Zhang, J.K. Yu, Advanced hydrogel material for meniscus repair. Adv. Funct. Mater. **34**, 2312276 (2024)
63. I. Doench, T.A. Tran, L. David, A. Montembault, E. Viguier, C. Gorzelanny, G. Sudre, T. Cachon, M. Louback-Mohamed, N. Horbelt, C. Peniche-Covas, A. Osorio-Madrazo, Cellulose nanofiber-reinforced chitosan hydrogel composites for intervertebral disc tissue repair. Biomimetics **4**, 19 (2019)
64. J. Joy, J. Pereira, R. Aid-Launais, G. Pavon-Djavid, A.R. Ray, D. Letourneur, A. Meddahi-Pellé, B. Gupta, Gelatin–oxidized carboxymethyl cellulose blend based tubular electrospun scaffold for vascular tissue engineering. Int. J. Biol. Macromol. **107**, 1922–1935 (2018)

65. L.R. Balahura, S. Dinescu, M. Balas, A. Cernencu, A. Lungu, G.M. Vlasceanu, H. Iovu, M. Costache, Cellulose nanofiber-based hydrogels embedding 5-fu promote pyroptosis activation in breast cancer cells and support human adipose-derived stem cell proliferation, opening new perspectives for breast tissue engineering. Pharmaceutics **13**, 1189 (2021)
66. J.J. Pavón, J.P. Allain, D. Verma, M. Echeverry-Rendón, C.L. Cooper, L.M. Reece, A.R. Shetty, V. Tomar, In situ study unravels bio-nanomechanical behavior in a magnetic bacterial nano-cellulose (mbnc) hydrogel for neuro-endovascular reconstruction. Macromol. Biosci. **19**, 1800225 (2019)
67. L. Lamboni, C. Xu, J. Clasohm, J.C. Yang, M. Saumer, K.H. Schäfer, G. Yang, Silk sericin-enhanced microstructured bacterial cellulose as tissue engineering scaffold towards prospective gut repair. Mater. Sci. Eng. C **102**, 502–510 (2019)

Chapter 7
Cellulose-Based Conductive Hydrogels for Emerging Intelligent Sensors

Xue Yao, Sufeng Zhang, Ning Wei, and Liwei Qian

Abstract Flexible intelligent sensing is a burgeoning field of study that covers various disciplines, including but not restricted to chemistry, physics, electronics and biology. However, the widespread use of flexible sensors remains challenging because of certain constraints, such as limited stretchability, poor biocompatibility, low responsivity, and the complexity of multifunctional integration. Conductive hydrogels with remarkable material properties are presently in the spotlight of flexible sensing. In the pursuit of high-performance and "green" conductive hydrogel-based sensors, cellulose is a promising candidate owing to its renewability, low cost, appealing mechanical properties, easy modification and other functional characteristics. Herein, cutting-edge progress in the fabrication of conductive cellulose hydrogels (CCHs) using cellulose and cellulose derivatives in terms of structural features, preparation approaches, functional properties, applications, and prospects for sensors is comprehensively summarized. The correlation between CCHs performances, reinforcement strategies and sensor properties is highlighted to gain insight into the process of developing smart sensors by utilizing CCHs. Besides, the state-of-the-art advances of CCHs toward emerging wearable sensors, including strain/pressure sensors, temperature sensors, humidity sensors, and biosensors, are systematically discussed. Finally, potential challenges and future outlooks of such attractive CCH-based flexible sensors are presented, providing valuable information for the development of next-generation cellulose-based electronic devices.

Keywords Cellulose · Conductive hydrogels · Biomass materials · Flexible sensors · Green electronics

X. Yao · S. Zhang (✉) · N. Wei · L. Qian
College of Bioresources Chemical and Materials Engineering, Shaanxi Provincial Key Laboratory of Papermaking Technology and Specialty Paper Development, Shaanxi University of Science and Technology, Xi'an, China
e-mail: zhangsufeng@sust.edu.cn

G. Wei and J. Zhang (eds.), *Cellulose-Based Hydrogels and Aerogels*, Lecture Notes in Nanoscale Science and Technology 41,
https://doi.org/10.1007/978-3-032-03008-5_7

7.1 Introduction

Flexible wearable sensors, which convert diverse external stimuli into electrical signals, and increasingly found in intelligent wearable systems, medical surveillance devices, soft robots, and human–machine interfaces [1]. Its flexible substrate and conductive material require higher stretchability, conformational recoverability, and biocompatibility to ensure accurate and non-delayed signal transmission, and meet the requirements of various application scenarios. Hydrogels are porous materials with high water contents and three-dimensional network, and they are formed by physical or chemical cross-linking of hydrophilic natural or synthetic polymers [2]. The flexibility and adjustable elastic moduli of hydrogels reduce mechanical mismatches with biological tissues, and they are promising materials for preparing sensing sensors [3].

Cellulose is a kind of crystalline natural polymer with good mechanical properties. Due to its rigid framework, it shows an attractive one-dimensional shape, making it an ideal material for the synthesis of conductive hydrogels. Cellulose based hydrogels have the advantages of abundance, regeneration, degradability, biocompatibility, low cytotoxicity and cost saving, which promote their widespread use.

The raw materials for the preparation of cellulose hydrogels include cellulose, cellulose nanocrystals (CNCs), cellulose nanofibers (CNFs) and commercial cellulose derivatives (e.g., hydroxypropyl cellulose [HPC] and carboxymethyl cellulose [CMC]). CCHs are typically engineered by adding conductive polymers, metal/carbon-based materials, salts or electrolytes to physically noncovalent or chemically covalent cross-linked 3D networks [4]. Hence, stable and strong bonds between the hydrogel network and the conductive filler are important in the construction of high-performance CCHs. To accurately translate human motion into electrical signals, hydrogels, which are wearable sensors that contact the skin, should exhibit good adhesion and biocompatibility with human skin or biological tissues. In addition, to withstand severe conditions, including extreme cold or heat, and exhibit long service lives, the hydrogels must resist drying and freezing and be capable of self-healing. To achieve this goal, CCHs would be exhibit unique merits. First, abundant hydroxyl groups or carboxyl acids are covalently immobilized at the periphery of the glucose unit, which provides a multifunctional platform for functionalization, grafting, and post-modification [5]. These merits endow the prepared CCHs with tunable features for tailored sensation, such as enhanced mechanical performance, conductivity, sensitivity, self-adhesion, self-repair, and freezing resistance. The second point is that nontoxic and biocompatible CCHs allow sensing equipment to come into direct contact with the human skin and can even be safely implanted in the body for long periods. The third advantage is the low-cost, biodegradability and renewability that enable large-scale preparation, commercialization and cyclic utilization of flexible sensors. Therefore, it is of great importance to review systematically the mechanisms of preparation, modification and cross-linking of the networks to tune the mechanical, electrical and multifunctional properties of the hydrogels.

In this chapter, the advanced development of flexible sensors from cellulose and cellulose-derived hydrogels is presented. In view of the preparation and engineering strategies, various CCHs design methods with inherently structural characteristics of cellulose were summarized. Subsequently, the key properties of CCHs enabling use in wearable sensors, including the conductivity, mechanical performance, self-repair, self-adhesion, and anti-freeze properties, are demonstrated, providing explicit design guidelines for the preparation of CCHs with tailor-made features. Thereafter, the cutting-edge progress made with CCHs for strain sensors, pressure sensors, temperature sensors, humidity sensors and biosensors is summarized. Finally, the future challenges and prospects of CCH based flexible sensors were outlined.

7.2 Fabrication and Engineering Strategies of CCHs

7.2.1 Molecular and Structural Features of Cellulose

Cellulose, the most plentiful natural polymer, is found in green wood, seafloor organisms, algae, and even bacteria. Cellulose is a linear polysaccharide consisting of anhydro-D-glucose molecules connected via β-1,4-glycosidic bonds; it has the molecular formula $(C_6H_{10}O_5)_n$ and molecular weights ranging from a few hundred to tens of thousands of Daltons. During the synthesis, van der Waals and interchain hydrogen bonds between the hydroxyl and oxygen atoms of adjacent chains cause the cellulose chains to stack in parallel, leading to the final cellulose fiber. Cellulose fibers demonstrate supramolecular structures composed of crystalline (regular molecular arrangements) and amorphous (disordered and loose arrangements) regions. Because the hydroxyl groups in the crystalline region form many hydrogen bonds, the mechanical rigidities of the crystalline zones are much higher than those of the amorphous zones. Due to these diverse characteristics, cellulose is an attractive building block for green sensors.

The relationships among the cellulose structure, performance, and advantages for designing conductive hydrogels are as follows: (1) Cellulose and its derivatives are promising replacements for current petrochemical materials due to their degradabilities and biocompatibilities. (2) The high aspect ratio of cellulose and its role as an additive in the construction of hydrogels facilitates reinforcement by effective stress transfer, making cellulosic materials attractive for use as gel sensing materials. (3) The hierarchical fibril structure of cellulose confers excellent structural tunability, leading to the development of multifunctional cellulose-based sensing sensors. (4) The many -OHs in cellulose make it susceptible to modification and hybridization, which facilitates material processing and ion mobility. (5) The plentiful intra- and intermolecular hydrogen bonds in cellulose lead to superior mechanical performance, which can be used to fabricate robust CCH-based sensors.

7.2.2 Fabrication Strategies of CCHs

Currently, there are two main strategies used in the preparation of cellulose hydrogels, i.e., bottom-up and top-down approaches. In a "bottom-u" approach, cellulose is converted into a hydrogel by processing it into a solution or suspension and then crosslinking or polymerizing it intramolecularly or with other polymers. With this method, it is possible to manufacture cellulose hydrogels utilizing dissolved cellulose, hydrosoluble cellulose derivatives or nanocellulose. In a top-down strategy, delignified wood is the initial substrate, which is transformed into a cellulose hydrogel via in situ polymerization, solution impregnation or polymer infiltration, etc. To synthesize hydrogels with multifunctional or enhanced properties, the cellulose can also be combined with other natural polymers (e.g., sodium alginate, chitosan, gelatin, and guar gum) or synthetic polymers (polyacrylamide, polyvinyl alcohol, and polyethylene glycol) to prepare a wide variety of cellulose-polymer hybrid hydrogels (Fig. 7.1a).

In addition, the design of the cross-linked networks is essential for the preparation of hydrogel-based electronic sensors. Physical cross-linking based on hydrogen

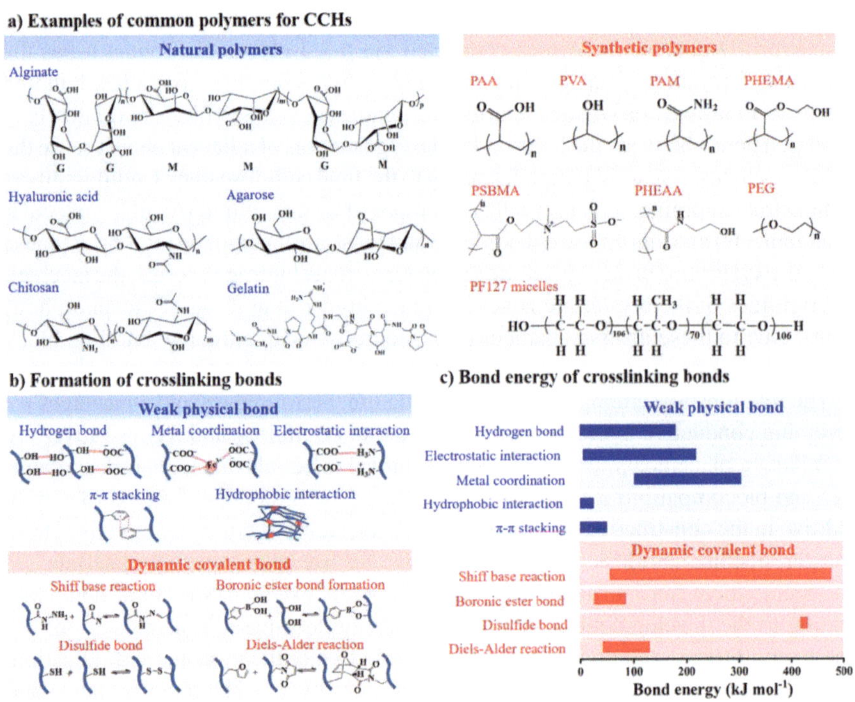

Fig. 7.1 Additional polymers commonly used for hybridization with cellulose to produce conductive hydrogels and the involved cross-linking strategies. **a** Examples of common polymers for CCHs. **b** Formation of crosslinking bonds. **c** Bond energy of crosslinking bonds. Reproduced with permission from Zhao et al. [6]; Copyright 2021, American Chemical Society

bonding, ionic, hydrophobic and host–guest interactions is generally weak but invertible. Chemical cross-linking is usually more robust but perpetual, except for the introduction of reversible covalent bonds (e.g., boronic ester bonds, Schiff base bonds, disulfide bonds, and Diels–Alder cycloaddition etc.) The energy of dynamic covalent cross-links ranges from 25 kJ mol^{-1} to 475 kJ mol^{-1}, depending on the type of bond (Fig. 7.1b, c).

7.3 Properties of CCHs

Flexible CCH-based sensors should meet a variety of functional requirements for different applications, including conductivity, self-repair, adhesion, and antifreeze requirements, the network design and the enhancement of these CCH features will be emphasized here.

7.3.1 Conductivity

Conductivity is critical for the use of cellulose hydrogels in flexible sensors. The strategy for CCH design is to embed conductive phases, including electronic conductive and ionic conductive materials, in the hydrogel matrix and to enable the conductive phases to form a permeable network, or in short, to form a permeable conductive phase. To optimize the electrochemical properties of CCHs, it is crucial to understand the basic mechanisms for conductivity in different situations.

In general, for electronically conducting materials, the conductive mechanism of hydrogels is associated with the conductive filler loading, which can be categorized into two main types: tunneling effect and ohmic conductivity [7]. The tunneling effect is mainly present in hydrogels with low filler loading, where most of the filler is covered by a layer of polymer during mixing, forming a barrier to free electron energy. According to classical mechanics, when the energy of a free electron is lower than the barrier height, it is improbable for it to jump over the barrier, thus hindering electronic conduction and making the conductivity of the hydrogel quite low. However, quantum mechanics demonstrate that microscopic particles with energies below the height of the barrier still have a certain probability of jumping over the barrier, which is commonly known as the tunneling effect. According to the tunneling conduction model, when the filler concentration increases to its percolation threshold, conductive networks distributed in the polymer matrix are constructed, forming relatively low-resistance paths for electron transport, and the ohmic conductance gradually becomes the main conductive mechanism in this case [8].

Enhancing the conductivity of CCHs will serve as an essential reference for building next-generation flexible sensing sensors. Conductivity enhancement strategies for electronic conduction and electrolyte/polyelectrolyte-based CCHs are discussed in this section.

There are usually two ways to enhance electron transport in hydrogels. First, when conductive polymers were introduced into the original cellulose frameworks by forming a backbone via in situ polymerization, the conductivity of the CCHs increased. Conjugated polymers are often used because their conductivity ($10^{-10} \sim 10^{-7}$ S cm^{-1}) is much greater than that of insulating cellulose. Conducting polymers contain alternating single and double bonds in the main chain, resulting in large conjugated π systems where the flow of π electrons increases the electrical conductivity. However, these conjugated polymers can only act as semiconductors due to their low conductivity.

The second method is to add conductive fillers (e.g., carbon particles, CNTs, graphene nanosheets, and metal particles) into cellulose hydrogels to form hybrid materials. Unlike conductive polymer-based CCHs, where the polymer chains are usually interconnected by cross-linking, these conductive fillers are isolated in the hydrogel, which makes it important to form an electrically percolating network [7]. There are three approaches that can be implemented to connect or penetrate a network. The first is simply increasing the amount of conductive filler so that it exceeds the percolation threshold. It has been reported that the conductivity of composite hydrogels increases dramatically once the filler-to-matrix ratio meets the percolation threshold, which is typically $0.1 \sim 10$ wt %, depending on the geometry of the filler. Furthermore, a facile percolation network can be constructed by designing the morphology of the filler. A third method is to form a preinterconnected scaffold or framework prior to forming a composite or hybrid hydrogel, which provides good electrical connectivity and thus confers high conductivity to the hydrogel [9].

Finally, according to the conduction mechanism, increasing the ion concentration and enhancing the ion mobility are the main strategies used to increase the conductivity of ionic conductive hydrogels. Specifically, there are three common ways to improve the ionic conductivity of ICCHs considering the hierarchical structure and adjustable chemistry of cellulose. The first approach is to design ionic nanochannels in a network of ICCHs to regulate and facilitate ion transport. For ICCHs fabricated via both bottom-up and top-down approaches, the channels can be easily tuned to the nanochannel size, which significantly enhances the ionic conductivity. For example, wood is one of the most representative cellulosic materials for the fabrication of CCHs through a top-down method, which can be easily engineered into nanoscale channels and modulate the transport behavior of ions to improve ionic conductivity due to its hierarchical porous structure and high anisotropy [5]. Second, by introducing polyelectrolytes, it is possible to improve the ion solubility and promote free ion migration. Specifically, the charged groups on the polyelectrolyte backbones exhibited higher ion solubilities and generated ion migration channels, thus improving ionic transport. For ICCHs, surface functionalization is the most straightforward approach to introduce charged groups into the backbone. Finally, cellulose itself is not conductive because of dense arrangement of cellulose chains and slow segmental relaxation impede ion transfer. Therefore, widening the molecular distances between the cellulose chains and uncoupling ion transportation from segmental relaxation of cellulose is an effective pathway. Recently, it was demonstrated that the interstitial spaces formed by the insertion of bentonite nanosheets into ICCHs provided fast cation transport channels, which improved the ionic conductivity [10]. These strategies open new horizons for designing high-performance ICCHs.

7.3.2 *Self-Healing Properties*

Malfunctions in wearable electronic sensors primarily stem from accidental mechanical deterioration resulting from reduplicative tears, cuts or scratches. Self-healing capabilities have the potential to confer damage mitigation and long-term reliability to materials [11, 12]. In general, the self-healing property of hydrogels is achieved by forming new and reversible cross-links, including non-covalent bonds, dynamic covalent bonds or bonds based on external stimuli, in the damaged regions (Fig. 7.2) [13].

Noncovalent bonds such as hydrogen bonds and ionic bonds are commonly available for CCHs owing to the unique structural characteristics of cellulose. It was found that CCHs based on pure cellulose exhibited fascinating self-healing by adjusting the hydrogen bonds between networks. As a typical example, Zhao et al. constructed a cellulose hydrogel by dissolving cellulose in an IL, and the hydrogel showed excellent self-healing (a self-healing efficiency of more than 95% within 15 min) due to invertible hydrogen bonding interactions [14]. In addition to hydrogen bonds, ionic bonds have also been employed to construct CCHs capable of self-repair.

Dynamic covalent bonding incorporates the stability of covalent bonding and the reversibility of noncovalent bonding to realize inherent dynamic balance between bond creation and detachment within gel networks, which also enables CCHs self-healing. A boronic ester bond is a dynamic covalent bond generated by a reaction between a diol and a boric acid. It has been demonstrated that cellulose with the

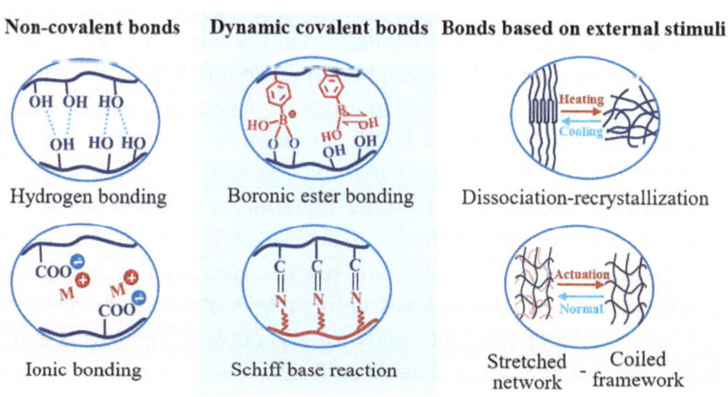

Fig. 7.2 Self-healing mechanism for CCHs, including non-covalent bonding (hydrogen bonding, ionic bonding), dynamic covalent bonds (boronic ester bonding, Schiff base reaction) and bonds based on external stimuli

advantages of abundant hydroxyl groups is often utilized in the fabrication of self-healing CCHs based on boronic ester bonds. This excellent self- repairing capability resulted from boronic ester bonding between the CNFs and PAM/PBA-IL within the hydrogel network. Schiff base bonds are reversible bonds formed between an amine group and an aldehyde group, and they have also been used to prepare CCHs. For example, Ling et al. prepared a self-healing CCH based on Schiff base bonds between dialdehyde carboxymethyl cellulose and amino groups of chitosan [15]. Because of the existence of the reversible Schiff base bonds, this CCH exhibited a self-healing rate of greater than 90% and a self-healing time of less than 10 min in water.

Aside from utilizing weak physical interactions and reversible covalent bonds, stimulus responses via temperature or solvents have also been devised to construct self-healing CCHs. A potential self-healing mechanism involved the dissociation of hydrogen bonds and melting of the crystalline domains during heating, as well as the recombination of the newly crystallized domains and hydrogen bonding at the junctions between two parts upon recooling. Furthermore, self-healing behavior was also achieved by a process of configuration switching under solvent stimulation. Specifically, in water and ethanol, the cellulose molecules exhibited linear stretched conformations, while the PAM molecules showed switching behavior by stretching in water and curling in ethanol. Therefore, the PAM chains were converted to coils by the ethanol stimulus, resulting in volume contraction, which obstructed the cracked areas and enabled self-healing.

7.3.3 Adhesive Properties

The signals from flexible wearable sensors are usually monitored with contact between the sensor and various human tissues (e.g., skin, organs). Hydrogel materials exhibit trade-offs between mechanical and electrical conductivity as well as the strong adhesion required for great dependability and sensitivity. Therefore, the preparation of hydrogels exhibiting rapid and reversible adhesion in different environments generally requires favorable interfacial connections between the hydrogels and substrates, including covalent bonds, physical cross-linking, and mechanical interlocking [16, 17]. The abundant hydroxyl groups on cellulose provide great possibilities for combinations with other materials, thus showing great potential for developing cellulose-based adhesive materials. Dynamic covalent bonds provide adhesion to hydrogels. For substrates with reactive functional groups, such as tissue surfaces, adhesion can be achieved by forming reversible covalent bonds. Coordination bonds are reversible covalent bonds formed between hydrogels and metals or metal oxides and can be used to provide adhesion.

Another preparation strategy is to penetrate a precursor solution of the hydrogel into the porous substrate to form a hydrogel that is mechanically interlocked with the matrix. This is because both biological tissues and hydrogels can be considered micro- or nanoporous polymer networks, where the adherent hydrogel is affixed to the surface of the tissue, and different groups react with each other to form a permeable network [18]. Specifically, polymer chains with multifunctional groups and triggerable cross-linking features are added to the bonding interface, and the polymer chains can diffuse into and cross-link within the hydrogel network. The in situ-formed polymer network stitches the hydrogel like a needle and thread suture by topologically entangling with the existing network without the need for other specific chemical designs of the hydrogel.

Mussel-inspired adhesive hydrogels exhibit superior properties and provide another strategy for designing self-adhesive CCHs. Mussels secrete sticky substances to adhere firmly to diverse surfaces, which is mainly attributed to the fact that 3,4-dihydroxy phenylalanine (DOPA) group and derivatives in the mussel adhesion proteins (e.g., dopamine (DA), catechol-containing polymers) form chemical and hydrogen bonds on the surface of substrates [19]. Therefore, other adhesion factors, such as tannic acid (TA) [20] and polyelectrolyte [21], have been introduced into CCHs to enable reversible adhesion of the hydrogels to a range of substrates via numerous interactions, which provided alternative adhesive strategies for CCH systems.

TA, abundant in edible fruits, has shown great potential for use in adhesives primarily owing to the abundant polyphenolic groups that may form multiple interactions and support strong adhesion to different substrates [22]. Our group reported all-biomass conductive hydrogels consisting of TA-coated CNCs, oxidized alginate and gelatin [23]. Due to the multiple interactions within the gel, particularly for the reversible connecting bridges mediated by borate bonding, the obtained hydrogels displayed great adhesion to diverse materials (e.g., wood, metal, and skin). Similar phenomena were observed with other reported TA-CNCs/CNF hydrogels [24, 25].

In general, the adhesive properties of hydrogels depend on the different mechanisms of adhesion [26]. As shown in Fig. 7.3, the strategies for the adhesion of various hydrogels are demonstrated. Covalent bonds possess high energy, so interfacial adhesion via covalent bonds is generally strong and stable but irreversible, hindering recyclability. Compared to irreversible covalent bonds, CCHs based on dynamic covalent bonding, such as boronic ester bonding, Schiff base bonding, and Diels–Alder reactions, are relatively easy to remove and exhibit recoverability. In addition, for interfacial adhesion between hydrogels and substrates through noncovalent interactions, such weak intermolecular chain interactions may be limited by potential delamination during long-term use. In response, the interfacial toughness of the hydrogel to the substrate can be improved by strong physical cross-linking (including crystalline domains, glassy nodules, and high-density hydrogen bonding). Since such strong physical interactions typically act as high-functional crosslinks, each crosslink can anchor multiple polymer chains on the substrate. For polymer entanglement and mechanical interlocking strategies, roughening or patterning of the substrate surface can further enhance the strength of the mechanical interlocking with the hydrogels [18]. Finally, while catechol compounds have been widely used to construct CCHs with adhesive properties, the interfacial toughness of adhesion achieved solely by catechol-based interfacial linkages is not high. Although CCHs have been widely used in preparing adhesive materials for sensing applications, harmonious combinations of adhesion with other features, such as electrical, mechanical and chemical

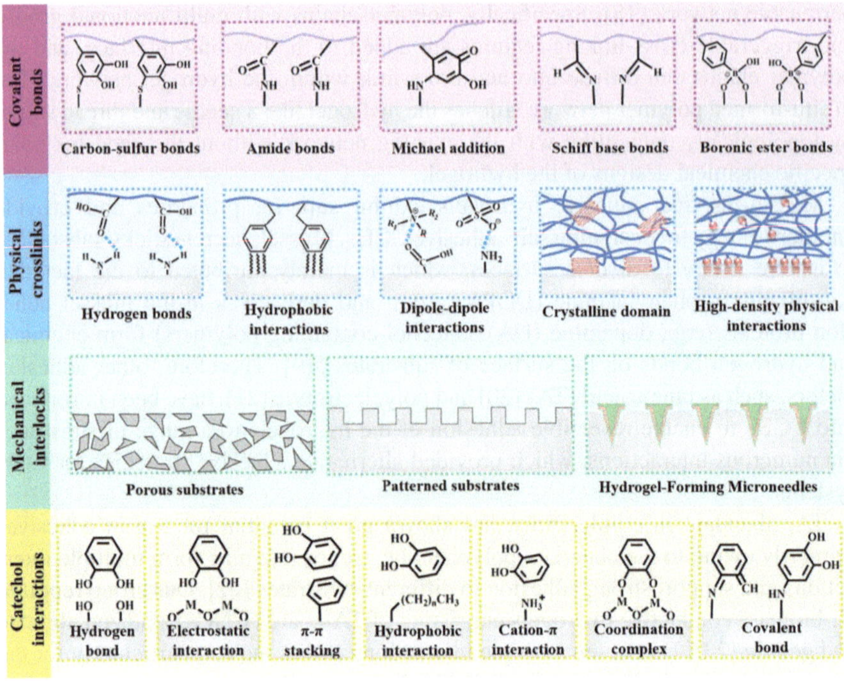

Fig. 7.3 Implementation strategies for the adhesion of hydrogels. The hydrogels are bonded on substrates via various types of interfacial linkages: covalent bonds, physical crosslinks, mechanical interlocks and catechol interactions. The catechol interactions section is reproduced with permission from Saiz-Poseu et al. [26]; Copyright 2018, Wiley–VCH

stability, remain challenging. Therefore, it is necessary to further develop and apply novel adhesive molecular mechanisms and microstructural engineering of hydrogel interfaces.

7.3.4 Low-Temperature Resistance

CCHs containing large amounts of water usually freeze below 0 °C, which leads to drastic reductions in mechanical performance, conductivity and functionality [7]. The development of freeze-resistant CCHs would overcome these difficulties and enable use in extreme environments. Understanding the principles of water freezing is a critical first step before designing of freezing-resistant CCHs. It is suggested that there are three states of water in hydrogels, including free water, weakly bound water (also defined as intermediate water), and unfrozen water (or bound water), which have disparate mobility and freezing points [27]. Since the interactions between free water and the polymer network are negligible, free water is practically equivalent to freely flowing water and crystallizes at temperatures below 0 °C. Intermediate water shows

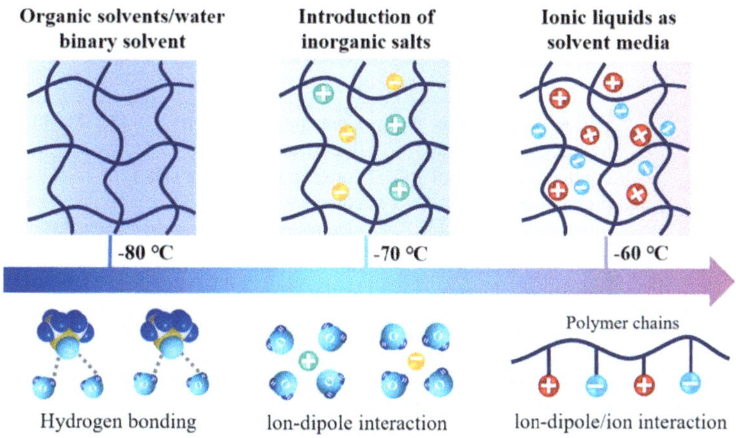

Fig. 7.4 Diverse strategies used to prepare freeze-resistant CCHs

few interactions with the hydrophilic moieties of the hydrogel, and freezes below 0 °C [28]. Furthermore, the residual water is tightly combined with the hydrogel network and is so-called "strongly bound water" or "unfrozen water", which remains fluid even below −100 °C. In general, the proportion of free water is higher than those of the weakly bound and unfrozen water [29]. Therefore, reducing the content of free water or altering its composition is an effective means to significantly improve the freeze resistances of CCHs. Based on these principles, several methods provide theoretical guidance for lowering the freezing temperatures of CCHs, such as the addition of salts [30, 31], or replacement with organic solvents or ionic liquids [32].

In this section, three methods for preparing freeze-resistant CCHs are introduced, and the mechanisms and characteristics of these methods are also described (Fig. 7.4).

Inspired by the freeze-resistance of ions (salts), inorganic salts were first incorporated into hydrogels to obtain freeze-resistant CCHs [33]. Salts (e.g., $CaCl_2$, NaCl, LiCl) often dissociate into anions and cations in aqueous solutions, which combine with some of the free water through hydration, resulting in water molecules that are less likely to aggregate into an ice crystals [34]. Zhang et al. synthesized freeze-tolerant PAM/PEDOT:PSS/CNF by introducing LiCl [35]. The proportion of nonfreezing water in the PAM/PEDOT:PSS/CNF hydrogel increased from 43.48% to 50.61% as the concentration of LiCl was increased from 0 to 45%. The as-prepared CCH exhibited stable mechanical properties at temperatures as low as -60 °C. Besides, other ions, such as NaCl, $ZnCl_2$ and $CaCl_2$, also depressed the freezing point of CCHs [36]. For example, Chen et al. added $ZnCl_2$ into a cellulose hydrogel, which kept the hydrogel from freezing at −33 °C [37].

The second strategy involves the introduction of organic solvents such as ethylene glycol (EG), glycerol, or dimethyl sulfoxide (DMSO) into CCHs to replace some of the aqueous solution. On the one hand, some organic solvents combine with H_2O to reduce the content of "free water" in the CCHs. On the other hand, the hydrogen

Table 7.1 Methods for constructing antifreeze CCHs and their advantages and disadvantages

Construction strategies	Advantages	Disadvantages	Refs.
Introduction of inorganic salts	Endowing cellulose hydrogels with good conductivity, ease of preparation, good biocompatibility	Unstable conductivity due to varying ionic concentrations, high salt corrodes metal electrodes	[30, 31, 33, 35–37]
Introduction of organic solvents (organohydrogels)	Ease of preparation; imparting cellulose hydrogels with good water retention	Some chemical inertness leading to reduced conductivity	[41–43]
Ionic liquids as solvent media (ionogels)	Outstanding resistance to low temperatures, good conductivity and stability over a wide temperature range	Risk of leakage, limited in applications with biological or environmental contact, relatively high cost	[32, 40, 44, 45]

bonding between functional groups such as amino, carboxyl or hydroxyl groups of organic molecules with H_2O weakens the interactions among the "free water" molecules, inhibiting the formation of ice crystals. To be precise, CCHs synthesized by this method are defined as conductive organohydrogels.

Third, the use of ILs as solvent media to prepare freeze-resistant CCHs has gained much attention from researchers in recent years owing to their excellent conductivity, mobility, and ease of non-evaporation [38, 39]. Strictly speaking, the gels acquired from crosslinked cellulose-IL solutions are not called hydrogels but ionogels. For example, Wang and coworkers dissolved cellulose in an aqueous solution of benzyltrimethylammonium hydroxide ($BzMe_3NOH$), and then prepared freeze-resistant cellulose gels directly by chemical cross-linking [40]. The reported gels maintained stable mechanical properties at −62.1 °C. Evidently, different preparation strategies for anti-freezing CCHs affect their properties. Therefore, the advantages and disadvantages of the various approaches are compared to enable the selection of a suitable strategy and to provide rational guidance for appropriate applications based on the principles of gel design (Table 7.1).

7.3.5 Water Retention

CCHs are susceptible to dehydration in normal or dry environments, leading to severe performance degradation and device dysfunction. Therefore, maintaining water in CCH structures and maintaining wet-soft activity in an open environment are the keys to further overcoming these issues. Generally, the proportion of free water molecules in conventional CCHs is relatively high, and once they are placed in an open environment, many free water molecules evaporate from the gel matrix into the air, and the residual bound and intermediate water in the hydrogel is insufficient to maintain its soft-wet properties. In essence, reducing the free water content and locking it up or introducing components that confer the ability to absorb free water from ambient air into the hydrogel is an effective strategy for improving the water retention of CCHs [46].

Dissociation of hygroscopic salts in water results in the formation of hydrated ions, which reduces the vapor pressure, and the salts further stabilize the water in the hydrogel [47]. CCHs containing soluble salts do not dry out in air but swell and deswell in response to fluctuations in ambient humidity, and the corresponding changes in water content will lead to alterations in their mechanical properties, electrical conductivity, and other functionalities. Alternatively, ethylene glycol or glycerol is usually mixed with water to form a binary solvent to reduce the amount of free water molecules in CCH networks. Jiang et al. prepared a conductive hydrogel consisting of tannic acid-coated cellulose nanofibers (TA@CNFs), Cu^{2+}, and PAM in a glycerol/water binary solvent system with excellent water retention that was able to withstand 1000% of the tensile strain when stored for 15 days at 25 °C and 30% relative humidity [24]. Since glycerin can form hydrogen bonds with water molecules, it promotes the adsorption and preservation of water molecules. Additionally, cellulose ionogels can be used since they are nonvolatile in a vacuum and have a wide range of operating temperatures, depending on the type of ionic liquid [38].

Despite significant progress in designing CCHs with water retention in the past decade, several serious challenges remain. The use of media such as water, organic solvents and ionic liquids may be limited by their narrow temperature windows or leakage issues. In addition, there are some problems to be solved for the method of designing hydrophobic coatings on hydrogel surfaces. First, it is very difficult to selectively modify only the exposed surface of a hydrogel without changing its internal structure or chemical composition. Second, the interfacial strength between the hydrophilic hydrogel surface and the hydrophobic coating is very weak, making it difficult to withstand external damage for practical applications. The solutions to these problems, along with the development of advanced CCHs, are widely available for exploration.

7.4 Applications of CCHs

CCHs have emerged as ideal candidates for intelligent sensors owing to the impressive characteristics mentioned above, including renewability, biocompatibility, great mechanical properties, high conductivity and environmental stability. This section provides a comprehensive overview of emerging CCH-based sensors, as well as their operational principles and pivotal figure of performance.

7.4.1 Strain/Pressure Sensors

Hydrogel-based flexible strain sensors sense external stress variations and convert mechanical stimuli into electrical signals to capture vital signs (e.g., pulse, body motion) of the human body. Depending on the sensing mechanism, strain sensors can operate in resistive, capacitive, piezoelectric and triboelectric modes for real-time detection. The most common CCH-based strain sensors use the resistive and capacitive modes, while piezoelectric and triboelectric transduction are not readily available on account of their intricate assembly and need for scarce materials.

7.4.1.1 Resistive Mode

The working of resistive strain/pressure sensors hinges on the conversion of geometrical deformation of materials under external force into the electrical signal in terms of change in resistances. Conductive hydrogels operate in two strain modes, i.e., tensile and pressing deformation.

(1) Strain Sensors

On account of the great stretchable and conductive properties, CCHs have been investigated for use as strain sensors in diversity of applications, including smart robotics, healthcare sensing and even artificial skin. Conductive hydrogels are generally categorized into electronic and ionic conductive hydrogels.

Compared to ICCHs, electronic conductive hydrogels are much more sensitive to compressive and tensile deformations since resistivity changes are realized through contact-resistance and tunneling effects [7]. The tunneling effect exists primarily within hydrogels with lower concentrations of electron-conducting materials. An electronically conductive network is formed in the hydrogel matrix, providing a relatively low-resistance path for electron transport as the concentration of the conductive filler is increased to the percolation threshold [48]. Contact resistance effect is the prevailing conductive mechanism, offering relative stability and high sensitivity for hydrogel-based resistive sensors. When stretched, the transformations of the conductive network cause the tunnel distances to change at small strains, resulting in a smaller signature output. The tunneling distance increases correspondingly as the applied strain is increased. Numerous direct linkages between the conductive substances are separated as the tunneling distance becomes greater than the cutoff distance, which generates a stronger signal output [49]. Thus, electronic CCH-based strain sensors typically exhibit high sensitivity and have been extensively used in the preparation of diverse wearable sensors. However, inorganic and metallic materials tend to aggregate in hydrophilic gel matrices due to their inherent stiffness, hydrophobicity and poor solubility, which limits the application range. Modifications of these materials with hydrophilic/charged phenolic hydroxyl, carboxylic acid and oxide groups usually solve this problem.

The hydrogel-based strain sensors described above meet the basic demands of conductivity and tensile properties. However, it is challenging to meet the multiple requirements for mechanical properties, conductivity, biocompatibility, transparency and a wide sensing range in practical applications. Similar to tissue activity generated via ionic conduction, ICCHs replicate long-distance ion movement instead of electron transport and show promise for biomedical applications. For ICCHs, stretching only alters the conformations of their polymer networks and water molecules and exhibits inappreciable changes in their ion conductivities. ICCH-based sensors can typically withstand large tensile deformations while detecting and monitoring a variety of human movements, due to their high stretchability, flexibility, and good compatibility. Yao and coworkers developed a CNFs-reinforced PAM/PBA-IL CCH for strain sensors [50]. The semi-interpenetrating network and presence of multifunctional PBA-IL resulted in a high sensitivity (GF = 8.36), quick response time (195 ms) and other multifunctionality.

(2) Pressure Sensors

CCHs can also be used to produce as pressure sensors by generating resistance changes through input pressures because of the great compressive elasticities. Typically, the available conductive channels in distorted hydrogels become denser when compressed, thereby reducing the resistance. Carbon materials, metal particles and conductive polymers have been used to design resistive pressure/compression sensors due to their high conductivity. In addition to carbon-based conductive materials, conductive polymers have been incorporated into hydrogel networks to construct dependable conductive networks.

Inspired by the ability of human skin to sense slight pressure changes, a resistive pressure sensor based on the CNFs/PAA/IL hydrogel was used to sense the slightest touch of a finger [51]. It showed an ionic conductivity of 0.70 S m^{-1} and a pressure sensitivity that decreased from 1.6 to 0.0004 kPa^{-1} over a pressure range of 0–2,000 kPa. On account of the high durability of CCHs, a hydrogel sensor also displayed satisfactory sensitivity in accurate real-time detection of pressure and human movement. In addition, as a proof-of concept, the CCH was employed for robot sensing that lit up LEDs on the robot's fingers.

7.4.1.2 Capacitive Mode

Capacitive sensors contain two parallel electrodes and a sandwiched conductive dielectric to form a parallel-plate capacitor. The principle of operation is conversion of the bilayer stress resulting from a thickness change into a capacitance signal. The advantages of this mechanism include fast responses, great linearity, low power consumption and easy processing, and they have led to extensive development of CCH-based capacitive sensors in wearable sensors.

In addition to being used as electrodes, CCHs also served as dielectric layers in the fabrication of flexible capacitive sensors. For instance, capacitive sensors prepared with CNT/PPy as the flexible electrode and CNFs-enhanced PAM hydrogel as the electrolyte showed excellent electrochemical performance and cycling stability [52]. In particular, the sensitivity was improved at the nanoscale and micron scale by incorporating low molecular weight cellulose and employing sandpaper as a template, respectively. The sensitivity was enhanced from 0.04 to 89.81 kPa^{-1}, and the system demonstrated remarkable durability, electrical transmission, and great potential for array applications.

7.4.1.3 Piezoelectric Mode

Piezoelectric sensors typically consist of a flexible substrate and piezoelectric materials such as polyvinylidene fluoride (PVDF), zinc oxide (ZnO) or zirconium titanate (PZT) [53]. As opposed to resistive and capacitive sensors, piezoelectric sensors can be self-powered, and their sensitivity is comparable to that of resistive sensors. Despite sensors based on piezoelectric materials have good prospects due to their high sensitivity and self-powered nature, they have limitations in healthcare sensing on account of poor biocompatibility or mechanical flexibility [54]. Flexible and biocompatible CCHs provide new solution ideas for fabricating the next generation of green and environmentally friendly piezoelectric sensors. The operating principle of a piezoelectric sensor is based on the piezoelectric effect, i.e., charge accumulation and voltage output under mechanical deformation [55]. More interestingly, the BC substrate of the piezoelectric sensor can be degraded by cellulose, and the embedded $ImClO_4$ can be solubilized with H_2O and recycled, leading to the development of a biodegradable and environmentally friendly flexible piezoelectric sensor.

Although many highly sensitive hydrogel-based strain/pressure sensors have been developed, the influence of the water content on the mechanical properties and the retention of electrical properties after repeated deformations are the main challenges. We believe that novel cellulose hydrogel networks inspired by the unique structures in nature, such as textured structures, micro/nanotransport channels, and layering mechanisms, is one of the effective strategies for future research.

7.4.2 Temperature Sensors

Temperature sensors are vital components of wearable sensors used to detect environmental and surface temperatures variations, which are crucial for the construction of smart human–machine interactions [56]. For example, the body temperature is one of the most important physiological parameters, and it affects the metabolic rate as well as the biochemical reactions in the human body [57]. Real-time and accurate monitoring of the local temperature is essential in allowing humans and intelligent robots to adjust their subsequent plans.

Current temperature sensing materials based on combinations of flexible substrates and temperature-sensitive materials generally present limited stretchability, low transparency, and cumbersome preparation processes, so it is difficult to meet the demand for high-performance, intelligent, and portable soft sensors. CCHs have shown promise in research on flexible temperature sensors due to their deformability, conductivity, transparency, and conformal contact between devices and curved surfaces [58]. The temperature sensing mechanism of a CCH relies on variations in the ionic conductivity with temperature. With rising temperatures, the conductivity of the CCH improves, and the resistance decreases on account of enhancement in both the ion concentration and mobility, resulting in an analog voltage output.

In addition, the introduction of inorganic salts is another strategy used in preparing CCH-based temperature sensors because the salts furnish free ions in the aqueous phase and lower the saturated vapor pressure and freezing point of the CCHs [59]. For instance, a second physically crosslinked tannic acid (TA)-NaCl network was introduced into a chemically crosslinked cellulose hydrogel to improve the hydrogel toughness and freezing tolerance.[60] The hydrogel showed a wider operating temperature window (20–100 °C) with a TCR of –0.0083 $°C^{-1}$, repeatability, and high stability, all of which allowed real-time monitoring of temperature changes during human exhalation and inhalation.

In addition to the salt solutions added to confer ion conductivity, the addition of polyelectrolytes in the matrix formed efficient ionic conductive pathways due to the mobilities of their charged side groups [61]. Additionally, adhesion between the zwitterionic active sites and the interface minimized interfacial sliding and delamination to ensure stable output of the electrical signal in real time. This provided the opportunity to construct a stable CCH-based temperature sensor with a long lifetime. The CCH-based sensor adhered to the finger for temperature monitoring and showed promised in the design of e-skins suitable for harsh temperature environments.

Organohydrogel-based temperature sensors with thermo-responsive networks were developed to provide sensors with high sensitivity and accuracy based on the lower heat capacities of organic solvents compared to that of water [62]. The introduction of DMSO into the PVA/CNF hydrogel allowed its use as a resistive sensor to detect temperatures between 0 and 60 °C [43]. The thermal sensitivities of the sensor at 0 to 6 °C and 6 to 60 °C were –0.0466 $°C^{-1}$ and –0.0063 $°C^{-1}$, respectively, which were ascribed to the H_2O/DMSO solvent system that facilitated ion transport.

7.4.3 Humidity Sensors

Humidity (relative humidity, RH) sensing plays an essential part in daily life since the humidity significantly impacts the quality of life and health by affecting everything from industry, agriculture, and food to breathing [63]. Traditional humidity sensors still suffer from problems in preparation and application, such as rigidity, high prices, cumbersome preparation processes and the generation of electronic waste [64].

The sensing principle of CCH-based humidity sensors is swelling of the hydrophilic groups within gel substrates stimulated by water, which increases the distance between the conductive materials and changes the resistance [65]. Pristine cellulose hydrogels serve as flexible substrates for sensors and as humidity-sensitive materials due to the surface hydroxyl functional groups of cellulose and their interactions with water that change the electrical conductivity [14]. However, they are not highly sensitive to humidity variations because of restricted moisture absorption. Therefore, the moisture sensitivity has been enhanced by introducing electronic fillers, inorganic salts or ILs.

Even though we have witnessed a growing interest in designing resistive humidity sensors, the problem of low sensitivity and accuracy still exist and impede their use in long-term monitoring. To overcome these problems, capacitive humidity sensors have been proposed because the dielectric constant of the sensor is more sensitive to humidity than that of a resistor; thus, capacitive humidity sensors exhibit higher sensitivity. For example, a capacitance-type humidity sensor fabricated from tannic acid (TA)/cellulose/NaCl exhibited high sensitivity, which enabled real-time monitoring of nasal airflow, the chest and abdomen, and the pulse during sleep [60]. Moreover, the sensor excluded interference problems arising from monitoring individual body parts or outputting a single electrical signal, and the results showed higher accuracy and confidence limits. This work provides an alternative to CCH-based sensors for respiratory monitoring and prediction of sleep disorder-derived health problems.

7.4.4 Biosensors

Biosensors are efficient analytical detection devices that provide specific physiological information via biological recognition [2]. Benefiting from their superior mechanical flexibility, conductivity, biocompatibility and adherence to human skin, CCHs have been integrated into biosensing electronics designed to monitor many biological parameters, including biochemical compounds (e.g., glucose) and biophysical indicators (e.g., pH, heart rate, electrocardiogram (ECG), and electroencephalogram (EEG)) [66].

Sweat contains a wealth of biochemical data, including electrolytes (e.g., potassium and sodium ions) and metabolites (e.g., glucose and lactate) [67]. Sweat sensors have attracted widespread use in disease detection and health monitoring due to the ease of collecting of these biomarkers [68, 69]. Generally, CCH-based wearable sweat sensors with the shapes of tattoos or patches are used to monitor the biomarkers in sweat to provide clinically relevant information on the health status.

CCH-based flexible sensors have also been used to detect glucose and pH. Shen et al. used carboxymethyl chitosan/oxidized CMC hydrogels to encapsulate gold nanoclusters and glucose oxidase [70]. These hydrogel-based biosensors exhibited high glucose sensing sensitivity, wide linear detection range (100 μM to 5 mM) and low detection limit (0.029 mM). Additionally, the hydrogels were able to perform blood fluorescence testing with diabetic mice due to the good biocompatibility, indicating promise for use in implantable biosensors. Furthermore, a BC/CMC hydrogel-based pH sensor was developed to monitor the colors produced by hydrogen peroxide and potassium iodide [71]. The sensor displayed a wide linear detection range, which facilitated use in urine pH monitoring to detect urological disorders.

Advanced implantable biosensors have the revolutionary potential to provide long-term effective functionality at the biological interface, enabling diagnostic and therapeutic capabilities for various diseases and disorders [72]. CCHs are used to coat and encapsulate bioelectronics like electrocardiography (ECG), electroencephalography (EEG), and electromyography (EMG) to bring interface adaptability and biocompatibility, and reduce interface impedance, because of their intrinsic softness, adhesion, mechanical compliance, and electrochemical stability. Lu et al. reported a regenerated cellulose/PAA-based ionogel with excellent fatigue resistance and abilities comparable to those of commercial silver/silver chloride electrodes for ECG monitoring [73]. Interestingly, an increase in the cellulose concentration reduced the noise wave and thus improved the accuracy. CCHs are ideal materials for constructing implantable biosensors due to their intrinsic characteristics and cross-linked networks that resemble the extracellular matrix. Furthermore, an efficient wound dressing that responded to electrical stimulation was also prepared [74]. The hydrogel promoted nerve repair, angiogenesis, and diabetic wound healing through electrical stimulation, which are highly important in clinical practice.

7.5 Summary and Outlook

Conductive hydrogels prepared from cellulose or its derivatives will be promising because of their unique features, including sustainability, degradability, nontoxicity, biocompatibility, abundant supply, low cost, outstanding mechanical performance, modifiability, natural nanochannels with tunable sizes, and ionic conductivity strengthening effects. These benefits make CCHs a desirable class of building blocks for designing and constructing flexible electronic devices in the future.

This chapter summarizes recent progress in preparing cellulose hydrogels for sensing applications. The structure-performance relationships of the cellulosic materials were briefly outlined. Design strategies for CCHs exhibiting reinforced mechanical performance, conductivity, self-healing, adhesion, frost resistance and water retention were discussed. Finally, the operation mechanisms for flexible sensors containing CCHs in wearable or implantable modes, including strain, pressure, temperature and biosensing were illustrated. Although great progress has been achieved in the development of advanced CCH-based sensors, some challenges remain.

Acknowledgements This work was supported by the National Natural Science Foundation of China (No. 22378247), International Joint Research Center for biomass chemistry and materials, Shaanxi international science and technology cooperation base (2018GHJD-19).

References

1. Q. He, Y. Cheng, Y. Deng, F. Wen, Y. Lai, H. Li, Conductive hydrogel for flexible bioelectronic device: current progress and future perspective. Adv. Funct. Mater. **34**(1), 2308974 (2024). https://doi.org/10.1002/adfm.202308974
2. A. Herrmann, R. Haag, U. Schedler, Hydrogels and their role in biosensing applications. Adv. Healthcare Mater. **10**(11), 2100062 (2021). https://doi.org/10.1002/adhm.202100062
3. J. Qu, Q. Yuan, Z. Li, Z. Wang, F. Xu, Q. Fan, M. Zhang, X. Qian, X. Wang, X. Wang, M. Xu, All-in-one strain-triboelectric sensors based on environment-friendly ionic hydrogel for wearable sensing and underwater soft robotic grasping. Nano Energy **111**(1), 108387 (2023). https://doi.org/10.1016/j.nanoen.2023.108387
4. Z. Wu, S. Chen, J. Li, B. Wang, M. Jin, Q. Liang, D. Zhang, Z. Han, L. Deng, X. Qu, H. Wang, Insights into hierarchical structure–property–application relationships of advanced bacterial cellulose materials. Adv. Funct. Mater. **33**(12), 2214327 (2023). https://doi.org/10.1002/adfm.202214327
5. D. Zhao, Y. Zhu, W. Cheng, W. Chen, Y. Wu, H. Yu, Cellulose-based flexible functional materials for emerging intelligent electronics. Adv. Mater. **33**(28), 2000619 (2021). https://doi.org/10.1002/adma.202000619
6. X. Zhao, X. Chen, H. Yuk, S. Lin, X. Liu, G. Parada, Soft materials by design: unconventional polymer networks give extreme properties. Chem. Rev. **121**(8), 4309–4372 (2021). https://doi.org/10.1021/acs.chemrev.0c01088
7. G. Li, C. Li, G. Li, D. Yu, Z. Song, H. Wang, X. Liu, H. Liu, W. Liu, Development of conductive hydrogels for fabricating flexible strain sensors. Small **18**(5), e2101518 (2022). https://doi.org/10.1002/smll.202101518
8. I. Alig, P. Pötschke, D. Lellinger, T. Skipa, S. Pegel, G.R. Kasaliwal, T. Villmow, Establishment, morphology and properties of carbon nanotube networks in polymer melts. Polymer **53**(1), 4–28 (2012). https://doi.org/10.1016/j.polymer.2011.10.063
9. L. Yan, T. Zhou, L. Han, M. Zhu, Z. Cheng, D. Li, F. Ren, K. Wang, X. Lu, Conductive cellulose bio-nanosheets assembled biostable hydrogel for reliable bioelectronics. Adv. Funct. Mater. **31**(17), 2010465 (2021). https://doi.org/10.1002/adfm.202010465
10. S. Wang, L. Yu, S. Wang, L. Zhang, L. Chen, X. Xu, Z. Song, H. Liu, C. Chen, Strong, tough, ionic conductive, and freezing-tolerant all-natural hydrogel enabled by cellulose-bentonite coordination interactions. Nat. Commun. **13**(1), 3408 (2022). https://doi.org/10.1038/s41467-022-30224-8
11. Y. Li, L. Shi, Y. Cheng, R. Wang, J. Sun, Development of conductive materials and conductive networks for flexible force sensors. Chem. Eng. J. **455**, 140763 (2023). https://doi.org/10.1016/j.cej.2022.140763
12. V.A. Glezakou, R. Rousseau, T. Lin, Introduction: self-healing in chemical systems. Chem. Rev. **123**(2), 555–557 (2023). https://doi.org/10.1021/acs.chemrev.2c00839
13. K. Zhang, Q. Feng, Z. Fang, L. Gu, L. Bian, Structurally Dynamic hydrogels for biomedical applications: pursuing a fine balance between macroscopic stability and microscopic dynamics. Chem. Rev. **121**(18), 11149–11193 (2021). https://doi.org/10.1021/acs.chemrev.1c00071
14. D. Zhao, Y. Zhu, W. Cheng, G. Xu, Q. Wang, S. Liu, J. Li, C. Chen, H. Yu, L. Hu, A dynamic gel with reversible and tunable topological networks and performances. Matter **2**(2), 390–403 (2020). https://doi.org/10.1016/j.matt.2019.10.020
15. Q. Ling, W. Liu, J. Liu, L. Zhao, Z. Ren, H. Gu, Highly sensitive and robust polysaccharide-based composite hydrogel sensor integrated with underwater repeatable self-adhesion and rapid self-healing for human motion detection. ACS Appl. Mater. Interfaces **14**(21), 24741–24754 (2022). https://doi.org/10.1021/acsami.2c01785
16. X. Chen, J. Zhang, G. Chen, Y. Xue, J. Zhang, X. Liang, I.M. Lei, J. Lin, B.B. Xu, J. Liu, Hydrogel bioadhesives with extreme acid-tolerance for gastric perforation repairing. Adv. Funct. Mater. **32**(29), 2202285 (2022). https://doi.org/10.1002/adfm.202202285

17. Y. Xue, J. Zhang, X. Chen, J. Zhang, G. Chen, K. Zhang, J. Lin, C. Guo, J. Liu, Trigger-detachable hydrogel adhesives for bioelectronic interfaces. Adv. Funct. Mater. **31**(47), 2106446 (2021). https://doi.org/10.1002/adfm.202106446
18. P. Rao, T.L. Sun, L. Chen, R. Takahashi, G. Shinohara, H. Guo, D.R. King, T. Kurokawa, J.P. Gong, Tough hydrogels with fast, strong, and reversible underwater adhesion based on a multiscale design. Adv. Mater. **30**(32), e1801884 (2018). https://doi.org/10.1002/adma.201801884
19. X. Ma, X. Zhou, J. Ding, B. Huang, P. Wang, Y. Zhao, Q. Mu, S. Zhang, C. Ren, W. Xu, Hydrogels for underwater adhesion: adhesion mechanism, design strategies and applications. J. Mater. Chem. A **10**(22), 11823–11853 (2022). https://doi.org/10.1039/d2ta01960d
20. W. Yan, M. Shi, C. Dong, L. Liu, C. Gao, Applications of tannic acid in membrane technologies: a review. Adv. Colloid Interface Sci. **284** (2020). https://doi.org/10.1016/j.cis.2020.102267
21. X. Pan, Q. Wang, D. Benetti, Y. Ni, F. Rosei, Polyelectrolyte hydrogel: a versatile platform for mechanical-electric conversion and self-powered sensing. Nano Energy **103**, 107718 (2022). https://doi.org/10.1016/j.nanoen.2022.107718
22. H. Sun, X. Fang, Y. Zhu, Z. Yu, X. Lu, J. Sun, Highly tough, degradable, and water-resistant bio-based supramolecular plastics comprised of cellulose and tannic acid. J. Mater. Chem. A **11**(13), 7193–7200 (2023). https://doi.org/10.1039/d3ta00351e
23. S. Cui, S. Zhang, F. Zhang, R. Lin, C. Tang, X. Jing, Tannic acid-coated cellulose nanocrystal-reinforced transparent multifunctional hydrogels with UV-filtering for wearable flexible sensors. Carbohydr. Polym. **323**, 121385 (2024). https://doi.org/10.1016/j.carbpol.2023.121385
24. S. Zong, H. Lv, C. Liu, L. Zhu, J. Duan, J. Jiang, Mussel inspired Cu-tannic autocatalytic strategy for rapid self-polymerization of conductive and adhesive hydrogel sensors with extreme environmental tolerance. Chem. Eng. J. **465**, 142831 (2023). https://doi.org/10.1016/j.cej.2023.142831
25. F. Han, X. Xie, T. Wang, C. Cao, J. Li, T. Sun, H. Liu, S. Geng, Z. Wei, J. Li, F. Xu, Wearable hydrogel-based epidermal sensor with thermal compatibility and long term stability for smart colorimetric multi-signals monitoring. Adv. Healthcare Mater. **12**(3), 2201730 (2022). https://doi.org/10.1002/adhm.202201730
26. J. Saiz-Poseu, J. Mancebo-Aracil, F. Nador, F. Busqué, D. Ruiz-Molina, The chemistry behind catechol-based adhesion, Angew. Chem. Int. Ed. **58**(3) 696–714 (2018). https://doi.org/10.1002/anie.201801063
27. Y. Jiana, S.H. Wang, J. Zhang, W. Lua, X. Zhou, T. Chen, Biomimetic anti-freezing polymeric hydrogels: keeping soft-wet materials active in cold environment Mater. Horiz. **8**(2) 351–369 (2020). https://doi.org/10.1039/D0MH01029D.1039/x0xx00000x
28. K. Lei, M. Chen, P. Guo, J. Fang, J. Zhang, X. Liu, W. Wang, Y. Li, Z. Hu, Y. Ma, H. Jiang, J. Cui, J. Li, Environmentally adaptive polymer hydrogels: maintaining wet-soft features in extreme conditions. Adv. Funct. Mater. **33**(41), 2303511 (2023). https://doi.org/10.1002/adfm.202303511
29. Y. Feng, S. Wang, Y. Li, W. Ma, G. Zhang, M. Yang, H. Li, Y. Yang, Y. Long, Entanglement in smart hydrogels: fast response time, anti-freezing and anti-drying. Adv. Funct. Mater. **33**(21), 2211027 (2023). https://doi.org/10.1002/adfm.202211027
30. H. Miao, Y. Liu, Y. Man, G. Huang, H. Huang, X. Fu, M. Ge, H. Liu, Y. Qian, High-efficiency and recyclable green molten salt hydrate solvent for cellulose hydrogels with high conductivity and freeze tolerance. ACS Sustainable Chem. Eng. **11**(24), 9225–9234 (2023). https://doi.org/10.1021/acssuschemeng.3c02175
31. Z. Li, H. Kumar, C. Guo, J. Shin, X. He, Q. Lu, H. Bai, K. Kim, J. Hu, Development of antifreezing, printable, adhesive, tough, biocompatible, high-water content hydrogel for versatile applications. ACS Appl. Mater. Interfaces **15**(12), 16034–16045 (2023). https://doi.org/10.1021/acsami.3c00152
32. H. Lee, A. Erwin, M.L. Buxton, M. Kim, A.V. Stryutsky, V.V. Shevchenko, A.P. Sokolov, V.V. Tsukruk, Shape persistent, highly conductive ionogels from ionic liquids reinforced with cellulose nanocrystal network. Adv. Funct. Mater. **31**(38), 2103083 (2021). https://doi.org/10.1002/adfm.202103083

33. S. Huang, L. Hou, T. Li, Y. Jiao, P. Wu, Antifreezing hydrogel electrolyte with ternary hydrogen bonding for high-performance zinc-ion batteries. Adv. Mater. **34**(14), 2110140 (2022). https://doi.org/10.1002/adma.202110140
34. S. Zhou, K. Guo, D. Bukhvalov, W. Zhu, J. Wang, W. Sun, M. He, H-bond/ionic coordination switching for fabrication of highly oriented cellulose hydrogels. J. Mater. Chem. A **9**(9), 5533–5541 (2021). https://doi.org/10.1039/d0ta10863d
35. M. Zhang, Y. Wang, K. Liu, Y. Liu, T. Xu, H. Du, C. Si, Strong, conductive, and freezing-tolerant polyacrylamide/PEDOT:PSS/cellulose nanofibrils hydrogels for wearable strain sensors. Carbohydr. Polym. **305**, 120567 (2023). https://doi.org/10.1016/j.carbpol.2023.120567
36. Z. Wang, X.-F. Zhang, L. Shu, J. Yao, Construction of MXene functionalized wood-based hydrogels using ZnCl2 aqueous solution for flexible electronics. J. Mater. Chem. A **11**(19), 10337–10345 (2023). https://doi.org/10.1039/d3ta01370g
37. D. Chen, X. Zhao, X. Wei, J. Zhang, D. Wang, H. Lu, P. Jia, Ultrastretchable, tough, antifreezing, and conductive cellulose hydrogel for wearable strain sensor. ACS Appl. Mater. Interfaces **12**(47), 53247–53256 (2020). https://doi.org/10.1021/acsami.0c14935
38. C. Zhao, Y. Wang, G. Tang, J. Ru, Z. Zhu, B. Li, C.F. Guo, L. Li, D. Zhu, Ionic flexible sensors: mechanisms, materials, structures, and applications. Adv. Funct. Mater. **32**(17), 2110417 (2022). https://doi.org/10.1002/adfm.202110417
39. L. Li, W. Li, X. Wang, X. Zou, S. Zheng, Z. Liu, Q. Li, Q. Xia, F. Yan, Ultra-tough and recyclable ionogels constructed by coordinated supramolecular solvents, Angew. Chem. Int. Ed. **61**(50) e202212512 (2022). https://doi.org/10.1002/anie.202212512
40. Y. Wang, L. Zhang, A. Lu, Transparent, antifreezing, ionic conductive cellulose hydrogel with stable sensitivity at subzero temperature. ACS Appl. Mater. Interfaces **11**(44), 41710–41716 (2019). https://doi.org/10.1021/acsami.9b15849
41. Y. Wei, L. Xiang, H. Ou, F. Li, Y. Zhang, Y. Qian, L. Hao, J. Diao, M. Zhang, P. Zhu, Y. Liu, Y. Kuang, G. Chen, MXene-based conductive organohydrogels with long-term environmental stability and multifunctionality. Adv. Funct. Mater. **30**(48), 2005135 (2020). https://doi.org/10.1002/adfm.202005135
42. X.F. Zhang, X. Ma, T. Hou, K. Guo, J. Yin, Z. Wang, L. Shu, M. He, J. Yao, Inorganic salts induce thermally reversible and anti-freezing cellulose hydrogels. Angew. Chem., Int. Ed. **58**(22) 7366–7370 (2019). https://doi.org/10.1002/anie.201902578
43. M. Li, D. Chen, X. Sun, Z. Xu, Y. Yang, Y. Song, F. Jiang, An environmentally tolerant, highly stable, cellulose nanofiber-reinforced, conductive hydrogel multifunctional sensor. Carbohydr. Polym. **284**, 119199 (2022). https://doi.org/10.1016/j.carbpol.2022.119199
44. Y. Liu, Q. Liu, L. Zhong, C. Chen, Z. Xu, Tough, antifreezing, and conductive double network zwitterionic-based hydrogel for flexible sensors. Chem. Eng. J. **452**, 139314 (2023). https://doi.org/10.1016/j.cej.2022.139314
45. A. Zhu, J. Huang, H. Xie, W. Yue, S. Qin, F. Zhang, Q. Xu, Use of a superbase/DMSO/CO2 solvent in order to incorporate cellulose into organic ionogel electrolyte for flexible supercapacitors. Chem. Eng. J. **446**, 137032 (2022). https://doi.org/10.1016/j.cej.2022.137032
46. C. Yang, Z. Suo, Hydrogel ionotronics. Nat. Rev. Mater. **3**(6), 125–142 (2018). https://doi.org/10.1038/s41578-018-0018-7
47. T. Zhu, Y. Ni, G.M. Biesold, Y. Cheng, M. Ge, H. Li, J. Huang, Z. Lin, Y. Lai, Recent advances in conductive hydrogels: classifications, properties, and applications. Chem. Soc. Rev. **52**(2), 473–509 (2022). https://doi.org/10.1039/d2cs00173j
48. H. He, H. Li, A. Pu, W. Li, K. Ban, L. Xu, Hybrid assembly of polymeric nanofiber network for robust and electronically conductive hydrogels. Nat. Commun. **14**(1), 759 (2023). https://doi.org/10.1038/s41467-023-36438-8
49. J. Tropp, C.P. Collins, X. Xie, R.E. Daso, A.S. Mehta, S.P. Patel, M.M. Reddy, S.E. Levin, C. Sun, J. Rivnay, Conducting polymer nanoparticles with intrinsic aqueous dispersibility for conductive hydrogels. Adv. Mater. 2306691 (2023). https://doi.org/10.1002/adma.202306691
50. X. Yao, S. Zhang, L. Qian, N. Wei, V. Nica, S. Coseri, F. Han, Super stretchable, self-healing, adhesive ionic conductive hydrogels based on tailor-made ionic liquid for high-performance

strain sensors. Adv. Funct. Mater. **32**(33), 2204565 (2022). https://doi.org/10.1002/adfm.202204565

51. Y. Ye, H. Oguzlu, J. Zhu, P. Zhu, P. Yang, Y. Zhu, Z. Wan, O.J. Rojas, F. Jiang, Ultrastretchable ionogel with extreme environmental resilience through controlled hydration interactions. Adv. Funct. Mater. **33**(2), 2209787 (2022). https://doi.org/10.1002/adfm.202209787
52. W. Ge, S. Cao, Y. Yang, O.J. Rojas, X. Wang, Nanocellulose/LiCl systems enable conductive and stretchable electrolyte hydrogels with tolerance to dehydration and extreme cold conditions. Chem. Eng. J. **408**, 127306 (2021). https://doi.org/10.1016/j.cej.2020.127306
53. M.N. Islam, R.H. Rupom, P.R. Adhikari, Z. Demchuk, I. Popov, A.P. Sokolov, H.F. Wu, R.C. Advincula, N. Dahotre, Y. Jiang, W. Choi, Boosting Piezoelectricity by 3D printing PVDF-MoS2 composite as a conformal and high-sensitivity piezoelectric sensor. Adv. Funct. Mater. 2302946 (2023). https://doi.org/10.1002/adfm.202302946
54. S. Chang, S. Hur, J. Park, D. Lee, J. Shin, H.S. Kim, S.E. Song, J.M. Baik, M. Kim, H.C. Song, C.Y. Kang, Optimization of piezoelectric polymer composites and 3D printing parameters for flexible tactile sensors. Addit. Manuf. **67**, 103470 (2023). https://doi.org/10.1016/j.addma.2023.103470
55. X. Liu, J. Tong, J. Wang, S. Lu, D. Yang, H. Li, C. Liu, Y. Song, BaTiO3/MXene/PVDF-TrFE composite films via an electrospinning method for flexible piezoelectric pressure sensors. J. Mater. Chem. C **11**(14), 4614–4622 (2023). https://doi.org/10.1039/d2tc05291a
56. Y. Li, L. Yang, S. Deng, H. Huang, Y. Wang, Z. Xiong, S. Feng, S. Wang, T. Li, T. Zhang, A machine learning-assisted multifunctional tactile sensor for smart prosthetics. InfoMat **5**(9), e12463 (2023). https://doi.org/10.1002/inf2.12463
57. P. Yao, Q. Bao, Y. Yao, M. Xiao, Z. Xu, J. Yang, W. Liu, Environmentally stable, robust, adhesive, and conductive supramolecular deep eutectic gels as ultrasensitive flexible temperature sensor. Adv. Mater. **35**(21), 2300114 (2023). https://doi.org/10.1002/adma.202300114
58. L. Shu, X. Zhang, Y. Wu, Z. Wang, J. Yao, Facile fabrication of strong and conductive cellulose hydrogels with wide temperature tolerance for flexible sensors. Int. J. Biol. Macromol. **240**, 124438 (2023). https://doi.org/10.1016/j.ijbiomac.2023.124438
59. L. Chen, J. Lou, X. Rong, Z. Liu, Q. Ding, X. Li, Y. Jiang, X. Ji, W. Han, Super-stretching and high-performance ionic thermoelectric hydrogels based on carboxylated bacterial cellulose coordination for self-powered sensors. Carbohydr. Polym. **321**, 121310 (2023). https://doi.org/10.1016/j.carbpol.2023.121310
60. J. Liu, H. Wang, T. Liu, Q. Wu, Y. Ding, R. Ou, C. Guo, Z. Liu, Q. Wang, Multimodal hydrogel-based respiratory monitoring system for diagnosing obstructive sleep Apnea syndrome. Adv. Funct. Mater. **32**(40), 2204686 (2022). https://doi.org/10.1002/adfm.202204686
61. A. Yasuda, A. Inagawa, N. Uehara, Charge-selective aggregation behavior of thermoresponsive polyelectrolytes having low charge density in aqueous solutions of organic counterions Langmuir **39**(5), 1730–1739 (2023). https://doi.org/10.1021/acs.langmuir.2c02286
62. Z. Chen, H. Liu, X. Lin, X. Mei, W. Lyu, Y. Liao, Competitive proton-trapping strategy enhanced anti-freezing organohydrogel fibers for high-strain-sensitivity wearable sensors. Mater. Horiz. **10**(9), 3569–3581 (2023). https://doi.org/10.1039/d3mh00459g
63. Y. Luo, J. Li, Q. Ding, H. Wang, C. Liu, J. Wu, Functionalized hydrogel-based wearable gas and humidity sensors. Nano-Micro Lett. **15**(1), 136 (2023). https://doi.org/10.1007/s40820-023-01109-2
64. I. Janica, V. Montes-García, F. Urban, P. Hashemi, A.S. Nia, X. Feng, P. Samorì, A. Ciesielski, Covalently functionalized mxenes for highly sensitive humidity sensors. Small methods **7**(8), 2201651 (2023). https://doi.org/10.1002/smtd.202201651
65. X. Wang, G. Wang, W. Liu, D. Yu, X. Liu, G. Li, Z. Song, H. Wang, Developing a carbon composite hydrogel with a highly conductive network to improve strain sensing performance. Carbon **216**, 118500 (2023). https://doi.org/10.1016/j.carbon.2023.118500
66. J.J. Quispe Haro, S.V. Wegner, an Adenosylcobalamin specific whole-cell biosensor, Adv. Healthcare Mater. **12**(25) 2300835 (2023). https://doi.org/10.1002/adhm.202300835
67. S. Zhang, R. Tan, X. Xu, S. Iqbal, J. Hu, Fibers/Textiles-based flexible sweat sensors: a review. ACS Mater. Lett. **5**(5), 1420–1440 (2023). https://doi.org/10.1021/acsmaterialslett.2c01207

68. J. Min, J. Tu, C. Xu, H. Lukas, S. Shin, Y. Yang, S.A. Solomon, D. Mukasa, W. Gao, Skin-interfaced wearable sweat sensors for precision medicine. Chem. Rev. **123**(8), 5049–5138 (2023). https://doi.org/10.1021/acs.chemrev.2c00823
69. L. Yin, M. Cao, K.N. Kim, M. Lin, J.M. Moon, J.R. Sempionatto, J. Yu, R. Liu, C. Wicker, A. Trifonov, F. Zhang, H. Hu, J.R. Moreto, J. Go, S. Xu, J. Wang, A stretchable epidermal sweat sensing platform with an integrated printed battery and electrochromic display. Nat. Electron. **5**(10), 694–705 (2022). https://doi.org/10.1038/s41928-022-00843-6
70. Y. Shen, Z. Wang, Y. Wang, Z. Meng, Z. Zhao, A self-healing carboxymethyl chitosan/oxidized carboxymethyl cellulose hydrogel with fluorescent bioprobes for glucose detection. Carbohydr. Polym. **274**, 118642 (2021). https://doi.org/10.1016/j.carbpol.2021.118642
71. T. Siripongpreda, B. Somchob, N. Rodthongkum, V.P. Hoven, Bacterial cellulose-based re-swellable hydrogel: facile preparation and its potential application as colorimetric sensor of sweat pH and glucose. Carbohydr. Polym. **256**, 117506 (2021). https://doi.org/10.1016/j.carbpol.2020.117506
72. Y. Xue, X. Chen, F. Wang, J. Lin, J. Liu, Mechanically-compliant bioelectronic interfaces through fatigue-resistant conducting polymer hydrogel coating. Adv. Mater. **35**(40) (2023). https://doi.org/10.1002/adma.202304095
73. X. Fu, Y. Chen, W. Wang, D. Yu, Self-adhesive and anti-fatigue cellulose-polyacrylate ionogels prepared by ultraviolet curing used as biopotential electrodes. Int. J. Biol. Macromol. **218**, 533–542 (2022). https://doi.org/10.1016/j.ijbiomac.2022.07.115
74. L. Guan, X. Ou, Z. Wang, X. Li, Y. Feng, X. Yang, W. Qu, B. Yang, Q. Lin, Electrical stimulation-based conductive hydrogel for immunoregulation, neuroregeneration and rapid angiogenesis in diabetic wound repair. Sci. China Mater. **66**(3), 1237–1248 (2022). https://doi.org/10.1007/s40843-022-2242-y

Chapter 8
Cellulose-Based Hydrogels for Wearable Devices

Ping Han

Abstract Cellulose-based hydrogels (CBHs) have shown important application value in the field of flexible sensing due to their excellent biocompatibility, degradability and controllability. This chapter focuses on the conductive properties, environmental stability and its application in wearable devices. By compositing conductive materials (such as MXene, conductive polymers), CBHs can achieve conductivity of 1–5 S/m and have sensor characteristics of fast response (response time <200 ms) and wide strain range (0–1000%). Their unique dual network structure design balances mechanical strength and conductivity, while endowing the material with self-healing, antifreeze (maintaining flexibility at −40 °C), and antibacterial properties to support long-term use in wearable devices. In terms of application, CBHs have been successfully used to monitor human movement (such as joint bending, pulse) and environmental parameters (such as temperature, sweat components). Most of cellulose raw materials are derived from agricultural waste, reflecting the concept of green and sustainable development. In the future, CBHs have broad prospects in fields such as smart medical care and human–computer interaction.

Keywords Cellulose hydrogels · Flexible sensor · Wearable device · Conductivity · Environmental stability

8.1 Introduction

Hydrogels are materials that have a porous 3D network structure with high moisture content, formed by hydrophilic polymers through physical or chemical crosslinking. In recent years, hydrogels have aroused extensive research interest and have been applied in many fields, given their excellent mechanical properties and adaptability. In particular, hydrogels based on natural polymers have found wide applications

P. Han (✉)
College of Physics, Qingdao University, Qingdao, China
e-mail: hanping828@qdu.edu.cn

G. Wei and J. Zhang (eds.), *Cellulose-Based Hydrogels and Aerogels*, Lecture Notes in Nanoscale Science and Technology 41,
https://doi.org/10.1007/978-3-032-03008-5_8

in the fields of flexible sensing and biomedical due to their unique advantages of degradability, environmental friendliness and low cytotoxicity.

Cellulose, as the most widely distributed and abundant natural polymer in nature, has gradually become the focus of academic and industrial circles in recent decades under the promotion of green chemistry and nanotechnology. Its biocompatibility, high strength and reinforcing ability give cellulose excellent adjustability and processability, making it an ideal choice for preparing diversified functional materials. In addition, cellulose surface is rich in hydroxyl functional groups, which makes it easy to modify, graft and crosslink by chemical or physical methods, thus showing great potential in sensor devices, optical sensing and biosensor materials preparation.

Cellulose hydrogel, as a natural hydrogel derived from biomass, inherits environmentally friendly and low toxicity properties and can replace traditional synthetic materials for the preparation of flexible sensing materials such as hydrogels [1]. In particular, cellulose-based hydrogels (CBHs) and membranes have been widely used in the fields of environmental protection, health monitoring and flexible sensing. In addition, the development of cellulose hydrogel sensing materials has promoted the high value-added utilization of biomass and the sustainable development of energy, thus verifying its huge application potential in the fields of biomedical, environmental monitoring and smart wearables [2].

As a green and environmentally friendly natural polymer material, CBH has shown broad application prospects in flexible sensing, biomedical and environmental governance due to its excellent biocompatibility, degradability and controllable physical and chemical properties. In the field of flexible sensing, CBH can impart high sensitivity and fast response characteristics through composite conductive materials (such as carbon nanotubes (CNTs), conductive polymers, etc.), and are suitable for wearable health monitoring devices, electronic skin, and human–computer interaction systems [3]. Its good flexibility and biocompatibility enable it to fit the human body for a long time, and monitor physiological signals such as heart rate, body temperature, and exercise status in real time, providing a safer and more sustainable material choice for smart wearable devices [4]. In terms of environmental governance, its 3D porous structure and rich active groups enable it to efficiently adsorb heavy metal ions, organic dyes and other pollutants in water. Meanwhile, its natural degradable properties avoid the secondary pollution of traditional synthetic adsorbents, which is in line with the development concept of green chemistry. In addition, the high value-added utilization of CBHs promotes the sustainable development of biomass resources [5]. By converting agricultural waste (such as straw, fruit shells, etc.) into high-performance functional materials, it not only improves resource utilization, but also reduces environmental pollution, providing innovative solutions for the circular economy. In the future, with the advancement of material modification technology and intelligent sensing technology, cellulose hydrogels will play a more important role in medical health, environmental monitoring and flexible electronics.

In this chapter, we present recent advance in the fabrication of CBH-based wearable devices for various applications. The thermal and electrical properties of CBHs are introduced, and their environmental stability is then analyzed. Additionally, the factors for affecting the sensor performance of the fabricated devices are discussed.

8.2 Thermal and Electrical Properties of CBHs for Wearable Devices

Resistivity test: The resistivity can be measured by the four-probe AC impedance method, and then the electrical conductivity ($\sigma = 1/\rho$) can be calculated, quantifying the ionic or electronic conduction ability of the material. Compared with the traditional two-probe method, the four-probe AC impedance method has become the gold standard for characterizing the electrical properties of CBHs by reducing contact resistance errors, separating bulk phase/surface effects, and providing frequency-dependent information. In the study of Wang et al., this method was used, and then the electrical conductivity of HPE-LiCl-7 was calculated to be 2.04 S/m [6]. Moreover, it can also be seen from the change in the resistivity of the hydrogel in the figure that its electrical conductivity first increased and then stabilized with the increase of LiCl concentration, reaching a plateau at 7 M.

Strain sensitivity test: Yi et al. measured the change in relative resistance $\Delta R/R_0$ of the hydrogel within the strain range of 0–1000% and calculated the strain coefficient GF [7]. At low strain levels ranging from 0 to 200%, GF = 3.13; At a higher strain of 400–600%, GF = 10.46. It has been demonstrated that the hydrogel can exhibit a linear response over a wide range, and the GF value increases as strain increases. It is suitable for multi-scenario monitoring from small deformations (such as pulse) to large movements (such as joint bending), demonstrating its dual sensitivity.

Conductivity test: In the application of flexible wearable devices, the high electrical conductivity of cellulose hydrogels can ensure a rapid response to strain signals. The electrical conductivity stability at low temperatures also expands the application scenarios of sensors in cold environments. Chen et al. measured the electrical conductivity of the hydrogel at different $ZnCl_2$ concentrations and temperatures using a digital source meter [8]. They found that at a 25% concentration, the electrical conductivity at room temperature reached 1.54 S/m, and remained at 1.21 S/m at −20 °C, which was attributed to the efficient migration of Zn^{2+} and Cl^-. In another case, Yu et al. measured the ionic conductivity at room temperature and low temperature using the four-probe method [9]. The room-temperature conductivity of SPCG-2 was 5.11 S/m, and it decreased to 4.85 S/m at −35 °C, with a retention rate of 95%.

Furthermore, MXene, as a conductive filler with good performance, is often used in the preparation of hydrogels. For example, in the study of Hasan et al., the room-temperature conductivity of CCH-5 was 1.11 S/cm, which was significantly higher than that of CH-2 (0.63 S/cm) and CCH-1 (0.78 S/cm) [10]. This improvement was attributed to the continuous conductive network of MXene and the ion transport channels of CMCs. When the temperature increased, the electrical conductivity increased linearly, which was due to the improvement of the ion mobility and the thermal excitation of MXene electrons.

Cyclic stability test: In the electrical performance test, the good cycling stability of hydrogels can improve their sensing life and signal accuracy. For instance, Zhao et al. conducted a 3000-s cycling stability test on the MXene-HC-PAM hydrogel at 150% strain, and the signal fluctuation was less than 5%, meeting the sensing requirements in wearable devices [11].

8.3 Environmental Stability of CBH Devices

Weight loss monitoring: Weight loss monitoring is to track the weight changes of CBHs under specific conditions, such as temperature, humidity, pH value, enzyme action, etc., by technical means, reflecting the process of water loss, degradation, or component release. Wang et al. exposed a CBH to room temperature (25 °C) and high temperature (80 °C) environments, measured the weight loss regularly, and evaluated the water retention capacity [6]. At room temperature, the weight loss of HPE-LiCl-0 (without LiCl) was 14.31%, while that of HPE-LiCl-7 was only 8.96%. At high temperatures, the loss of HPE-LiCl-0 was 24.42% and that of HPE-LiCl-7 was 19.76%, which was significantly better than that of traditional hydrogels (Fig. 8.1a). It has been confirmed that LiCl and EG work in synergy to suppress evaporation through ionic hydration shells and hydrogen bond networks, ensuring the maintenance of ionic conductivity and mechanical properties at high temperatures or during long-term use.

Frost resistance performance test: Frost resistance performance testing refers to the ability of CBHs to maintain structural integrity, mechanical properties and functional stability in low-temperature environments, such as freezing or repeated freeze–thaw cycles. Wang et al. placed the hydrogel in an environment of −40 °C for 6 h, and then observed the morphological changes and measured the tensile/compressive properties [6], and found that the resistivity of HPE-LiCl-7 was stable (Fig. 8.1b). Moreover, HPE-LiCl-7 remained soft and could be stretched to 5 times its original length. The retention rate of tensile strength was >90%, and the rebound rate after compression was >95%. It has been verified that the hydrogel can maintain flexibility and electrical conductivity at extremely low temperatures, showing high potential for monitoring in low-temperature environments.

Optical transmissivity: Light transmittance refers to the percentage of the intensity of transmitted light to that of incident light after light (usually visible light or light of a specific wavelength) passes through CBHs, reflecting the transparency of the material. Hydrogels with high transmittance usually have a more regular network structure and fewer scattering centers, reflecting the uniformity of the material's microstructure. It has potential impacts on optical applications such as transparent sensors, photo-responsive devices, and biological imaging carriers. As shown in Fig. 8.1c, Gao et al. measured the visible light transmittance of the hydrogel. The visible light transmittance at 800 nm was 98%, and the text could be seen clearly [12].

Antimicrobial property test: Antibacterial performance is an indispensable property for the application of hydrogels in wearable devices. Wei et al. measured the diameter of the inhibition zone and the inhibition rate of hydrogels against

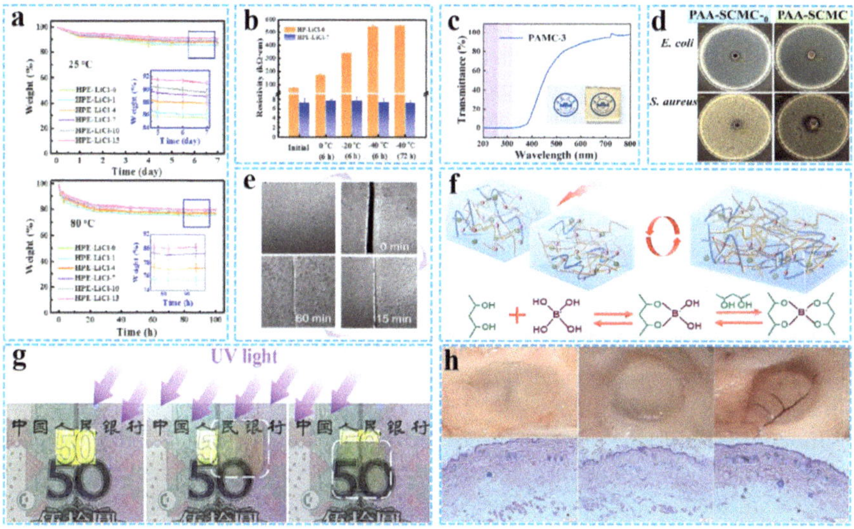

Fig. 8.1 Environmental stability tests of CBH devices: **a** The mass loss of hydrogel when stored at 25 and 80 °C for 100 h. **b** The resistivity of hydrogels at different temperatures. Reprinted with the permission from Reference [6], Copyright 2024, American Chemical Society. **c** The UV spectrum of hydrogel (the illustration photo shows high transparency). Reprinted with the permission from Reference [12], Copyright 2024, Elsevier Ltd. **d** The antibacterial ability of PAA-SCMC-0 and PAA-SCMC hydrogels. Reprinted with the permission from Reference [13], Copyright 2024, Elsevier Ltd. **e** The self-healing process of hydrogel. Reprinted with the permission from Reference [7], Copyright 2025, Springer. **f** Self-healing mechanism hydrogel. Reprinted with the permission from Reference [14], Copyright 2025, Elsevier Ltd. **g** The UV filtering effect of 2-mm-thick hydrogel on RMB banknotes. Reprinted with the permission from Reference [12], Copyright 2024, Elsevier Ltd. **h** Biocompatibility of PPNGC hydrogel. Reprinted with the permission from Reference [15], Copyright 2024, Elsevier Ltd

Escherichia coli and Staphylococcus aureus [13]. The diameters of the antibacterial zones were 10.6 and 20.7 mm, respectively, and the antibacterial rate was >99% (Fig. 8.1d). Antibacterial properties prevent infection, biocompatibility meets medical-grade application requirements, and can be used for direct skin contact monitoring.

Self-healing performance test: Self-healing ability can extend the service life of wearable devices and is suitable for scenarios prone to wear and tear. In the study of Huang et al. [7], when the hydrogel was cut, it could heal after 60 min (Fig. 8.1e), and the tensile strength recovered to 70.45%, the elongation at break recovered to 91.21%, the recovery rate of electrical conductivity reached 97%, and still maintained the initial value of 88.69% after three cuts. The healing mechanism relies on the coordination between Ga^{3+} and -COO $^{-}$, as well as the hydrogen bond reconstitution between MXene/CNF and PAA. In addition, Zhuang et al. discovered that in the P/P:P/C conductive hydrogel, PVA form a dynamic boric acid bond cross-linking network with borax (Fig. 8.1f), which are reversible and prone to breakage and recombination, endow the hydrogel with excellent self-healing ability [14].

UV resistance test: UV rays can trigger the photo-oxidative degradation of cellulose chains, leading to a decrease in the mechanical strength of hydrogels, yellowing of color, and destruction of network structures. Testing its UV resistance can determine the service life of materials in outdoor, medical light, or industrial ultraviolet environments. In addition, cellulose hydrogels with excellent UV resistance can be used in sun protection coatings, outdoor environment remediation materials or photo-stabilizing drug carriers. The corresponding specific applications include UV-resistant soil water-retaining agents and drugs that prevent deterioration caused by ultraviolet rays. UV resistance tests can prove the anti-aging ability of materials during long-term outdoor use and are suitable for scenarios such as outdoor sensors. Gao et al. measured the ultraviolet shielding effect of hydrogels [12]. The transmittance of 200–400 nm ultraviolet light is less than 20%, which can effectively shield ultraviolet rays (Fig. 8.1g).

Biocompatibility test: To verify the biocompatibility and safety of hydrogels and provide ethical and biological evidence for their application in implantable sensors. Jiang et al. made the hydrogel into 1 cm cylinders and implanted them into the subcutaneous tissue on the back of rats for staining and observation [15]. Within 14 days of implantation, there was no redness, swelling, inflammation or necrosis in the tissues around the hydrogel. Staining showed that the cell morphology was normal and no immune rejection reaction occurred (Fig. 8.1h). Combined with antibacterial performance, it can enhance the safety of wearable devices on the human body surface and subcutaneous tissues.

8.4 Sensor Performance Test

Strain sensing test: Strain sensing test of hydrogels is essential for their application in flexible wearable devices. In the tests conducted by Mogli et al. [16], the corresponding resistances increased linearly at different bending angles of the fingers (30°–90°), and the signal repeatability of repeated bending cycles was >95%. Similar tests were also conducted in the experiments of Zhou and co-workers [17]. When hydrogels were applied to human motion responses, it was found that actions such as finger (30°–90°), wrist and knee bending could all trigger significant and unique electrical signals, and the signal amplitudes were positively correlated with the motion amplitudes (Fig. 8.2a). This will facilitate the distinction of response signals from different body parts and provide reliable data input for subsequent machine learning.

Response-recovery time test: Cellulose hydrogel, as a stimulus-responsive material, is sensitive to stimuli such as temperature, pH, electric field and light. Its response time and recovery time are the core indicators for measuring its dynamic performance. The fast response and recovery time of cellulose hydrogels is conducive to efficient monitoring of human movement. The response time of the hydrogel sensor studied was 153 ms and the recovery time was 165 ms [7], meeting the requirements of real-time monitoring and suitable for full-range human motion monitoring (Fig. 8.2b).

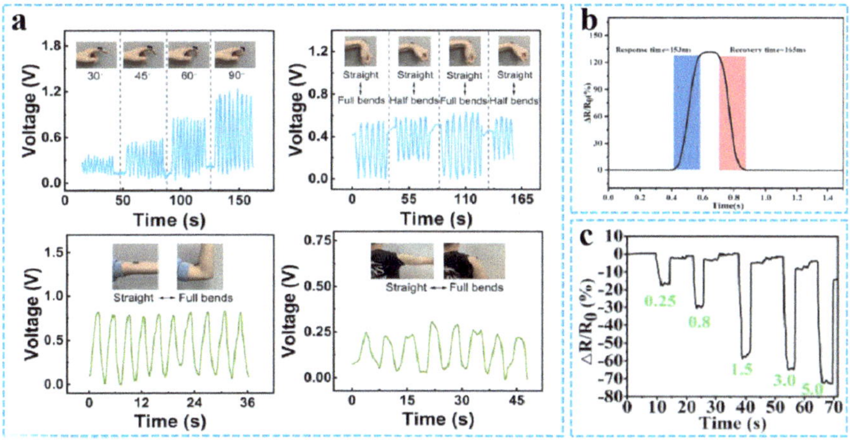

Fig. 8.2 Sensor performance tests of CBH sensors: **a** The output voltage of the hydrogel sensor as a function of time. Reprinted with the permission from Reference [17], Copyright 2024, American Chemical Society. **b** The response and recovery time of hydrogels. Reprinted with the permission from Reference [7], Copyright 2025, Elsevier Ltd. **c** $\Delta R/R_0$ at different sodium chloride concentrations. Reprinted with the permission from Reference [13], Copyright 2024, Elsevier Ltd

Materials with fast response and short recovery time are suitable for reusable sensors, such as rapid recovery after repeated deformation to maintain detection accuracy.

Humoral sensing test: The changes in the components of body fluids are directly related to the health status of the human body. CBH sensors achieve real-time detection of glucose, pH, electrolytes, metabolites, etc. through chemical/physical signal conversion. By exploring the biocompatibility, modifiability, and signal conversion ability of CBH sensors, real-time and non-invasive monitoring of human physiological states can be achieved. In a previous study [13], $\Delta R/R_0$ was measured in artificial sweat with different NaCl concentrations (0.25–5 mg/mL) and pH (4.0–8.0). It was found that $\Delta R/R_0$ decreased with the increase of NaCl concentration, and the response was stable at pH = 5.5, which could distinguish different sweating states (Fig. 8.2c). This can be combined with the detection of signals such as heart rate to comprehensively analyze the body's movement status, laying the foundation for more intelligent wearable devices. In the future, the combination of micro-nano processing and intelligent algorithms can further enhance its application potential in the fields of personalized medicine and sports health.

8.5 Applications of CBHs for Wearable Devices

CBH-based flexible sensors can be divided into the following categories in terms of application: biological motion monitoring and flexible wearable devices.

8.5.1 Biological Motion Detection

CBH flexible sensing materials have features including human–computer interaction [18], e-skin [19], and health tracking [20], as well as a series of medical biomedical applications. In biomotion sensing applications, CBH sensing materials primarily achieve their diverse functions through the fabrication of pressure, strain, biomolecule, and pH sensors. Pressure and strain sensors are capable of monitoring human joint movements, while biological and chemical sensors are well-suited for medical diagnostics and detection applications. CBH sensing materials exhibit broad application prospects in the biomedical field by offering effective tools and methodologies for disease diagnosis and therapeutic monitoring. As flexible and biocompatible materials, CBHs represent a new frontier in biological motion detection. With the growing demand for highly sensitive and biocompatible flexible sensors, CBHs have emerged as promising alternatives to traditional rigid sensors. Owing to their unique physical and chemical properties, cellulose hydrogels have successfully overcome the limitations of conventional materials, positioning themselves as next-generation smart sensing platforms for biomedical applications [21].

Yang et al. explored the feasibility of PPZ (PVA/HPEI/CNCs@PA/PA/$ZnCl_2$, defined as PPZ hydrogel) nanocomposite hydrogels as flexible sensors for human motion detection at sub-zero temperatures (Fig. 8.3) [21]. To explore the feasibility of PPZ nanocomposite hydrogels as flexible sensors for human motion detection at subzero temperatures, we applied the hydrogel sensors to various body parts of a volunteer. It was found show that the hydrogel-based sensor could sensitively and stably detect subtle joint movements in the eyebrows, mouth, and fingers (Fig. 8.3a–c). In addition, the CBH sensor could sensitively and stably capture larger joint movements of the elbow and knee (Fig. 8.3d–e). Figure 8.3f demonstrates that the hydrogel sensors could rapidly and precisely detect this series of movements. Hydrogel sensors show quick, accurate responses and can detect minor human movements in low temperatures. This makes them promising for wearable and flexible technology applications. Therefore, the fabricated hydrogel sensors showed fast, accurate responses and could detect tiny human movements at low temperatures, making them promising for wearable and flexible technology applications [23]. In another study, Lee et al. began by focusing on temperature, one of the most commonly perceived physiological parameters in humans and animals [24]. Body temperature served as a critical indicator of various organ functions and can reflect a wide range of health conditions. For instance, spatiotemporal imaging or thermometric analysis of skin temperature holds significant experimental and diagnostic value in numerous medical fields, including cardiovascular health assessment, evaluation of cognitive function, diagnosis of pulmonary diseases, breast cancer detection, and the monitoring of other syndromes.

CBHs can be applied for monitoring mechanical matching and wearable health. By tuning crosslink density, polymer concentration, or incorporating nano-reinforcing phases such as cellulose nanocrystals (CNCs), the mechanical modulus of hydrogels can be finely adjusted, typically within the kilopascal to megapascal range, to match

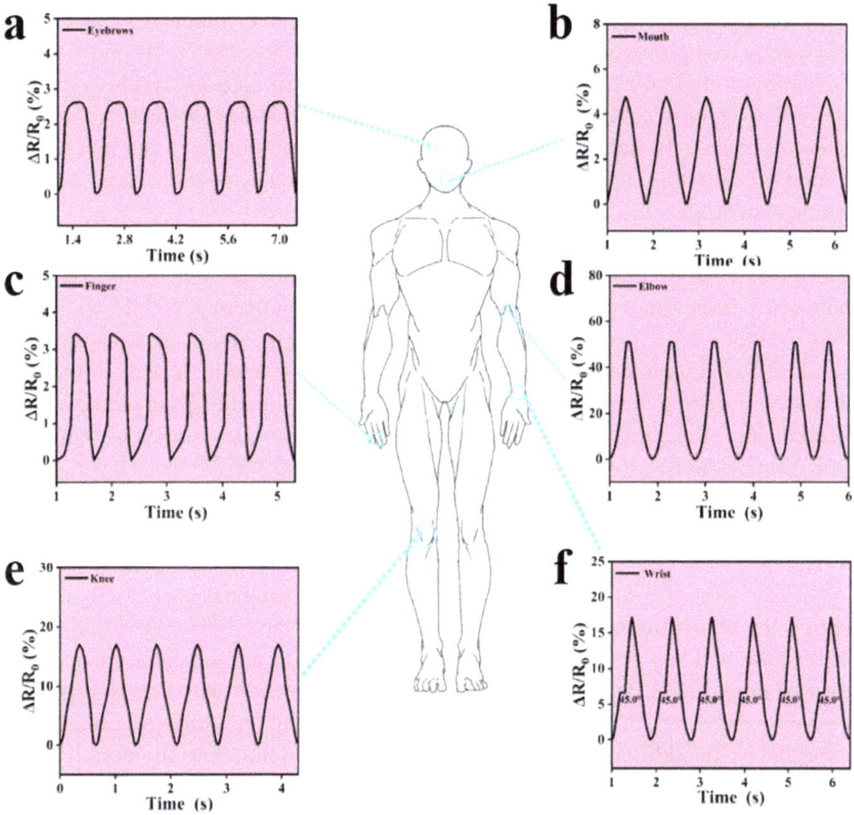

Fig. 8.3 CBH-based sensors for detecting movements of various parts of the human body. **a** Frowning. **b** Opening and closing of the mouth. **c** Finger flexion. **d** Elbow flexion. **e** Knee flexion. **f** Wrist flexion at angles of 45.0° and 90.0°. Reprinted with the permission from Reference [22], Copyright 2024, Elsevier Ltd

that of various biological tissues, from soft skin to robust tendons [25]. This facilitates the formation of non-invasive, stress-free conformal interfaces. In practical applications, such flexible sensors have been mounted on human hand mannequins to monitor finger joint flexion and extension. Flexible sensor arrays, such as AgNWs/bacterial cellulose hydrogel composites, can be attached to joints to capture real-time data on gait, gesture dynamics, and spinal curvature, supporting motion analysis, rehabilitation assessment, and posture anomaly detection. When used as electronic skin, these hydrogels could be integrated into smart textiles or wearable devices (e.g., wristbands) to accurately monitor physiological signals such as pulse waves, respiration patterns, and even subtle facial micro-expressions. These capabilities provide valuable data for cardiovascular health monitoring and mood recognition.

In addition to electrical conductivity, wearable electronics should also have desirable mechanical properties. The equilibrium between ionic conductivity and mechanical strength is often explored, as salt ions easily diffuse into the semi-crystalline/crystalline region of cellulose and weaken the hydrogen bonding interactions between cellulose chains. To balance the constraints between conductivity and mechanical properties, a feasible approach was applied to create rational and robust dual-network cellulose hydrogels [26].

Polyvinyl alcohol (PVA) is an ideal material for the preparation of wearable sensors due to its good safety, high flexibility, transparency and scalability. Recently, there have been some studies on the application of molecular weight-controlled PVA to improve the performance of cellulose hydrogels. Shu et al. investigated a cellulose-based ionic conductive hydrogel (ICHs) for wearable devices in the field of flexible electronics and microelectronics [26]. The formed ICHs could be assembled into multifunctional sensors for human motion and temperature monitoring, and the sensors exhibited long-lasting sensitivity and cycling stability over a wide operating temperature. Their study contributes to the design of high-performance CBHs with good ionic conductivity and mechanical properties under extreme conditions.

The study by Liang et al. proposed a feasible strategy to design a robust and anisotropic wood-based hydrogel through cell wall nanoengineering [27]. By manipulating the microstructure of the wood cell wall, flexible CBHs were obtained by Schiff base bonding through polyacrylamide and cellulose molecular chains. This resulted in excellent flexibility and mechanical properties of wood hydrogels, with tensile strengths of 22.3 and 6.1 MPa in the longitudinal and transverse directions, respectively. In addition, the aqueous salt electrolyte confined to the porous structure yields anisotropic ionic conductivity (longitudinal and transverse directions of 19.5 and 6.02 S/m, respectively). Through cell wall nanoengineering, combined with cellulose scaffolds and polyacrylamide hydrogels, a dual-network wood-based hydrogel was successfully realized. The prepared ionically conductive Na-TDW/PAM-75 hydrogels exhibited highly anisotropic mechanical properties, excellent flexibility, and excellent ionic conductivity in wet state.

The hydrogel strain sensor designed by Jia and co-workers showed high sensing sensitivity and was able to stably monitor a wide range of joint movements and subtle facial expressions [28]. They provided a promising approach for the design of biomass-based multifunctional hydrogels with excellent mechanical properties and energy harvesting and storage capabilities for emerging applications of self-powered flexible sensors.

Zhang et al. proposed a stretchable, water-retaining, antifreeze-resistant, conductive hemicellulose/polypyrrole (H/PPy) composite hydrogel for epidermal strain sensors by freeze–thaw method. Hemicellulose was used as the hydrophilic carrier and PPy as the conductive matrix to prepare composite hydrogels. The introduction of glycerol and sodium chloride decreases the freeze–thaw temperature of the hydrogel and increases the conductivity. After 7 days of storage, the hydrogel had good mechanical properties, remodeling ability and water retention. In addition, the strain sensor assembled from the hydrogel can accurately monitor the human motion signal and has good strain sensing durability [29].

Using these functional nanocomposite hydrogels, Pei et al. designed strain-sensitive, wearable, retractable sensors to monitor finger joint movements, knee bending, and even slight pulse beats [30]. Surprisingly, this flexible sensor could apparently sense the movement of the body, from large movements (knee bending) to tiny signals (pulse beating). In addition, it exhibited excellent durability after repeated cycles. The self-healing nanocomposite hydrogels prepared by this method had potential prospects in the design of biomedicine, biosensors, and flexible electronic devices.

Zhong et al. manipulated the internal structure of the hydrogel by prestretching synergistic ion cross-linking, and constructed a carboxymethyl cellulose (CMC)-based double-network-oriented hydrogel similar to muscle [31]. The introduction of prestretching increased the tensile strength of the dual-network hydrogel from 1.45 to 4.32 MPa, and its light transmittance increased from 67.3% to 84.5%. In addition, the thermal stability and electrical conductivity of hydrogels have also been improved to a certain extent. Its good mechanical properties and electrical conductivity could be converted into stable electrical signal output during the deformation process. A dual-network oriented hydrogel based on CMC was used as a flexible substrate to further assemble a flexible sensor device. Hydrogel sensors could monitor simple joint movements as well as complex spatial movements, which makes them potentially useful in the field of research of smart responsive electronic devices such as flexible wearables, intelligent strain sensing, and soft robotics.

8.5.2 *Conductive Wearable Sensors*

Du et al. reviewed the latest research progress of cellulose-based conductive hydrogels [32]. The applications of cellulose-based conductive hydrogels in multiple fields were introduced. In wearable sensors, they can effectively and realistically monitor human physiological signals. In smart biomedicine, they contribute to wound healing, tissue engineering, and nerve regeneration. In flexible supercapacitors, they offer the potential for green and sustainable energy storage. In recent years, flexible wearable smart sensing materials have considerable prospects, and people are increasingly focusing on renewable hydrogel flexible sensing materials for biomass to meet the growing demand for sustainability and eco-friendliness. Cellulose, as a superior biomass material, has great advantages in terms of use in flexible sensors. This chapter mainly introduces the application of cellulose hydrogels in flexible sensors, and focuses on the properties of cellulose-based hydrogels such as tensile, pH and conductive, introduces the future development trend of cellulose hydrogels, and looks forward to the future development of cellulose hydrogel sensing materials [31].

After connecting the circuit, the LED emits dazzling green light, and the LED was darkened after stretching (Fig. 8.4a), which was attributed to the increase of ion transport resistance during the stretching so that the hydrogel exhibits more excellent resistance during tensile deformation. The assembled hydrogel sensor was continuously bending (Fig. 8.4b) and twisting (Fig. 8.4c), which showed a stable relative resistance change, indicating that the hydrogel strain sensor showed a specific relative resistance change during different deformation processes, which further indicates the application potential of CPDA-P hydrogel with $\lambda = 4$ in strain transducers. Hydrogel output regular relative resistance change curve (Fig. 8.4d) during 150 times load-unload processes, indicating that the hydrogel sensor shows good fatigue resistance [31].

Gao et al. introduced a novel regulatory system based on natural ice structural proteins (ISPs) and CNFs into a hydrogel network composed of a chemical cross-linking network and a physically cross-linked polyvinyl alcohol chain copolymerized with acrylamide and 2-acrylamide-2-methylpropanesulfonic acid, and obtained a freeze-resistant hydrogel with high conductivity (2.63S/m). These hydrogels exhibit excellent adhesion to a variety of substrates, including aluminum, glass, porcine skin, and plastic. With the increase of CNF content, the mechanical properties of the composites were significantly improved (tensile strength was 106.4 kPa and elastic modulus was 133.8 kPa). In addition, the ISP inhibits the growth of ice. This imparts frost resistance to hydrogels and allows them to maintain satisfactory mechanical, conductive, and sensing properties at sub-zero temperatures [33].

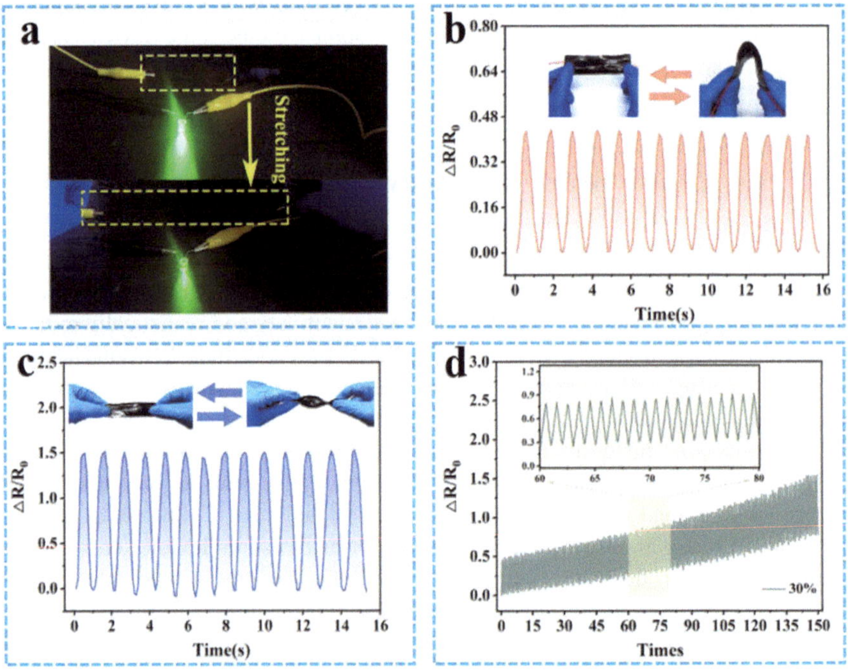

Fig. 8.4 **a** Photographs of LED brightness changes after the hydrogel was connected to the circuit and stretched; Changes in the relative resistance under tensile-release. **b** bending. **c** and twisting. **d** Schematic diagram of the strain sensor assembly; Changes in relative resistance during 150 cycles. Reprinted with the permission from Reference [31], Copyright 2023, Elsevier Ltd.

Additionally, Ix-Cy-PAA/PVA hydrogel was used as a conductor, and the strain-dependent resistance changes were observed by stretching or compressing the hydrogel. When the hydrogel was stretched, the resistance increased due to lengthening of the ion transport path, and the brightness of the bulb in the circuit decreased. Contrarily, the bulb in the circuit became bright again when the hydrogel was compressed due to shortening of the ion transfer path, indicating reduced reactance. The above results suggested that Ix-Cy-PAA/PVA hydrogel has potential application as a multi-functional sensor [33].

Zhu et al. improved the microstructure and mechanical properties of CBHs by dissolving cellulose in an $AlCl_3/ZnCl_2$ complex solution and inducing interactions between cellulose molecules through the regeneration process. Cellulose undergoes a process of dissolution and regeneration, from the original cellulose I to cellulose II. The composition of regenerated solvent had a significant effect on the crystallinity of regenerated cellulose, and the crystallinity of water regenerated cellulose (61.42%) was higher than that of absolute ethanol regenerated cellulose (40.17%). By adjusting the mass ratio of ethanol/aqueous solution, the tensile strength and elongation at break of cellulose hydrogels can reach 2.38 MPa and 196%, respectively. In addition, regenerated cellulose hydrogels exhibited excellent electrical conductivity, making a promising material in flexible sensing devices [34].

The preparation process of the cellulose hydrogel is simple, environmentally friendly, and relatively cheap. Therefore, the cellulose hydrogel prepared in this work has great potential in manufacturing cheap, green, and efficient personal wearable monitoring devices. The battery-powered closed-loop circuits of the LED indicator was constructed using ICH and RCHs as a conductor, respectively (Fig. 8.5a). The RCH-60/40 was attached to the finger to assemble a sensor, and the bending motion of the finger was recorded as a fluctuating current signal (Fig. 8.5b). Similarly, placing a weight above the sensor can sense the pressure of the code. The strain sensitivity can satisfactorily meet the requirements of personal wearable monitoring equipment (Fig. 8.5c) [34].

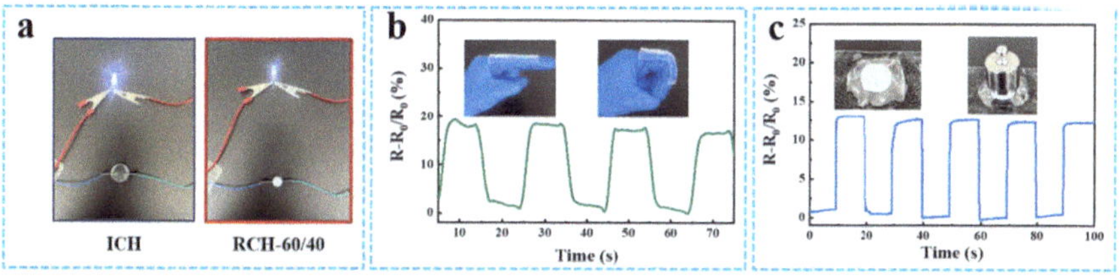

Fig. 8.5 **a** ICH and RCH-60/40 as circuit conductors. **b** The relative resistance variation of the sensor under the bending and loosening process. **c** The relative resistance variation of the sensor under the loading and unloading heavy objects process. Reprinted with the permission from Reference [34], Copyright 2024, Elsevier Ltd.

8.6 Conclusions

As a new generation of flexible sensing materials, CBHs exhibit great potential in wearable device applications due to their unique structural characteristics and outstanding performance. In practical settings, CBHs have been successfully integrated into wearable technologies such as electronic skins and smart textiles, enabling real-time monitoring of physiological parameters including joint motion, body temperature, and pulse. They have also demonstrated excellent performance in biochemical sensing, particularly in the analysis of sweat components. Notably, CBH-based materials offer significant value in terms of sustainability. Their raw materials are primarily derived from agricultural waste, and the preparation processes adhere to the principles of green chemistry. This ensures environmental sustainability across the entire life cycle of the material, aligning closely with the current goals of green manufacturing and sustainable development.

Looking ahead, the deep integration of CBHs with Internet of Things (IoT) technology is poised to drive the innovative development of intelligent medical monitoring systems. CBHs also hold significant promise in emerging fields such as human–computer interaction and virtual reality. With continued advancements in scalable fabrication techniques and breakthroughs in multifunctional integration, CBHs are expected to offer intelligent, eco-friendly material solutions for next-generation flexible electronic devices. These developments will promote the evolution of wearable technologies toward greater comfort, precision, and sustainability. As a class of smart materials derived from natural sources, CBHs are redefining the boundaries of human–machine interfaces and catalyzing transformative innovations in medical health monitoring and smart wearable systems.

References

1. H. Shi, H. Huo, H. Yang, H. Li, J. Shen, J. Wan et al., Cellulose-based dual-network conductive hydrogel with exceptional adhesion. Adv. Funct. Mater. **34**, 2408560 (2024)
2. H. Zhang, X. Gan, Z. Song, J. Zhou, Amphoteric cellulose-based double-network hydrogel electrolyte toward ultra-stable zn anode. Angew. Chem. Int. Ed. **62**, e202217833 (2023)
3. J.-X. Ren, J.-L. Zhu, S.-C. Shi, M.-Q. Yin, H.-D. Huang, Z.-M. Li, In-situ structuring a robust cellulose hydrogel with zno/sio2 heterojunctions for efficient photocatalytic degradation. Carbohydr. Polym. **296**, 119957 (2022)
4. S. Hu, J. Han, Z. Shi, K. Chen, N. Xu, Y. Wang et al., Biodegradable, super-strong, and conductive cellulose macrofibers for fabric-based triboelectric nanogenerator. Nano-Micro Lett. **14**, 115 (2022)
5. S. Choi, J. Jo, J. Park, S. Kim, S. Jeong, S.Y. Jeong et al., Stimuli-responsive, methyl cellulose-based, interpenetrating network hydrogels: Non-covalent design, injectability, and controlled release. Carbohydr. Polym. **347**, 122689 (2024)
6. Y. Wang, H. Liu, J. Yu, H. Liao, L. Yang, E. Ren et al., Ionic conductive cellulose-based hydrogels with superior long-lasting moisture and antifreezing features for flexible strain sensor applications. Biomacromol **25**, 838–852 (2024)

7. S. Yi, X. Lv, X. Zhang, J. Chen, D. Fang, H. Jia et al., Multifunctional composite hydrogels based on hydroxypropyl cellulose for catalytic degradation of organic pollutants and high-performance wearable sensors. Int. J. Biol. Macromol. **307**, 141689 (2025)
8. D. Chen, X. Zhao, X. Wei, J. Zhang, D. Wang, H. Lu et al., Ultrastretchable, tough, antifreezing, and conductive cellulose hydrogel for wearable strain sensor. ACS Appl. Mater. Interfaces **12**, 53247–53256 (2020)
9. K. Yu, L. Yang, S. Zhang, N. Zhang, H. Liu, Stretchable, tough, self-healing, antifreezing, and multifunctional nanocellulose-based hydrogel for wearable monitoring of human motion. MSEB. **317**, 118175 (2025)
10. H.M. Zahid, X. Chuanghua, L. Lei, K.Z.M.A. Motaleb, A. M. Foysal, T. Sirui et al., Facile fabrication of robust, recyclable, and multifunctional composite hydrogel sensor with MXene and cellulose microcrystals. Ind. Crops Prod. **231**, 121224 (2025)
11. L. Zhao, B. Luo, S. Gao, Y. Liu, C. Lai, D. Zhang et al., Stretchable, self-adhesive, and conductive hemicellulose-based hydrogels as wearable strain sensors. Int. J. Biol. Macromol. **282**, 137313 (2024)
12. J. Gao, X. Li, L. Xu, M. Yan, H. Bi, Q. Wang, Transparent multifunctional cellulose-based conductive hydrogel for wearable strain sensors and arrays. Carbohydr. Polym. **329**, 121784 (2024)
13. J. Wei, C. Liu, L. Shi, Y. Liu, H. Lu, High-performance conductive double-network hydrogel base on sodium carboxymethyl cellulose for multifunctional wearable sensors. Carbohydr. Polym. **350**, 122943 (2024)
14. J. Zhuang, X. Zhang, W. Jin, F. Mei, Y. Xu, L. He et al., Conductive, self-healing and adhesive cellulose nanofibers-based hydrogels as wearable strain sensors and supercapacitors. Ind. Crops Prod. **225**, 120547 (2025)
15. W. Jiang, Y. Ma, Q. Wang, T. Zhu, Y. Gao, G. Gao et al., Biocompatible and 3d-printable conductive hydrogels driven by sodium carboxymethyl cellulose for wearable strain sensors. Polymer **395**, 126763 (2024)
16. J.T. Carvalho, I. Cunha, J. Coelho, E. Fortunato, R. Martins, L. Pereira, Carbon-yarn-based supercapacitors with in situ regenerated cellulose hydrogel for sustainable wearable electronics. ACS Appl. Energy Mater. **5**, 11987–11996 (2022)
17. Y. Zhou, G. Huang, Z. Zhang, L. Qin, H. Cai, J. Sha, Flexible, stretchable, and antifreezing triboelectric nanogenerators based on cellulose hydrogel for energy harvesting and wearable electronics. ACS Appl Nano Mater. **7**, 20866–20876 (2024)
18. F. Han, S. Chen, F. Wang, M. Liu, J. Li, H. Liu et al., High-conductivity, self-healing, and adhesive ionic hydrogels for health monitoring and human-machine interactions under extreme cold conditions. Adv. Sci. **12**, 2412726 (2025)
19. H. Shi, F. Kuang, H. Huo, Y. Gao, X. Duan, J. Shen et al., Temperature-responsive cellulose-based janus hydrogel as underwater electronic skin. Nano Lett. **25**, 7835–7844 (2025)
20. X. Wang, M. Geng, X. Pan, Y. Wang, T. Zhan, Y. Liu et al., Skin-adherent, cellulose-based photonic patch for visual strain mapping. ACS Mater. Lett. **7**, 854–861 (2025)
21. C. Yang, J. Liu, P. Liu, W. Wang, H. Chen, L. Bai et al., Phytic acid extracted cellulose nanocrystals for designing self-healing and anti-freezing hydrogels' flexible sensor. Chem. Eng. J. **493**, 152276 (2024)
22. S. Swingler, A. Gupta, H. Gibson, M. Kowalczuk, W. Heaselgrave, I. Radecka, Recent advances and applications of bacterial cellulose in biomedicine. Polymers-Basel. **13**, 412 (2021)
23. N. Wang, X. Yang, X. Zhang, Ultrarobust subzero healable materials enabled by polyphenol nano-assemblies. Nat. Commun. **14**, 814 (2023)
24. T.Q. Trung, N.E. Lee, Flexible and stretchable physical sensor integrated platforms for wearable human-activity monitoringand personal healthcare. Adv. Mater. **28**, 4338–4372 (2016)
25. T. Yang, S.-H. Lu, H. Zhu, A. Patetsos, E. McDonald, M.D. Mellor et al., Tough and elastic cellulose composite hydrogels/films for flexible wearable sensors. ACS Appl. Mater. Interfaces **16**, 40018–40029 (2024)
26. L. Shu, X.-F. Zhang, Y. Wu, Z. Wang, J. Yao, Facile fabrication of strong and conductive cellulose hydrogels with wide temperature tolerance for flexible sensors. Int. J. Biol. Macromol. **240**, 124438 (2023)

27. S. Liang, Q. Ji, R. Wang, G. Hu, W. Li, L. He et al., Wood cell wall nanoengineering toward anisotropic, strong, and flexible cellulosic hydrogel sensors. Nano Lett. **24**, 12741–12749 (2024)
28. L. Jia, L. Zhang, Z. Cai, F. Zhu, B. Huang, Q. Lu, Ultra-flexible anti-freezing cellulose conductive hydrogel for energy storage and self-powered sensors. Ind. Crops Prod. **232**, 121222 (2025)
29. W. Zhang, J.-Y. Wen, M.-G. Ma, M.-F. Li, F. Peng, J. Bian, Anti-freezing, water-retaining, conductive, and strain-sensitive hemicellulose/polypyrrole composite hydrogels for flexible sensors. J Mater. Res. Technol. **14**, 555–566 (2021)
30. Z. Pei, Z. Yu, M. Li, L. Bai, W. Wang, H. Chen et al., Self-healing and toughness cellulose nanocrystals nanocomposite hydrogels for strain-sensitive wearable flexible sensor. Int. J. Biol. Macromol. **179**, 324–332 (2021)
31. L. Zhong, Y. Zhang, F. Liu, L. Wang, Q. Feng, C. Chen et al., Muscle-inspired anisotropic carboxymethyl cellulose-based double-network conductive hydrogels for flexible strain sensors. Int. J. Biol. Macromol. **248**, 125973 (2023)
32. Z. Du, N. Wang, J. Du, Recent progress in cellulose-based conductive hydrogels. Polymers-Basel. **17**, 1089 (2025)
33. X. Gao, J. Wu, Y. Wang, Y. Wang, Y. Zhang, T.T. Nguyen et al., Anti-freezing hydrogel regulated by ice-structuring proteins/cellulose nanofibers system as flexible sensor for winter sports. Int. J. Biol. Macromol. **265**, 131118 (2024)
34. Y. Zhu, S. Song, Y. Yang, R. Zhang, D. Xie, C. Zhang et al., Regenerated cellulose hydrogel with excellent mechanical properties for flexible sensors. Ind. Crops Prod. **210**, 118026 (2024)

Chapter 9
Cellulose-Based Hydrogels for Sustainable Environmental Science

Yu Lei, Xiaoxuan Fan, Qi Sun, and Lei Guo

Abstract Cellulose-based hydrogels (CBHs) are natural hydrogels with excellent biocompatibility, degradability, and environmental friendliness, making them widely applied in environmental science. Through modification and optimization, CBHs exhibit enhanced adsorption capacities and efficient removal of pollutants, particularly in removing harmful substances such as pharmaceuticals, dyes, and heavy metals from water, as well as their use in soil remediation and environmental sensor design. This chapter focuses on the preparation methods and modification strategies of CBHs, exploring their application prospects in environmental protection, particularly in water pollution control and soil remediation. Additionally, a comparative analysis of the sustainability of CBHs with gelatin-based hydrogels and polyurethane hydrogels is presented, evaluating the performance and application value of different materials in environmental science. This chapter provides essential theoretical support and practical guidance for the development of efficient and sustainable environmental remediation materials.

Keywords Cellulose-based hydrogels · Water pollution · Soil remediation · Environmental sensors · Sustainability

9.1 Introduction

Cellulose is a natural linear polysaccharide composed of D-glucose units connected by $\beta(1 \rightarrow 4)$ glycosidic bonds, with reactive hydroxyl groups present at the C2, C3, and C6 positions. The unique molecular structure of cellulose imparts significant reactivity, enabling it to interact with various materials through a range of chemical modifications, promoting the formation of hydrogels [1, 2]. As one of the

Y. Lei · X. Fan · Q. Sun · L. Guo (✉)
Institute of Biomedical Engineering, College of Life Science, Qingdao University, Qingdao, PR China
e-mail: guoleibio@qdu.edu.cn

G. Wei and J. Zhang (eds.), *Cellulose-Based Hydrogels and Aerogels*, Lecture Notes in Nanoscale Science and Technology 41,
https://doi.org/10.1007/978-3-032-03008-5_9

most abundant natural polymers globally, cellulose's distinctive chemical and physical properties have shown great application potential in the field of environmental science. The inherent characteristics of cellulose, including its abundance, renewability, biocompatibility, and biodegradability, make it an ideal material for hydrogel preparation, offering remarkable environmental friendliness.

Compared to other bio-based materials, cellulose exhibits significant advantages in hydrogel development, particularly in environmental protection fields, such as water pollution control and soil remediation, where it has received increasing attention. The reactive hydroxyl groups in cellulose molecules are crucial for hydrogel synthesis, as they allow the hydrogel's structure and properties to be adjusted through processes such as crosslinking and functionalization. Through these chemical modifications, the water absorption capacity, mechanical strength, degradation rate, and responsiveness to external environments of hydrogels can be precisely controlled, endowing cellulose-based hydrogels (CBHs) great potential in various fields [3, 4].

In environmental science, CBHs have shown excellent capabilities in removing drugs, dyes, and heavy metal ions from water, providing an efficient and sustainable method for water pollution control. Moreover, CBHs can also be applied in soil remediation, improving soil quality by adsorbing and immobilizing pollutants, thus reducing the spread of harmful substances. In recent years, increasing research has focused on the application of CBHs in environmental sensors, which can effectively monitor changes in water quality, offering real-time feedback for water pollution control [5–7]. The sustainability of cellulose is reflected not only in its renewability and biodegradability but also in its low environmental load. Compared to traditional synthetic hydrogel materials, CBHs have lower production energy consumption and a smaller environmental impact, with degradation products that are harmless to ecosystems.

Therefore, CBHs are considered a green, environmentally friendly material, aligning with today's sustainable development goals. With advancements in technology and ongoing research, the application prospects of CBHs in environmental science are increasingly promising, particularly in water resource management and soil remediation.

9.2 Preparation and Modification of CBHs

CBHs, due to their excellent biodegradability and renewability, are widely used in fields such as environmental protection, drug delivery, tissue engineering, and water treatment. The main preparation methods include physical crosslinking, chemical crosslinking, and radiation crosslinking [8]. The physical crosslinking method forms a three-dimensional (3D) network through non-covalent interactions, such as hydrogen bonds, van der Waals forces, or charge interactions, with a relatively low environmental burden and a simple preparation process. Common physical crosslinking methods include freeze–thaw, thermal gelation, and ionic crosslinking. Among these, the freeze–thaw method promotes intermolecular crosslinking through repeated freezing and thawing, providing good biocompatibility and tunability [9].

Chen et al. proposed a cellulose-based ionic thermoelectric hydrogel, which achieved excellent thermoelectric performance through physical cross-linking [10], as shown in Fig. 9.1a. By the sub-nanoscale confinement between the oriented cellulose molecular chains and conductive ions, the selective migration of ions under thermal gradients was successfully regulated, significantly enhancing the thermoelectric potential. The chemical cross-linking method, on the other hand, involved the reaction between crosslinking agents (such as glutaraldehyde, isocyanates, etc.) and the hydroxyl groups on the cellulose molecular chains to form stable covalent bonds, significantly enhancing the mechanical strength and thermal stability of the hydrogel. This method is suitable for applications requiring high stability and durability. As shown in Fig. 9.1b, Lu and co-workers introduced catechol functional groups through chemical crosslinking and combined them with dialdehyde cellulose, significantly enhancing the tissue adhesion, sensing capabilities, and hemostatic properties of the hydrogel [11]. Through chemical cross-linking, a stable 3D network structure was formed, which not only improved the mechanical properties of the hydrogel but also endowed it with excellent biocompatibility and antibacterial properties, making it potentially valuable for wound healing applications. However, the choice and concentration of cross-linking agents have a significant impact on the biocompatibility of the hydrogel.

Excessive use of certain cross-linking agents can lead to potential toxicity, especially limiting the use of CBHs in biomedical applications. Therefore, researchers focus on finding low-toxicity and environmentally friendly agents and optimizing cross-linking conditions to ensure performance and safety. Radiation cross-linking, on the other hand, involves irradiating cellulose with γ-rays, electron beams, or ultraviolet light to induce free radical reactions for crosslinking, eliminating the need for chemical crosslinking agents and avoiding potential toxicity issues. Hydrogels prepared by this method exhibit excellent mechanical properties and transparency, making them especially suitable for 3D printing technology. However, the effectiveness of this method relies on precisely controlling the radiation source, dose, and exposure time. As shown in Fig. 9.1c, Lai et al. used radiation cross-linking to prepare a nanocomposite hydrogel based on cationic cellulose nanocrystals (CNCs), successfully achieving a 3D printable capacitive sensor [12]. Through radiation cross-linking, a robust cross-linked network structure was formed inside the hydrogel, which not only enhanced its mechanical strength but also maintained excellent transparency. This method effectively improved the electrical properties of the hydrogel, making it highly applicable in the sensor field, particularly in scenarios that require high transparency and mechanical strength. Different cross-linking methods directly influence the network structure of the hydrogel, thus determining its mechanical strength, elasticity, and response to environmental stimuli. Selecting the appropriate crosslinking method is crucial to ensuring the performance and applicability of hydrogels in various applications. With the increasing demand for environmental protection and biocompatibility, developing efficient and eco-friendly fabrication technologies will be a key focus of future research.

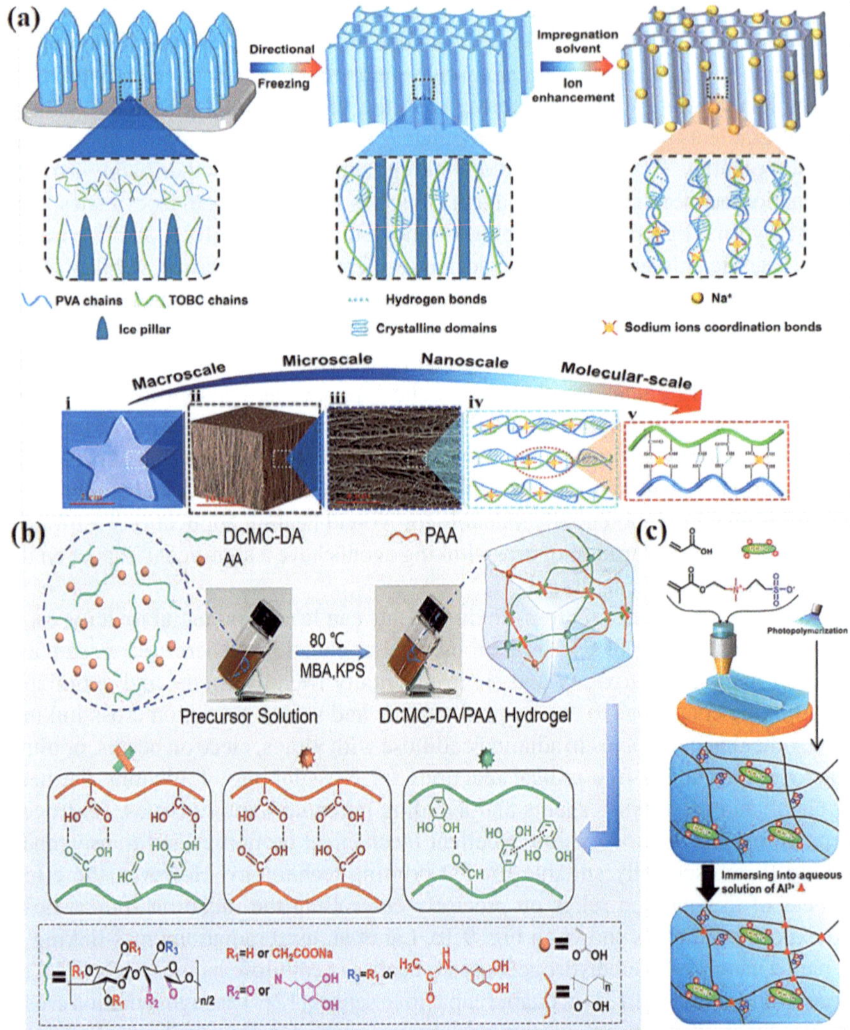

Fig. 9.1 Preparation of CBHs: **a** Preparation of CBHs via physical cross-linking. Reproduced with the permission from Reference [10]. Copyright 2024, Elsevier. **b** Preparation of CBHs via chemical cross-linking. Reproduced with the permission from Reference [11]. Copyright 2024, Elsevier. **c** Preparation of CBHs via radiation cross-linking. Reproduced with the permission from Reference [12]. Copyright 2021, MDPI

The modification methods of CBHs include chemical, physical, and biological modifications, aimed at enhancing their performance and applications [13]. Chemical modification introduces different functional groups through methods such as etherification, esterification, and amination, adjusting hydrophilicity, hydrophobicity, and reactivity. Etherification improves solubility and thermal stability; esterification imparts antimicrobial properties and hydrophobicity; amination enhances adsorption properties and biocompatibility, making it suitable for drug delivery and tissue engineering. Physical modifications, such as heat treatment and ultraviolet radiation, improve molecular alignment and cross-linking reactions, enhancing mechanical properties and thermal stability. Nanofiller modification improves mechanical strength and conductivity, expanding its applications in sensors and energy storage. Biological modification, using natural polysaccharides, proteins, or biological enzymes, improves biodegradability and compatibility, enhancing the effectiveness of CBHs in tissue engineering. The diverse modification methods of CBHs promotes their widespread applications in various fields.

9.3 Applications of CBHs in Environmental Science

9.3.1 Removal of Medicine from Water

Driven by industrialization and urbanization, water pollution has become a critical environmental issue that needs urgent resolution worldwide. Over the past two decades, pharmaceutical pollution has increasingly highlighted its harmful effects on the water environment. With characteristics such as persistence, strong bioaccumulation, and limited biodegradability, pharmaceutical pollutants pose a serious threat to water resource security, gradually evolving into a new challenge for water environmental management [14]. Hydrogels are three-dimensional colloidal network structures formed by the interweaving of hydrophilic polymer chains, which possess excellent water absorption and retention properties. Due to their large specific surface area, ideal polymer-water interface characteristics, and the ability to flexibly adjust their microstructure, hydrogels exhibit unique advantages and application potential in wastewater purification [15].

Cellulose is commonly used as a cross-linking agent in hydrogels for aqueous phase remediation [16]. Tie et al. constructed a bio-based nanocellulose/polyacrylic acid (PAA/NC) composite hydrogel adsorption system with a 3D hierarchical pore structure by precisely controlling the geometry (aspect ratio), interface properties (surface charge density), and crystal structure (crystallinity) of nanocellulose (NC) [17]. Figure 9.2a shows the process of preparing PAA/NC hydrogels via free radical polymerization of acrylic acid (AA) and interfacial hydrogen bonding crosslinking between NC and PAA. The self-assembled PAA chains form a homogeneous network with uniformly distributed carboxyl side chains. The carboxyl groups (-COOH) of PAA and the hydroxyl groups (-OH) of NC form hydrogen bonds and physical entanglements, building a multi-level crosslinking structure. When NC with different morphologies is added, cellulose nanofibers (CNFs) with a high aspect ratio (TCNF) can maintain a layered network. The transformation of hydroxymethyl groups into carboxyl groups enhanced the surface negative charge density of the material. Shorter and stiffer CNCs acted as reinforcing fillers, while longer and more flexible CNFs enhanced the network density. In this system, the composite hydrogel (PAA/TCNF) formed by TEMPO-oxidized TCNF and PAA increased the adsorption capacity for diclofenac to 559.8 mg·g^{-1}, which was 2.1 times higher than the original PAA hydrogel (293.5 mg·g^{-1}). The 3D porous structure, high specific surface area, and pore size regulation of the PAA/NC hydrogel efficiently captured pollutants, making it suitable for pharmaceutical wastewater treatment and drinking water purification.

9.3.2 Dye Removal from Water

The wastewater containing dyes discharged from the papermaking, printing, and textile dyeing industries can cause severe water pollution. Due to their high stability, dyes tend to persist in aquatic environments for long periods, disrupting the ecological balance (e.g., inhibiting phytoplankton photosynthesis). They can also accumulate through the food chain, leading to chronic health risks in humans, such as liver and kidney damage and cancer risk [18].

Natural polymers (such as chitosan and cellulose) are commonly used to prepare dye-adsorbing hydrogels due to their low-cost characteristics. However, traditional materials often suffer from insufficient mechanical strength. Tu et al. successfully developed a chitosan/cellulose composite hydrogel through structural modification strategies [19], which exhibited both high dye adsorption capacity (with a 3.2-fold increase in adsorption efficiency) and high mechanical strength (with a fracture stress of 1.2 MPa). As shown in Fig. 9.2b, the composite hydrogel possessed high flexibility (Fig. 9.2b, capable of being tied and twisted without breaking) and high elasticity (Fig. 9.2b1, fully recovering after compression), while also demonstrating efficient adsorption ability for Congo Red dye (Fig. 9.2b2: the solution turns from red to transparent, and the hydrogel changes color). This hydrogel is biodegradable and easily recyclable, and after dye adsorption, it can be quickly collected through physical separation methods (such as filtration, centrifugation, or magnetic field-responsive separation), completely avoiding the consumption of chemical reagents and the generation of solid waste, thus preventing secondary pollution risks.

Carboxymethyl cellulose (CMC), a water-soluble anionic polymer derived from cellulose, has high hydrophilicity, biocompatibility, and the ability to undergo chemical modification, making it an ideal natural base material for preparing hydrogels [20]. Li et al. constructed an oxidation–reduction responsive green hydrogel system by activating the surface of carboxymethyl CNCs (C-CNCs) and dynamically covalently cross-linking them with L-cysteine [21]. This material demonstrated excellent adsorption performance for methylene blue (MB), with a maximum adsorption capacity of 756 mg/g, significantly outperforming other cellulose-based adsorbents reported in the literature. Under the reducing condition of glutathione (GSH), the disulfide bonds (-S–S-) in the hydrogel could be specifically cleaved, enabling the controlled release of MB. Therefore, the enrichment-release cycle of pollutants could be regulated by adjusting the redox conditions. This study provides a new strategy

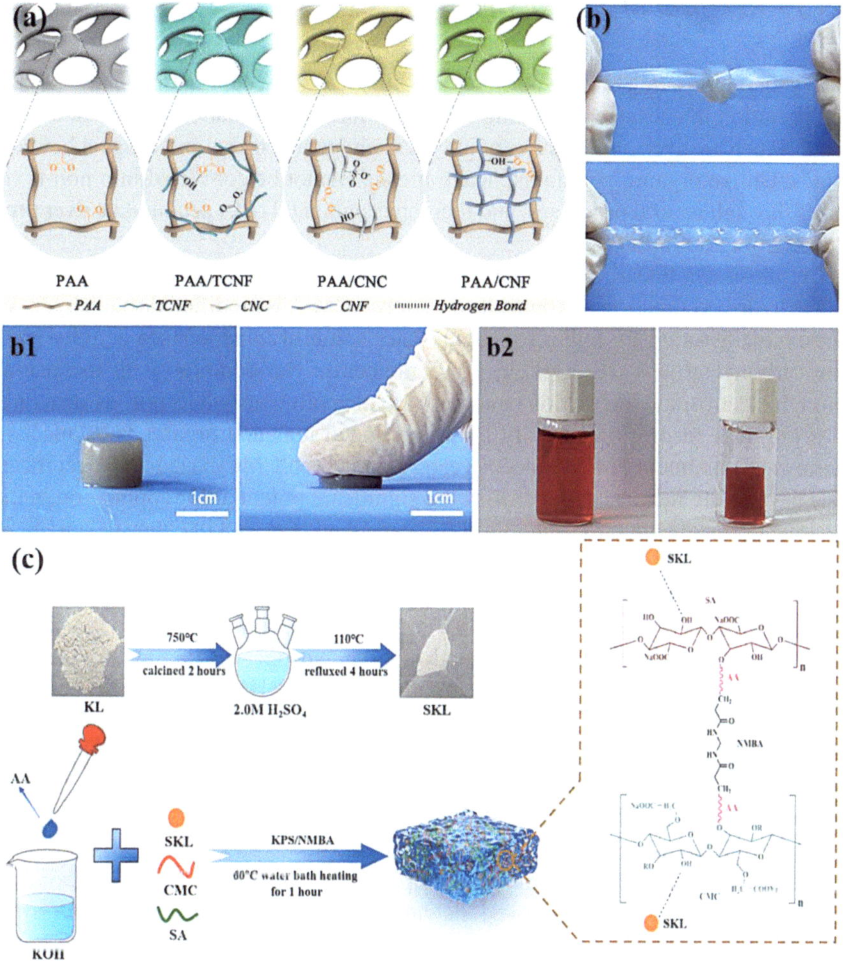

Fig. 9.2 Applications of CBHs in water pollution remediation. **a** Synthesis process of PAA/NC hydrogels. Reproduced with the permission from Reference [17]. Copyright 2022, Elsevier. **b** Images of hydrogels with chitosan contents: cellulose = 3: 1, 5% REC under **b** knotting and twisting: **b1** compression ability and (b2) CR adsorption ability. Reproduced with the permission from Reference [19]. Copyright 2017, Royal Society of Chemistry. **c** Synthesis of SKL and CMC/SA/SKL. Reproduced with the permission from Reference [25]. Copyright 2025, Elsevier

for the application of environmentally responsive hydrogels in intelligent wastewater treatment.

9.3.3 Removal of Heavy Metals

Heavy metal ions, due to their non-degradability and bioaccumulation, cause persistent pollution in ecosystems, posing long-term threats to aquatic organisms, human health (such as liver and kidney toxicity, carcinogenic risks), and biogeochemical cycles. Unlike organic pollutants, heavy metals cannot be converted into non-toxic forms through natural processes, and their environmental risks accumulate over time with ongoing pollution [22].

Cellulose is a linear polysaccharide connected by β-1,4-glucosidic bonds and is rich in hydroxyl functional groups. Its derivative, CMC, is modified through hydroxy carboxymethylation to obtain excellent water solubility and surface activity [23]. The carboxyl groups (-COOH) in CMC molecules can combine with metal ions through electrostatic interaction or adsorb organic pollutants via its porous structure, showing dual adsorption capacity for heavy metal ions and organic substances in wastewater treatment [24]. In recent years, research on CMC-based hydrogel materials has rapidly increased, demonstrating significant potential for application in the treatment of heavy metal-polluted wastewater. Chen et al. sed inexpensive kaolin as a raw material, modified with sulfuric acid, and polymerized it with sodium alginate (SA) and CMC via free radical polymerization to prepare the SA/CMC/SKL composite hydrogel (Fig. 9.2c) [25]. After the optimization by orthogonal experiments, the maximum adsorption capacities of the hydrogel for Pb^{2+}, Cu^{2+}, and MB reached 879.84, 543.50, and 805.16 mg/g, respectively (pH = 5.0, adsorbent amount = 0.01 g), which were significantly higher than most cellulose-based adsorbents reported in the literature. After 5 adsorption–desorption cycles, the hydrogel maintained over 90% of its adsorption efficiency, showing excellent recyclability and stability.

Zeng et al. introduced fish gelatin and bamboo shoot particles (BSP) into CMC hydrogel to construct a high-strength composite hydrogel [26]. Hydrogen bonding interactions significantly improved its mechanical properties (the critical strain reached 214.68%), achieving efficient adsorption of Cd^{2+}, Hg^{2+}, and Pb^{2+} (maximum adsorption capacities of 147.7, 88.62, and 163.89 mg/g, respectively). After 5 cycles of adsorption, the material still maintained over 90% efficiency, demonstrating its potential for low-cost, recyclable heavy metal pollution treatment. The construction of the composite network structure endowed the hydrogel with excellent cycle stability, allowing it to maintain efficient heavy metal removal capability even after multiple adsorption–desorption cycles. This characteristic provides key support for its sustainable application in long-term pollution treatment.

9.3.4 Soil Repair

CBHs have attracted widespread attention in the academic community for their application in soil remediation, particularly in the areas of heavy metal contamination, organic pollutant degradation, and soil structure improvement. Over-fertilization in agriculture and the untreated discharge of industrial wastewater have led to excessive levels of heavy metals and organic pollutants, posing serious threats to the ecological environment and human health [27].

Traditional remediation methods, such as physical, chemical, and biological treatments, although effective, are often associated with high costs, complex operations, and potential issues of secondary pollution [28]. Hydrogels, due to their excellent water absorption and retention properties, can effectively enhance the soil's moisture retention capacity, reduce water evaporation, and thus provide a more stable moisture environment for plant growth.

As a natural polymer material, cellulose has good biocompatibility and biodegradability, which can improve the soil's pore structure, enhance aeration and permeability, and promote the healthy growth and development of plant roots. Furthermore, by chemically modifying cellulose, it can be endowed with the ability to adsorb heavy metal ions and organic pollutants, thus enabling the remediation of polluted soil. These properties make CBHs highly promising for applications in soil remediation and agricultural production [29]. In the remediation of heavy metal pollution, CBHs possess abundant carboxyl (-COOH) and amino (-NH_2) functional groups, which can capture and immobilize heavy metal ions such as Pb^{2+}, Cu^{2+}, and other ions in the soil through electrostatic adsorption, ion exchange, and complexation. This effectively reduces their concentration in the soil and lowers the toxicity to plants [30].

For example, Shaghaleh et al. prepared NC using the TEMPO oxidation method, which was then compounded with lignin and MIL-100(Fe) to form a hydrogel with excellent adsorption properties [31]. This composite hydrogel demonstrated outstanding lead and copper removal capabilities in soil remediation, effectively capturing and immobilizing heavy metal ions, thereby reducing their bioavailability in the soil (Fig. 9.3a). Its environmentally friendly preparation process and efficient remediation effect make it an ideal material for in-situ remediation of soil heavy metal pollution. In another study by Shaghaleh et al. CBHs were prepared by compounding aminated nanocellulose with metal–organic frameworks (MOFs) [32]. These hydrogels exhibited good mechanical strength and hydration properties, and were effective in removing Cr(VI) contaminants from the soil (Fig. 9.3b). Figure 9.3c shows the effects of different soil treatments on wheat growth and heavy metal absorption. The combination of soil washing and the hydrogel composite treatment significantly promoted the elongation of wheat roots and seed germination index, while significantly reducing Cr(VI) concentration. This strategy effectively maintained soil/crop quality and soil health, and exhibited high chemical/environmental stability after remediation, demonstrating its sustainability and feasibility for agricultural soil applications.

Xu et al. developed a CBH, which was compounded with natural zeolite to form a recyclable hydrogel material [33]. This hydrogel effectively removed cadmium ions from the soil, reducing the bioavailability of cadmium and minimizing its harm to the environment and crops. Its composite properties enhanced adsorption performance

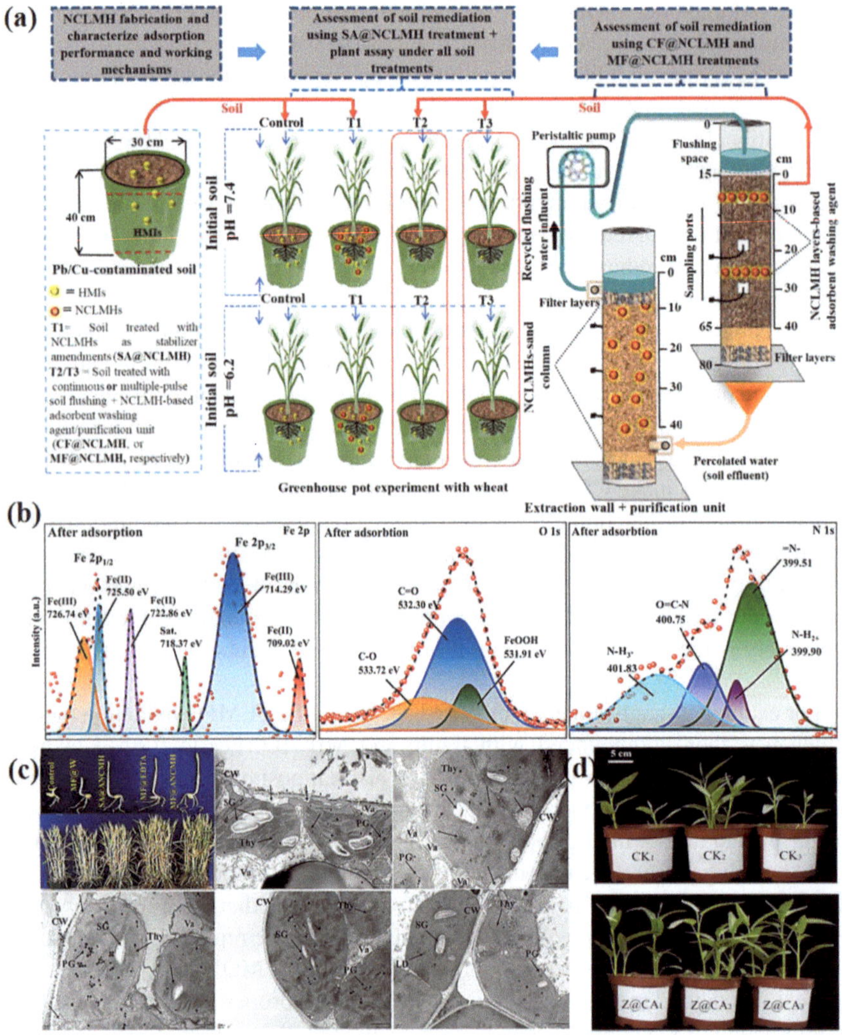

Fig. 9.3 Application of CBHs for soil repair. **a** Illustration of the three soil remediation procedures based on the NCLMH adsorbent concept. Reproduced with the permission from Reference [31]. Copyright 2024, Elsevier. **b** ANCMH composite hydrogel for adsorbing Cr(VI). Reproduced with the permission from Reference [32]. Copyright 2025, Elsevier. **c** Growth and biological performance of wheat plants under different soil treatments. Reproduced with the permission from Reference [32]. Copyright 2025, Elsevier. **d** Digital photos of the water spinaches after planting for 21 days. Reproduced with the permission from Reference [33]. Copyright 2024, Elsevier

and mechanical stability. As shown in Fig. 9.3d, the hydrogel treatment significantly promoted the growth of water spinach, increasing its height, fresh weight, and chlorophyll content, improving its growth in cadmium-contaminated environments, and alleviating the toxic effects of cadmium. The CBH plays an important role in improving soil structure. Its excellent water retention and holding capacity make it highly valuable in arid regions. The swelling properties of the hydrogel could effectively reduce soil moisture evaporation, maintain soil humidity, and promote plant growth. In addition, the cellulose hydrogel can optimize the soil's pore structure, improving soil aeration and permeability, which is beneficial for the healthy development of plant roots [34].

9.3.5 Environmental Sensors

CBHs, due to their excellent biocompatibility, degradability, and tunability, are widely applied in the field of environmental sensors [35]. These hydrogels not only exhibit good water absorption and swelling properties, but can also achieve a sensitive response to environmental changes through chemical modification or the incorporation of conductive materials, making them ideal sensing materials [36]. In the monitoring of gases, temperature, humidity, and heavy metal ions, CBHs demonstrate outstanding detection performance. The abundant functional groups in cellulose molecules enable them to composite with conductive materials, further enhancing their sensitivity and conductivity. Therefore, CBA-based materials with high stretchability and complex environmental sensitivity have become an important driving force for the future development of sensor technologies.

Shi et al. designed a thermo-responsive cellulose-based Janus hydrogel, which is formed by copolymerizing cellulose carbon dots (CDs) with acrylic acid (AA) and Poly (*N*-isopropylacrylamide) (PNIPAM), resulting in a dual-responsive structure [37]. This material not only exhibited excellent strain sensing ability and temperature regulation performance in underwater environments but also maintained stability under fluctuating temperature and humidity conditions (Fig. 9.4a). Its unique Janus structure enabled high sensitivity to gases, particularly showing a more significant response to certain harmful gases, demonstrating its broad potential for applications in environmental monitoring and gas detection.

Hydroxypropyl cellulose (HPC) is a cellulose derivative with anisotropic liquid crystal (LC) properties and thermochromic characteristics. Shi et al. introduced HPC into poly(2-acrylamido-2-methylpropane sulfonic acid) (PAMPS) and formed anisotropic hydrogels through mechanical shear (Fig. 9.4b) [38]. The hydrogel exhibited uniform interference colors under crossed polarizers and demonstrates responsiveness to temperature, pressure, and stretching forces. As a temperature sensor, the HPC/PAMPS hydrogel responded to a wide temperature range without light transmission loss. The HPC aqueous solution is temperature-sensitive, with its long-range order decreasing as the temperature rises, until phase separation occurs at the critical solution temperature (LCST). When HPC was combined with the PAMPS

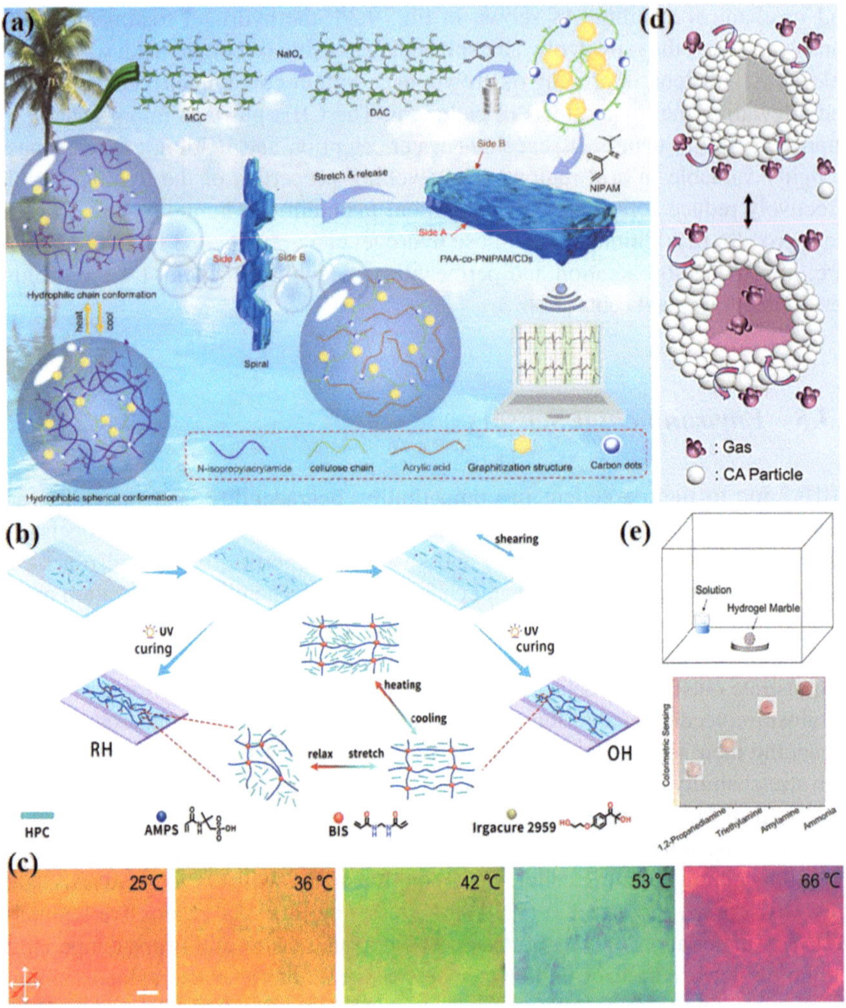

Fig. 9.4 Application of CBHs in environmental sensors. **a** Fabrication of a temperature-responsive cellulose-based Janus hydrogel (PAA-co-PNIPAM/CDs) for underwater electronic skin. Reproduced with the permission from Reference [37]. Copyright 2025, American Chemical Society. **b** Synthesis of HPC/PAMPS hydrogel and **c** POM images of hydrogel samples at different temperatures. Reproduced with the permission from Reference [38]. Copyright 2023, American Chemical Society. **d** CBH particles for gas sensing. and **e** reaction of hydrogel particles with ammonia and organic amine gases. Reproduced with the permission from Reference [39]. Copyright 2023, Elsevier

hydrogel, the thermal response effect was retained. As shown in Fig. 9.4c, the CBA displayed a change in interference colors from orange, yellow, yellow-green, blue, to magenta at different temperatures. This study provides new insights into the application of anisotropic hydrogels and advances the development of multifunctional CBH sensors.

Li et al. developed a CBH marble by dissolving cellulose in a 68% zinc chloride solution and crosslinking it with calcium ions to form a 3D network structure [39]. Its porous shell allowed gases to enter the core, providing excellent gas sensing performance (Fig. 9.4d). When ammonia or organic amine gases (such as 1,2-propanediamine, triethylamine, and pentylamine) were introduced, the hydrogel particles underwent a noticeable color change, especially rapidly turning pink in the presence of ammonia, indicating high sensitivity to ammonia (Fig. 9.4e). Therefore, the hydrogel particles served as a highly sensitive gas sensor, particularly suitable for the detection of harmful gases. This material was not only environmentally friendly and cost-effective but also exhibited excellent stability and reusability, enabling it to function stably under different humidity and temperature conditions, making it suitable for gas monitoring in complex environments.

Mercury ions (Hg^{2+}) are one of the most toxic heavy metals, posing severe threats to both human health and environmental ecosystems. To effectively detect and remove Hg^{2+} from polluted water, Li et al. developed a new cellulose-based fluorescent hydrogel using a proportional probe, which exhibited significant fluorescence changes in the presence of Hg^{2+}, providing a sensitive detection method for harmful substances in water [40]. This CBH not only revealed excellent monitoring performance, but also effectively removed Hg^{2+}, demonstrating its potential applications in environmental protection and sensing technology.

9.4 Sustainability Analysis

To better assess the sustainability of CBHs in the field of environmental science, this paper conducts a multi-factor analysis using the Overall Sustainability Footprint (OSF) method [41]. Based on the three core principles of sustainability, including ecological sustainability, economic sustainability, and social sustainability, eight specific indicators were selected for evaluation, including environmental friendliness, waste utilization, ease of preparation, public acceptance, recyclability, degradability, cost-effectiveness, and application versatility. The performance of each evaluation was rated on a three-level scale: low (L, $i = 1$), medium (M, $i = 2$), and high (H, $i = 3$). It should be noted that for the "cost" factor, the ratings L, M, and H correspond to ranking values $i = 3$, $i = 2$, and $i = 1$, respectively, because in the assessment of material sustainability, cost has a negative impact on its sustainability. Then, the Overall Sustainability Footprint (OSF) is calculated through the sustainable impacts reflected in the following summarized formula:

$$OSF = 100\% \times \sum_{j=1}^{8} \left(\frac{i}{3}\right)_j \times \frac{1}{8}$$

in which j represents the aspect of sustainability, and i represents the corresponding ranking score.

To better understand the advantages of CBHs, a comparison was made with widely used polyurethane (PU) hydrogels and gelatin hydrogels. PU hydrogels are a class of hydrogels formed by cross-linking PU to create a hydrated 3D network structure [42, 43], with adjustable mechanical properties and biocompatibility. They are characterized by excellent elasticity, controllable hydration, and strong durability, making them widely used in drug delivery, wound dressings, and tissue engineering. In environmental science, PU hydrogels, as high-performance materials, can be used in water treatment, pollutant adsorption, and wastewater treatment. Gelatin hydrogels are natural hydrogels formed by cross-linking collagen from animal sources [44, 45], offering good biodegradability and biocompatibility. They are characterized by excellent hydrophilicity, degradability, and high biocompatibility, making them suitable for biomedical and environmental applications. In environmental science, gelatin hydrogels can be used as green materials for pollutant removal, wastewater treatment, and ecological restoration, especially demonstrating significant potential in the development of sustainable materials.

According to the ranking values, the OSF values of CBHs, PU hydrogels, and Gelatin hydrogels are 87.5%, 66.7%, and 79.2%, respectively, as shown in Fig. 9.5. CBHs and gelatin hydrogels are widely accepted by the public due to their natural origin, good environmental friendliness, and biodegradability, and they perform excellently in environmental protection and waste utilization. Cellulose hydrogels have better recyclability, while Gelatin hydrogels, due to their biocompatibility, are primarily applied in the biomedical field. In contrast, the chemically synthesized nature of PU hydrogels leads to a greater environmental burden, lower waste utilization, and poor biodegradability. Nevertheless, their adjustable properties make them widely applicable in environmental management, drug delivery, and tissue engineering. Overall, cellulose hydrogels outperform other hydrogels in environmental performance and public acceptance, Gelatin hydrogels hold an important position in the biomedical field, and PU hydrogels need further improvements in their environmental friendliness to enhance their overall competitiveness.

9.5 Conclusion and Outlooks

The preparation methods of CBHs mainly include physical, chemical, and radiation cross-linking. By adopting different crosslinking techniques, their mechanical properties, biocompatibility, and stability can be adjusted. Furthermore, through chemical, physical, and biological modifications, the application performance of hydrogels in

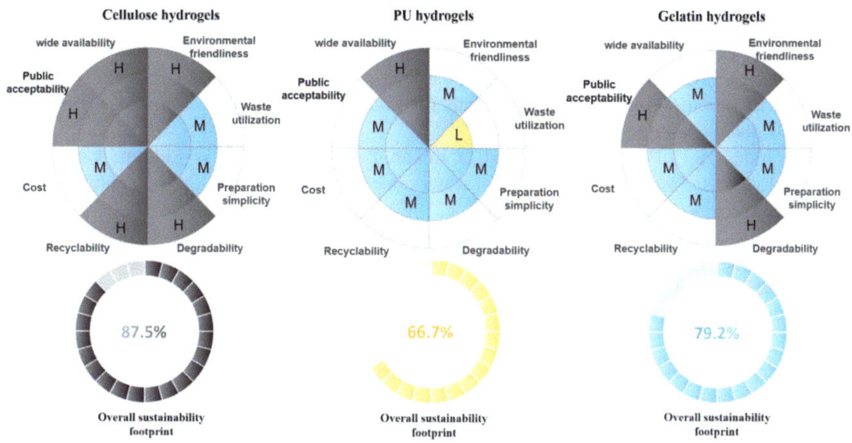

Fig. 9.5 Sustainability analysis of the Cellulose hydrogels by comparing with PU and Gelatin hydrogels

drug release, tissue engineering, and sensors can be further enhanced. Water pollution, including pharmaceuticals, dyes, and heavy metals, has become a global environmental issue. Due to their excellent water absorption properties and adjustable structure, CBHs are widely applied in water purification. In addition to water purification, CBHs also have broad applications in soil remediation and environmental sensors. Through structural regulation and functional modification, they can achieve real-time monitoring and remediation of environmental pollution, providing innovative solutions for environmental protection. As environmental pollution becomes an increasingly severe issue, CBHs are expected to play an increasingly important role in future environmental science.

Future research can be focused on optimizing the synthesis methods of CBHs, exploring more efficient crosslinking techniques and functional modifications to enhance their adsorption capacity and degradability. In addition, the development of intelligent CBHs materials that can respond to environmental changes, self-repair, or actively remove pollutants will be an important direction in the future. By integrating nanotechnology, green chemistry, and the principles of a circular economy, CBHs are expected to provide more precise and sustainable solutions in water resource management, soil remediation, and environmental monitoring. With the advancement of technology, CBHs has a broad application prospect in environmental pollution control, offering significant support for global environmental protection.

References

1. P. Alexandridis, M. Ghasemi, E.P. Furlani, M. Tsianou, Solvent processing of cellulose for effective bioresource utilization. Curr. Opin. Green Sustain. Chem. **14**, 40–52 (2018)

2. D.H. Tran, M. Ulbricht, Cellulose-cellulose composite membranes for ultrafiltration. J. Membr. Sci. **672**, 121426 (2023)
3. K. Sharma, P. Choudhary, A. Majeed, S. Guleria, M. Kumar, A.K. Rana, G. Rajauria, Cellulose based membranes, hydrogels and aerogels for water treatment application. Ind. Crops Prod. **225**, 120474 (2025)
4. A.A. Aslam, S.U. Hassan, M.H. Saeed, O. Kokab, Z. Ali, M.S. Nazir, W. Siddiqi, A.A. Aslam, Cellulose-based adsorbent materials for water remediation: harnessing their potential in heavy metals and dyes removal. J. Clean. Prod. **421**, 138555 (2023)
5. Q.-F. Guan, H.-B. Yang, Z.-M. Han, Z.-C. Ling, C.-H. Yin, K.-P. Yang, Y.-X. Zhao, S.-H. Yu, Sustainable cellulose-nanofiber-based hydrogels. ACS Nano **15**, 7889–7898 (2021)
6. L.C. Wong, C.P. Leh, C.F. Goh, Designing cellulose hydrogels from non-woody biomass. Carbohydr. Polym. **264**, 118036 (2021)
7. D.M. Nascimento, Y.L. Nunes, M.C.B. Figueirêdo, H.M.C. de Azeredo, F.A. Aouada, J.P.A. Feitosa, M.F. Rosa, A. Dufresne, Nanocellulose nanocomposite hydrogels: technological and environmental issues. Green Chem. **20**, 2428–2448 (2018)
8. Y. Tang, Z. Fang, H.-J. Lee, Exploring applications and preparation techniques for cellulose hydrogels: a comprehensive review. Gels **10**, 365 (2024)
9. Z. Peng, F. Chen, Hydroxyethyl cellulose-based hydrogels with various pore sizes prepared by freeze-drying. J. Macromol. Sci. Part B **50**, 340–349 (2010)
10. L. Chen, X. Rong, Z. Liu, Q. Ding, X. Li, Y. Jiang, W. Han, J. Lou, Negative thermopower anisotropic ionic thermoelectric hydrogels based on synergistic coordination and hydration for low-grade heat harvesting. Chem. Eng. J. **481**, 148797 (2024)
11. S. Lu, S. Bian, Y. Jia, Y. Guo, H. Xiao, M. Zhang, K. Liu, L. Huang, L. Chen, Y. Ni, H. Wu, Catechol-functionalised dialdehyde cellulose-containing hydrogels with tissue adhesion, sensing and haemostatic properties for wound healing. Cellulose **31**, 2355–2377 (2024)
12. P.-C. Lai, S.-S. Yu, Cationic cellulose nanocrystals-based nanocomposite hydrogels: achieving 3d printable capacitive sensors with high transparency and mechanical strength. Polymers **13**, 688 (2021)
13. F. Wang, A. Borjas, A. Bonto, A.V. Ursu, M. Dupont, J. Roche, C. Delattre, Exploring novel applications for hydrogels derived from modified celluloses. Polymers **16**, 530 (2024)
14. H.G.R. Younis, H.R.S. Abdellatif, F. Ye, G. Zhao, Tuning the physicochemical properties of apple pectin films by incorporating chitosan/pectin fiber. Int. J. Biol. Macromol. **159**, 213–221 (2020)
15. O. Duman, T.G. Polat, C.Ö. Diker, S. Tunç, Agar/κ-carrageenan composite hydrogel adsorbent for the removal of Methylene Blue from water. Int. J. Biol. Macromol. **160**, 823–835 (2020)
16. J. Shojaeiarani, D. Bajwa, A. Shirzadifar, A review on cellulose nanocrystals as promising biocompounds for the synthesis of nanocomposite hydrogels. Carbohydr. Polym. **216**, 247–259 (2019)
17. L. Tie, Y. Ke, Y. Gong, W.X. Zhang, Z. Deng, Nanocellulose fine-tuned poly(acrylic acid) hydrogel for enhanced diclofenac removal. Int. J. Biol. Macromol. **213**, 1029–1036 (2022)
18. P. Saravanan, J. Josephraj, P. Thillainayagam Bhagavathi, G. Ravindiran, B. Laura, Biochar for removal of dyes in contaminated water: an overview. Biochar **4**, 2–16 (2022)
19. H. Tu, Y. Yu, J. Chen, X. Shi, J. Zhou, H. Deng, Y. Du, Highly cost-effective and high-strength hydrogels as dye adsorbents from natural polymers: chitosan and cellulose. Polym. Chem. **8**, 2913–2921 (2017)
20. K. Enoch, A.A. Somasundaram, Rheological insights on carboxymethyl cellulose hydrogels. Int. J. Biol. Macromol. **253**, 127481 (2023)
21. Y. Li, X. Hou, Y. Pan, L. Wang, H. Xiao, Redox-responsive carboxymethyl cellulose hydrogel for adsorption and controlled release of dye. Eur. Polym. J. **123**, 109447 (2020)
22. M. Zhang, L. Song, H. Jiang, S. Li, Y. Shao, J. Yang, J. Li, Biomass based hydrogel as an adsorbent for the fast removal of heavy metal ions from aqueous solutions†. J. Mater. Chem. A **5**, 3434–3446 (2017)

23. M. Ren, O.A. Fakayode, F. Kong, C. Zhou, L. Chen, X. Fan, J. Liang, H. Li, Characterization of cellulose nanocrystals prepared by different delignification methods and application of ultra-light, hydrophobic aerogels as oil absorbent in food systems. Ind. Crops Prod. **197**, 116653 (2023)
24. Y. Chen, Y. Long, Q. Li, X. Chen, X. Xu, Synthesis of high-performance sodium carboxymethyl cellulose-based adsorbent for effective removal of methylene blue and Pb (II). Int. J. Biol. Macromol. **126**, 107–117 (2018)
25. J. Chen, X. Wang, X. Huang, Z. Tong, J. Zhou, Y. Shen, C. Hao, Construction of cellulose-based hydrogel compounded with modified kaolin and its removal performance for heavy metal ions and dyes in water. Int. J. Biol. Macromol. **306**, 141398 (2025)
26. R. Zeng, J. Zheng, Y. Zuo, C. Xiao, Y. Zhu, Synergistic and simultaneous removal of heavy metal ions over waste bamboo shoot particles encapsulated carboxymethyl cellulose/gelatin composite hydrogel. Int. J. Biol. Macromol. **283**, 137578 (2024)
27. S. Gupta, A. Nighojkar, N. Mayilswamy, B. Kandasubramanian, Recent trends in the application of silk-based composites for remediation of toxic contaminants from wastewater. J. Polym. Environ. **31**, 2243–2272 (2022)
28. T. Zhou, Y. Wang, S. Huang, Y. Zhao, Synthesis composite hydrogels from inorganic-organic hybrids based on leftover rice for environment-friendly controlled-release urea fertilizers. Sci. Total. Environ. **615**, 422–430 (2017)
29. J. Dhiman, S.O. Prasher, E. ElSayed, R.M. Patel, C. Nzediegwu, A. Mawof, Effect of hydrogel based soil amendments on heavy metal uptake by spinach grown with wastewater irrigation. J. Clean. Prod. **311**, 127644 (2021)
30. H. Shaghaleh, X. Xu, S. Wang, Current progress in production of biopolymeric materials based on cellulose, cellulose nanofibers, and cellulose derivatives. RSC Adv. **8**, 825–842 (2018)
31. H. Shaghaleh, Y. Alhaj Hamoud, Q. Sun, Effective and green in-situ remediation strategies based on TEMPO-nanocellulose/lignin/MIL-100(Fe) hydrogel nanocomposite adsorbent for lead and copper in agricultural soils. Environ. Pollut. **360**, 124623 (2024)
32. H. Shaghaleh, Y. Alhaj Hamoud, Q. Sun, M.S. Sheteiwy, H. AbdElgawad, Soil flushing coupled with aminated-nanocellulose/MOF hydrogel nanocomposites adsorbents: a novel sustainable remediation strategy for Cr(VI)-contaminated agricultural soils. Sep. Purif. Technol. **353**, 128440 (2024)
33. H. Xu, Z. Ou, W. Li, T. Hu, Y. Zhang, H. Xu, J. Wang, Y. Li, Cadmium (II) adsorption by recyclable zeolite-loaded hydrogel: extension to the removal of cadmium (II) from contaminated soil. Chem. Eng. J. **492**, 151842 (2024)
34. H. Shaghaleh, Y. Alhaj Hamoud, Q. Sun, Functionalized nanocellulose nanocomposite hydrogels for soil and water pollution prevention, remediation, and monitoring: a critical review on fabrication, application properties, and potential mechanisms. J. Environ. Chem. Eng. **12**, 111892 (2024)
35. N. Al-Shamery, T. Benselfelt, P.S. Lee, Melanin and polypyrrole-coated nanocellulose hydrogel networks for environmental sensing and energy storage. ACS Appl. Mater. Interfaces **15**, 25966–25979 (2023)
36. M. Li, D. Chen, X. Sun, Z. Xu, Y. Yang, Y. Song, F. Jiang, An environmentally tolerant, highly stable, cellulose nanofiber-reinforced, conductive hydrogel multifunctional sensor. Carbohydr. Polym. **284**, 119199 (2022)
37. H. Shi, F. Kuang, H. Huo, Y. Gao, X. Duan, J. Shen, J. Wan, Y. Li, G. Du, L. Yang, Temperature-responsive cellulose-based Janus hydrogel as underwater electronic skin. Nano Lett. **25**, 7835–7844 (2025)
38. H. Shi, Y. Deng, Y. Shi, Cellulose-based stimuli-responsive anisotropic hydrogel for sensor applications. ACS Appl. Nano Mater. **6**, 11524–11530 (2023)
39. N. Li, H. Wanyan, S. Lu, H. Xiao, M. Zhang, K. Liu, X. Li, B. Du, L. Huang, L. Chen, Y. Ni, H. Wu, Robust cellulose-based hydrogel marbles with excellent stability for gas sensing. Carbohydr. Polym. **306**, 120617 (2023)
40. L. Meng, L. Xiaoning, X. Mengwen, L. Bowen, Y. Mengqing, C. Zhijun, G. Tianchu, D.J. Tony, W. Lidong, X. Huining, A ratiometric fluorescent hydrogel of controlled thickness prepared

continuously using microtomy for the detection and removal of Hg(II). Chem. Eng. J. **426**, 131296 (2021)
41. M. Peydayesh, R. Mezzenga, Protein nanofibrils for next generation sustainable water purification. Nat. Commun. **12**, 3248 (2021)
42. A.T. Speidel, P.R.A. Chivers, C.S. Wood, D.A. Roberts, I.P. Correia, A.S. Caravaca, Y.K.V. Chan, C.S. Hansel, J. Heimgärtner, E. Müller, J. Ziesmer, G.A. Sotiriou, P.S. Olofsson, M.M. Stevens, Tailored biocompatible polyurethane-poly(ethylene glycol) hydrogels as a versatile nonfouling biomaterial. Adv. Healthc. Mater. **11**, 2201378 (2022)
43. K. Li, C. Zhou, S. Liu, F. Yao, G. Fu, L. Xu, Preparation of mechanically-tough and thermo-responsive polyurethane-poly(ethylene glycol) hydrogels. React. Funct. Polym. **117**, 81–88 (2017)
44. X. Yuan, Z. Zhu, P. Xia, Z. Wang, X. Zhao, X. Jiang, T. Wang, Q. Gao, J. Xu, D. Shan, B. Guo, Q. Yao, Y. He, Tough gelatin hydrogel for tissue engineering. Adv. Sci. **10**, 2301665 (2023)
45. J. Ren, X. Wang, L. Zhao, M. Li, W. Yang, Effective removal of dyes from aqueous solutions by a gelatin hydrogel. J. Polym. Environ.Polym. Environ. **29**, 3497–3508 (2021)

Chapter 10
Cellulose-Based Bioactive Hydrogels for Tumor Therapy

Rongqiu Mu and Gang Wei

Abstract Bioactive cellulose-based hydrogels (CBHs) have shown significant potential in cancer treatment through various therapeutic strategies. This chapter highlights recent advances in the functionalization of CBHs to impart key physical, chemical, and biological properties, such as injectability, biocompatibility, biocatalytic activity, and targeted recognition, which facilitate their application in tumor chemotherapy, chemodynamic therapy, photothermal therapy, photodynamic therapy, and immunotherapy. Additionally, the underlying mechanisms by which CBHs exert therapeutic effects are discussed to emphasize their advantages in achieving high-performance cancer treatment. This chapter aims to provide readers with a comprehensive understanding of the design and modification strategies used to enhance CBH functionality, thereby broadening their biomedical applications in areas such as tissue engineering, antibacterial coatings, and biosensing.

Keywords Cellulose · Hydrogels · Bioactive materials · Drug delivery · Tumor therapy

10.1 Introduction

Cellulose is the most abundant renewable polymer in nature. Its outstanding biocompatibility, biodegradability, mechanical strength, and chemical modifiability endow it with significant potential in the field of biomedical materials [1, 2]. Cellulose hydrogels are three-dimensional (3D) network structures formed by physical or chemical cross-linking of cellulose or its derivatives, capable of maintaining structural stability in high water-content environments [3, 4]. Based on the structural features

R. Mu · G. Wei (✉)
School of Polymer Science and Engineering, Qingdao University of Science and Technology, Qingdao, China
e-mail: wei-lab@qust.edu.cn

R. Mu
College of Chemistry and Chemical Engineering, Qingdao University, Qingdao, China

G. Wei and J. Zhang (eds.), *Cellulose-Based Hydrogels and Aerogels*, Lecture Notes in Nanoscale Science and Technology 41,
https://doi.org/10.1007/978-3-032-03008-5_10

and sources of cellulose, commonly used forms include bacterial cellulose (BC), cellulose nanofibers (CNFs), and cellulose nanocrystals (CNCs). These structural units impart desirable mechanical properties and functionalization potential to the hydrogel systems [5].

In recent years, increasing attention has been paid to the design of cellulose nanofiber-based bioactive compounds or drug delivery carriers with stimuli-responsive release capabilities. Such systems aim to enhance therapeutic efficacy, improve targeting precision, and reduce side effects in cancer treatment [6, 7]. Bioactive hydrogels function not merely as passive drug carriers but also actively participate in therapeutic processes by incorporating drugs, biologics, immune factors, or inorganic nanoparticles, thereby achieving both therapeutic activity and functional responsiveness [8]. Their unique features, such as injectability, controlled release behavior, and responsiveness to the tumor microenvironment (TME) make them ideal platforms for precision oncology [9–11]. Traditional cancer chemotherapy is often hindered by poor targeting, severe systemic toxicity, and drug resistance [12]. Bioactive hydrogels offer novel strategies to overcome these limitations through localized delivery, smart release control, combination therapy, and immune activation. These features contribute to improved therapeutic efficacy and reduced systemic toxicity, paving the way for personalized and low-side-effect cancer treatments [13–15].

Compared to other natural or synthetic polymers, cellulose-based hydrogels (CBHs) offer several distinctive advantages in cancer therapy. Firstly, their excellent biocompatibility and degradability ensure safer in vivo applications, effectively minimizing immune rejection and toxic accumulation [10]. Secondly, the abundance of hydroxyl groups in the cellulose molecular structure provides convenient sites for chemical modification, allowing for the grafting of various functional groups and the integration of targeting ligands, therapeutic agents, and responsive moieties [5, 15]. Moreover, CBHs possess favorable mechanical properties and rheological behavior, making them suitable for injectable local gelation, especially advantageous for postoperative management and recurrence prevention in solid tumors [16]. In addition, CBHs demonstrate broad application potential in areas such as sustained drug release, immune activation, TME modulation, and multimodal therapy. These advantages underscore their strategic value in the development of precise anticancer treatments [17, 18]. Taken together, CBHs represent a high-performance bioactive platform, offering a robust material foundation and broad application prospects for achieving personalized and intelligent cancer therapy.

This chapter presents recent advance in the synthesis and functionalization of bioactive CBHs, as well as their potential applications in tumor therapy.

10.2 Functionalization Strategies of Bioactive CBHs for Cancer Therapy

Bioactive CBHs for cancer therapy leverage multifunctional strategies, such as the integration of photothermal and photodynamic modalities, to achieve TME-responsive and precisely targeted drug delivery. These hydrogels significantly reduce systemic toxicity and help overcome drug resistance through multimodal synergistic treatment approaches. Their injectability, biocompatibility, and smart design offer promising potential for clinical translation. However, further optimization is required in terms of stability, safety, and large-scale production to fully advance the development of personalized and comprehensive cancer therapies.

10.2.1 Incorporation of Chemotherapeutic Agents

Traditional chemotherapy is often limited in clinical application due to its high toxicity and severe side effects on healthy cells. CBHs, with their rich 3D network structures, offer an effective platform for the localized and sustained release of chemotherapeutic agents. By encapsulating drugs within these hydrogels, drug stability can be enhanced, therapeutic efficacy prolonged, and systemic toxicity significantly reduced [19].

Lim and colleagues developed a highly biocompatible, porous soft hydrogel for the targeted delivery of the chemotherapeutic drug doxorubicin (DOX) by employing an inverse electron demand Diels–Alder (IEDDA) click-cross-linking strategy. This system combined norbornene-functionalized carboxymethyl cellulose (CMC-Nb) with a redox-responsive diselenide-containing crosslinker (DSe-DPEG-DTz) (Fig. 10.1a) [20]. In the TME, the elevated levels of glutathione (GSH) trigger the cleavage of diselenide bonds, thereby enabling targeted and stimuli-responsive drug release. Compared to blank hydrogels, DOX-loaded hydrogels exhibited superior antitumor activity. Under 10 mM GSH conditions, the hydrogel demonstrated a rapid and efficient release of DOX, with 99% of the drug released within 12 h. This study presents a promising strategy for the development of targeted drug delivery systems that respond intelligently to the TME.

Cisplatin is a widely used chemotherapeutic agent. However, its systemic administration is often associated with severe side effects. Localized drug delivery systems offer a promising strategy to minimize off-target toxicity by confining drug action to the tumor site. Therefore, the development of a delivery platform capable of prolonged retention and sustained release of cisplatin at the tumor site is critical to addressing this limitation. Lavasanifar and colleagues designed an adhesive hydrogel composed of CNC grafted with poly (acrylic acid) (PAA), referred to as CNC-g-PAA, for the localized and sustained delivery of cisplatin [21]. By optimizing the grafting ratio between CNC and PAA, the hydrogel exhibited improved adhesiveness,

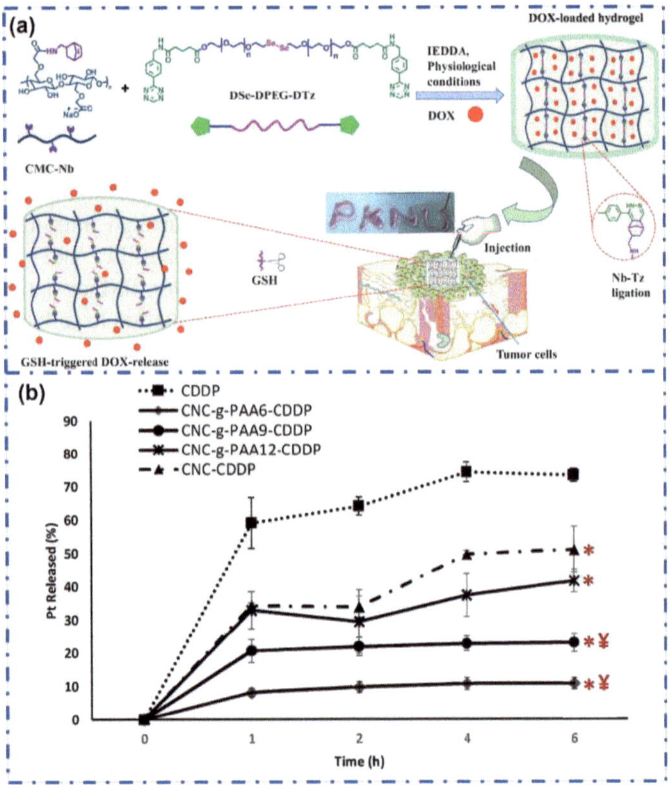

Fig. 10.1 **a** Schematic illustration of the development and drug release mechanism of CMC-based hydrogels. Adapted from Reference [20], Copyright 2022 Elsevier B.V. **b** Release profile of cisplatin from different formulations reported as mean ± SD. Adapted from Reference [21], Copyright 2021 Elsevier B.V

swelling behavior, and drug loading capacity. Among the tested formulations, CNC-g-PAA9 demonstrated the most favorable release profile and drug release control (Fig. 10.1b). In vitro studies confirmed that this system significantly prolonged the release of cisplatin, increased the IC50 value in tumor cells, and showed lower cytotoxicity and better biocompatibility compared to free cisplatin. This CNC-based hydrogel not only achieved long-term retention and sustained release of cisplatin at the tumor site, thereby minimizing systemic toxicity, but also significantly enhanced the overall antitumor efficacy. These findings offer a promising strategy for localized cisplatin delivery and improved cancer treatment outcomes.

10.2.2 Loading of Biomolecules

Biological agents such as small interfering RNA (siRNA), DNA, growth factors, and cytokines are playing increasingly important roles in cancer therapy [22, 23]. These biomolecules function by regulating gene expression, promoting cell growth, or inhibiting tumor proliferation and metastasis, making them essential components in modern oncological treatment. However, their clinical application is often limited by poor stability, low targeting efficiency, and rapid clearance in vivo. Incorporating these agents into CBHs can effectively address these challenges by enhancing their stability, prolonging retention, and enabling controlled, localized delivery.

In cancer treatment, DNA serves not only as a gene regulatory molecule but also as a structural component, targeting moiety, and stimulus-responsive element in hydrogel design. Zheng et al. developed an intelligent nanohydrogel system based on CMC as the matrix and DNA as both a cross-linking and targeting element [24]. This system integrates DNA-templated CdTeSe quantum dots (QDs) and the anti-nucleolin aptamer AS1411, enabling DOX delivery through microRNA (miRNA) responsiveness and GSH-triggered release (Fig. 10.2a). The hydrogel exhibited multifunctionality, combining structural integrity, specific targeting, gene regulation, and imaging capabilities. It also showed excellent performance in controlled drug release and cellular uptake, exemplifying an integrated approach to utilizing DNA-based biologics for CBH functionalization.

Growth factors hold significant therapeutic value in tissue regeneration and tumor-associated injury repair. However, their clinical use is hampered by instability and dose-dependent side effects. To address this, Zhang et al. designed a composite injectable hydrogel system composed of thiolated chitosan (TCS) and CMC for the sustained delivery of transforming growth factor-β1 (TGF-β1) [25]. The hydrogel exhibited high structural stability, good mechanical strength, and sustained release behavior, with only 18.6% of the drug released over 21 d, thereby significantly extending the biological activity of TGF-β1. In vitro studies showed enhanced proliferation of bone marrow mesenchymal stem cells (BMSCs), while in vivo experiments demonstrated effective cartilage defect repair. This study highlights the feasibility and advantages of using CBHs for the delivery of biologics, offering a novel strategy for the development of efficient and safe platforms for tissue regeneration and adjuvant cancer therapy.

10.2.3 Integration of Immunotherapeutic Agents

With the rapid advancement of cancer immunotherapy, agents such as immune checkpoint inhibitors (PD-1/PD-L1 antibodies) and chimeric antigen receptor T (CAR-T) cells have demonstrated remarkable efficacy in various malignancies [26, 27]. However, their clinical outcomes are often hampered by rapid degradation, short in vivo retention, and significant systemic toxicity, which limit the sustainability and

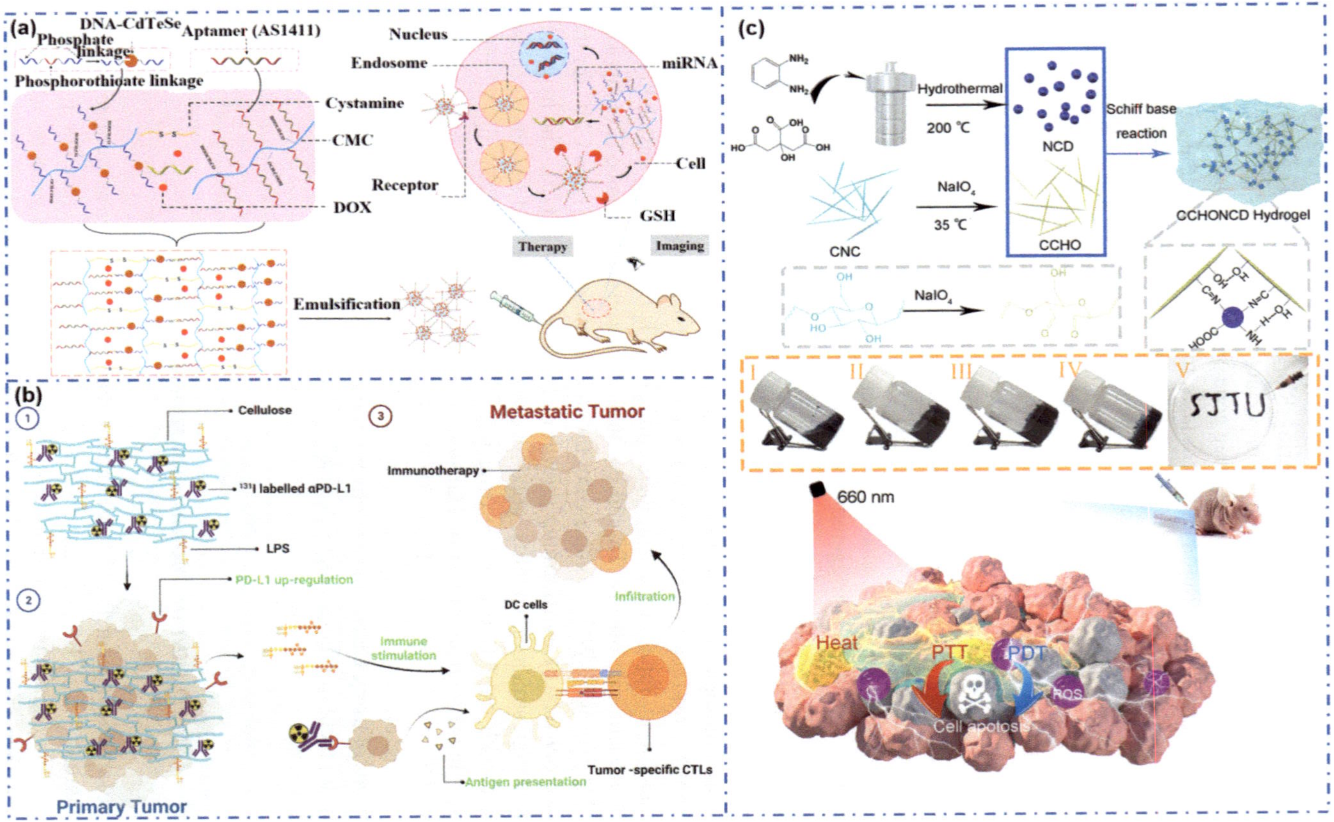

Fig. 10.2 **a** CBHs for tumor imaging and therapy. Adapted from Reference [24], Copyright 2022 Royal Society of Chemistry. **b** Schematic diagram of 131I-αPD-L1/BC for enhanced radio-immunotherapy. Adapted from Reference [29], Copyright 2022 Elsevier B.V. **c** Synthetic route of CCHONCD hydrogel for cancer treatment. Adapted from Reference [32], Copyright 2021 Wiley–VCH

safety of therapeutic effects. Integrating immunotherapeutic agents into CBHs offers a promising strategy to overcome these limitations. Leveraging the inherent biocompatibility, tunable drug release profiles, and TME responsiveness of CBHs enables localized delivery, sustained release, and enhanced immune activation, ultimately improving the specificity and efficacy of immunotherapy [28].

To address the limitations of radioimmunotherapy (RIT) and immune checkpoint blockade (ICB), including nonspecific distribution, low response rates, and immune-related adverse effects, Yang et al. developed a multifunctional bacterial cellulose (BC)-based hydrogel platform incorporating ^{131}I-labeled αPD-L1 antibodies for synergistic radiotherapy and immunotherapy following local injection [29]. In vivo studies demonstrated that this hydrogel significantly inhibited both primary and distant tumor growth in CT26 colon cancer and 4T1 breast cancer models, completely eradicated pulmonary metastases, and achieved a single-dose tumor inhibition rate exceeding 90%, with 100% survival in treated animals. Mechanistically, the hydrogel system elicited a potent therapeutic response through a combination of immunogenic cell death (ICD) induction, PD-L1/PD-1 pathway blockade, and dendritic cell maturation. This enabled a "local radiotherapy–immune activation–systemic antitumor" tri-modal synergistic effect (Fig. 10.2b). This study highlights the powerful potential of CBHs as integrated carriers for immunotherapeutic agents and radiopharmaceuticals. The platform offers a novel and highly translatable strategy for combinational cancer therapy, paving the way for precise, localized, and systemic tumor suppression.

10.2.4 Inorganic Nanoparticle Functionalization

Inorganic nanoparticle functionalized cellulose hydrogels combine the multifunctionality of inorganic materials with the biocompatibility of cellulose matrices. As a natural polymer, cellulose can effectively encapsulate and stabilize nanoparticles, preventing aggregation, reducing toxicity, and improving the overall safety and stability of the system. The incorporation of inorganic nanoparticles such as QDs, metals, and metal oxides endows the hydrogels with additional functionalities, such as magnetism, photothermal effects, and radiosensitization, thereby enhancing their potential for precision therapy and diagnosis [30].

In recent studies, the integration of inorganic nanoparticles has emerged as a pivotal strategy for functionalizing CBHs, particularly in enhancing the targeting capability, controlled release behavior, and bioactivity of drug delivery systems. For instance, Ostovar et al. developed a pH-responsive hydrogel system based on chitosan/carboxymethyl cellulose (CS/CMC) loaded with zinc oxide nanoparticles (ZnO NPs) and graphene quantum dots (GQDs) for the targeted delivery of quercetin (QUR), a natural anticancer agent, in brain tumor therapy [31]. The synergistic effects of ZnO and GQDs significantly enhanced both encapsulation efficiency (87.5%) and drug loading capacity (46.75%). Moreover, the hydrogel exhibited excellent pH sensitivity, releasing up to 95% of QUR at acidic pH 5.4, while limiting release to 56% at physiological pH 7.4. GQDs contributed potential fluorescence imaging capabilities, while ZnO offered supplementary anticancer effects, enabling a theranostic approach. In vitro studies confirmed that the composite hydrogel significantly inhibited the proliferation of U-87 MG glioblastoma cells, while maintaining low cytotoxicity against normal fibroblasts (L929), demonstrating good biosafety.

In another innovative study, Chen et al. proposed a multifunctional hydrogel fabricated via Schiff base reactions between amino-functionalized carbon dots (NCDs) and aldehyde-modified CNC (CCHOs), enabling integrated photothermal and photodynamic therapy (PTT/PDT) within a single platform (Fig. 10.2c) [32]. In this system, NCDs served triple roles: as crosslinkers, photothermal agents (with a photothermal conversion efficiency of 77.6%), and photosensitizers (with a singlet oxygen quantum yield of 0.37). Remarkably, both PTT and PDT could be simultaneously activated using a single 660 nm light source, greatly simplifying the conventional multicomponent treatment process. The porous architecture of CCHOs enhanced light scattering and reactive oxygen species diffusion, while their shear-thinning behavior facilitated injectability and prolonged retention at tumor sites. In vitro results showed that the hydrogel reduced the viability of B16F10 melanoma cells to 11% under light exposure, with negligible cytotoxicity. In vivo, a single injection combined with light irradiation led to complete tumor suppression without recurrence. Hematological and histological analyses confirmed excellent biocompatibility and safety.

These findings underscore the versatility of inorganic nanoparticle-functionalized cellulose hydrogels and offer new insights into their design. By harnessing the synergistic effects of a single nanocomponent, such systems hold great promise for the clinical translation of near-infrared (NIR)-responsive biomaterials in precision cancer therapy.

10.2.5 Combination Therapy Strategies

With the continuous advancement of cancer treatment technologies, single-modality therapies often face limitations such as drug resistance, tumor recurrence, and suboptimal therapeutic efficacy. As a result, the development of multimodal treatment strategies has become a major focus in oncology research [33]. Combination therapies can effectively overcome the shortcomings of monotherapies by offering enhanced targeting capability and therapeutic efficacy [34].

One common approach involves combining PTT with chemotherapy. To enhance therapeutic outcomes while minimizing drug-related side effects, Chen et al. developed a multi-stimuli-responsive hydrogel (CNF TRIH@DOX) based on TEMPO-oxidized cellulose nanofibers (TOCNFs) for non-invasive imaging and chemo-photothermal combination therapy [35]. In this system, pH- and temperature-responsive polymers (PEI-NIPAM) were covalently grafted onto TOCNFs, followed by co-loading of DOX and citrate-stabilized Prussian blue nanoparticles (PBNPs), enabling a highly integrated and controllable drug delivery platform. The hydrogel's 3D nanocage structure exhibited exceptional loading capacities for DOX (415 mg/g) and PBNPs (2 mg/g). Simulated TME studies revealed a progressive, environment-triggered release of DOX, with a sharp release enhancement under 808 nm laser irradiation, achieving a cumulative release of 56.8%. This demonstrated a synergistic tri-responsive mechanism, sensitive to temperature, pH, and NIR light, supporting efficient drug delivery, rapid tumor diagnosis, and dual-modality therapy, thereby advancing the development of smart theranostic systems.

You et al. designed a multifunctional hydrogel system based on CNCs to enable combined chemotherapy and chemodynamic therapy (CDT) for effective tumor treatment [36]. By incorporating 5-aminolevulinic acid (ALA) and polydopamine (PDA)-modified CNCs with Fe^{3+} ions through a pH-responsive design, they constructed a biocompatible, tumor-targeted nanohydrogel (Fig. 10.3a). This system triggered hydroxyl radicals (•OH) radical generation via Fenton reactions to induce oxidative stress in tumor cells while delivering the chemotherapeutic agent paclitaxel (PTX) to enhance the anticancer effect. Under acidic conditions mimicking the TME, the hydrogel achieved rapid PTX release and robust reactive oxygen species (ROS) production. In MCF-7 breast cancer cells, the PTX-PDA/ALA/Fe@CNC hydrogel reduced cell viability to 2.7%, demonstrating potent cytotoxicity and apoptosis-inducing effects mediated by ROS. Meanwhile, the hydrogel showed low toxicity toward normal mammary epithelial cells, highlighting its potential to minimize the side effects of chemotherapy.

In chemoradiotherapy applications, CBHs have emerged as ideal drug carriers due to their favorable biocompatibility, biodegradability, and tunable physicochemical properties. Xi et al. proposed a novel nanogel platform based on CMC and bovine serum albumin (BSA) assembled via electrostatic interactions, enabling a green synthesis approach with excellent biocompatibility and biodegradability (Fig. 10.3b) [37]. This nanogel was capable of co-delivering the radioisotope ^{131}I and camptothecin, achieving dual-mode chemoradiotherapy. It exhibited pH-responsive release behavior, minimizing damage to normal tissues. Furthermore, its high cellular uptake efficiency and tumor-targeting accumulation significantly improved therapeutic outcomes, offering a promising strategy for functionalizing bioactive hydrogels in combinational cancer therapy.

10.3 Action Mechanisms of CBHs in Tumor Therapy

10.3.1 Sustained and Targeted Drug Release

In conventional cancer chemotherapy, many drugs suffer from low bioavailability, nonspecific biodistribution, and rapid systemic clearance, making it difficult to maintain effective drug concentrations in tumor tissues. This often results in reduced therapeutic efficacy and increased side effects. To address these challenges, Chatap et al. developed a magnetic nanocomposite hydrogel (CON-A-CU-MFNCs) based on wheat cellulose nanofibers, enabling sustained and targeted delivery of the

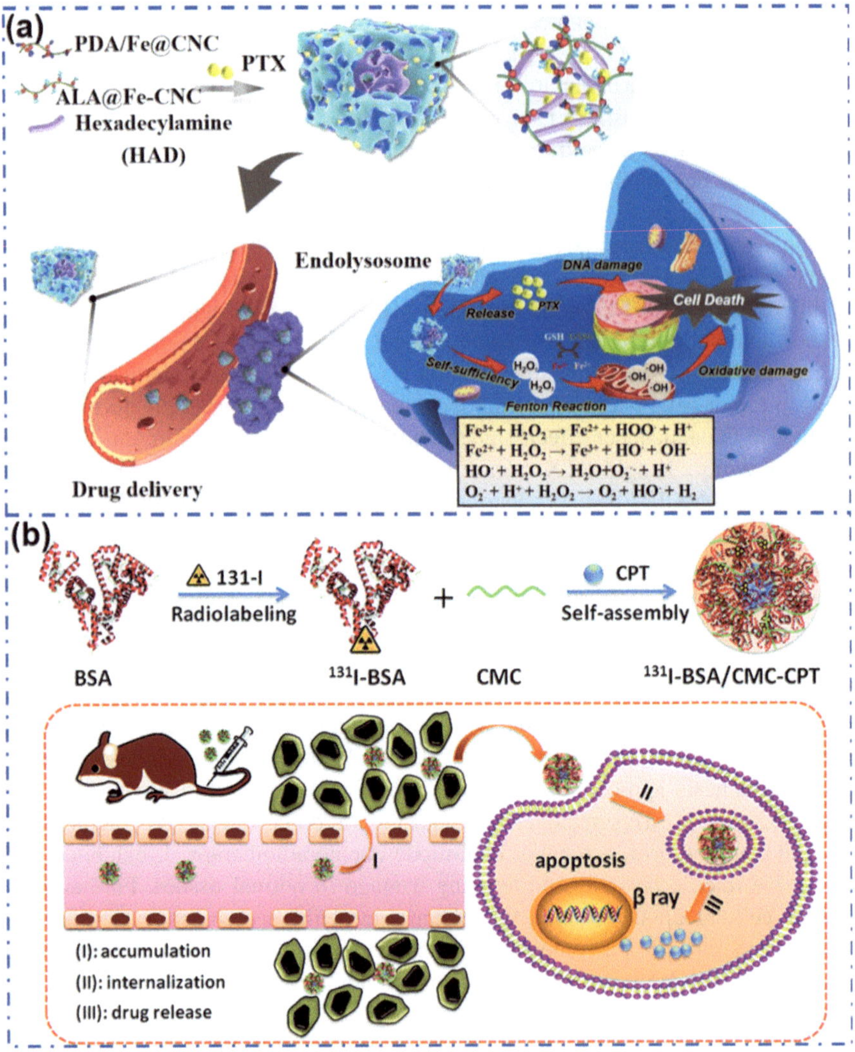

Fig. 10.3 **a** Schematic illustration of the PTX-PDA/ALA/Fe@CNC nanohydrogel, and Fe^{3+}-activated Fenton reaction to increase ROS concentration in tumor cells. Adapted from Reference [36], Copyright 2021 Elsevier BV. **b** Schematic illustration to show the preparation of ^{131}I-BSA/CMC-CPT nanogels for in vivo combined chemo-radioisotope therapy. Adapted from Reference [37], Copyright 2018 Royal Society of Chemistry

chemotherapeutic agent curcumin through multiple synergistic mechanisms [38]. In this system, concanavalin A was incorporated to actively target tumor cells by recognizing glycoproteins on their surfaces. Meanwhile, magnetic Fe_3O_4 nanoparticles endowed the hydrogel with magnetic responsiveness, allowing drug accumulation at the tumor site under external magnetic field guidance. The hydrogel also exhibited sensitivity to the acidic TME, enabling pH-triggered drug release and preventing premature leakage in normal tissues. Drug release profiles demonstrated clear pH dependency and magnetothermal responsiveness, with high loading efficiency (up to 83.87%) and a stable, prolonged release pattern. This system highlights a promising strategy for enhancing targeted and sustained chemotherapy using CBHs.

In another study, Zhang et al. designed a composite hydrogel system by combining CMC with a thermo-responsive polymer P(NIPAM-co-AH), and embedding gold nanorods (GNRs) to achieve NIR-induced photothermal-chemo combination therapy [39]. This hydrogel system exhibited excellent capabilities for sustained and site-specific drug delivery in tumor treatment. The porous network structure of the hydrogel allowed for controlled and prolonged release of DOX, achieving a release rate of up to 89% under tumor-relevant conditions (pH 5.4 and 42 °C), thus increasing tumor cell exposure to the drug and reducing dosing frequency. Additionally, the hydrogel was responsive to the acidic TME, promoting targeted degradation and drug release at the tumor site. In vivo experiments demonstrated that the system not only significantly inhibited tumor growth but also reduced DOX-induced toxicity in non-target tissues. These results underscore the hydrogel's potential as an effective and safe platform for spatiotemporally controlled chemotherapy.

10.3.2 PH/Temperature/Enzyme-Responsive Drug Delivery

Stimuli-responsive hydrogel-based drug delivery systems have shown great promise in improving tumor treatment efficacy while minimizing the adverse effects of chemotherapeutic agents. The unique characteristics of the TME, such as acidic pH, localized hyperthermia, and specific enzymatic activities, offer ideal triggers for the design of intelligent hydrogels.

In tumor therapy, the acidic pH and elevated local temperature of the TME provide key stimuli for the controlled release of drugs from smart hydrogels. Cai et al. developed a dual-responsive nanohydrogel (CMC/POM) based on CMC, thermo-sensitive monomer MEO_3MA, and pH-sensitive monomer MAA, designed for precise DOX release within the TME (Fig. 10.4a) [40]. This system exhibited rapid responsiveness under acidic (pH 3.0) and elevated temperature (45 °C) conditions, achieving a drug release efficiency of up to 92.2% within 24 h, while significantly reducing release rates at neutral or alkaline pH. The combined pH- and thermo-stimuli induced structural changes in the hydrogel, enabling highly selective, diffusion-controlled drug release in tumor tissues. Besides excellent thermal stability, biocompatibility, and drug-loading capacity, this hydrogel system realized spatiotemporally precise drug delivery.

Moreover, Qi et al. designed a multi-responsive hydrogel (Au NCs/5-FU@Carr-LA) based on CMC, integrating dual-modal pH/temperature stimuli to achieve precise release of the chemotherapeutic drug 5-fluorouracil (5-FU) [41]. Under tumor-mimicking acidic conditions (pH 5.0), the hydrogel significantly accelerated drug release. Additionally, gold nanoclusters (Au NCs) embedded within the

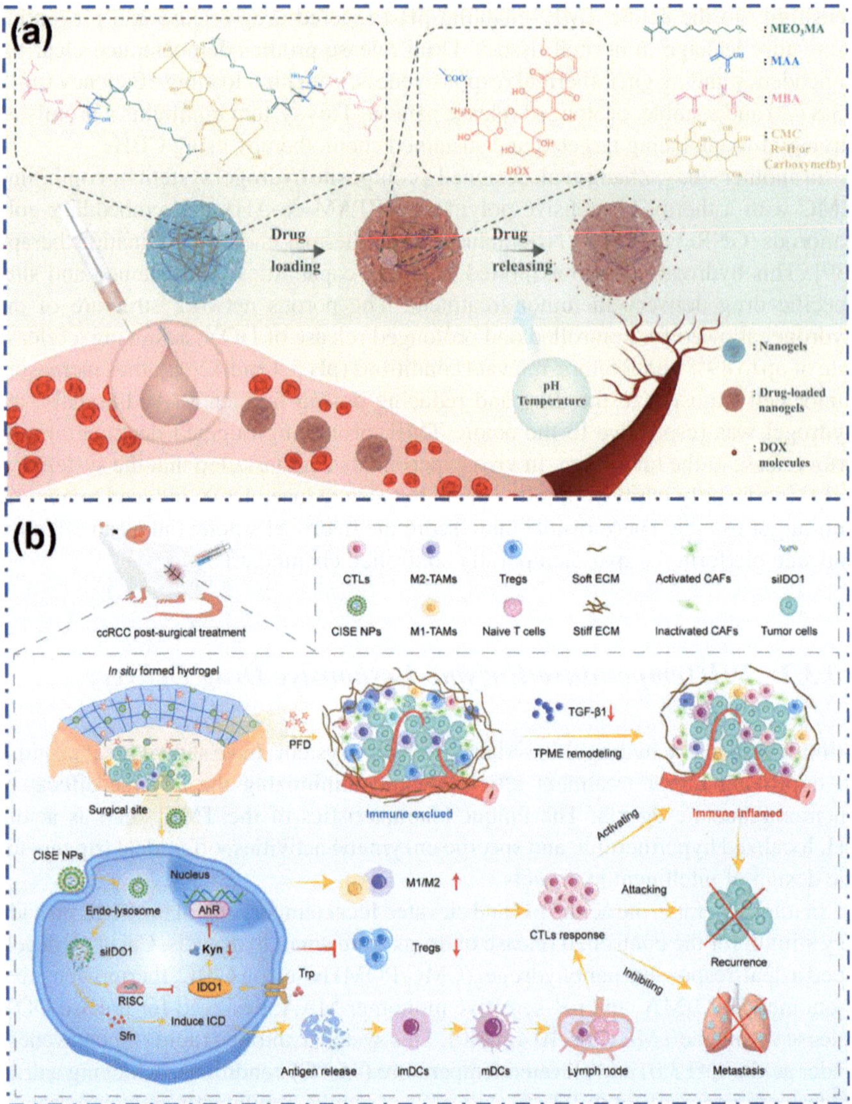

Fig. 10.4 **a** The structure of CMC/POM nnogel and the drug loading and releasing process by intravenous injection. Adapted from Reference [40], Copyright 2025 Springer. **b** Schematic of the in situ biomimetic hydrogel system for postoperative ccRCC mechano-immunometabolic therapy. Adapted from Reference [45], Copyright 2024 Wiley–VCH

hydrogel generated a photothermal effect under NIR irradiation, further promoting the breakdown of the drug-loaded network, thereby enhancing release rate and treatment depth. This study highlights the potential of CBHs incorporating multiple responsive mechanisms for on-demand drug release, targeted delivery, and synergistic therapy within tumor sites.

10.3.3 Tumor Microenvironment Modulation

TME is typically characterized by high reductive conditions, enhanced oxidative stress barriers, acidic pH, hypoxia, immunosuppression, abnormal metabolism, and a dense extracellular matrix (ECM) barrier. These features hinder drug penetration, immune cell infiltration, and the release of antitumor agents [42]. Recently, researchers have introduced GSH-responsive moieties, loaded oxidases or metal ions, and modulated ECM mechanical properties to make CBHs not only effective drug delivery platforms but also unique modulators of the TME [43, 44].

The rigidity of the ECM and immune-metabolic imbalance within the TME are key factors contributing to therapeutic failure and tumor immune evasion. Dong et al. developed a biomimetic hydrogel system based on fibrinogen (CISE-PFD@Gel) for in situ post-surgical modulation of the TME to inhibit renal clear cell carcinoma recurrence [45]. This hydrogel forms in situ after local injection and allows sequential drug release. By releasing pirfenidone (PFD), it suppresses the TGF-β1 signaling pathway, reduces collagen deposition and cancer-associated fibroblast (CAF) activation, softens tumor tissue, improves perfusion, and alleviates hypoxia, thereby promoting immune cell infiltration. Meanwhile, siRNA-loaded nanocarriers silence the IDO1/Kynurenine/AhR immunometabolic axis, reversing immunosuppression mediated by regulatory T cells and M2-type macrophages. Combined with sorafenib-induced ICD, this system enhances dendritic cell maturation and T cell activation (Fig. 10.4b). By remodeling the TME, this approach significantly improves immunotherapy efficacy and demonstrates promising anti-recurrence and anti-metastasis effects in animal models.

Zhou et al. reported an injectable multi-enzyme mimetic hydrogel (MELH) based on thiolated CMC (CMC-SH) that synergistically modulates the TME through multiple mechanisms [46]. This hydrogel selectively degrades in the GSH-rich tumor milieu, releasing 5-fluorouracil (5-FU), glucose oxidase (GOx), and Fe^{3+} ions. GOx catalyzes the oxidation of glucose to generate hydrogen peroxide (H_2O_2), which reacts with Fe^{2+} via the Fenton reaction to produce highly toxic •OH, disrupting the redox homeostasis of tumor cells. Additionally, GOx-induced glucose depletion creates a 'starvation' microenvironment, further inhibiting tumor metabolism. The disulfide bonds within the CMC-SH backbone confer high GSH sensitivity, enabling targeted drug release at the tumor site while minimizing systemic toxicity. Furthermore, this hydrogel downregulates GPX4 expression and induces ferroptosis, effectively overcoming chemotherapy resistance. Overall, MELH employs a multifaceted approach, GSH depletion, enhanced ROS generation, and glucose starvation, to remodel the TME, markedly increasing tumor sensitivity to chemotherapy. This highlights the promising application of CBHs in precise and responsive tumor therapy.

10.3.4 PTT/PDT Enhancement

The combined application of PTT and PDT has been demonstrated as a synergistic strategy to enhance antitumor efficacy. The thermal effect generated by PTT not only directly destroys tumor cells but also increases local blood flow and alleviates hypoxia, thereby improving the efficiency of ROS generation during PDT [47, 48]. Meanwhile, ROS produced by PDT further activates apoptosis pathways at the cellular level, amplifying the cytotoxic effects of PTT. This dual assault mechanism effectively eradicates primary tumor cells and clears residual microscopic lesions post-surgery, significantly reducing the risk of tumor recurrence.

Zhang et al. developed a smart multifunctional hydrogel system based on CMC to synergistically enhance the effects of PTT and PDT for efficient tumor recurrence suppression [49]. Dopamine (DA) was covalently grafted onto the CMC backbone, which then self-assembled under the oxidizing agent sodium periodate ($NaIO_4$) to form a CDI hydrogel containing polydopamine (PDA) and sodium iodide (NaI). The PDA component endowed the system with excellent NIR absorption and photothermal conversion efficiency (up to 45.7%), while the incorporation of the photosensitizer Ce6 enabled efficient singlet oxygen ($^1O_2$) generation. Under laser irradiation, the synergistic action of PDA and Ce6 rapidly increased the local temperature causing thermal damage and simultaneously produced ROS to induce apoptosis, thereby realizing an enhanced dual-modal PTT/PDT treatment effect. In vitro studies showed that the cell viability in the combined treatment group dropped significantly to 4.65%. In vivo, the combined therapy completely inhibited tumor recurrence, outperforming monotherapy groups. This work provides a novel strategy for designing efficient and safe photoresponsive antitumor materials.

Additionally, Grijalvo et al. developed a multifunctional drug delivery platform based on CNCs, co-loading the photosensitizer Ce6, the photothermal agent indocyanine green (ICG), and the chemotherapeutic drug PTX, with folic acid (FA) modification to enhance tumor targeting and achieve multimodal synergistic therapy [50]. Upon NIR laser irradiation, the system simultaneously generated heat and released ROS, significantly increasing apoptosis rates and demonstrating excellent photoresponsiveness and synergistic cytotoxicity. In vitro results showed that the combined therapy group exhibited a much higher cell death rate than any single treatment group, while FA modification effectively enhanced cancer cell uptake, further improving therapeutic selectivity. Notably, the CNCs used not only possess excellent biocompatibility and biodegradability but also serve as a green and efficient drug carrier, exhibiting great potential for clinical translation.

10.3.5 Immune System Activation and Tumor Suppression

CBHs can enhance antitumor immune responses by loading antigens, immune adjuvants, or immune checkpoint inhibitors to activate immune effector cells such as dendritic cells (DCs) and T cells. Their excellent biocompatibility and controllable release properties also help remodel the tumor immune microenvironment, enabling sustained tumor suppression.

Thermo-sensitive injectable hydrogels, due to their superior local sustained release and tissue compatibility, have become important carriers to boost tumor immunotherapy. Wang et al. constructed an OxP/R848@PLEL hydrogel for a colorectal cancer peritoneal metastasis model, achieving synergistic chemo-immunotherapy by co-delivering oxaliplatin and the immune adjuvant R848 [51]. Oxaliplatin induces ICD, promoting calreticulin (CRT) exposure, and the release of HMGB1 and ATP, which enhance dendritic cell recognition and activation of tumor antigens. R848 activates the TLR7/8 pathway, further promoting DC maturation, $CD8^+$ T cell expansion, and inducing tumor-associated macrophages polarization toward the pro-inflammatory M1 phenotype, thereby reversing the immunosuppressive microenvironment (Fig. 10.5). This hydrogel significantly increases local drug concentration and retention time, showing excellent antitumor efficacy and immune memory induction in vivo. This study demonstrates the potential of hydrogels as immune-activating platforms in cancer therapy.

CBHs exhibit excellent multifunctional integration potential in tumor treatment. Chen et al. developed an injectable hydrogel assembled from aldehyde-modified CNC (mCNCs), carbon dots (CDs), and Fe_3O_4 nanoparticles to realize synergistic PTT, PDT, and CDT [52]. Under NIR irradiation, CDs generate heat and $^1O_2$, while Fe_3O_4 induces Fenton reactions producing •OH, collectively triggering tumor cell death. High levels of ROS and apoptotic signals induce ICD, enhancing antigen release and immune activation. This platform combines injectability, magnetic responsiveness, and treatment visualization, providing new design concepts for CBHs in tumor immunotherapy.

Additionally, Shi et al. designed a membrane-targeted self-assembling short peptide hydrogel (PepABS-py) that forms a nanofiber network triggered by tumor cell surface alkaline phosphatase, physically blocking tumor cell membranes and effectively inhibiting exosome release and spread [6, 53]. This hydrogel captures immunosuppressive exosomes, reducing T cell exhaustion and M2 macrophage polarization risks, while enhancing $CD8^+$ T cell infiltration and activation. Furthermore, loaded carbonic anhydrase IX (CA IX) inhibitors alleviate tumor hypoxia, further promoting immune activation. In breast cancer models, PepABS-py combined with chemotherapy reverses paclitaxel-induced immunosuppression and metastatic tendencies, significantly suppressing tumor growth and lung metastasis. This study proposes a novel hydrogel mechanism integrating exosome blockade and microenvironment remodeling, offering new insights for cellulose-based immune hydrogels in antitumor therapy.

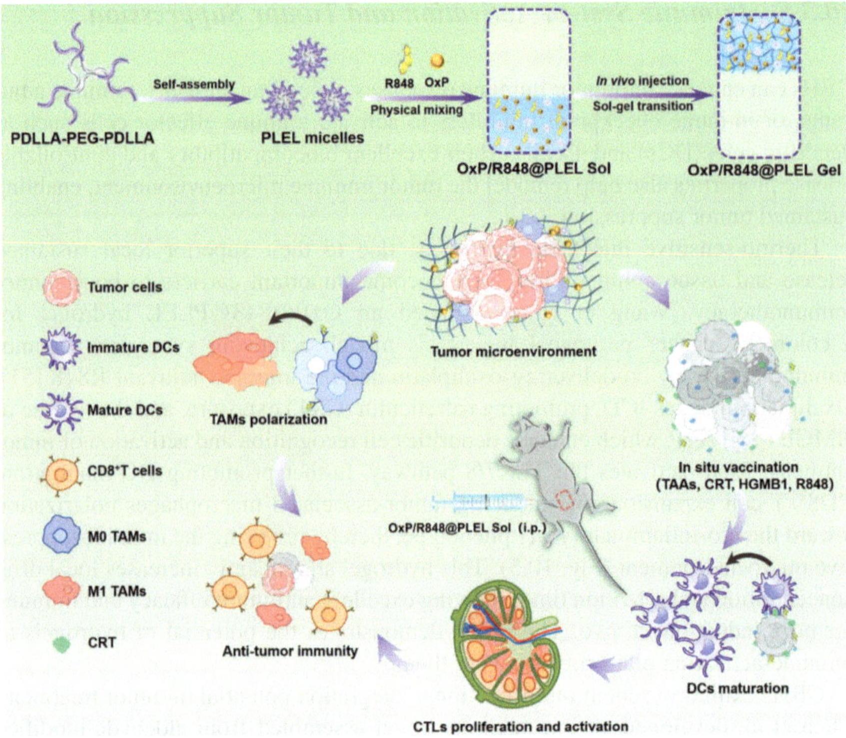

Fig. 10.5 Schematic illustration of the preparation and mechanism of OxP/R848@PLEL injectable thermo-sensitive hydrogel. Adapted from Reference [51], Copyright 2023 Wiley–VCH

10.4 Conclusion

In summary, bioactive CBHs exhibit excellent overall performance and hold great promise as multifunctional platforms for tumor therapy. Their injectability, stimulus-responsiveness, biodegradability, and capacity for multimodal therapeutic integration have enabled significant advances in chemotherapy, immunotherapy, phototherapy, and targeted treatment. With a growing understanding of the TEM and immune mechanisms, alongside the continuous advancement of material science, CBHs are poised to play an increasingly important role in personalized therapy and precision drug delivery. Nevertheless, challenges remain, particularly in large-scale production, in vivo stability, and long-term biosafety, which should be addressed to facilitate clinical translation. Looking ahead, the integration of synthetic biology, artificial intelligence-driven design, and 3D printing technologies is expected to propel CBHs toward safer, more efficient, and highly precise solutions for future cancer therapies.

References

1. Y.W. Li, W. Liu, Y.Y. Wang, S. Lu, Cellulose based nano-scaffolds for targeted cancer therapies: current status and future perspective. Int. J. Nanomedicine **20**, 199 (2025)
2. D.Y. Xia, P. Wang, X.F. Ji, N.M. Khashab, J.L. Sessler, F.H. Huang, Functional supramolecular polymeric networks: The marriage of covalent polymers and macrocycle-based host-guest interactions. Chem. Rev. **120**, 6070 (2020)
3. Y.S. Ai, Z.X. Lin, W.Q. Zhao, M. Cui, W. Qi, R.L. Huang, R.X. Su, Nanocellulose-based hydrogels for drug delivery. J. Mater. Chem. B **11**, 7004 (2023)
4. P.F. Zou, J.X. Yao, Y.N. Cui, T. Zhao, J.W. Che, M.Y. Yang, Z.P. Li, C.S. Gao, Advances in cellulose-based hydrogels for biomedical engineering: a review summary. Gels **8**, 364 (2022)
5. W.K. Cheng, Y. Zhu, G.Y. Jiang, K.Y. Cao, S.Q. Zeng, W.S. Chen, D.W. Zhao, H.P. Yu, Sustainable cellulose and its derivatives for promising biomedical applications. Prog. Mater. Sci. **138**, 101152 (2023)
6. A.P.A. De Carvalho, R. Értola, C. Conte-Junior, Nanocellulose-based platforms as a multi-purpose carrier for drug and bioactive compounds: from active packaging to transdermal and anticancer applications. Int. J. Pharm. **652**, 123851 (2024)
7. L. Shi, H.J. Lv, C.T. Chen, F.M. Cui, L. Zhang, J.P. Cao, R.P. Zaccaria, Q. Zhang, D.P. Sun, Regulation of gut microbiome with redox responsible bacterial cellulose hydrogel for precision chemo-radiotherapy of intestinal cancer. Chem. Eng. J. **446**, 137340 (2022)
8. H. Zhang, L.S. Hu, W. Xiao, Y.Q. Su, D.L. Cao, An injectable, in situ forming and NIR-responsive hydrogel persistently reshaping tumor microenvironment for efficient melanoma therapy. Biomater. Res. **27**, 118 (2023)
9. G.X. Xu, K. Liu, X.W. Chen, Y. Lin, C.C. Yu, X.X. Nie, W.X. He, N. Karin, Y.X. Luan, Hydrogel-mediated tumor T cell infiltration and immune evasion to reinforce cancer immunotherapy. Nanoscale Horiz. **9**, 295 (2024)
10. M. Tang, J.Z. Song, S.Y. Zhang, X.L. Shu, S. Liu, M. Ashrafizadeh, Y.N. Ertas, Y. Zhou, M. Lei, Innovative theranostic hydrogels for targeted gastrointestinal cancer treatment. J. Transl. Med. **22**, 970 (2024)
11. J.W. Gong, L.L. Hou, Y.C. Ching, K.Y. Ching, N.D. Hai, C.H. Chuah, A review of recent advances of cellulose-based intelligent-responsive hydrogels as vehicles for controllable drug delivery system. Int. J. Biol. Macromol. **264**, 130525 (2024)
12. J.Q. Zhou, Z.S. Zhang, Z.L. Zhang, T. Wu, H. Li, X.J. Hao, X. Liu, T.X. Gong, D. Liu, S.B. Wei, A facile dual-drug delivery system using cellulose-based microgel/hydrogel for enhanced gastric cancer therapy. Colloid. Surface. B **254**, 114813 (2025)
13. Y. Cheng, H.T. Zhang, H. Wei, C.Y. Yu, Injectable hydrogels as emerging drug delivery platforms for tumor therapy. Biomater. Sci. **12**, 1151 (2024)
14. Z.P. Wang, B.T. Zhai, J. Sun, X.F. Zhang, J.B. Zou, Y.J. Shi, D.Y. Guo, Recent advances of injectable in situ-forming hydrogels for preventing postoperative tumor recurrence. Drug Deliv. **31**, 2400476 (2024)
15. X.Y. Jiang, F.W. Zeng, X.F. Yang, C. Jian, L.A. Zhang, A.X. Yu, A. Lu, Injectable self-healing cellulose hydrogel based on host-guest interactions and acylhydrazone bonds for sustained cancer therapy. Acta Biomater. **141**, 102 (2022)
16. S. Selvaraj, A. Chauhan, V. Dutta, R. Verma, S.K. Rao, A. Radhakrishnan, S. Ghotekar, A state-of-the-art review on plant-derived cellulose-based green hydrogels and their multifunctional role in advanced biomedical applications. Int. J. Biol. Macromol. **265**, 130991 (2024)
17. Y.X. Zhou, K.Z. Chen, H.W. Cheng, S.S. Zhang, Recent advances in polysaccharide-based hydrogels for tumor immunotherapy. Gels-Basel **11**, 152 (2025)
18. F. Ghorbani, B. Ghalandari, Z.C. Liu, D.J. Li, B.Q. Yu, Injectable light-assisted thermo-responsive methylcellulose-sodium humate hydrogel proposed for photothermal ablation and localized delivery of cisplatin. Front. Bioeng. Biotech. **10**, 967438 (2022)
19. F.F. Wang, J.S. Chen, J.F. Liu, H.B. Zeng, Cancer theranostic platforms based on injectable polymer hydrogels. Biomater. Sci. **9**, 3543 (2021)

20. I. Ali, M. Gulfam, S.H. Jo, J.W. Seo, A. Rizwan, S.H. Park, K.T. Lim, Reduction-responsive and bioorthogonal carboxymethyl cellulose based soft hydrogels cross-linked via IEDDA click chemistry for cancer therapy application. Int. J. Biol. Macromol. **219**, 109 (2022)
21. M.R. Vakili, W. Mohammed-Saeid, A. Aljasser, J. Hopwood-Raja, B. Ahvazi, Y. Hrynets, M. Betti, A. Lavasanifar, Development of mucoadhesive hydrogels based on polyacrylic acid grafted cellulose nanocrystals for local cisplatin delivery. Carbohydr. Polym. **255**, 117332 (2021)
22. B.H. Shan, F.G. Wu, Hydrogel-based growth factor delivery platforms: strategies and recent advances. Adv. Mater. **36**, 2210707 (2024)
23. W.Q. Li, J. Hu, C. Chen, X.Y. Li, H.H. Zhang, Y.R. Xin, Q.C. Tian, S.L. Wang, Emerging advances in hydrogel-based therapeutic strategies for tissue regeneration. Regen. Ther. **24**, 459 (2023)
24. Y. Zheng, S.Q. He, P.H. Jin, Y.B. Gao, Y. Di, L.M. Gao, J.D. Wang, Construction of multifunctional carboxymethyl cellulose nanohydrogel carriers based on near-infrared DNA-templated quantum dots for tumor theranostics. RSC Adv. **12**, 31869 (2022)
25. Z.F. Zhang, S.F. Lin, Y.P. Yan, X.X. You, H. Ye, Enhanced efficacy of transforming growth factor-β1 loaded an injectable cross-linked thiolated chitosan and carboxymethyl cellulose-based hydrogels for cartilage tissue engineering. J. Biomat. Sci-Polym. E. **32**, 2402 (2021)
26. Z.L. Li, X.Q. Lai, S.Q. Fu, L. Ren, H. Cai, H. Zhang, Z.W. Gu, X.L. Ma, K. Luo, Immunogenic cell death activates the tumor immune microenvironment to boost the immunotherapy efficiency. Adv. Sci. **9**, 2201734 (2022)
27. M. Kundu, R. Butti, V.K. Panda, D. Malhotra, S. Das, T. Mitra, P. Kapse, S.W. Gosavi, G.C. Kundu, Modulation of the tumor microenvironment and mechanism of immunotherapy-based drug resistance in breast cancer. Mol. Cancer **23**, 92 (2024)
28. P.P. Ji, W.Q. Sun, S.Y. Zhang, Y.Q. Xing, C. Wang, M.Y. Wei, Q.Y. Li, G. Ji, G.D. Yang, Modular hydrogel vaccine for programmable and coordinate elicitation of cancer immunotherapy. Adv. Sci. **10**, 2301789 (2023)
29. Z.Y. Qi, P. Pei, Y.X. Zhang, H. Chen, S. Yang, T. Liu, Y.J. Zhang, K. Yang, 131I-αPD-L1 immobilized by bacterial cellulose for enhanced radio-immunotherapy of cancer. J. Control. Release **346**, 240 (2022)
30. X.N. Zhang, Q. Zhao, J.J. Yang, T.X. Wang, F.B. Chen, K. Zhang, Tumor microenvironment-triggered intratumoral in-situ biosynthesis of inorganic nanomaterials for precise tumor diagnostics. Coord. Chem. Rev. **484**, 215115 (2023)
31. S. Ostovar, M. Pourmadadi, M.A. Zaker, Co-biopolymer of chitosan/carboxymethyl cellulose hydrogel improved by zinc oxide and graphene quantum dots nanoparticles as pH-sensitive nanocomposite for quercetin delivery to brain cancer treatment. Int. J. Biol. Macromol. **253**, 127091 (2023)
32. T.X. Chen, T.T. Yao, H. Peng, A.K. Whittaker, Y. Li, S.M. Zhu, Z.Y. Wang, An injectable hydrogel for simultaneous photothermal therapy and photodynamic therapy with ultrahigh efficiency based on carbon dots and modified cellulose nanocrystals. Adv. Funct. Mater. **31**, 2106079 (2021)
33. L. Chen, G.J. Chen, K. Hu, L.X. Chen, Z.Y. Zeng, B. Li, G.H. Jiang, Y. Liu, Combined photothermal and photodynamic therapy enhances ferroptosis to prevent cancer recurrence after surgery using nanoparticle-hydrogel composite. Chem. Eng. J. **468**, 143685 (2023)
34. C.T. Liu, Y.Y. Liao, L. Liu, L.Y.J. Xie, J.B. Liu, Y.M. Zhang, Y.Z. Li, Application of injectable hydrogels in cancer immunotherapy. Front. Bioeng. Biotech. **11**, 1121887 (2023)
35. Z.P. Chen, R.M. Chen, C. Zhao, Z.Y. Quan, H.X. Zhu, L. Wang, Q. Bu, Y.H. He, H. He, A novel medically imageable intelligent cellulose nanofibril-based injectable hydrogel for the chemo-photothermal therapy of tumors. Chem. Eng. J. **431**, 133255 (2022)
36. C.Q. You, L.K. Ning, H.S. Wu, C.B. Huang, F. Wang, A biocompatible and pH-responsive nanohydrogel based on cellulose nanocrystal for enhanced toxic reactive oxygen species generation. Carbohydr. Polym. **258**, 117685 (2021)
37. K.F. Liu, D. Zheng, J.Y. Zhao, Y.H. Tao, Y.S. Wang, J. He, J.D. Lei, X.J. Xi, PH-Sensitive nanogels based on the electrostatic self-assembly of radionuclide I labeled albumin and

carboxymethyl cellulose for synergistic combined chemo-radioisotope therapy of cancer. J. Mater. Chem. B **6**, 4738 (2018)
38. V. Chatap, P. Vanjari, N.V. Bhilare, Curcumin loaded magnetic nanocellulose fiber composites with con-a cap for theranostics application in breast cancer. Cellulose **32**, 3855 (2025)
39. L.M. Chang, X.J. Liu, J.J. Zhu, Y.X. Rao, D.Y. Chen, Y. Wang, Y.L. Zhao, J.L. Qin, Cellulose-based thermo-responsive hydrogel with NIR photothermal enhanced DOX released property for anti-tumor chemotherapy. Colloid. Surface. B **218**, 112747 (2022)
40. Y.P. Cai, D.Y. Li, S.H. Peng, H. Liu, Synthesis of dual-responsive carboxymethyl cellulose-based nanogels for drug delivery applications. Colloid Polym. Sci. **303**, 287 (2025)
41. C.X. Qi, A. Li, B.M. Wu, P.S. Wang, Multi-sensitive Au NCs/5-FU@Carr-LA composite hydrogels for targeted multimodal anti-tumor therapy. Molecules **29**, 4051 (2024)
42. Z.M. Li, W.G. Xu, J.Z. Yang, J. Wang, J.L. Wang, G. Zhu, D. Li, J.X. Ding, T.M. Sun, A tumor microenvironments-adapted polypeptide hydrogel/Nanogel composite boosts antitumor molecularly targeted inhibition and immunoactivation. Adv. Mater. **34**, 2200449 (2022)
43. Y.S. Liu, Y.Y. Geng, B.L. Yue, P.C. Lo, J. Huang, H.L. Jin, Injectable hydrogel as a unique platform for antitumor therapy targeting immunosuppressive tumor microenvironment. Front. Immunol. **13**, 832942 (2022)
44. S.L. Gong, X.Q. Liang, M.M. Zhang, L. Li, T. He, Y. Yuan, X.C. Li, F.R. Liu, X. Yang, M.L. Shen, Q.J. Wu, C.Y. Gong, Tumor microenvironment-activated hydrogel platform with programmed release property evokes a cascade-amplified immune response against tumor growth, metastasis and recurrence. Small **18**, 2107061 (2022)
45. Y.Z. Dong, J. Luo, M.L. Pei, S. Liu, Y.C. Gao, H.M. Zhou, Y. Nueraihemaiti, X.C. Zhan, T.C. Xie, X.D. Yao, X. Guan, Y.F. Xu, Biomimetic hydrogel-mediated mechano-immunometabolic therapy for inhibition of ccRCC recurrence after surgery. Adv. Sci. **11**, 2308734 (2024)
46. J.Q. Zhou, D. Jia, L. Jin, W.L. He, Z.L. Zhang, Z.S. Zhang, T. Wu, U.O. Häfeli, X. Liu, X. Han, S.B. Wei, T.X. Gong, An injectable hydrogel for synergistic therapy in colorectal cancer by targeting glutathione. Cell Rep. Phys. Sci. **6**, 102399 (2025)
47. S. Yang, J.L. Wu, Z.Y. Wang, Y. Cheng, R. Zhang, C. Yao, D.Y. Yang, A smart DNA hydrogel enables synergistic immunotherapy and photodynamic therapy of melanoma. Angew. Chem. Int. Ed. **63**, 2319073 (2024)
48. S. Kim, J. Lee, H. Byun, J. Kim, W.J. Kim, Postoperative tumor suppression ROS scavenging, VEGF capture, and photothermal therapy using a split aptamer-loaded hydrogel. J. Control. Release **384**, 113883 (2025)
49. Y. Zhang, C.P. Zhu, Z.R. Zhang, J.L. Zhao, Y.K. Yuan, S.G. Wang, Oxidation triggered formation of polydopamine-modified carboxymethyl cellulose hydrogel for anti-recurrence of tumor. Colloid. Surface. B **207**, 112025 (2021)
50. T.T.A. Do, S. Grijalvo, T. Imae, M.J. Garcia-Celma, C. Rodríguez-Abreu, A nanocellulose-based platform towards targeted chemo-photodynamic/photothermal cancer therapy. Carbohydr. Polym. **270**, 118366 (2021)
51. M. Wang, D.R. Hu, Y. Yang, K. Shi, J.A. Li, Q.Y. Liu, Y.C. Li, R. Li, M. Pan, D. Mo, W. Chen, X.C. Li, Z.Y. Qian, Enhanced chemo-immunotherapy strategy utilizing injectable thermosensitive hydrogel for the treatment of diffuse peritoneal metastasis in advanced colorectal cancer. Adv. Sci. **10**, 2303819 (2023)
52. T.X. Chen, T.T. Yao, H. Pan, H. Peng, A.K. Whittaker, Y. Li, S.M. Zhu, Z.Y. Wang, One-step nanoarchitectonics of a multiple functional hydrogel based on cellulose nanocrystals for effective tumor therapy. Nano Res. **15**, 8636 (2022)
53. K.J. Shi, W.J. Fu, Z.F. Sabet, J.M. Ye, S.J. Liang, T. Liu, Q.L. Liu, M.Y. Guo, M. You, J.G. Wu, R. Bai, Y. Liu, B. Hu, X.J. Cui, J.Y. Li, C.Y. Chen, Hydrogel-mediated jamming of exosome communications that counter tumor adaption in the tumor immune microenvironment. ACS Nano **18**, 33042 (2024)

Chapter 11
Preparation and Properties of Cellulose-Based Aerogels

Yunxiao Liu and Jianming Zhang

Abstract Cellulose-based aerogels (CBAs) represent a sustainable class of materials with ultra-low density, high surface area, and exceptional porosity, offering significant potential for diverse applications including thermal insulation, environmental remediation, and energy storage. This chapter comprehensively examines the preparation pathways and resultant properties of CBAs derived from various cellulose sources (plant biomass, bacterial cellulose, and nanocellulose). It details the critical processing stages, which include raw material selection and pretreatment (delignification, nanocellulose production, dissolution), dissolution/dispersion in solvent systems (ionic liquids, alkali/urea, nanocellulose suspensions), gelation via physical (hydrogen bonding, freeze–thaw) or chemical crosslinking, and drying (freeze-drying, supercritical drying). The synthesis-structure–property relationships are emphasized, highlighting how processing choices impact the final aerogel's porosity, morphology, mechanical integrity, and functionality. Furthermore, the fabrication of specialized dimensional forms—films/membranes, continuous fibers (via wet spinning, freeze-spinning), and microspheres (via emulsification or simple dropping methods) is explored, demonstrating their application-specific advantages and tailored production routes. By integrating fundamental material science with process engineering, this chapter provides a foundation for designing high-performance, sustainable CBAs for advanced technological solutions.

Keywords Cellulose · Aerogels · Preparation · Properties · Sustainable materials

Y. Liu · J. Zhang (✉)
Key Laboratory of Rubber-Plastics, Ministry of Education, Shandong Provincial Key Laboratory of RubberPlastics, Qingdao University of Science and Technology, Qingdao, China
e-mail: zjm@qust.edu.cn

G. Wei and J. Zhang (eds.), *Cellulose-Based Hydrogels and Aerogels*, Lecture Notes in Nanoscale Science and Technology 41,
https://doi.org/10.1007/978-3-032-03008-5_11

11.1 Introduction

The relentless pursuit of sustainable advanced materials has propelled aerogels to the forefront of scientific and technological innovation. Characterized by their ultra-low density, high specific surface area, open-cell porosity (>90%), and exceptionally low thermal conductivity, these unique materials hold immense promise across diverse fields such as thermal insulation, environmental remediation (adsorption, filtration), catalysis, biomedicine, and energy storage.

This landscape presents a compelling opportunity for cellulose, the planet's most abundant, renewable, biocompatible, and biodegradable natural polymer, to emerge as a cornerstone for the next generation of high-performance, "green" aerogels. Sourced from wood, cotton, agricultural residues (e.g., bagasse, straw), industrial waste streams, or even microbial biosynthesis (bacterial cellulose), cellulose offers an unparalleled combination of sustainability and versatile chemical functionality [1]. The plethora of hydroxyl groups on its molecular chains facilitates extensive chemical modification and the engineering of intricate hierarchical structures. Cellulose-based aerogels (CBAs) leverage the inherent advantages of cellulose or its nanoscale derivatives, including cellulose nanocrystals (CNCs), cellulose nanofibrils/nanofibers (CNFs), and bacterial cellulose (BC), to construct three-dimensional (3D) nanoporous networks [2]. The fabrication pathway, as detailed extensively in this chapter, is critical to realizing the unique structural features and resultant properties of CBAs.

This chapter provides a comprehensive and in-depth examination of the science and engineering underpinning CBAs, with a primary focus on their preparation and the resulting properties. We systematically dissect each critical stage of the fabrication process-from the initial selection and pre-processing of cellulose sources, through the complexities of dissolution/dispersion and network formation (gelation and cross-linking), to the decisive drying methodologies (Fig. 11.1). Crucially, we explore how the choices made at each step profoundly shape the aerogel's structure and, consequently, its key physical, mechanical, thermal, and sportive properties. By establishing these critical structure-processing-property relationships, this chapter aims to serve as a foundational resource for researchers and engineers seeking to harness the remarkable potential of CBAs for sustainable technological solutions across a broad spectrum of applications. The discussion on performance sets the stage for understanding the practical implications of the presented fabrication strategies.

11.2 Raw Material Selection and Pretreatment

The development of CBAs begins with the selection of suitable cellulose sources and essential pretreatment steps to yield processable precursors. The most common raw materials are plant-derived cellulose from traditional sources like wood pulp, cotton linters, and bamboo, known for their established supply, and increasingly

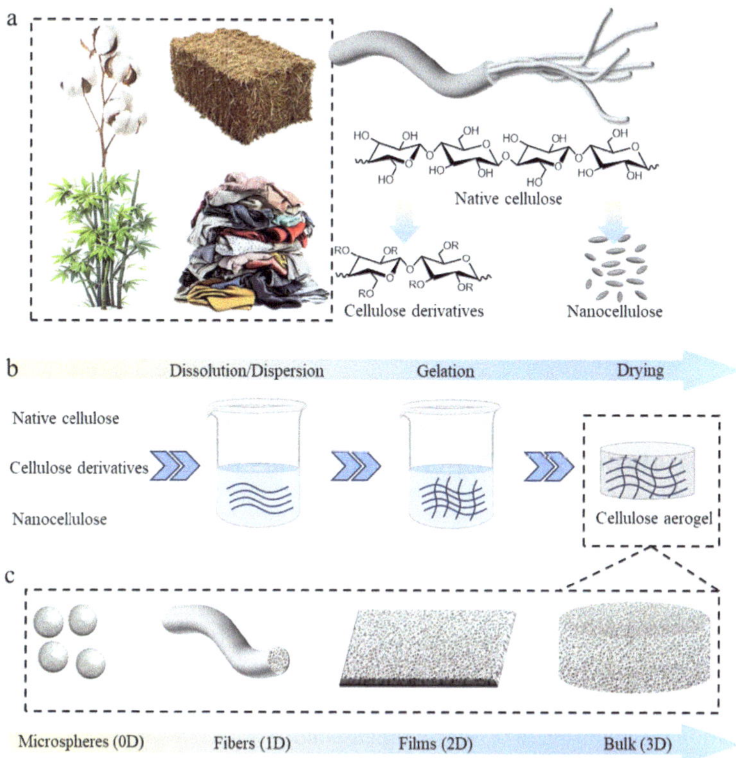

Fig. 11.1 Structural evolution and dimensional versatility of cellulose-based materials. **a** Natural cellulose sources, molecular structures highlighting functional groups (hydroxyl), and processed forms (cellulose derivatives, nanocellulose). **b** Fabrication workflow for cellulose aerogels: dissolution/dispersion of cellulose sources → gelation → drying → porous aerogel structure. **c** Tunable architectures: 0D microspheres, 1D fibers, 2D films, and 3D bulk materials for multifunctional applications

agro-industrial wastes such as sugarcane bagasse, corn stalks, and rice straw, valued for sustainability and cost-effectiveness. BC, produced microbially (e.g., using fruit waste substrates), offers distinct advantages including high purity, an inherent 3D nanofibrillar structure, and excellent mechanical strength, making it ideal for demanding applications like biomedicine. It is worth noting that the variety in cellulose sources affects the morphology of the gels and aerogels. One clear example is that for low to medium-DP cellulose, the raw material source generally does not significantly impact the properties and morphology of the resulting aerogel. Therefore, low and medium-DP cellulose can be readily utilized to manufacture cellulose aerogels. However, when preparing aerogels from high-DP cellulose, distinct structural features arise during drying processes such as vacuum freeze-drying and supercritical CO_2 drying [3]. In these cases, the aerogels form vertically aligned channels and large pores, often referred to as "fingers".

Pretreatment is mandatory to overcome the recalcitrance of raw cellulose and transform it into a form conducive to aerogel formation. Key processes include delignification via alkaline or acid hydrolysis to remove lignin/hemicellulose from plant biomass; fiber liberation and size reduction through mechanical means; production of nanocellulose, specifically flexible CNFs or rod-like CNCs, often enhanced by chemical treatments like TEMPO-oxidation to improve dispersion and gelation; and dissolution using solvents such as ionic liquids (ILs) or alkali/urea aqueous systems for regenerated aerogels, or the use of soluble derivatives like carboxymethyl cellulose (CMC). These fundamental pretreatment steps collectively tailor the physical and chemical properties of the cellulose precursor, thereby critically governing the achievable structure and performance of the final cellulose aerogel.

11.3 Dissolution and Dispersion Systems for CBA Fabrication

The fabrication of CBAS begins with dissolving or dispersing cellulose into a homogeneous solution or suspension, followed by gelation, solvent removal, and drying. The choice of dissolution or dispersion systems critically determines the regenerated cellulose structure, aerogel porosity, and mechanical properties. The following section examines key solvent and dispersion systems alongside their dissolution and stabilization mechanisms in cellulose processing.

11.3.1 Traditional Solvent Systems

Cellulose is insoluble in water and most organic solvents due to the large number of hydrogen bonds in the cellulose molecular chain, dispersion interactions and hydrophobic effects. To obtain the gel shape, cellulose should first be dissolved in a solvent to produce a gel. This is the starting point of aerogel production. *N*-methylmorpholine-*N*-oxide (NMMO) is a direct solvent for cellulose, capable of dissolving high-molecular-weight cellulose without derivatization [4–7]. The dissolution mechanism involves the disruption of inter- and intramolecular hydrogen bonds via the strong electron-donating ability of N–O groups. NMMO-based solutions enable the production of high-strength regenerated cellulose fibers (e.g., Lyocell) and aerogels. Other complex solvents such as *N*, *N*-dimethylacetamide/lithium chloride (DMAc/LiCl), *N*, *N*-dimethylformamide/dinitrogen tetroxide (DMF/N_2O_4), and dimethyl sulfoxide/tetrabutylammonium fluoride (DMSO/TBAF) have also been used to dissolve cellulose [8]. However, its high viscosity, thermal instability, and environmental concerns (e.g., solvent recovery costs) limit large-scale applications. Imidazolium-based ILs are currently considered highly potential solvents for cellulose and have been the subject of extensive research.

These ILs displayed a range of desirable characteristics such as nonvolatility, a high solvency power for cellulose, and exceptional thermal stability. These are superior to those of other cellulose solvents. [Bmim]Cl and [Emim]Ac have been widely considered at present for direct dissolution of cellulose, and had the capacity to dissolve cellulose up to approximately 15–20% [9–11].

11.3.2 Green and Emerging Solvent Systems

Developed as a low-cost and environmentally friendly alternative, alkali/urea systems (e.g., NaOH/urea/H_2O) dissolve cellulose via the formation of a cellulose-NaOH-urea inclusion complex at low temperatures (–10 to –5 °C) [12]. This system avoids toxic solvents and produces transparent cellulose solutions suitable for regenerated membranes, fibers, and aerogels. The rapid gelation kinetics (~minutes) and ice-templating compatibility enable the fabrication of anisotropic aerogels with aligned pores. Drawbacks include limited solubility for high-degree-of-polymerization cellulose and sensitivity to temperature fluctuations.

Furthermore, it is noteworthy that cellulose can be converted into more readily soluble derivatives via chemical reactions such as esterification and etherification. For instance, derivatives like cellulose nitrate[13, 14], cellulose acetate[15, 16], and carboxymethyl cellulose [17] can be rendered water-soluble by controlling their degree of substitution. The dissolution of these cellulose derivatives in water facilitates their subsequent gelation process, making them highly suitable materials for the preparation of cellulose aerogels.

11.3.3 Nanocellulose Dispersion Systems

Nanocellulose, as the smallest physical structural unit of cellulose, has a diameter of nanometer scale and a certain length-diameter ratio. As shown in Fig. 11.2, nanocellulose mainly consists of short rod-like CNCs (diameter 5–70 nm and length 100–250 nm), long fibrous CNFs (diameter 2–20 nm and length > 1 μm). Nanocellulose boasts numerous exceptional characteristics, including a large specific surface area, superior mechanical strength, high purity, and significant adsorption capacity [18, 19].

Typically, CNFs are produced from lignocellulosic materials through 2,2,6,6-tetramethylpiperidine-1-oxylradical (TEMPO)-mediated oxidation. This process selectively converts surface hydroxyl groups to carboxylates (–COO^-), creating electrostatic repulsion that enables stable dispersion in water and facilitates low-energy mechanical fibrillation into individualized nanofibrils [20]. This method employs two primary systems: an alkaline approach using catalytic TEMPO and sodium bromide (NaBr) in water at pH 10–11, with sodium hypochlorite (NaClO) as the primary oxidant [20]; and an acidic approach utilizing TEMPO, NaClO, and sodium

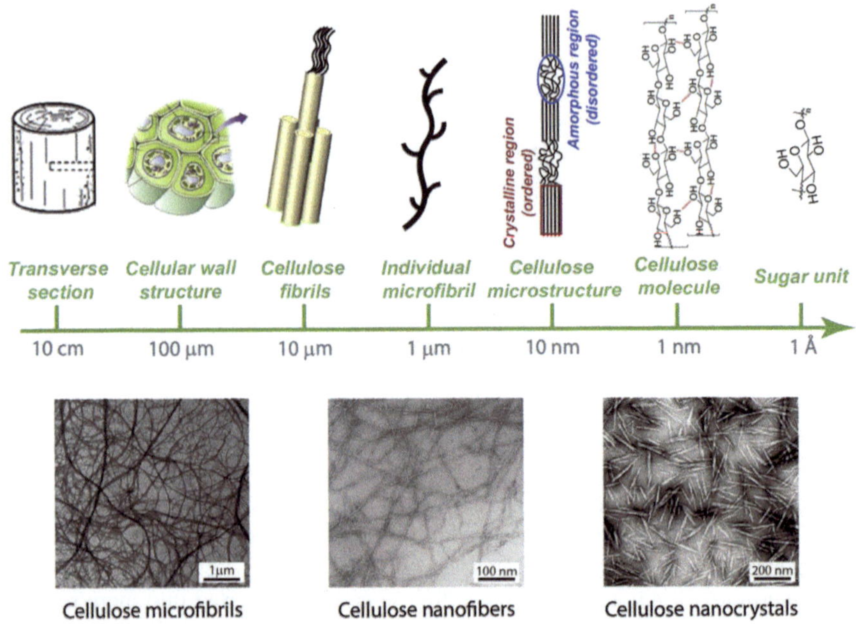

Fig. 11.2 Hierarchical structure of cellulose and TEMs of cellulose microfibrils, nanofibers, and nanocrystals. Reprinted with the permission from Calvino et al. [19], Copyright 2020, Elsevier Ltd

chlorite ($NaClO_2$) in water at pH 4.8–6.8 under elevated temperatures (typically 40–60 °C) for one to three days [21]. In both cases, oxidation selectively converts the C6 primary hydroxyl groups of cellulose to carboxylate anions ($-COO^-$), generating significant electrostatic repulsion between fibrils. This repulsion facilitates efficient separation into individualized CNFs during mild mechanical disintegration. Crucially, the oxidation process primarily targets accessible fibril surfaces and amorphous regions, thereby preserving the underlying fibrous morphology and core crystalline structure of the cellulose [21].

The most well-studied and most widely used protocol to produce CNCs uses concentrated sulfuric acid to hydrolyse and esterify cellulose in less than 2 h with yields in the 20–75% range. Sulfuric acid is a strong acid, and the high concentration of protons attacks the glycosidic bonds faster in less ordered regions of cellulose. As these bonds are broken, the cellulose chains become shorter and the average degree of polymerization decreases until the disordered regions have been fully degraded and the levelling off degree of polymerization (LODP) is reached. Sulfuric acid hydrolysis introduces sulfate half-ester groups ($–OSO_3^-$) on crystalline domains [22]. The resulting negative charges prevent aggregation via electrostatic stabilization, allowing homogeneous aqueous suspensions critical for nanomaterial applications. In recent years, advanced oxidation technologies have driven significant interest in the oxidative preparation of CNCs. Strong oxidizing agents, including persulfates, periodates, permanganates, and hydrogen peroxide, are employed to generate CNCs

[23–25]. During the hydrolysis of cellulose, these oxidants perform in situ oxidation of surface hydroxyl groups, producing carboxyl groups (-COOH). These carboxyl groups subsequently ionize in aqueous systems, conferring electrostatic repulsion that facilitates nanocrystal dispersion. The surface charges on both CNCs and CNFs provide sites for gelation through ionic crosslinking or electrostatic interactions, making them versatile precursors for cellulose aerogels.

11.4 Gelation and Cross-Linking

The gelation of cellulose is an important step in preparing cellulose aerogel. Gelation generally refers to interactions between molecules (hydrogen bonding, hydrophobic interactions, etc.) to form interconnected polymer networks, which can also be called polymer solutions becoming solid. During gelation, three-dimensional networks of cellulose are constructed, enhancing the gel's strength with their network frameworks. The process of gel formation includes both physical and chemical cross-linking mechanisms [26]. Physical cross-linking primarily results from hydrogen bonding within or between molecules, as well as physical entanglement among molecules [27]. Chemical cross-linking usually requires the addition of extra cross-linking agents to help create the cross-linked matrix [28]. The common physical cross-linking approaches are achieved mainly by solvent exchange, freeze–thaw cycling and adding metal salt ions ($CaCl_2$, etc.). For solution systems of cellulose and its derivatives, solvent exchange is the most effective gelation technique. Non-solvent-induced phase separation occurs when a material is in solution prior to the solvent exchange. Nanocellulose suspensions can create gels through freeze–thaw cycling, a process in which the expansion of ice crystals aids in the development of solid physical networks [29]. Furthermore, the pore structure and appearance of the aerogel are significantly influenced by the freezing temperature. Additionally, electrostatic attraction serves as an alternative method for creating physical gels through cross-linking [30]. Unlike hydrogen bonding, electrostatic interaction can enhance the hygroscopic strength of porous nanomaterials prepared from nanocellulose materials.

Chemical cross-linking method refers to the cross-linking of cellulose and crosslinking agent (epoxy compound, aldehyde compound, poly-amino compounds, and acids) under corresponding reaction conditions. The aim of crosslinking is to create network structure between cellulose molecular chains, maintaining a good three-dimensional structure and improving the physical and mechanical properties. In general, chemical crosslinking is faster than physical crosslinking, so chemical crosslinking results in a more stable configuration.

11.5 Drying Process

Drying represents the final and most critical step in aerogel fabrication, essential regardless of the initial preparation route [28, 31]. Its primary objective is to prevent network collapse-characterized by pore structure degradation-caused by capillary forces at air–liquid interfaces during solvent removal. Two principal techniques address this challenge: freeze-drying (FD) and supercritical drying (SCD) [32–34] (Fig. 11.3).

11.5.1 Freeze-Drying

FD removes solvent while preserving low density, high porosity, and high surface area. This technique avoids liquid/vapor interfaces through a three-stage process: (1) reducing temperature below the solvent's freezing point to induce solidification; (2) applying vacuum to lower pressure; and (3) elevating temperature under low pressure to sublime the solid solvent directly into vapor, yielding aerogels (or cryogels when FD-derived). Careful execution is vital, as uncontrolled ice crystal growth during freezing can expand and damage the gel network, leading to large pores (up to hundreds of microns), thick non-porous walls, reduced density, and diminished surface area [31, 34]. Rapid freezing techniques-such as immersion in liquid nitrogen

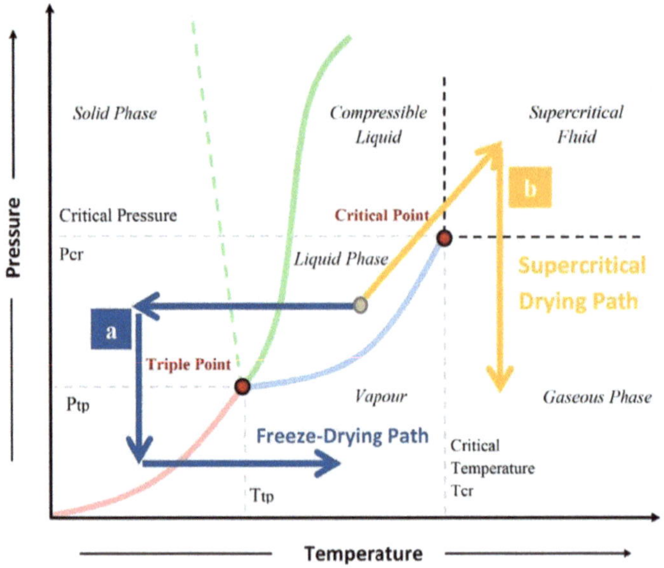

Fig. 11.3 Phase diagram of water, with the process of removal through FD and *SCD*. Reprinted with the permission from Zaman et al. [34], Copyright 2020, Elsevier Ltd

(unidirectional freezing) or ultra-cold freezers-minimize crystal growth and preserve nanostructure integrity. For instance, rapid freezing of CNF gels in liquid ethanol (–114 °C) yielded 1–60 nm pores, while slower freezing in benzonitrile (–13 °C) produced larger voids [32]. Employing *tert*-butanol (TBA)/water mixtures near their eutectic point also suppresses crystal growth. This approach yielded denser aerogels (~0.114 kg/m^3) compared to pure water systems (~0.044 kg/m^3) and facilitated cellulose microfibril aerogels with surface areas of ~ 100 m^2/g [31, 35].

11.5.2 Supercritical Drying

SCD eliminates capillary forces by bypassing the liquid–vapor phase boundary. The process involves: (1) replacing the gel's pore solvent (e.g., water) with a fluid miscible with supercritical CO_2 ($scCO_2$), typically an alcohol, due to water's immiscibility with $scCO_2$ and oxidizing properties under supercritical conditions; (2) introducing CO_2 and elevating pressure (>7.38 MPa) and temperature (>31.3 °C) to achieve supercritical state-favored for CO_2's accessibility, low cost, and safety; (3) depressurizing slowly once pores contain only supercritical fluid, allowing gas-phase CO_2 exit without interface formation or structural damage [31, 32, 36]. SCD typically minimizes shrinkage and enables high surface areas (e.g., 605 m^2/g at 10 MPa) [37]. Studies confirm SCD yields lower shrinkage than FD [38], though its adoption is constrained by expensive high-pressure equipment and operational complexity [28, 32, 36].

Comparative studies underscore performance differences. Buchtová and Budtova evaluated microcrystalline cellulose gels (3–11 wt% in DMSO/[Emim]Ac) dried via: FD (with/without pre-unidirectional freezing in liquid N_2 at –196 °C), $scCO_2$, and oven vacuum drying [39]. FD with pre-freezing minimized shrinkage and density while maximizing porosity; $scCO_2$ drying delivered the highest surface area; vacuum drying performed poorest. Separately, Wang et al. (2017) utilized TBA/water exchange (replacing $CaCl_2$ gelling solution) prior to FD for nanocellulose, achieving high surface area (164.97 m^2/g) and low shrinkage (5.89%), highlighting the efficacy of solvent modification in FD protocols [40].

11.6 Various Dimensional CBAs: Specialized Forms and Fabrication

Beyond conventional monolithic forms (e.g., cylinders), CBAs can be engineered into functional microspheres, continuous fibers and thin films/membranes [41]. These specialized geometries open unique application pathways, demanding tailored fabrication approaches derived from bulk aerogel processes.

11.6.1 CBA Films/Membranes

CBA films/membranes, characterized by their reduced thickness relative to bulk aerogels, are fabricated using core principles similar to those for bulk structures but often incorporate modifications suited to their form (Fig. 11.4). The inherent fine-scale structure of aerogels translates well into films/membranes, often leading to enhanced gas filtration performance compared to their bulk counterparts [42]. Preparation routes leverage established techniques:

Sol–Gel Casting with Post-Drying: A common approach involves forming a hydrogel via sol–gel chemistry, often followed by functionalization (e.g., hydrophobic modification) and subsequent freeze-drying. Zhang et al. fabricated a 3D Janus all-cellulose aerogel via sol–gel processing and freeze-drying, creating controllable asymmetric wettability to circumvent the membrane "Trade-off" effect for sustainable oil–water separation. This pathway successfully produces membranes with excellent adsorption and filtration capabilities [43] (Fig. 11.4a).

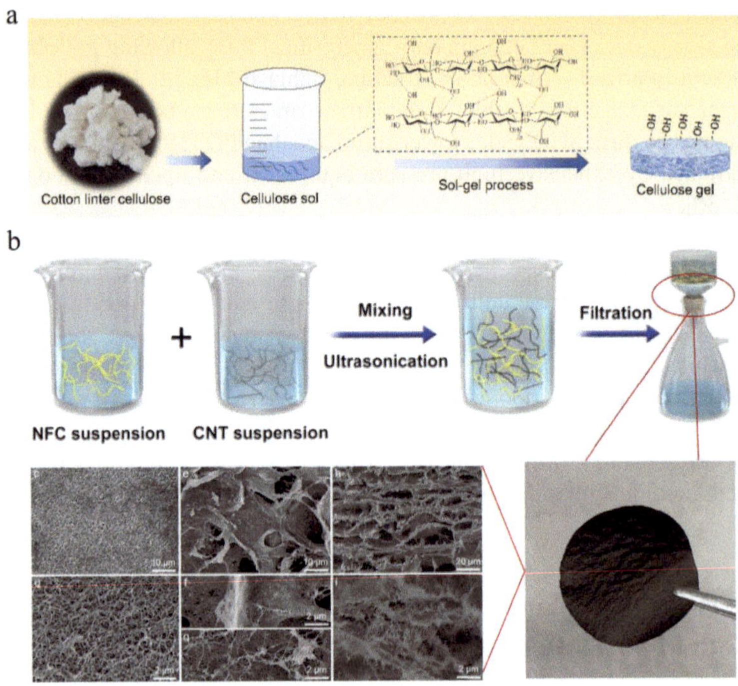

Fig. 11.4 Preparation methods for CBA membranes: **a** Hydrophobic cellulose aerogel membrane produced via sol–gel processing and freeze-drying. Reprinted with the permission from Zhang et al. [43], Copyright 2023, Elsevier Ltd. **b** Electrically conductive CBA film fabricated through filtration and freeze-drying. Reprinted with the permission from Bai et al. [45], Copyright 2021, American Chemical Society

Filtration/Deposition Followed by Drying: Thin layers of CNF dispersions, often mixed with functional additives (e.g., MXene, CNTs), can be deposited via filtration or casting into molds. Subsequent drying-employing freeze-drying, supercritical CO_2 drying, or ambient drying techniques-yields aerogel films. Surface modifications can further enhance functionality. For example, Xin et al. produced freeze-dried CNF/MXene composite films; subsequent silicon coating imparted hydrophobicity, photothermal properties, and effective EMI shielding, making them suitable for humid-environment protective gear [44]. Similarly, Bai et al. fabricated high-aspect-ratio, asymmetrically porous NFC/CNT composite aerogel films by filtering suspensions and freeze-drying the resulting hydrogels [45] (Fig. 11.4b). These films demonstrated benefits for Li–S batteries by promoting electrolyte penetration, mitigating polysulfide shuttling, and facilitating rapid electron transport.

11.6.2 CBA Fibers

Aerogel fibers represent a distinct class of materials, merging the ultralight, porous architecture of aerogels with the processability and slenderness of fibers. This combination significantly expands their potential application scope (Fig. 11.5). The fabrication methods primarily include these aspects.

Wet Spinning: This dominant method involves dissolving cellulose (e.g., in ionic liquids like [BMIM]Cl or urea/alkali aqueous systems) to form a spinning dope, potentially incorporating functional fillers (e.g., CNTs). The solution is extruded through a spinneret into a coagulating bath containing non-solvent(s), inducing rapid solidification/gelation into fibers. Parameters like solution concentration, flow rate, nozzle size, and bath composition critically influence fiber diameter, uniformity, and mechanical properties [46]. Post-spinning, fibers undergo drying (typicallySCD or FD) to achieve the aerogel structure. This process yields fibers with excellent mechanical characteristics; Liu et al. fabricated super tough CBA fibers via wet-spinning and SCD of ionic-liquid-dissolved cellulose, achieving record-breaking toughness (21.85 MJ/m^3) that exceeds biological ligaments while maintaining high porosity, wearability, and thermal insulation for advanced wearable textiles [47] (Fig. 11.5a).

Freeze-Spinning: Inspired by polar bear hair, this technique involves extruding a spinning dope onto a cold plate, inducing directional freezing as the fiber forms. Ice crystal growth aligns within the freezing direction, creating a porous template. Subsequent freeze-drying removes the ice, leaving behind an aerogel fiber with aligned porosity [48] (Fig. 11.5b). This method enables continuous production. Wu et al. further innovated by producing bionic encapsulated aerogel fibers via freeze-spinning and coating, creating a unique core–shell structure: a porous, insulating core locked within a dense, protective shell, decoupling and optimizing insulation and mechanical properties [49].

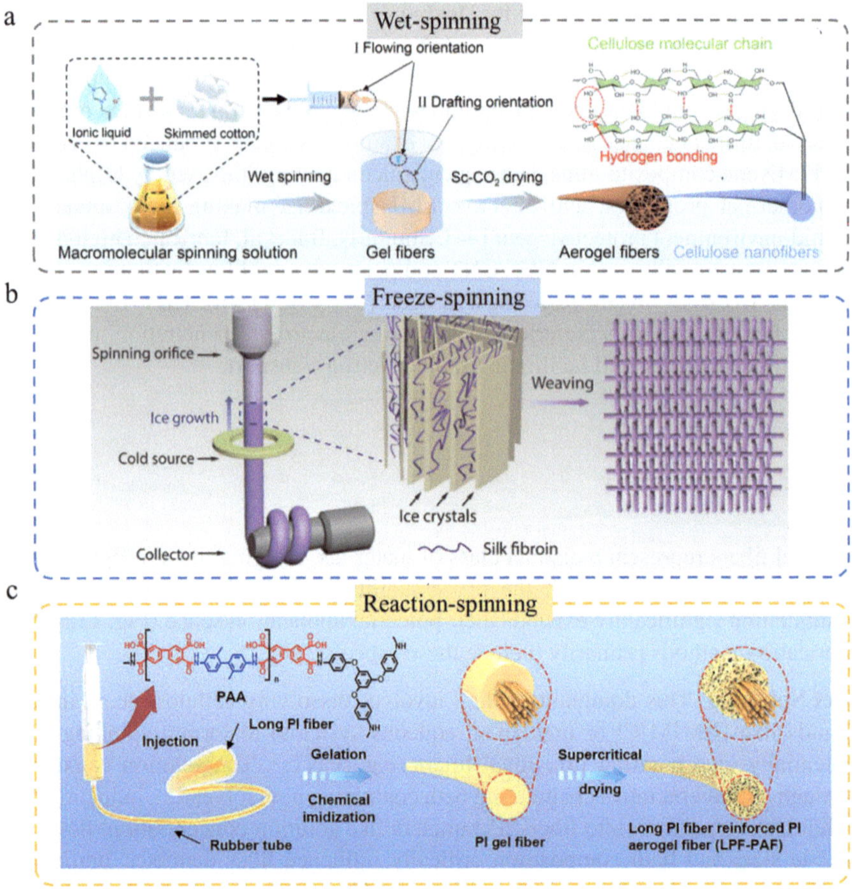

Fig. 11.5 The synthesis technologies for CBA fibers: **a** Wet-spinning. Reprinted with the permission from Liu et al. [47], Copyright 2023, American Chemical Society. **b** Freeze-spinning. Reprinted with the permission from Cui et al. [48], Copyright 2018, Wiley. **c** Reaction spinning. Reprinted with the permission from Zhu et al. [54], Copyright 2023, American Chemical Society

Reaction Spinning: Here, polymerization or cross-linking reaction kinetics control the solidification process rapidly after extrusion from the spinneret [50]. While less common for pure cellulose, it offers potential for achieving specific properties like transparency. Recent advances include sophisticated strategies like microfluidic-assisted dynamic dual-cross-linking, yielding fibers with unconventional structures like a dense inner core surrounded by a tunable porous sponge-like layer, contrasting with the classic skin–core structure of wet-spun fibers [51–54] (Fig. 11.5c).

11.6.3 CBA Microspheres

CBA microspheres in the size range of 1–3000 μm can be prepared by vigorously stirring a water-in-oil emulsion, where an aqueous cellulose solution (sol dispersed phase) is emulsified within an oil phase (continuous phase), promoting gelation of the dispersed phase [55, 56]. Following gelation, the microspheres are separated via filtration or centrifugation, washed, and dried under suitable conditions [57]. The size distribution of these microspheres is demonstrably influenced by agitation intensity, precursor concentrations, surfactant presence (if used), and the aqueous-to-oil phase volume ratio [58, 59]. For producing millimeter to centimeter-scale aerogel beads, a droplet extrusion method is typically employed: the cellulose solution is extruded through a syringe nozzle into a solution containing a crosslinker or gelation promoter (such as specific ions, controlled temperature/pH environments) to induce immediate gelation upon droplet contact [60]. Critically, the size of the beads produced by this approach is primarily governed by the orifice diameter of the syringe nozzle used during extrusion. The primary goal of droplet-based methods is to control droplet morphology, size, and size distribution. Two dominant approaches are simple dropping and emulsification [61] (Fig. 11.6).

Simple dropping utilizes devices such as syringes or syringe pumps to extrude cellulose solutions through a nozzle [62]. Droplets form when ejection forces (gravity and pressure) overcome the solution's surface tension and capillary forces at the outlet [63]. The droplets then fall into a gelation bath (e.g., crosslinker solution or controlled pH/temperature environment). Uniformity is significantly influenced by flow rate control: manual extrusion produces non-uniform beads, whereas syringe pump extrusion enables precise flow regulation for highly uniform cellulose gel beads [64]. This method readily achieves particles of 0.2–10 mm, but requires thorough parameter optimization via experimental design and prior characterization of system properties (viscosity, gelation kinetics, temperature dependencies). Despite being the simplest production approach, syringe extrusion remains constrained by solution viscosity limitations and comparatively low throughput [65].

Emulsification disperses a cellulose aqueous phase (dispersed phase) into an immiscible oil phase (continuous phase, e.g., silicone oil), stabilized by surfactants (e.g., Span80, Tween80) [66, 67]. Mechanical agitation breaks droplets into finer sizes until energy input balances interfacial tension [68, 69]. Subsequent physical (e.g., cooling) or chemical (e.g., crosslinkers) triggers induce gelation. The emulsification process enables large-scale production of polymeric microspheres with controllable particle sizes and distributions, providing an efficient route to customizable microspherical aerogels when combined with appropriate drying techniques. This approach facilitates the synthesis of CNFs aerogel microspheres with tunable particle and pore sizes. Zhai et al. developed a novel W/O emulsification and freeze-drying method to produce size-controlled PVA/CNF aerogel microspheres, strategically selecting oil phases and cooling agents to ensure separation of frozen ice microspheres from the emulsion [70]. Similarly, glutaraldehyde-crosslinked PVA/CNF aerogel microspheres with distinct size distributions were successfully produced using combined

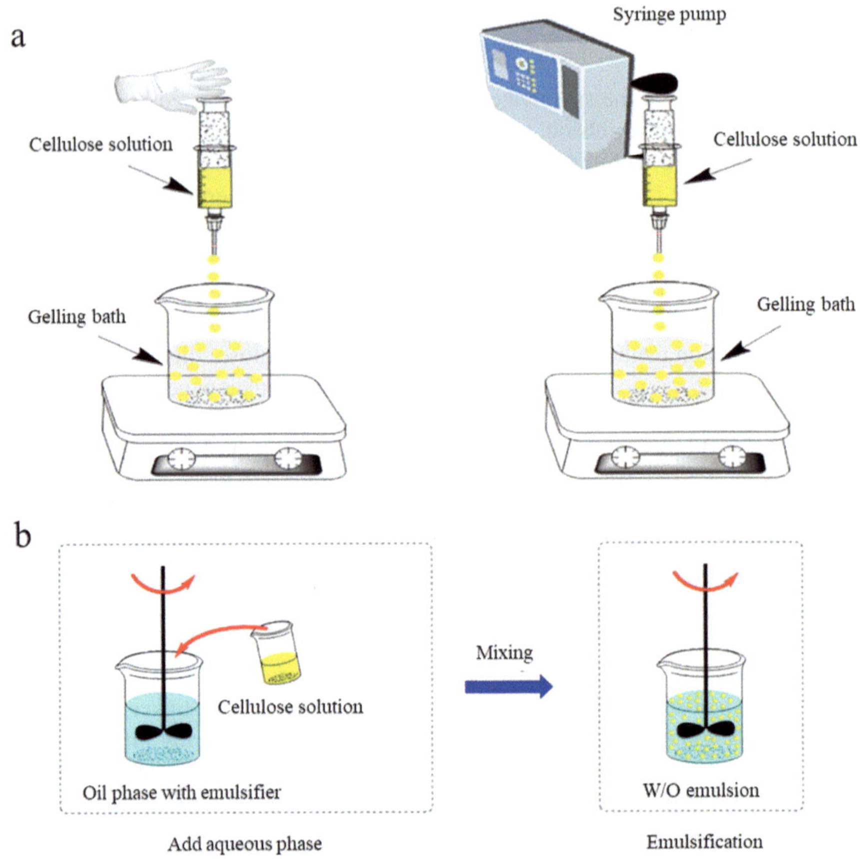

Fig. 11.6 Comparative schematic of cellulose aerogel microsphere fabrication methods. **a** Simple dropping approach: Manual extrusion *vs.* automated extrusion injecting cellulose solution into a gelation bath. **b** Emulsification process: Sequential steps: (i) Addition of aqueous cellulose phase to surfactant-containing oil phase; (ii) Mixing under mechanical agitation; (iii) Formation of stabilized W/O emulsion with discrete microdroplets. Reprinted with the permission from Shi et al. [61], Copyright 2021, Elsevier Ltd

W/O emulsification and FD [71]. This emulsion-gelation approach achieves particle sizes of 1–1500 μm without specialized equipment, where spherical morphology is maintained through regulated stirring and surfactant selection. Compared to dropping methods, this technique demonstrates superior reproducibility for small aerogel particles while eliminating complex equipment requirements.

The selection of appropriate bead production techniques is contingent on the target particle size and scale requirements. Simple dropping is economical for millimeter-scale beads but limited by solution viscosity and low throughput. Emulsification allows scalable production of micrometer-scale microspheres with tunable sizes via surfactant/stirring control, though droplet coalescence may cause size heterogeneity without optimized parameters.

11.7 Conclusion and Outlooks

CBAs have emerged as a transformative class of sustainable, high-performance materials, capitalizing on cellulose's renewable abundance, biocompatibility, and versatile chemical functionality. This chapter comprehensively detailed the critical fabrication pathway-encompassing raw material selection and pretreatment, dissolution/dispersion in tailored solvent systems (ionic liquids, alkali/urea, nanocellulose suspensions), controllable gelation (physical/chemical cross-linking), and optimized drying (FD and SCD). Each processing step profoundly shapes the final aerogel's ultralight, nanoporous structure and resultant properties. Furthermore, the engineering of dimensional forms-films/membranes for enhanced filtration, wet-spun or freeze-spun fibers for textile integration, and microspheres via emulsification for adsorption-demonstrates CBAs' exceptional versatility. These attributes underpin their significant promise across thermal insulation, environmental remediation (oil/water separation, contaminant adsorption), energy storage, catalysis, and biomedicine.

Despite substantial progress, challenges remain in scaling production cost-effectively, enhancing mechanical robustness under humid conditions, standardizing nanocellulose isolation, and reducing the energy footprint of drying processes. Future research should prioritize integrating CBAs with functional nanomaterials (e.g., MXenes, CNTs) to engineer composites with enhanced properties like electrical conductivity, EMI shielding, or catalytic activity. Developing truly multifunctional CBAs exhibiting stimuli-responsiveness, asymmetric wettability (e.g., Janus structures), or hierarchical porosity will unlock next-generation smart applications. Advancing greener synthesis routes, such as low-energy photo-gelation or recyclable solvent systems, is crucial for improved sustainability. Concurrently, optimizing high-throughput techniques like continuous freeze-spinning or scalable emulsion processing is essential to translate laboratory innovations into commercially viable products particularly for large-scale applications in atmospheric water harvesting, carbon capture, personal thermal management, and advanced filtration. Realizing this potential requires deeper fundamental understanding of structure–property relationships across scales.

References

1. Y.S. Ai, L. Zhang, M. Cui, R.L. Huang, W. Qi, Z.M. He, R.X. Su, Toward cleaner production of nanocellulose: a review and evaluation. Green Chem. **24**, 6406–6434 (2022)

2. S. Sen, A. Singh, C. Bera, S. Roy, K. Kailasam, Recent developments in biomass derived cellulose aerogel materials for thermal insulation application: a review. Cellulose **29**, 4805–4833 (2022)
3. M. Négrier, E. El Ahmar, R. Sescousse, M. Sauceau, T. Budtova, Upcycling of textile waste into high added value cellulose porous materials, aerogels and cryogels. RSC Sustain. **1**, 335–345 (2023)
4. H.F. Peng, G.L. Dai, S.P. Wang, H.Y. Xu, The evolution behavior and dissolution mechanism of cellulose in aqueous solvent. J. Mol. Liq. **241**, 959–966 (2017)
5. R. Protz, A. Lehmann, J. Ganster, H.P. Fink, Solubility and spinnability of cellulose-lignin blends in aqueous NMMO. Carbohydr. Polym. **251**, 117027 (2021)
6. Y.X. Guo, J. Cai, T.Q. Sun, L.Q. Xing, C.Z. Cheng, K.D. Chi, T. Li, The purification process and side reactions in the N-methylmorpholine-N-oxide (NMMO) recovery system. Cellulose **28**, 7609–7617 (2021)
7. B. Medronho, B. Lindman, Competing forces during cellulose dissolution: from solvents to mechanisms. Cur. Opin. Colloid Interf. Sci. **19**, 32–40 (2014)
8. A.J. Sayyed, N.A. Deshmukh, D.V. Pinjari, A critical review of manufacturing processes used in regenerated cellulosic fibres: viscose, cellulose acetate, cuprammonium, LiCl/DMAc, ionic liquids, and NMMO based lyocell. Cellulose **26**, 2913–2940 (2019)
9. C. Verma, A. Mishra, S. Chauhan, P. Verma, V. Srivastava, M.A. Quraishi, E.E. Ebenso, Dissolution of cellulose in ionic liquids and their mixed cosolvents: a review. Sustain. Chem. Pharm. **13**, 100162 (2019)
10. N. Luo, Y.X. Lv, D.X. Wang, J.M. Zhang, J. Wu, J.S. He, J. Zhang, Direct visualization of solution morphology of cellulose in ionic liquids by conventional TEM at room temperature. Chem. Commun. **48**, 6283–6285 (2012)
11. Y. Li, J.J. Wang, X.M. Liu, S.J. Zhang, Towards a molecular understanding of cellulose dissolution in ionic liquids: anion/cation effect, synergistic mechanism and physicochemical aspects. Chem. Sci. **9**, 4027–4043 (2018)
12. X. Chen, C. Burger, D. Fang, D. Ruan, L. Zhang, B.S. Hsiao, B. Chu, X-ray studies of regenerated cellulose fibers wet spun from cotton linter pulp in NaOH/thiourea aqueous solutions. Polymer **47**, 2839–2848 (2006)
13. Y.A. Gismatulina, V.V. Budaeva, G.V. Sakovich, Cellulose nitrates from intermediate flax straw. Russ. Chem. Bull. **65**, 2920–2924 (2016)
14. S.Y. Li, Y. Li, Y.J. Ding, Y. Yuan, Z.L. Xiao, Chemical modification of nitrocellulose by grafting sodium carboxymethyl. Cellulose **29**, 8103–8115 (2022)
15. Y. Watabe, Y. Suzuki, S. Koike, S. Shimamoto, Y. Kobayashi, Cellulose acetate, a new candidate feed supplement for ruminant animals: In vitro evaluations. J. Dairy Sci. **101**, 10929–10938 (2018)
16. S.J. Feng, L. Yao, M.X. Feng, H.R. Cai, X. He, M. He, Y.M. Zhou, Regeneration of pea-pod-like cellulose acetate fibers as aerogel-derived boards for building thermal regulation and carbon reduction. Advanced Fiber Materials **6**, 570–582 (2024)
17. Z.M. Liu, S.F. Zhang, B. He, S.J. Wang, F.G. Kong, Synthesis of cellulose aerogels as promising carriers for drug delivery: a review. Cellulose **28**, 2697–2714 (2021)
18. C.M. Clarkson, S.M. El Awad Azrak, E.S. Forti, G.T. Schueneman, R.J. Moon, J.P. Youngblood, Recent developments in cellulose nanomaterial composites. Adv. Mater. **33**, 2000718 (2020)
19. C. Calvino, N. Macke, R. Kato, S.J Rowan, Development, processing and applications of bio-sourced cellulose nanocrystal composites. Prog. Polym. Sci. **103** (2020)
20. A. Isogai, T. Saito, H. Fukuzumi, TEMPO-oxidized cellulose nanofibers. Nanoscale **3**, 71–85 (2011)
21. A. Isogai, Y.X. Zhou, Diverse nanocelluloses prepared from TEMPO-oxidized wood cellulose fibers: nanonetworks, nanofibers, and nanocrystals. Cur. Opin. Solid State Mate. Sci. **23**, 101–106 (2019)
22. B. Sun, M. Zhang, Q.X. Hou, R. Liu, T. Wu, C.L. Si, Further characterization of cellulose nanocrystal (CNC) preparation from sulfuric acid hydrolysis of cotton fibers. Cellulose **23**, 439–450 (2016)

23. L.J. Zhou, N. Li, J. Shu, Y.X. Liu, K.T. Wang, X. Cui, J.M. Zhang, One-pot preparation of carboxylated cellulose nanocrystals and their liquid crystalline behaviors. ACS Sustain. Chem. Eng. **6**, 12403–12410 (2018)
24. Y.X. Liu, H.X. Wei, Z.W. Liu, X.R. Liu, Y.J. Fang, M. Jiang, J.M. Zhang, Ultrafast and energy-saving extraction of cellulose nanocrystals. Green Chem. **24**, 6823–6829 (2022)
25. P.W. Liu, B. Pang, S. Dechert, X.C. Zhang, L.B. Andreas, S. Fischer, K. Zhang, Structure selectivity of alkaline periodate oxidation on lignocellulose for facile isolation of cellulose nanocrystals. Angew. Chem. Int. Ed. **59**, 3218–3225 (2020)
26. Y.M. Chen, L. Zhang, Y. Yang, B. Pang, W.H. Xu, G.G. Duan, K. Zhang, Recent progress on nanocellulose aerogels: preparation, modification, composite fabrication. Applications. Adv. Mater. **33**, 2005569 (2021)
27. T.M. Zhang, Y. Zhang, X.Y. Wang, S. Liu, Y. Yao, Characterization of the nano-cellulose aerogel from mixing CNF and CNC with different ratio. Mater. Lett. **229**, 103–106 (2018)
28. L.Y. Long, Y.X. Weng, Y.Z. Wang, Cellulose aerogels: synthesis, applications, and prospects. Polymers **10**, 623 (2018)
29. D. Ciftci, A. Ubeyitogullari, R.R. Huerta, O.N. Ciftci, R.A. Flores, M.D.A. Saldaña, Lupin hull cellulose nanofiber aerogel preparation by supercritical CO_2 and freeze drying. J. Supercrit. Fluids **127**, 137–145 (2017)
30. D.H. Li, Y. Wang, Y.Y. Sun, Y. Lu, S. Chen, B.B. Wang, D.J. Yang, Turning gelidium amansii residue into nitrogen-doped carbon nanofiber aerogel for enhanced multiple energy storage. Carbon **137**, 31–40 (2018)
31. T. Budtova, Cellulose II aerogels: a review. Cellulose **26**, 81–121 (2019)
32. N. Lavoine, L. Bergström, Nanocellulose-based foams and aerogels: processing, properties, and applications. J. Mater. Chem. A **5**, 16105–16117 (2017)
33. H. Al Abdallah, J.H. Tannous, B. Abu-Jdayil, Cellulose and nanocellulose aerogels, their preparation methods, and potential applications: a review. Cellulose **31**, 2001–2029 (2024)
34. A. Zaman, F. Huang, M. Jiang, W. Wei, Z. Zhou, Preparation, properties, and applications of natural cellulosic aerogels: a review. Energy Built Environ. **1**, 60–76 (2020)
35. M. Fumagalli, D. Ouhab, S.M. Boisseau, L. Heux, Versatile gas-phase reactions for surface to bulk esterification of cellulose microfibrils aerogels. Biomacromol **14**, 3246–3255 (2013)
36. M.A. Dahlem, C. Borsoi, B. Hansen, A.L. Catto, Evaluation of different methods for extraction of nanocellulose from yerba mate residues. Carbohydr. Polym. **218**, 78–86 (2019)
37. L. Heath, W. Thielemans, Cellulose nanowhisker aerogels. Green Chem. **12**, 1448–1453 (2010)
38. S. Lee, K.Y. Kang, M.J. Jeong, A. Potthast, F. Liebner, Evaluation of supercritical CO_2 dried cellulose aerogels as nano-biomaterials. J. Korean Phy. Soc. **71**, 483–486 (2017)
39. N. Buchtová, T. Budtova, Cellulose aero-, cryo- and xerogels: towards understanding of morphology control. Cellulose **23**, 2585–2595 (2016)
40. X.Y. Wang, Y. Zhang, H. Jiang, Y.X. Song, Z.B. Zhou, H. Zhao, Tert-butyl alcohol used to fabricate nano-cellulose aerogels via freeze-drying technology. Mater. Res. Exp. **4**, 065006 (2017)
41. Q.L. Zhang, L.X. Li, H. Wu, Y.L. Cheng, C.K. Liu, C.Q. Fang, Recent advances in cellulose based aerogels with various dimensions: design, functionalization, and applications. Cellulose **32**, 1–27 (2025)
42. M.S. Toivonen, A. Kaskela, O.J. Rojas, E.I. Kauppinen, O. Ikkala, Ambient-dried cellulose nanofibril aerogel membranes with high tensile strength and their use for aerosol collection and templates for transparent. Flexible Devices. Adv. Funct. Mater. **25**, 6618–6626 (2015)
43. J.J. Zhang, C.C. Wang, Z.B. Lin, Q. Fu, C.X. Niu, Z.W. Zhao, L.B. Lu, Asymmetric wettable three-dimensional Janus all-cellulose aerogel: Controllable construction with a similar phase self-assembling strategy and oil-water separation performance. J. Environ. Chem. Eng. **11**, 110776 (2023)
44. W. Xin, M.G. Ma, F. Chen, Silicone-coated MXene/Cellulose nanofiber aerogel films with photothermal and joule heating performances for electromagnetic interference shielding. ACS Appl. Nano Mater. **4**, 7234–7243 (2021)

45. L.L. Bai, J.S. Ma, H.Q. Song, Y. Yang, C.Y. Zhi, S.Y. Lee, W.S. Chen, Flexible, electrically conductive, nanostructured, asymmetric aerogel films for lithium-sulfur batteries. Acs Appl. Mater. Interfaces **13**, 59174–59184 (2021)
46. M. Rostamitabar, A.M. Abdelgawad, S. Jockenhoevel, S. Ghazanfari, Drug-eluting medical textiles: from fiber production and textile fabrication to drug loading and delivery. Macromol. Biosci. **21**, 2100021 (2021)
47. Z.S. Liu, Z.Z. Sheng, Y.Q. Bao, Q.Q. Cheng, P.X. Wang, Z.W. Liu, X.T. Zhang, Ionic liquid directed spinning of cellulose aerogel fibers with superb toughness for weaved thermal insulation and transient impact protection. ACS Nano **17**, 18411–18420 (2023)
48. Y. Cui, H.X. Gong, Y.J. Wang, D.W. Li, H. Bai, A thermally insulating textile inspired by polar bear hair. Adv. Mater. **30**, 1706807 (2018)
49. M.R. Wu, Z.Y. Shao, N.F. Zhao, R.Z. Zhang, G.D. Yuan, L.L. Tian, H. Bai, Biomimetic, knittable aerogel fiber for thermal insulation textile. Science **382**, 1379–1383 (2023)
50. Y. Du, X.H. Zhang, J. Wang, Z.W. Liu, K. Zhang, X.F. Ji, X.T. Zhang, Reaction-spun transparent silica aerogel fibers. ACS Nano **14**, 11919–11928 (2020)
51. G.Y. Li, D. Fang, G. Hong, A. Eychmueller, X.T. Zhang, W.H. Song, Assembling graphene aerogel hollow fibres for solar steam generation. Compos. Commun. **35**, 101302 (2022)
52. M. Li, C.X. Tang, S.H. Fu, K.C. Tam, Y.Z. Zong, Cellulose-based aerogel beads for efficient adsorption-reduction-sequestration of Cr(VI). Int. J. Biol. Macromol. **216**, 860–870 (2022)
53. Q.H. Li, Z.H. Yuan, C. Zhang, S.Q. Hu, Z.M. Chen, Y.Z. Wu, D.D. Ye, Tough, highly oriented, super thermal insulating regenerated All-cellulose sponge-aerogel fibers integrating a graded aligned nanostructure. Nano Lett. **22**, 3516–3524 (2022)
54. C.Y. Zhu, T.T. Xue, Z.C. Ma, W. Fan, T.X. Liu, Mechanically strong and thermally insulating polyimide aerogel fibers reinforced by prefabricated long polyimide fibers. ACS Appl. Mater. Interfaces **15**, 12443–12452 (2023)
55. M. Alnaief, I. Smirnova, In situ production of spherical aerogel microparticles. J. Supercrit. Fluids **55**, 1118–1123 (2011)
56. C.A. García-González, J.J. Uy, M. Alnaief, I. Smirnova, Preparation of tailor-made starch-based aerogel microspheres by the emulsion-gelation method. Carbohydr. Polym. **88**, 1378–1386 (2012)
57. C.A. García-González, M. Alnaief, I. Smirnova, Polysaccharide-based aerogels-Promising biodegradable carriers for drug delivery systems. Carbohydr. Polym. **86**, 1425–1438 (2011)
58. M. Alnaief, M.A. Alzaitoun, C.A. García-González, I. Smirnova, Preparation of biodegradable nanoporous microspherical aerogel based on alginate. Carbohydr. Polym. **84**, 1011–1018 (2011)
59. T. Mehling, I. Smirnova, U. Guenther, R.H.H. Neubert, Polysaccharide-based aerogels as drug carriers. J. Non-Cryst. Solids **355**, 2472–2479 (2009)
60. H. Maleki, L. Duraes, C.A. García-González, P. del Gaudio, A. Portugal, M. Mahmoudi, Synthesis and biomedical applications of aerogels: possibilities and challenges. Adv. Colloid Interface Sci. **236**, 1–27 (2016)
61. W. Shi, Y.C. Ching, C.H. Chuah, Preparation of aerogel beads and microspheres based on chitosan and cellulose for drug delivery: a review. Int. J. Biol. Macromol. **170**, 751–767 (2021)
62. I. Preibisch, P. Niemeyer, Y. Yusufoglu, P. Gurikov, B. Milow, I. Smirnova, Polysaccharide-based aerogel bead production via jet cutting method. Materials **11**, 1287 (2018)
63. S.H. Ching, N. Bansal, B. Bhandari, Alginate gel particles-a review of production techniques and physical properties. Crit. Rev. Food Sci. Nutr. **57**, 1133–1152 (2017)
64. B.B. Lee, P. Ravindra, E.S. Chan, Size and shape of calcium alginate beads produced by extrusion dripping. Chem. Eng. Technol. **36**, 1627–1642 (2013)
65. S.M.K. Mohamed, K. Ganesan, B. Milow, L. Ratke, The effect of zinc oxide (ZnO) addition on the physical and morphological properties of cellulose aerogel beads. RSC Adv. **5**, 90193–90201 (2015)
66. G.J. Chen, Y.Y. Wang, R.S. Xie, S.Q. Gong, A review on core-shell structured unimolecular nanoparticles for biomedical applications. Adv. Drug Del. Rev. **130**, 58–72 (2018)

67. Y.C. Ching, T. Gunathilake, C.H. Chuah, K.Y. Ching, R. Singh, N.S. Liou, Curcumin/Tween 20-incorporated cellulose nanoparticles with enhanced curcumin solubility for nano-drug delivery: characterization and in vitro evaluation. Cellulose **26**, 5467–5481 (2019)
68. V. Baudron, P. Gurikov, I. Smirnova, A continuous approach to the emulsion gelation method for the production of aerogel micro-particle. Colloid. Surf. A **566**, 58–69 (2019)
69. J. Viganó, A.A.D. Meirelles, G. Náthia-Neves, A.M. Baseggio, R.L. Cunha, M.R. Marostica, J. Martínez, Impregnation of passion fruit bagasse extract in alginate aerogel microparticles. Int. J. Biol. Macromol. **155**, 1060–1068 (2020)
70. T.L. Zhai, Q.F. Zheng, Z.Y. Cai, H.S. Xia, S.Q. Gong, Synthesis of polyvinyl alcohol/cellulose nanofibril hybrid aerogel microspheres and their use as oil/solvent superabsorbents. Carbohydr. Polym. **148**, 300–308 (2016)
71. C.M. Zhang, T.L. Zhai, L.S. Turng, Aerogel microspheres based on cellulose nanofibrils as potential cell culture scaffolds. Cellulose **24**, 2791–2799 (2017)

Chapter 12
Functional Regulation of Cellulose-Based Aerogels

Yuchen Jin and Zhiqiang Su

Abstract Cellulose-based aerogels (CBAs) are advanced functional materials characterized by ultra-low density, high porosity, biocompatibility, biodegradability, and renewability. Their versatile porous architecture and abundant hydroxyl groups enable precise functional tailoring. This chapter reviews key functionalization strategies for CBAs, including physical modifications, chemical modifications (such as oxidation, esterification, etherification, grafting, and cross-linking), in situ hybridization with functional nanoparticles, and compound functionalization approaches. These techniques enable multi-component synergy, enhanced reactivity, and improved biological activity, transforming CBAs into intelligent platforms with extended high-performance functionalities. Applications of functionalized CBAs are diverse, spanning thermal insulation, oil–water separation and adsorption, atmospheric water harvesting, CO_2 capture, electromagnetic interference shielding, energy storage and conversion, catalysis, and biomedical uses. Despite these advancements, several challenges remain, including the development of stimuli-responsive systems, green and scalable fabrication methods, in-depth biocompatibility assessments, standardized clinical translation pathways, and integration with AI-assisted design tools. Addressing these issues is essential to fully unlock the potential of CBAs in environmental, energy, electronic, and biomedical applications.

Keywords Cellulose-based aerogels · Functional regulation · Sol–gel method · Multicomponent synergy · Composite functionalization

---

Y. Jin · Z. Su (✉)
State Key Laboratory of Chemical Resource Engineering, Key Laboratory of Advanced Functional Polymer Composites, Beijing University of Chemical Technology, Beijing, China
e-mail: suzq@mail.buct.edu.cn

G. Wei and J. Zhang (eds.), *Cellulose-Based Hydrogels and Aerogels*, Lecture Notes in Nanoscale Science and Technology 41,
https://doi.org/10.1007/978-3-032-03008-5_12

12.1 Introduction

The functional regulation of cellulose-based aerogels (CBAs) has been a focus of research since the early twenty-first century. Driven by the growing emphasis on renewable materials and green technologies, as well as the increasing demand for advanced material performance, the customization of CBAs has received substantial attention. Owing to the inherently high crystallinity of cellulose, CBAs not only exhibit excellent mechanical properties but also possess a highly porous architecture, endowing them with considerable potential as adsorbent materials. By optimizing pore structure, modulating surface roughness, and employing directional alignment techniques, it is possible to precisely tailor their mechanical strength, adsorption capacity, hydrophobicity/hydrophilicity, and anisotropic thermal or electrical conductivity.

Moreover, the abundance of free hydroxyl groups on cellulose chains provides a high density of graftable sites, offering extensive possibilities for chemical modification and multicomponent integration [1]. Such functional moieties can interact physically or chemically with nanomaterials (e.g., graphene oxide [2], Mxenes [3]), siloxanes [4], ions [5], or functional molecules, through non-covalent or covalent bonding, thereby enhancing the performance of CBAs. These versatile modification strategies not only enable their unique application potential in environmental remediation, smart devices, energy storage, and biomedical fields but also promote the transformation of aerogels from single-functional materials toward intelligent and multifunctionally integrated platforms [6].

In this chapter, we systematically expound the core objective of functional regulation of CBAs, that is, to precisely control its physical properties through physical and chemical means and endow it with extended functions such as intelligent responsiveness and biological activity, thereby upgrading it from a single material to an intelligent multi-functional platform. CBAs, with their highly porous nano/micron structure, abundant free hydroxyl, and the renewable and biodegradable properties derived from cellulose, have laid the foundation for their role as high-performance and sustainable green materials. The precise functional regulation strategies break through the limitations of traditional aerogels, greatly expanding the application prospects of CBAs in cutting-edge fields such as environmental remediation, energy storage, biomedicine, and smart devices, while promoting green and sustainable development by leveraging renewable resources.

12.2 Physical Modifications

CBAs have garnered significant attention in the field of green functional materials owing to their renewability, biodegradability, and tunable multiscale porous architectures. Through physical modification strategies, the pore morphology and hierarchical structure of aerogels can be precisely engineered, enabling controlled modulation of pore size from the nanometer to the micrometer scale. Studies have demonstrated that the synergistic regulation of process parameters—such as emulsifier concentration gradients, ionic liquid/co-solvent systems, and supercritical drying—can effectively optimize porosity, specific surface area, and mechanical strength. This structure–property co-design endows CBAs with outstanding thermal insulation, adsorption capacity, and electromagnetic shielding properties, highlighting their promising applications in environmental remediation, biomedicine, and smart sensing technologies.

12.2.1 Pore Structure Optimization

The pore structure plays a critical role in determining the thermal insulation, adsorption efficiency, and mechanical performance of aerogels. Based on the morphological features of the porous network, CBAs can be categorized into disordered isotropic porous structures, hierarchical porous architectures, directionally aligned channel structures, and bioinspired natural-channel structures. By precisely controlling the pore size, distribution, and orientation, the aerogels can be tailored to meet the specific requirements of various application domains. Detailed information on pore structure parameters, including their corresponding physical quantities and characterization methods, is provided in Table 12.1.

The sol–gel method remains the predominant approach for the preparation of aerogels. Among the critical factors influencing the formation of the gel network, solution homogeneity and viscosity play a decisive role. The uniformity of the gel network directly determines the pore size, distribution, and connectivity of the resultant aerogel after drying, which are core parameters for pore structure optimization. Emulsifiers facilitate the formation of stable emulsions between two immiscible phases, such as oil and water, by reducing interfacial tension. This allows monomers to disperse uniformly in the aqueous phase as emulsified droplets, thereby promoting the homogeneous formation of the aerogel network. By adjusting the amount of emulsifier used, and employing emulsions as templating agents, the porosity, density, pore diameter, and microstructural morphology of the aerogel can be precisely controlled. Notably, the density of aerogels initially decreases and then increases with increasing emulsifier concentration. The optimal performance is achieved at a concentration of 0.003 $mol{\cdot}L^{-1}$.

Furthermore, the type and content of ionic liquids and co-solvents exert significant effects on the initial dissolution state and dissolution process of cellulose. These

Table 12.1 Fundamental physical parameters and characterization methods for pore structure features of CBAs

Physical properties	Characterization techniques	Ref
Density	Drainage method Formula derivation	[7]
Porosity	Characterization method	[8]
Specific surface area	BET	[9]
Average aperture	Mercury compression method SAXS	[8]
Aperture distribution	Mercury compression method	[8]
Hole volume	BET	[10]
Microscopic morphology	SEM AFM TEM	[11]
Surface chemical structure	NMR FTIR NMR XPS	[12]

processes are primarily governed by the hydrogen bond basicity and the microscopic structural features of the ionic liquid/co-solvent systems. It has been reported that acetate-based ionic liquids exhibit superior dissolution capabilities compared to chloride-based ionic liquids when paired with co-solvents. However, an excessive amount of co-solvent may substantially impair the overall dissolution capacity of the system [13].

The drying procedure is another critical step in tailoring the pore structure of aerogels. By modulating the physical conditions during drying—such as temperature distribution, solvent properties within the hydrogel, and the application of directional drying techniques—it is possible to fine-tune the pore architecture to optimize the aerogel's performance in thermal insulation, gas adsorption, and electromagnetic wave absorption applications. Drying involves the removal of solvent from the wet gel in the form of vapor, while maintaining the integrity of the three-dimensional aerogel network. During solvent evaporation, menisci form at the pore openings, and due to the nanoscale-to-microscale nature of the pores, the skeleton is subjected to capillary forces at the liquid–vapor interface. These capillary pressures can be quantitatively described by the Laplace Eq. (12.1).

$$\mathrm{P} = \frac{2\gamma\cos\theta}{\mathrm{R}} \tag{12.1}$$

Here, r denotes the pore radius, γ the surface tension of the solvent, θ the contact angle between the solvent and the pore wall, and P the capillary pressure. Excessive capillary forces may result in structural collapse or fracturing of the aerogel network. Currently, the most commonly employed drying techniques for aerogel fabrication include supercritical drying, freeze-drying, and ambient pressure drying. Among these, supercritical drying eliminates capillary forces during solvent removal, enabling the production of aerogels with high porosity, large specific surface area, and low density. Freeze-drying, on the other hand, is less suitable for aerogels with nanoscale pores due to the formation of ice crystals, which can disrupt fine structures. In the case of ambient pressure drying, careful control over the drying temperature, implementation of gradient heating, or the design of surface wettability (hydrophilic or hydrophobic) is required to preserve the desired pore architecture and prevent collapse, depending on the mechanical robustness of the aerogel framework.

For CBAs, the incorporation of cellulose introduces flexible polymer chains and additional crosslinking or branching sites into the otherwise rigid inorganic network. This enhances intermolecular interactions, thereby improving the mechanical strength and toughness of the composite aerogel. Moreover, the inclusion of cellulose imparts biodegradability and biocompatibility, which are crucial for biomedical applications such as drug delivery.

In disordered porous aerogels, the pore structure can be tailored through modulation of the drying temperature, the crystallization rate of ice templates, or solvent exchange. During ambient pressure drying, stepwise temperature ramping helps minimize shrinkage and maintains the integrity of the aerogel skeleton. A rapid ice crystallization rate yields smaller pore diameters due to reduced crystal size, whereas a slower rate results in larger pores. Solvent exchange, which replaces the initial solvent with another of lower surface tension, can also effectively reduce capillary-induced shrinkage.

Additionally, emulsion templating methods allow for precise design of the aerogel's porous morphology through the regulation of emulsion characteristics (e.g., droplet size and stability), followed by freeze-drying. CBAs prepared via Pickering emulsion templating exhibit distinctive and tunable porous morphologies, including alveolar-like or two-dimensional lamellar structures [14]. Further reduction of pore size can be achieved via post-treatment with solvent exchange [15]. Due to the templating effect of emulsions, CNF aerogels fabricated by emulsion techniques can form interconnected microscale pores between the entangled nanofibers (Fig. 12.1a).

The hierarchical porous architecture of CBAs significantly enhances their adsorption performance. By processing cellulose into nanofibers and incorporating them with other functional materials, a multilevel porous structure comprising macropores, mesopores, and micropores can be constructed. For example, in TEMPO-oxidized cellulose nanofiber (TOCN) aerogels, the incorporation of halloysite nanotubes (hNTs) results in a hybrid porous system featuring both large pores (~20 μm) and mesopores. This structure enables efficient water absorption from blood, leading to localized concentration of blood cells and coagulation factors. Moreover, the increased number of active sites enhances the adsorption and adhesion of red blood cells and platelets, ultimately facilitating blood clot formation (Fig. 12.1b).

In addition, 3D printing allows for the rational design of scaffolds to achieve multiscale porosity [17]. By fabricating scaffolds with pores in the micrometer to millimeter range, the aerogels exhibit both high water absorption/release capacities and the ability to adsorb lithium chloride (LiCl) particles (Fig. 12.1c). The hydrophilic

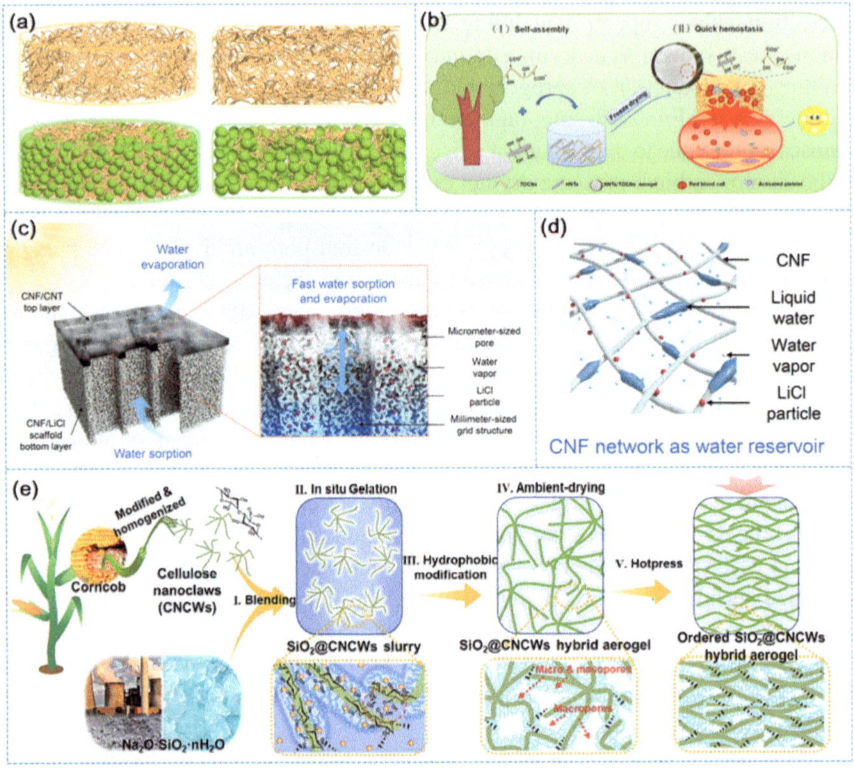

Fig. 12.1 **a** Structure of traditional CNF aerogels and CNF/emulsion composite aerogels. Reproduced with permission from Song et al. [15]. Copyright 2020, American Chemical Society **b** CBA with hierarchical porous architecture for hemostatic applications. Reproduced with permission from Ref. [16]. Copyright 2024, Elsevier B.V **c** Illustration of atmospheric water harvesting using CNF/CNT hybrid aerogels **d** Structural schematic of the CNF/CNT aerogel after water absorption. Reproduced with permission from Ref. [17]. Copyright 2023, Wiley-VCH GmbH **e** Synthesis of SiO_2@CNCWs hybrid aerogels. Reproduced with permission from Ref. [9]. Copyright 2022, Elsevier Ltd

CNF network within the scaffold serves as a reservoir, effectively storing the absorbed water, while the micron-sized pores physically confine LiCl particles, minimizing leakage and aggregation (Fig. 12.1d). The hierarchical porous architecture significantly increases the porosity and specific surface area of CBAs, enabling superior adsorption performance. These materials have shown great promise in applications such as hemostatic treatment [16], moisture-driven transport [18], and atmospheric water harvesting [17].

Furthermore, the abundant hydrogen bonding network and unique "claw-like" molecular chain configuration within the aerogel matrix (Fig. 12.1e) endow the material with outstanding mechanical performance, making it highly suitable for applications in architecture, industrial manufacturing, and aerospace engineering [9].

Gradient pore structure design enables spatial functionalization within a single material, allowing for regional synergy through either gradual pore size reduction or layered arrangements of ordered and disordered structures. Such designs are particularly beneficial in applications like broadband noise absorption and electromagnetic interference (EMI) shielding. Gradient pore structures are primarily achieved by controlled gradients in solute concentration or drying conditions. For instance, stepwise dilution of a cellulose nanocrystal-containing solution followed by freeze-drying can yield an open porous structure with a gradient of large, medium, and small pores [19]. Similarly, combining directional freeze-drying with ambient pressure drying in a layered manner has been shown to achieve ultrahigh EMI shielding efficiency [1].

In ordered porous structures, the aerogel skeleton grows along specific directions, forming architecturally organized pore networks with high degrees of programmability [20]. Directional freezing techniques enable the alignment of ice crystals along the freezing direction, resulting in anisotropic, oriented channel structures. When combined with uniaxial or biaxial thermal pressing, the pore walls can be induced to bend inward, giving rise to negative Poisson's ratio superstructures [21]. Furthermore, bidirectional freezing can produce three-dimensional interpenetrating lamellar networks in CBAs, which significantly enhance the specific surface area and improve adsorption efficiency. This structural advantage facilitates their effective use in applications such as marine oil spill remediation [22].

12.2.2 *Surface Roughness Modulation*

Precise control of the surface roughness of CBAs is a critical factor for enhancing their functional performance and expanding their application scope. Various strategies—such as chemical vapor deposition (e.g., deposition of nanoscale silica or organosilane coatings), physical adsorption, and polysiloxane-based surface modification—have been employed to finely tailor the micro- and nanoscale roughness of the aerogel surface. Such structural modulation can be leveraged to tune the wettability of the aerogel, rendering it either hydrophilic or oleophilic/hydrophobic. For instance, specially treated superhydrophobic surfaces, characterized by water contact angles greater than 150°, exhibit remarkable anti-fouling properties and enhanced oil–water separation efficiency. Conversely, hydrophilic surface modifications improve the aerogel's adsorption capacity for polar substances and strengthen interfacial bonding with composite matrices.

This regulation of surface roughness not only optimizes the wettability, mechanical stability, and functional selectivity of CBAs, but also significantly broadens their potential applications. In the environmental field, modified cellulose aerogels serve as highly efficient filtration media for air and water purification. In energy storage systems, their superior properties make them promising candidates for use as high-performance electrode or separator materials in supercapacitors and batteries. In biomedical applications, surface-modified cellulose aerogels have demonstrated excellent biocompatibility and functional adaptability, making them suitable for tissue engineering scaffolds, drug delivery carriers, and related uses. Therefore, surface roughness modulation represents a key technological pathway for the functional design of cellulose aerogels and provides a strong theoretical and practical foundation for the development of high-performance, multifunctional cellulose-based materials.

Silane-based CVD techniques have been employed to modify nanocellulose surfaces, yielding superhydrophobic aerogels. These aerogels exhibit excellent hydrophobicity and oleophilicity, making them suitable for oil–water separation in marine environments (Fig. 12.2a). By combining bidirectional freeze-drying with organosilane deposition, aerogels with dual-directional oleophilic channels can be fabricated, which show enhanced mechanical properties and adsorption performance (Fig. 12.2b). Utilizing the inherent thermal stability of deposited silanes, the resulting aerogels exhibit a water contact angle of 156.5° and can withstand exposure to open flame at 1300 °C for up to 70 min without structural degradation (Fig. 12.2c).

The hierarchical porous structures of CBAs also enable multilevel adsorption and anchoring of diverse single-atom metal species, facilitating the stable immobilization of single-atom catalysts (SACs) within the aerogel matrix [25]. For example, cadmium (Cd), a low-boiling-point metal, can be adsorbed by the negatively charged functional groups on CNFs and graphene oxide (GO) surfaces. During pyrolysis,

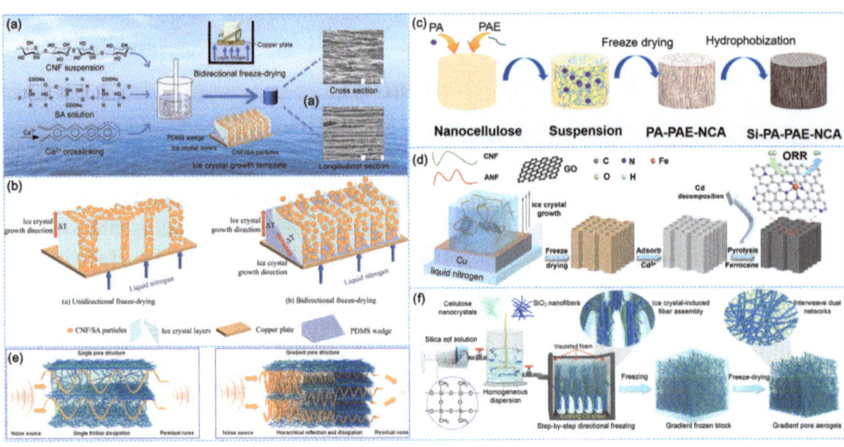

Fig. 12.2 **a** CNF/SA aerogels for oil spill remediation in marine environments. **b** Diagram of bidirectional freeze-drying for the preparation of CNF/SA aerogels. Reproduced with permission from Ref. [22]. Copyright 2022, Elsevier B.V. **c** Synthesis of superhydrophobic nanocellulose aerogels. Reproduced with permission from Ref. [23]. Copyright 2024, Elsevier B.V **d** Schematic representation of the preparation of atomically dispersed FeN_4 active sites in porous nitrogen-doped carbon aerogels (Fe-SA@PNC). Reproduced with permission from Ref. [24]. Copyright 2023, Elsevier B.V. **e** Sound absorption mechanisms of conventional versus gradient-structured aerogels. **f** Diagram of the fabrication process of elastic ceramic nanofiber aerogels with gradient pore structures incorporating cellulose nanonets. Reproduced with permission from Ref. [19]. Copyright 2023, Wiley–VCH GmbH

ferrocene is introduced, leading to the evaporation of Cd and the subsequent formation of iron-based single atoms via chemical vapor deposition. The resulting $Fe–N_4$-rich active sites confer excellent catalytic activity and stability in oxygen reduction reactions [23].

Nanocellulose can also enhance its hydrophobicity through the physical adsorption of cationic surfactants and copolymers via electrostatic interactions, van der Waals forces, and hydrogen bonding. Under interfacial electrostatic attraction, cationic surfactants self-assemble with nanocellulose, further enhancing the surface hydrophobicity. However, factors such as imbalanced adsorption–desorption, low critical micelle concentration of surfactants, large pore size of the nanocellulose matrix, and weak binding interactions may result in residual surfactants within the aerogel matrix, ultimately compromising mechanical performance. To address this, 2,3-epoxypropyltrimethylammonium chloride (EPTMAC) has been used to modify CNFs. EPTMAC binds to the negatively charged groups on nanocellulose via electrostatic interaction, introducing cationic functional groups and thereby enhancing hydrophobicity. This approach has proven effective for capturing microplastic particles with small diameters from aqueous environments [24].

12.2.3 Directional Alignment Techniques

Directional alignment techniques for CBAs enable precise control over ordered structures, significantly enhancing the material's multidimensional performance. By inducing anisotropic pore architectures through tailored fabrication processes, aerogels can be endowed with directionally dependent microstructures that optimize thermal management capabilities, improve mechanical strength and toughness, and facilitate the construction of multiscale conductive networks for enhanced EMI shielding. This structure-oriented design strategy overcomes the inherent performance homogeneity of conventional aerogels and offers innovative solutions for advanced applications, including thermal protection in aerospace, integration of flexible electronics, and mitigation of high-frequency electromagnetic pollution.

The principal methods for preparing directionally aligned CBAs include directional freeze-drying, magnetic or electric field-assisted alignment, 3D printing with ink writing, template-guided growth, mechanical stretching, chemical crosslinking combined with self-assembly, and biomimetic or bio-templated structural design.

For instance, in aerogels fabricated via directional freeze-drying, regularly aligned channels are formed along a specific axis. This anisotropic pore orientation alters the propagation path of electromagnetic waves—reducing shielding effectiveness along the aligned direction (Fig. 12.2e)—while in the perpendicular direction, the more complex pore structures increase wave scattering and internal reflection along the pore walls, thereby significantly enhancing EMI shielding. This approach imparts directional dependence to the shielding effect, allowing fine-tuning of electromagnetic response through ordered structural design [26].

Moreover, gradient pore structures obtained by stepwise directional freeze-casting can integrate large, medium, and small pores along the direction of acoustic wave propagation. This gradient configuration enables continuous sound wave entry and dissipation, while the high specific surface areas of nanofibers ($\approx$300 nm in diameter) and nanonets ($\approx$30 nm in diameter) provide extensive contact interfaces for sound absorption. The resulting material exhibits outstanding broadband noise attenuation capabilities (Fig. 12.2f).

The incorporation of magnetic nanoparticles (e.g., Fe_3O_4) into cellulose solutions enables the formation of ordered structures through external magnetic field guidance. Under the influence of a magnetic field, the nanoparticles align into chains along a specific direction, which in turn drives the orientation of cellulose molecular chains. Subsequent processing, such as low-temperature calcination, immobilizes these magnetic nanostructures within the aerogel framework, resulting in composite aerogels with gradient porous architectures and continuous conductive networks. This structural design not only optimizes the electromagnetic parameter matching but also enhances microwave absorption by increasing the multiple reflection, scattering, and polarization loss of electromagnetic waves [27].

Regulating the pore structure through physical modification can significantly improve the overall performance of biomass-derived aerogels. Ordered honeycomb-like or lamellar structures can evenly distribute applied stress, thereby enhancing mechanical strength. In contrast, disordered structures improve toughness through entanglement of fibrous components. The synergistic effect of nanoscale and microscale hierarchical pores increases the specific surface area and expands the material's fluid storage capacity. Meanwhile, the introduction of aligned channels facilitates mass transport, further boosting the functional performance and application potential of CBAs.

12.3 Chemical Modifications

Chemical modification enhances the physicochemical properties of CBAs and introduces specific functionalities by incorporating reactive groups from other molecules or initiating chemical reactions. Such modifications significantly expand the application potential of cellulose aerogels across various fields.

12.3.1 Surface Oxidization

Through oxidation, etherification, esterification, graft copolymerization, and chemical crosslinking, cellulose can be transformed into a wide range of derivatives with distinct structural and functional attributes. These derivatives serve as precursors for aerogels with tailored properties and extended application domains. For example, chemical modification alters the solubility of cellulose derivatives: carboxymethyl cellulose and hydroxypropyl methyl cellulose are water-soluble; ethyl cellulose dissolves in dichloromethane; and cellulose acetate dissolves in acetone. These solubility enhancements help circumvent time-consuming solvent exchange processes and improve the efficiency of aerogel fabrication.

The abundant hydroxyl groups on cellulose chains provide accessible sites for functionalization via esterification, etherification, and oxidation. Such surface chemical modifications lower the surface energy of cellulose aerogels, which helps prevent structural collapse during drying and improves interfacial compatibility when combined with polymer matrices. Functional polymer chains or small molecules can be grafted onto the cellulose framework through radical polymerization, crosslinker-assisted strategies, or click chemistry. Targeted functional groups may also be introduced directly onto the cellulose backbone. Furthermore, biomolecules can be immobilized on cellulose surfaces via covalent bonding or physical adsorption, enabling the integration of bioactive functionalities for applications in drug delivery systems and tissue engineering scaffolds.

For instance, TEMPO-mediated oxidation can convert over 70% of surface hydroxyl groups on CNFs into carboxyl groups (Fig. 12.3a). These carboxyls can undergo amide crosslinking with various small-molecule amines, resulting in amine-functionalized CNFs with enhanced performance as triboelectric materials. The resulting aerogels can be utilized in flexible triboelectric nanogenerators (TENGs) (Fig. 12.3b) [28]. Additionally, grafting amino-functional polymers onto cellulose provides active adsorption sites for carbon dioxide, enabling selective CO_2 capture (Fig. 12.3c) [29].

From a kinetic perspective, increasing the temperature enhances molecular motion, activating more adsorption sites that interact with CO_2, thereby improving adsorption capacity. However, excessively high temperatures weaken intermolecular forces, leading to a reduction in the fixed amount of CO_2 on the adsorbent. Thermodynamically, since the reaction between amine groups and CO_2 is exothermic, elevated temperatures favor the desorption process and shift the adsorption–desorption equilibrium toward desorption (Fig. 12.3d).

12.3.2 Introduction of Functional Groups

Chemical modification of cellulose aerogels through the incorporation of specific functional groups is a highly effective approach to optimize their physicochemical properties and expand their application scope.

12.3.2.1 Surface Chemical Modification

The hydroxyl groups (–OH) on cellulose molecular chains serve as the primary reactive sites for chemical functionalization. Specific functional groups can be introduced

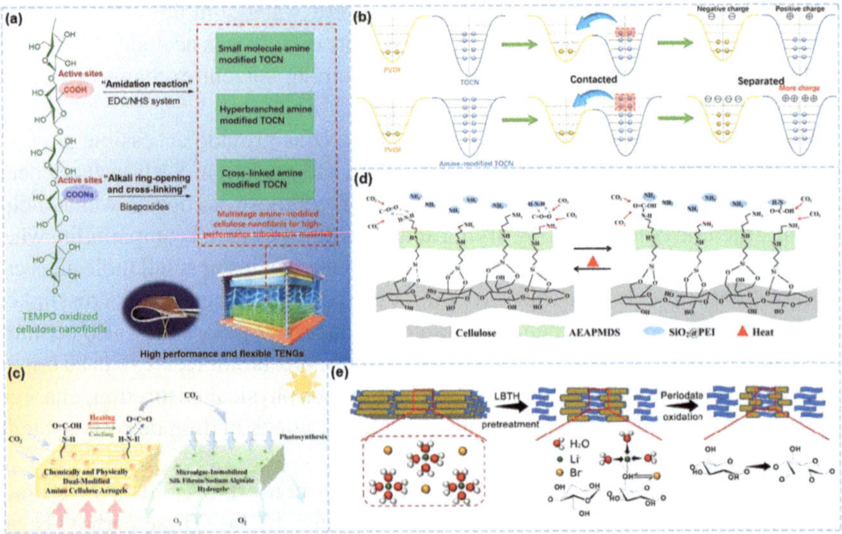

Fig. 12.3 **a** Fabrication of TOCN-based TENG device. **b** Electron cloud potential well model of the contact–separation process between a dual TOCN-based triboelectric layer and a PVDF membrane. Reproduced with permission from Song et al. [28]. Copyright 2022, Elsevier. **c** Chemically and physically dual-modified aminated cellulose aerogel composite for CO_2 adsorption and desorption. **d** Schematic representation of the CO_2 capture mechanism of the dual-modified aminated cellulose aerogel composite. Reproduced with permission from Li et al. [29]. Copyright 2024, American Chemical Society. **e** Mechanistic illustration of the role of $LiBr·3H_2O$ (LBTH) in insolubilization pretreatment and enhanced periodate oxidation. Reproduced with permission from Zhang et al. [30]. Copyright 2023, American Chemical Society

via esterification, etherification, and oxidation reactions, enabling control over the surface properties and functionalities of aerogels.

Esterification reactions: Ester groups (–O–CO–R) can be introduced by reacting hydroxyl-containing compounds with acid anhydrides or acyl chlorides. For instance, cellulose can be heated with an excess of carboxylic acid under acidic conditions or reacted with acyl chlorides in suitable solvents to graft carboxyl groups. A representative example involves the reaction between cellulose and acetic anhydride, leading to the formation of acetyl groups ($–O–CO–CH_3$) and the synthesis of cellulose derivatives such as cellulose acetate [31]. Moreover, reaction with stearoyl chloride ($C_{17}H_{35}COCl$) introduces long-chain hydrophobic alkyl groups [32].

Etherification reactions: In etherification, hydroxyl groups react with alkylating agents (e.g., chloroacetic acid, ethylene oxide) to form ether linkages (–O–R). One notable example is the synthesis of carboxymethyl cellulose (CMC) from cellulose and chloroacetic acid [33]. Similarly, the reaction of cellulose with propylene oxide introduces hydroxypropyl groups ($–O–CH_2–CHOH–CH_3$).

Oxidation reactions: Oxidation of hydroxyl groups leads to the formation of carboxyl (–COOH) or aldehyde (–CHO) groups [34]. In the presence of a NaClO/NaBr system, TEMPO-mediated oxidation selectively converts C6 primary hydroxyl groups into carboxyl groups (–COOH) [35]. Alternatively, periodate oxidation selectively cleaves vicinal diols in the cellulose backbone, generating aldehyde groups (–CHO) [30] (Fig. 12.3e).

12.3.2.2 Graft Copolymerization

Graft copolymerization is a chemical technique used to introduce functional monomers or polymer chains onto the cellulose backbone, aiming to impart specific functionalities and broaden the applicability of cellulose-based materials. This method has been extensively employed in cellulose modification for the preparation of functionalized cellulose aerogels.

Grafting of hydrophobic/hydrophilic groups: Hydrophobic or hydrophilic groups can be grafted onto cellulose chains via radical polymerization, click chemistry, and other approaches. For example, polymethyl methacrylate (PMMA), which contains ester groups, can be reinforced with CNFs, resulting in composites that maintain high optical transparency while exhibiting significantly improved mechanical strength and stiffness [36]. Moreover, the grafted polymer chains can interpenetrate the cellulose backbone to form network structures, thereby enhancing the mechanical integrity of the aerogel.

Grafting of functional monomers: Grafting functional monomers endows cellulose with unique properties. For instance, fluorescent carbon dots grafted onto cellulose can create cellulose aerogels with excellent photoluminescent properties, which are applicable in sensing, anti-counterfeiting, and bioimaging fields [37]. Magnetic nanoparticles such as Fe_3O_4, when surface-modified with carboxyl groups, can form covalent bonds with amino-functionalized cellulose, imparting magnetic responsiveness to the resulting material [38] (Fig. 12.4a). Similarly, nanocellulose integrated with highly oriented nickel chains can form a complex interpenetrating network, resulting in multifunctional carbonized cellulose aerogels with superior microwave absorption properties [27]. Moreover, in situ growth of Zr-based metal–organic frameworks (Zr-MOFs) on bacterial cellulose (BC) microspheres produces composite materials with inherent fluorescence (Fig. 12.4b). The microporous structure of Zr-MOFs allows for molecular sieving, and their ligands interact with target compounds such as tetracycline (TC), thereby enhancing selectivity in detection and adsorption processes [39].

12.3.2.3 Chemical Cross-Linking

Chemical cross-linking involves the reaction of crosslinkers (e.g., glutaraldehyde, epichlorohydrin) with hydroxyl groups on cellulose to form a stable crosslinked network. In CBAs, crosslinking reduces the aggregation of hydrogen bonds between fibrils, thereby maintaining the porous structure and improving both specific surface area and thermal insulation properties. For instance, the three-dimensional porous

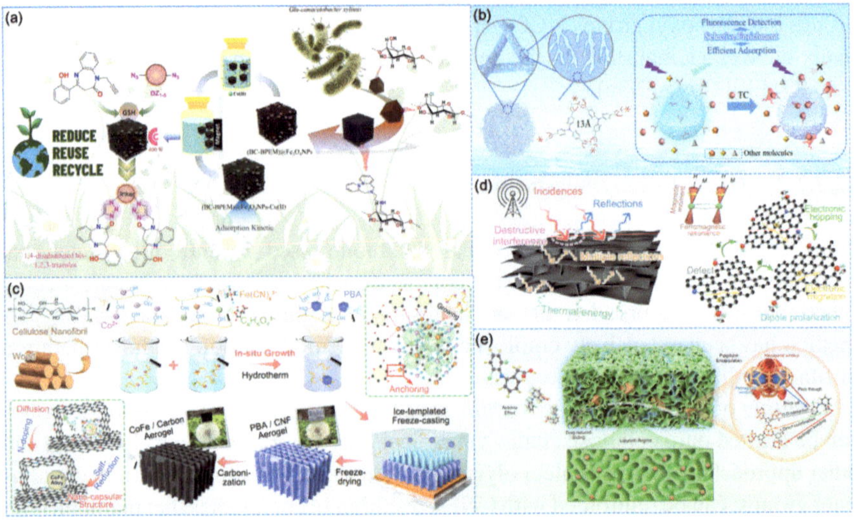

Fig. 12.4 **a** Synthesis of a TOCN-based triboelectric nanogenerator (TENG) device. Reproduced with permission from Chaabane et al. [38]. Copyright 2025, Elsevier. **b** Schematic representation of dual-functional nanocellulose@MOF composite aerogels for selective fluorescence detection of tetracycline. Reproduced with permission from Wang et al. [39]. Copyright 2025, Elsevier. **c** Schematic of the in-situ growth of Prussian blue analogues (PBAs) on CNFs to fabricate biomimetic CoFe/carbon ordered aerogels. **d** Schematic illustration of the electromagnetic wave absorption mechanism of CoFe/carbon aerogels. Reproduced with permission from Qiao et al. [40]. Copyright 2024, Wiley–VCH GmbH. **e** Schematic diagram of the cascade mechanism of diclofenac (DCF) encapsulation driven by MIL@TOCN diffusion. Reproduced with permission from Hu et al. [41]. Copyright 2025, Elsevier

structure of cellulose aerogels has been used as a scaffold for KOH etching and chemical crosslinking, enabling the successful incorporation and uniform distribution of carbon nitride (CN) throughout the aerogel network [42].

12.4 In-Situ Hybridization

Functional nanoparticles (e.g., silica, carbon nanotubes) or metal oxides (e.g., TiO_2) can be embedded into the cellulose matrix via physical blending or in situ synthesis. Incorporation of nanoparticles enhances the thermal insulation performance of aerogels, as these nanoparticles act as scattering centers within the porous channels, thereby suppressing heat conduction. In addition, doping with magnetic Fe_3O_4 nanoparticles imparts magnetic responsiveness to the aerogel, facilitating its recyclability. In situ generated functional CBAs effectively prevent nanoparticle agglomeration and ensure uniform dispersion of fillers. Strong interfacial interactions—mediated by hydrogen bonding or covalent bonding—enhance the synergy between components. Moreover, the templating effect of cellulose enables morphological control over the embedded functional materials, including nanoparticle size and the crystallinity of metal–organic frameworks (MOFs). This technique not only improves dispersion and interfacial compatibility, but also allows precise morphological control, ultimately enhancing the performance and applicability of the composite materials.

For instance, metal salts such as $AgNO_3$ and $HAuCl_4$ can be dissolved in cellulose solutions (e.g., NaOH/urea systems), and subsequently reduced using reducing agents such as ascorbic acid or $NaBH_4$, or via thermal reduction, to form metallic nanoparticles. The resulting silver nanoparticles are distributed throughout the nanofiber networks, embedded within the cell walls and pore spaces of the cellulose matrix [43].

Furthermore, carbon nanotubes (CNTs), GO, or biochar can be mixed with cellulose solutions and processed via freeze-drying and carbonization to yield carbon–cellulose aerogel composites. The hydroxyl groups on the surface of nanocellulose interact with graphene, forming interconnected networks that inhibit graphene aggregation. The surface chemistry and physicochemical properties of CBAs are tunable, enabling effective dispersion of carbonaceous materials and improving the specific capacitance of electrodes—thus enhancing the electrochemical performance of supercapacitors. Carboxyl-rich CNFs can be synthesized via TEMPO-mediated oxidation, and then combined with CNTs to produce well-dispersed CNFs/MWCNTs suspensions. Freeze-drying and subsequent high-temperature carbonization of these suspensions yield CNFs/MWCNTs carbon aerogels, further expanding the potential of cellulose-derived carbon aerogels as electrode materials for high-performance supercapacitors [44].

In situ hydrothermal synthesis of MOFs on CNFs enables the MOF particles to become tightly embedded and uniformly dispersed within the carbon matrix. This architecture significantly enhances interfacial bonding quality, thereby improving both the mechanical and electromagnetic properties of the composite aerogels [40] (Fig. 12.4c). CoFe nanocapsules exhibit excellent dispersion and interfacial connectivity within the carbon matrix, and their biomimetic ordered porous structures markedly improve impedance matching, thereby facilitating the penetration of electromagnetic waves into the aerogel interior. Additionally, the layered microstructures induce multiple reflection effects that extend the propagation path of the waves, further enhancing electromagnetic attenuation performance (Fig. 12.4d).

When used as a scaffold, cellulose aerogels allow for the in-situ hybridization of functional nanoparticles within their porous frameworks. The high specific surface area of the aerogel significantly increases contact with target analytes or fields of action, thereby boosting specific performance metrics. For example, in situ growth of Zr-based MOFs on bacterial cellulose microspheres produces composites with fluorescent characteristics. The microporous network of Zr-MOFs enables molecular sieving, while their ligands interact with tetracycline, improving both selectivity and efficiency in detection and adsorption applications [39]. Moreover, cellulose aerogels derived from wood with three-dimensional hierarchical structures and compressible, ordered channel arrays have been demonstrated as effective scaffolds for MOF nanocrystals, and have been employed, for the first time, in iodine capture and recycling.

In addition to nanoparticle incorporation, polymer chains can also be grafted onto CNFs (Fig. 12.4e). Along the radial direction, increased inter-fiber spacing, introduction of reinforcing polymers, and formation of strong interfacial bonding synergistically endow the aerogel with compressible elasticity. The anisotropic aerogel matrix provides structural support, while polydimethylsiloxane (PDMS) polymers prevent fiber slippage and impart both rigidity and flexibility to the composite aerogel [41].

12.5 Composite Functionalization

Composite functionalization of cellulose aerogels integrates chemical modification, structural regulation, and incorporation of functional components through multidimensional design strategies. This approach overcomes the limitations of conventional single-function materials, transforming cellulose aerogels from passive structural substrates into advanced multifunctional platforms with intelligent responsiveness and bioactivity. At the core of this design philosophy lies the concept of "structure–function integration," which focuses on three principal directions: synergistic effects of multicomponent systems, responsiveness to dynamic environments, and compatibility with biological interfaces. By incorporating nanomaterials, stimuli-responsive molecules, and bioactive components, cellulose aerogels retain their intrinsic lightweight and porous characteristics while acquiring new functionalities such as electrical conductivity, self-healing, antibacterial activity, and tissue regeneration.

12.5.1 Multicomponent Synergy

In composite systems, two or more components interact via physical or chemical mechanisms, such as hydrogen bonding, electrostatic attraction, or covalent bonding, to produce synergistic effects. The functional complementarity and performance superposition of individual components result in an overall enhancement of material properties beyond what could be achieved by any single constituent alone.

The incorporation of graphene, MXenes, or CNTs significantly improves the electrical conductivity and mechanical strength of cellulose aerogels. For instance, the addition of multi-walled carbon nanotubes (MWCNTs) markedly enhances the thermal stability of aerogel precursors. Under identical high-temperature carbonization conditions, CBAs containing MWCNTs exhibit superior morphological integrity, greater microstructural stability, and higher degrees of graphitization (Fig. 12.5a). MWCNTs, as reinforcing phases, effectively improve the physicochemical properties of cellulose-derived carbon aerogels [45].

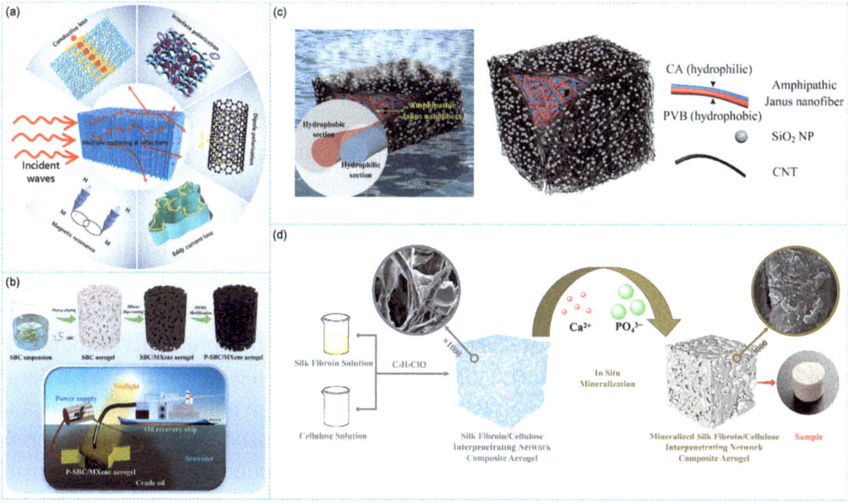

Fig. 12.5 **a** MWCNTs serving as reinforcing agents and functional components within a cellulose scaffold. Reproduced with permission from Zhang et al. [45]. Copyright 2025, Elsevier. **b** Schematic of an eco-friendly bacterial cellulose/MXene aerogel for efficient crude oil/seawater separation. Reproduced with permission from Chen et al. [46]. Copyright 2024, Elsevier. **c** Schematic of a Janus-structured eco-friendly bacterial cellulose/MXene aerogel for efficient crude oil/seawater separation. Reproduced with permission from Wang et al. [47]. Copyright 2023, The Authors. **d** Schematic illustration of the self-assembly process of a mineralized silk fibroin (SF)/cellulose interpenetrating network (M-S-C) composite aerogel. Reproduced with permission from Chen et al. [48]. Copyright 2022, Elsevier

CNFs can also serve as intercalating agents to increase the interlayer spacing of MXenes. The abundant hydroxyl groups on CNFs form strong interactions with the hydrophilic surface groups of MXenes, thereby enhancing the mechanical properties of MXene/CNF hybrid membranes without significantly compromising their electrical conductivity. For example, the synergistic integration of $Ti_3C_2T_x$ MXene, carboxylated single-walled carbon nanotubes (SWCNTs), and cellulose aerogels resulted in the formation of a fibrous conductive network within aerogel film electrodes. These networks provided a greater number of electroactive sites and more efficient ion transport and diffusion pathways, offering significant potential for lightweight and wearable smart electronic devices [49].

Moreover, cellulose can interact with polyurethane primarily through hydrogen bonding between surface hydroxyl groups on cellulose and functional groups on the polyurethane chains. This strong interfacial adhesion contributes to significant improvements in the mechanical strength and thermal stability of the resulting composites. The amphiphilic nature of cellulose also facilitates homogeneous dispersion of fillers within the polyurethane matrix. Due to these characteristics, cellulose nanofiber/polyurethane composites have shown wide-ranging applications in areas such as sensing, 3D printing, self-healing materials, and flame retardants.

12.5.2 Dynamic Response Design

Dynamic response design enables CBAs to react adaptively to external environmental stimuli (e.g., temperature, pH, light) by incorporating stimuli-responsive components, thereby promoting their applications in fields such as smart controlled release systems and soft robotics.

The integration of photothermal phase-change materials imparts light-responsive energy storage and heat release capabilities to cellulose aerogels, enabling their use in crude oil absorption, solar thermal energy storage devices, and aerogel evaporators. MXenes exhibit localized surface plasmon resonance (LSPR) effects, and when composited with cellulose aerogels, their light absorption performance is significantly enhanced. This enhancement arises from the generation of a large number of hot electrons via the LSPR effect, which undergo non-radiative decay and generate thermal energy (Fig. 12.5b). The heat produced can be effectively transferred through the composite aerogel to crude oil in contact, reducing the oil's viscosity and thereby improving its absorption efficiency. Furthermore, due to the metallic-like electrical conductivity of MXenes, the composite aerogel exhibits excellent and stable electrothermal conversion capabilities, allowing for electrically assisted crude oil recovery [46].

Gallium–indium alloys are also promising photothermal materials. Through vacuum impregnation, these phase-change materials can be embedded into modified cellulose aerogels, where they form thermally conductive networks and enhance thermal conductivity for solar thermal energy storage applications [50]. Recently, Janus-structured modified CBAs have been developed for photothermal interfacial evaporation, enabling seawater desalination and removal of organic pollutants (Fig. 12.5c). These aerogels can produce fresh water using only sunlight, offering a green, safe, efficient, and renewable solution for seawater purification [47].

Incorporation of thermosensitive materials enables aerogels to achieve thermal regulation functions. For example, CNFs can encapsulate paraffin droplets to form shell-like structures that stabilize Pickering emulsions without phase separation before or after the phase transition. CNF/PCM composite aerogels maintain structural integrity even at temperatures as high as 190 °C [51]. Furthermore, by combining thermos-responsive and photothermal materials, the photothermal conversion of light to heat triggers thermal contraction of phase-change materials, which alters the spatial configuration of imprinted sites, enabling controlled and reversible desorption of dysprosium ions (Dy^{3+}) under green, energy-efficient conditions [52].

12.5.3 Bioactivity Integration

Endowing CBAs with bioactivity is one of the key strategies for advancing their biomedical applications. By introducing natural or synthetic bioactive molecules, the interactions between the material and biological systems (e.g., cells, tissues, body fluids) can be precisely regulated, thereby enabling enhanced biocompatibility, tissue regeneration, and antimicrobial or anti-inflammatory properties.

The release of bioactive ions (e.g., Ca^{2+}, Zn^{2+}, Sr^{2+}) can modulate the local microenvironment to promote the regeneration of specific tissues such as bone or skin. For instance, mineralized silk fibroin (SF)/cellulose interpenetrating network (IPN) composite aerogels leverage biomolecular templates to control the nucleation and growth of nano-hydroxyapatite (n-HA) crystals, resulting in the formation of a mineralized IPN aerogel material (Fig. 12.5d). Such materials hold great promise for applications in bone repair and regeneration [48].

Additionally, by combining partial anisotropic freeze-drying, crosslinking, and sacrificial templating strategies, nanofibrous aerogel 3D-printed grids have been developed using electrospun short fiber segments composed of poly(glycolide-co-lactide) (PGLA)/gelatin and poly(p-dioxanone) (PDO)/gelatin. These scaffolds exhibit nanoscale fibrous architecture and well-defined micro-/macrochannel networks, which facilitate the migration and proliferation of keratinocytes and dermal fibroblasts, thereby accelerating wound healing in diabetic ulcer models.

CBAs are characterized by low density, high specific surface area, and large pore volume, along with a loose and highly porous structure. These features allow them to effectively absorb and retain moisture, making them excellent carriers for antimicrobial agents. A novel aerogel-based fresh-keeping pad was developed by combining cellulose aerogels with polyvinyl alcohol (PVA), whey protein isolate, and cinnamon essential oil (CEO). The incorporation of CEO endowed the resulting aerogel with outstanding antibacterial and antioxidant properties. Specifically, it effectively inhibited the growth and proliferation of bacteria on the surface of beef, suppressed lipid oxidation, and preserved the meat's color during storage [53].

12.6 Conclusion and Outlooks

Composite functionalization represents a transformative strategy in the design of CBAs, enabling the integration of structural modulation, chemical modification, and multifunctional components into a unified material system. Through multicomponent synergy, dynamic responsiveness, and biointerface compatibility, these aerogels evolve from passive porous matrices into intelligent, multifunctional platforms. The incorporation of conductive nanomaterials enhances mechanical integrity and electronic conductivity, while stimuli-responsive components endow them with tunable properties under external triggers such as light or temperature. Furthermore, the introduction of bioactive ions, molecules, and scaffolding architectures facilitates tissue regeneration, antimicrobial efficacy, and wound healing, thus expanding their applicability to advanced fields such as flexible electronics, environmental remediation, and biomedical engineering. This section underscores the pivotal role of composite functionalization in pushing the boundaries of CBA performance and versatility.

Although CBAs have made significant progress in functional regulation and application expansion, the following key challenges still need to be overcome in future research: (i) Intelligent response upgrade: Develop stimuli-responsive aerogels (such as light/magnetic and pH/enzyme synergistic triggering) to achieve environmentally adaptive behavior regulation. (ii) Green manufacturing innovation: Explore low-energy consumption processes such as supercritical CO_2 assisted drying and bio-enzyme catalytic modification to reduce reliance on organic solvents. (iii) Accelerated clinical transformation: Deepen the research on long-acting biocompatibility in vivo and establish a standardized performance evaluation system (such as ISO 10993). (iv) Artificial intelligence empowerment: By integrating machine learning to optimize the design of multi-level pore structures, predict the synergistic effects of functional components, and shorten the material research and development cycle.

References

1. Y. Mao, Y. Sheng, Z. Fan, J. Yang, J. Liu, C. Tang, S. Fu, Atmospheric pressure dried discontinuous pore gradient structured CNF-Based Aerogel for Ultra-Low Reflection, Broadband, and Super-High EMI Shielding. Adv. Funct. Mater. 2421492 (2025)
2. Q. Zhou, W. Hong, K. Sheng, M. Pan, C. Liu, C. Mei, T. Ma, Direct ink writing 3D printed graphene oxide nanocomposite aerogel for intelligent fire-warning and exceptional fire-shielding. Chem. Eng. J. **499**, 156325 (2024)
3. J. Li, C. Liang, C. Lin, Y. Wang, Y. Liang, D. Dong, Lightweight, superhydrophobic, and superelastic MXene/carboxymethylcellulose sodium composite aerogels for efficient and multifunctional electromagnetic interference shielding. Chem. Eng. J. **506**, 160274 (2025)
4. K. Dhandapani, G. Ham, H. Fattahimoghaddam, I.H. Kim, H. Cha, X. Tang, D.W. Kim, T.K. An, Y.J. Jeong, Polymer semiconductor-coated 3D bilayer cellulose nanofiber/polydimethylsiloxane aerogel for enhanced solar-driven interfacial water evaporation. Desalination **611**, 118947 (2025)
5. B. Schroeter, S. Holst, I. Jung, L. Gibowsky, R. Subrahmanyam, P. Gurikov, I. Smirnova, Transparent cellulose aerogels from concentrated salt solutions: synthesis and characterization. Adv. Funct. Mater. **34**, 2407547 (2024)
6. L. Chen, X. Yu, M. Gao, C. Xu, J. Zhang, X. Zhang, M. Zhu, Y. Cheng, Renewable biomass-based aerogels: from structural design to functional regulation. Chem. Soc. Rev. **53**, 7489–7530 (2024)
7. S. Zong, L. Zhu, F. Lei, J. Jiang, J. Duan, Low cost Non freeze-drying construction strategy of cellulose aerogels for efficient solar desalination. Chem. Eng. J. **500**, 156859 (2024)
8. Q. Bi, Q. Zhao, Z. Qin, W. Gao, Z. Li, S. Deng, L. Mo, Lightweight, robust, and antimicrobial cellulose nanofibril aerogels decorated with TiO2@PDA core–shell nanoparticles for highly efficient oil–water separation. Chem. Eng. J. **507**, 159989 (2025)
9. Q. Peng, Y. Lu, Z. Li, J. Zhang, L. Zong, Biomimetic, hierarchical-ordered cellulose nanoclaw hybrid aerogel with high strength and thermal insulation. Carbohydr. Polym. **297**, 119990 (2022)
10. Y. Huang, Y. Sun, H. Liu, Fabrication of chitin nanofiber-PDMS composite aerogels from Pickering emulsion templates with potential application in hydrophobic organic contaminant removal. J. Hazard. Mater. **419**, 126475 (2021)
11. M. Cao, J. Zhu, G. Miao, J. Sha, D. Li, J. Li, C. Wang, C. Li, J. Zhang, Y. Xu, S. Chen, F. Xu, Ambient-Dried Nanocellulose Composite Aerogels for Enhanced Hydrovoltaic Electricity Generation. Adv. Funct. Mater. 2418823 (2024)
12. Y. Liang, H. Zhu, L. Wang, H. He, S. Wang, Biocompatible smart cellulose nanofibres for sustained drug release via pH and temperature dual-responsive mechanism. Carbohydr. Polym. **249**, 116876 (2020)

13. Y. Zhou, X. Zhang, D. Yin, J. Zhang, Q. Mi, H. Lu, D. Liang, J. Zhang, The solution state and dissolution process of cellulose in ionic-liquid-based solvents with different hydrogen-bonding basicity and microstructures. Green Chem. **24**, 3824–3833 (2022)
14. C. Jiménez-Saelices, B. Seantier, Y. Grohens, I. Capron, Thermal Superinsulating Materials Made from Nanofibrillated Cellulose-stabilized pickering emulsions. ACS Appl. Mater. Interfaces **10**, 16193–16202 (2018)
15. M. Song, J. Jiang, H. Qin, X. Ren, F. Jiang, Flexible and Super Thermal Insulating Cellulose Nanofibril/Emulsion Composite Aerogel with Quasi-Closed Pores. ACS Appl. Mater. Interfaces **12**, 45363–45372 (2020)
16. M. Long, X. Yang, T. Shi, Y. Yang, All-natural aerogel of nanoclay/cellulose nanofibers with hierarchical porous structure for rapid hemostasis. Int. J. Biol. Macromol. **278**, 134592 (2024)
17. P. Zhu, Z. Yu, H. Sun, D. Zheng, Y. Zheng, Y. Qian, Y. Wei, J. Lee, S. Srebnik, W. Chen, G. Chen, F. Jiang, 3D Printed Cellulose Nanofiber Aerogel Scaffold with Hierarchical Porous Structures for Fast Solar-Driven Atmospheric Water Harvesting. **36**, 2306653 (2024)
18. R. Xiao, X. Zhou, T. Yang, Z. Liu, S. Han, J. Wang, H. Wang, D. Ye, Biomimetic gradient aerogel fibers for sustainable energy harvesting from human sweat via the hydrovoltaic effect. Nano Energy **136**, 110759 (2025)
19. D. Zong, W. Bai, X. Yin, J. Yu, S. Zhang, B. Ding, Gradient Pore Structured Elastic Ceramic Nanofiber Aerogels with Cellulose Nanonets for Noise Absorption. **33**, 2301870 (2023)
20. Y. He, Z. Zhang, H. Ma, Y. Zhou, X. Tian, D. Sheng, X. Liu, Y. Yang, Waterborne polyurethane/cellulose composite aerogel with passive cooling and thermal insulation properties. **141**, e56158 (2024)
21. Q. Xu, M. Biancalana, J.C. Grant, H. Chiu, L. Jaroszewski, M.W. Knuth, S.A. Lesley, A. Godzik, M. Elsliger, A.M. Deacon, I.A. Wilson, Structures of single-layer β-sheet proteins evolved from β-hairpin repeats. Protein Sci. **28**, 1676–1689 (2019)
22. Q. Liu, Y. Liu, Q. Feng, C. Chen, Z. Xu, Preparation of antifouling and highly hydrophobic cellulose nanofibers/alginate aerogels by bidirectional freeze-drying for water-oil separation in the ocean environment. J. Hazard. Mater. **441**, 129965 (2023)
23. M. Shen, J. Qi, K. Gao, C. Duan, J. Liu, Q. Liu, H. Yang, Y. Ni, Chemical vapor deposition strategy for inserting atomic FeN4 sites into 3D porous honeycomb carbon aerogels as oxygen reduction reaction catalysts in high-performance Zn-air batteries. Chem. Eng. J. **464**, 142719 (2023)
24. J. Zhuang, N. Rong, X. Wang, C. Chen, Z. Xu, Adsorption of small size microplastics based on cellulose nanofiber aerogel modified by quaternary ammonium salt in water. Sep. Purif. Technol. **293**, 121133 (2022)
25. S. Chen, X. Li, C. Zhang, L. Long, X. Zhao, C. Zhu, B. Li, X. Yi, Superhydrophobic nanocellulose aerogel for thermal insulation and thermal resistance up to 1300 °C. Chem. Eng. J. **499**, 156043 (2024)
26. X. Dai, X. Chen, X. Jing, Y. Zhang, M. Pan, M. Li, Q. Li, P. Liu, C. Fan, X. Liu, DNA Origami-Encoded Integration of Heterostructures. Angew. Chem. Int. Ed. **61**, e202114190 (2022)
27. B. Wang, K. Nan, H. Rao, Y. Chen, R. Pei, Y. Wang, Integrated design of multifunctional lightweight magnetic cellulose-based aerogel with 1D/2D/3D hierarchical network for efficient microwave absorption. Compos. Commun. **49**, 101987 (2024)
28. Y. Song, J. Bao, Y. Hu, Y. Liu, Y. Wen, M. Xu, Q. Yang, Z. Sun, C. Xiong, Z. Shi, Multistage amine-modified cellulose nanofibrils as high-performance positive triboelectric materials. Cell Rep. Phys. Sci. **5**, 101939 (2024)
29. S. Li, Y. Yu, J. Chang, Z. Zheng, G. Li, X. Wang, D.L. Kaplan, Enhanced CO2 Capture and Utilization through Chemically and Physically Dual-Modified Amino Cellulose Aerogels Integrated with Microalgae-Immobilized Hydrogels. ACS Environ. Sci. Technol. Eng. **5**, 551–565 (2025)
30. Y. Zhang, W. Deng, C. Liu, F. Yan, M. Wu, Q. Cui, S. Willför, C. Xu, B. Li, Preparation of Antibacterial Dialdehyde Nanocellulose Using LiBr·3H2O Non-Dissolving Pretreatment Promoted Periodate Oxidation. ACS Sustain. Chem. Eng. **11**, 6641–6651 (2023)

31. T. Nakamura, T. Ishiyama, Molecular dynamics study of hydrogen bond structure and tensile strength for hydrated amorphous cellulose. Biomacromol **25**, 7249–7259 (2024)
32. Q. Yang, W. Su, J. Hu, Y. Xu, Z. Liu, L. Hui, Synthesis of Superhydrophobic Cellulose Stearoyl Ester for Oil/Water Separation. **12**, 1964 (2022)
33. L. Xiong, W. Yu, X. Lu, J. Teng, C. Chen, B. Li, L. Shen, H. Lin, A novel dual-function anti-scalant carboxymethyl cellulose-graft-protocatechuic acid for efficient scaling and biofouling prevention. J. Membr. Sci. **713**, 123357 (2025)
34. Y. Yang, Z. Zhao, Y. Wang, L. Yang, D. Liu, X. Yang, W. Pan, A novel asymmetric membrane osmotic pump capsule with in situ formed delivery orifices for controlled release of gliclazide solid dispersion system. Int. J. Pharm. **506**, 340–350 (2016)
35. Y. Song, J. Bao, Y. Hu, Y. Liu, Y. Wen, M. Xu, Q. Yang, Z. Sun, C. Xiong and Z. Shi, Multistage amine-modified cellulose nanofibrils as high-performance positive triboelectric materials. Cell Rep. Phys. Sci. **5** (2024)
36. Y. Duan, H. Yang, K. Liu, T. Xu, J. Chen, H. Xie, H. Du, L. Dai, C. Si, Cellulose nanofibril aerogels reinforcing polymethyl methacrylate with high optical transparency. Adv. Compos. Hybrid Mater. **6**, 123 (2023)
37. C. Zhang, S. Zhao, W. Liu, X. Han, M. Wu, P. Zhao, A.J. Ragauskas, X. Song, Advances in multi-dimensional cellulose-based fluorescent carbon dot composites. Compos. Part B Eng. **260**, 110752 (2023)
38. L. Chaabane, Z. Jaafar, M. Chaaben, S. Chaaben, A.E. Ghali, M. Msaddek, E. Beyou, M.H.V. Baouab, Dual-function advanced magnetic bacterial cellulose materials: From enhanced adsorption phenomena to an unprecedented circular green catalytic strategy. J. Colloid Interface Sci. **686**, 1215–1229 (2025)
39. Z. Yan, L. Meng, S. Jiang, Y. Deng, J. Xi, L. Zhang, P. Li, H. Xiao, W. Wu, Bifunctional Nanocellulose@MOF composite aerogel for selective fluorescent detection and efficient removal of tetracycline. Carbohydr. Polym. **347**, 122697 (2025)
40. J. Qiao, Q. Song, X. Zhang, S. Zhao, J. Liu, G. Nyström, Z. Zeng, Enhancing interface connectivity for multifunctional magnetic carbon aerogels: an in situ growth strategy of metal-organic frameworks on cellulose Nanofibrils. **11**, 2400403 (2024)
41. X. Hu, J. Zhou, Z. Deng, W.-X. Zhang, MIL-100-Fe self-assembled cellulose nanofibers sponge for Diclofenac cascade encapsulation. Carbohydr. Polym. **352**, 123182 (2025)
42. Z. Sun, X. Zhao, Z. Chen, Y. Jin, F. Ren, P. Ren, Facile synthesis of carbon nitride-doped cellulose aerogels with high selective adsorption of cationic dyes from wastewater. Ind. Crops Prod. **221**, 119384 (2024)
43. J. Garemark, X. Yang, X. Sheng, O. Cheung, L. Sun, L.A. Berglund, Y. Li, Top-Down approach making anisotropic cellulose aerogels as universal substrates for multifunctionalization. ACS Nano **14**, 7111–7120 (2020)
44. J. Miao, X. Dong, Y. Xu, Z. Zhai, L. Zhang, B. Ren, Z. Liu, Preparation and electrochemical performance of 1,4-naphthaquinone-modified carbon nanotubes as a supercapacitor material. Org. Electron. **73**, 304–310 (2019)
45. B. Zhang, Z. Qu, C. Ruiz-Agudo, L. Yang, B. Chi, Y. Kong, F. Wang, Lightweight magnetic carbon nanotube/cellulose nanofibre aerogels with microstructure engineering for enhanced microwave absorption. Carbon **234**, 120020 (2025)
46. Z. Chen, B. Wang, J. Qi, T. Liu, Y. Feng, C. Liu, C. Shen, Eco-friendly bacterial cellulose/MXene aerogel with excellent photothermal and electrothermal conversion capabilities for efficient separation of crude oil/seawater mixture. Carbohydr. Polym. **336**, 122140 (2024)
47. R. Wang, J. Deng, P. Wu, Q. Ma, X. Dong, W. Yu, G. Liu, J. Wang, L. Liu, Amphipathic Janus nanofibers aerogel for efficient solar steam generation. **7**, e12667 (2024)
48. Z.-J. Chen, Y. Zhang, L. Zheng, H. Zhang, H.-H. Shi, X.-C. Zhang, B. Liu, Mineralized self-assembled silk fibroin/cellulose interpenetrating network aerogel for bone tissue engineering. Biomater. Adv. **134**, 112549 (2022)
49. S. Lyu, H. Chang, L. Zhang, S. Wang, S. Li, Y. Lu, S. Li, High specific surface area MXene/SWCNT/cellulose nanofiber aerogel film as an electrode for flexible supercapacitors. Compos. Part B Eng. **264**, 110888 (2023)

50. Z. Wei, Y. Zhang, C. Cai, H. Qu, Y. Fu, S.C. Tan, Wood Lamella-Inspired Photothermal Stearic Acid-Eutectic Gallium-Indium-Based Phase Change Aerogel for Thermal Management and Infrared Stealth. **19**, 2302886 (2023)
51. M. Song, J. Jiang, J. Zhu, Y. Zheng, Z. Yu, X. Ren, F. Jiang, Lightweight, strong, and form-stable cellulose nanofibrils phase change aerogel with high latent heat. Carbohydr. Polym. **272**, 118460 (2021)
52. T. Xu, X. Zheng, X. Zhang, G. Li, J. Mei, Z. Li, Photothermal switchable cellulose nanocrystals-graphene oxide ionic imprinted aerogel for selective adsorption and controlled desorption of dysprosium ions. ACS Sustain. Chem. Eng. **12**, 9833–9844 (2024)
53. K. Yu, L. Yang, N. Zhang, S. Wang, H. Song, H. Liu, Superabsorbent, antibacterial, and antioxidant nanocellulose aerogels: Preparation, characterization, and application in beef preservation. Food Chem. **466**, 142251 (2025)

Chapter 13
Cellulose-Based Aerogels for Thermal Insulation

Benkang Zhu, Boxiao Li, Xiaofang Zhang, and Jianming Zhang

Abstract Cellulose aerogel, as a lightweight, highly porous, and renewable biomass-derived material, has witnessed rapid development in thermal insulation applications. This chapter reviews its research progress in the field of thermal insulation. Based on the macroscopic morphology, cellulose-based aerogels (CBAs) are categorized into one-dimensional (1D) aerogel fibers, two-dimensional (2D) aerogel films, and three-dimensional (3D) bulk aerogels. Here, their preparation methods are summarized first, followed by a review of their application areas. Their primary applications include building insulation, textile filling, and infrared stealth thermal management. Research indicates that structural regulation and functional modifications (e.g., hydrophobic treatment, compositing with nanoparticles) significantly enhance the mechanical strength and environmental adaptability of these three material types. Future efforts need to focus on optimizing scalable manufacturing processes and reducing costs to facilitate industrial application.

Keywords Cellulose-based aerogels · Thermal insulation · Fibers · Films · Scaffolds

13.1 Introduction

The global imperative to mitigate climate change and achieve energy security has thrust the built environment into sharp focus. In this critical context, thermal insulation emerges as a cornerstone, offering one of the most cost-effective and immediate pathways to drastically reduce operational energy demands and associated carbon footprints. For example, effective insulation minimizes unwanted

B. Zhu · B. Li · X. Zhang · J. Zhang (✉)
Key Laboratory of Rubber-Plastics, Ministry of Education, Shandong Provincial Key Laboratory of RubberPlastics, Qingdao University of Science and Technology, Qingdao, China
e-mail: zjm@qust.edu.cn

X. Zhang
e-mail: zhangxf@qust.edu.cn

G. Wei and J. Zhang (eds.), *Cellulose-Based Hydrogels and Aerogels*, Lecture Notes in Nanoscale Science and Technology 41,
https://doi.org/10.1007/978-3-032-03008-5_13

heat transfer through building envelopes, stabilizing indoor temperatures, enhancing occupant comfort [1]. Consequently, the selection and performance of insulation materials are pivotal in the pursuit of sustainable construction and energy-efficient retrofitting. Traditionally, the insulation market has been dominated by materials derived from fossil fuels (e.g., phenol phenol formaldehyde (PF) [2], polystyrene (PS) [3], polyurethane (PU) [4] foams). While often effective thermally, these conventional materials present substantial environmental challenges. Their production typically involves high embodied energy, reliance on non-renewable petrochemical feedstocks, and the potential release of persistent pollutants or hazardous dust during installation and disposal. Furthermore, end-of-life management poses significant problems, as many synthetic insulations are not readily biodegradable and contribute to long-term landfill burdens. These drawbacks increasingly conflict with the principles of a circular economy and the growing demand for genuinely sustainable building solutions that prioritize human and environmental health throughout their lifecycle.

This limitation, coupled with persistent environmental concerns regarding synthetic materials' embodied carbon, resource depletion, and end-of-life management, has catalyzed intense research into ultra-efficient, multifunctional, and truly sustainable alternatives. Within this landscape, cellulose-derived aerogels, foams, and engineered porous materials have emerged as revolutionary candidates, promising to redefine the performance boundaries of bio-based insulation through advanced nano- and meso-structuring. Cellulose, Earth's most abundant natural polymer, offers an unparalleled foundation for sustainable material design. Its renewability, biodegradability, low toxicity, and compatibility with waste-stream valorization (e.g., recycled paper, agricultural residues) align perfectly with circular economy principles [5, 6]. However, transforming this linear polysaccharide into materials with ultra-low thermal conductivity (< 0.025 W/m·K) and superior mechanical robustness requires sophisticated engineering beyond simple fiber compaction.

Cellulose-based aerogels (CBAs), produced by creating a nanostructured cellulose network within a solvent or dispersion followed by their removal under conditions preventing pore collapse (e.g., supercritical CO_2 drying, freeze-drying), achieve remarkably low densities (down to 0.01 g/cm^3) and extraordinarily high porosities ($> 95\%$). This intricate nano-architecture, comprising cellulose fibrils surrounding nanoscale air pockets, drastically minimizes solid conduction, gas conduction, and convection—the three primary heat transfer pathways. The result is thermal conductivity potentially rivalling or even surpassing silica aerogels, the current gold standard for high-performance insulation. CBA foams, often fabricated via directional freezing, templating, or gas-foaming techniques, provide a more scalable route to hierarchical porosity, combining micro-, meso-, and macro-pores to optimize tortuosity and phonon scattering while offering improved mechanical resilience for structural applications. The porous materials also contain one-dimensional (1D) porous cellulose fibers [7], two-dimensional (2D) porous cellulose films [8], and three-dimensional (3D) cellulose aerogel bulk [9, 10].

The allure of these advanced CBAs extends far beyond exceptional thermal resistance. Their inherently tunable structure allows precise optimization for specific insulation contexts—from cryogenic applications requiring nanopores smaller than the mean free path of air molecules (≈70 nm at ambient) to building envelopes demanding balanced hygrothermal performance. Furthermore, cellulose's rich chemistry enables facile functionalization or hybridization to further improve the thermal insulation properties and other properties, for example, moisture resilience through hydrophobic modification (e.g., silanization [11–14] and esterification [15]) while preserving the porous structure critical for low thermal conductivity. Mechanical enhancement can be reinforced with polymers, cross-linkers, or nanofillers to improve compressive strength and elasticity of CBAs, addressing the brittleness often associated with aerogels. Despite their immense promise, significant challenges remain in translating laboratory breakthroughs into widespread commercial adoption. Scalable, energy-efficient, and cost-effective manufacturing processes, particularly for aerogels avoiding intensive supercritical drying, are crucial. Ensuring long-term stability under thermal cycling, humidity fluctuations, and mechanical load of CBAs is paramount.

13.2 Fabrication and Thermal Insulation Applications of 1D CBAs

13.2.1 Fabrication of 1D CBAs

Different cellulose sources and manufacturing methods can impart distinct microstructures and properties to nanocellulose aerogels. Typically, nanocellulose-based aerogels are fabricated in three steps: (1) dispersion of nanocellulose, (2) gelation of nanocellulose, and (3) drying of the gel. Various techniques can assemble individual fibers or aggregates into three dimensional cellulose aerogels with diverse structural characteristics, where mechanical and functional properties are highly dependent on microstructure. Primary fabrication strategies for one-dimensional cellulose aerogels include electrospinning, solution blow spinning, freeze-drying, and template-directed methods [16].

Electrospinning is a technique that utilizes high-voltage electric fields to produce ultrafine fibers from polymer solutions or melts. The core principle involves using a high-voltage power supply (typically thousands to tens of thousands of volts) to charge the polymer solution, forming a Taylor cone. When the electric field force overcomes the solution's surface tension, a jet ejects from the cone tip and solidifies into nano- or micro-scale fibers as the solvent evaporates during travel to the collector. Fiber diameter and morphology are governed by solution parameters (e.g., concentration, viscosity, conductivity), process parameters (voltage, flow rate, collection distance), and environmental factors (temperature, humidity). For instance, high humidity may induce phase separation, creating fluffy 3D structures, while high polymer concentration increases fiber diameter. The unique advantage of electrospinning lies in its ability to produce fibers with a wide diameter range (tens of nanometers to micrometers) and complex structures such as porous or core–shell configurations through parameter optimization [16].

Solution blow spinning employs high-speed airflow (rather than an electric field) to stretch polymer solutions into fibers, combining features of meltblown and electrospinning technologies. The principle involves: high-pressure gas generates shear force through an annular nozzle, forming solution jets that solidify into fibers upon solvent evaporation. Key parameters include gas pressure (typically 0.2–0.5 MPa), solution viscosity (affecting fiber continuity), and collection distance (determining fiber deposition morphology). Fibers produced by this method generally range from hundreds of nanometers to micrometers in diameter, forming 3D networks through physical entanglement or chemical crosslinking. For example, TiO_2 ceramic aerogels exhibit densities as low as 8 mg/cm^3, porosity > 99.7%, and maintain superelasticity at high temperatures (e.g., 1100 °C). The advantages of solution blow spinning include high-throughput production (e.g., kilogram-scale polystyrene fibers) and broad material compatibility (e.g., cellulose, metal oxides), though fiber uniformity is slightly inferior to electrospinning [16].

In addition to electrospinning and high-speed air-jet spinning, wet-spinning methods are the most commonly used approach for preparing cellulose aerogel fibers. This method first involves preparing a cellulose solution or dispersion, which is then extruded into a coagulation bath via a metering pump. The resulting nascent fiber bundles (filaments) formed in the coagulation bath are subsequently drawn and extracted, followed by washing and drying to obtain cellulose aerogel fibers (As shown in Fig. 13.1). During the spinning process, four speeds have to be considered for the spinning kinetics, that is, extrusion speed, building block diffusion rate, gelling agent diffusion rate, and collecting speed. The spinnability of the desired dope should be related to the viscosity, velocity, and surface tension of the spinning dope [17]. Therefore, it is also a complex process.

Freeze-drying is a dehydration process where frozen solvents sublimate directly from solid to gas under reduced pressure, creating oriented pores. During freezing, uniformly dispersed nanofibers become enriched between solution crystals. As ice crystals grow, nanofibers are confined between them and entangled due to crystal expansion, forming a 3D solid network. After drying, the solidified solvent sublimates, yielding a three-dimensional fluffy aerogel. Xu et al. employed a bottom-up unidirectional freezing process to construct vertically porous CNF/Ti_3CNT_x and CNF/rGO hybrid aerogel electrodes with coupled nanochannels [18]. The vertical porous structure significantly shortens ion diffusion paths, enhancing charge/ion transport/diffusion kinetics, while inserted cellulose nanofibers (CNFs) prevent nanosheet restacking and expand interlayer nanochannels, improving electrolyte ion accessibility. Long et al. developed an anisotropic composite aerogel (SPCCAM) using directional freezing and interfacial crosslinking, featuring a 1D building block at the nanoscale and a honeycomb-like anisotropic structure at the microscale [19]. The core of carboxylated cellulose nanofibers (CNFs-C) exhibits substantially higher strength than the intermediate polybenzoxazine (PBOZ) shell layer, while interfacial

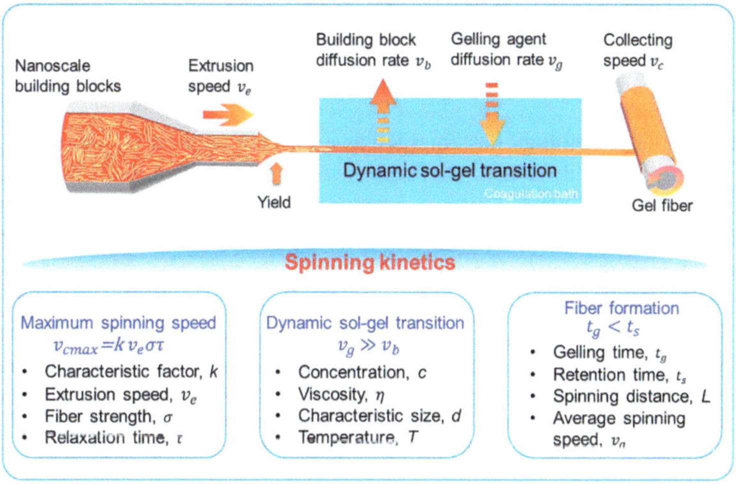

Fig. 13.1 Spinning dynamics of aerogel fibers. During the dynamic sol–gel transition, the gelling agent diffusion rate competes with the building block diffusion rate to form the gel fiber. Reprinted with the permission from Sheng et al. [17]**,** Copyright 2023, Wiley–VCH

crosslinking via amino groups and hydrogen bonds forms a highly stable 1D cellulose aerogel building block. Wang et al. fabricated multidimensional, gradient porous, electrically and magnetically conductive carbon nanocellulose aerogels composed of highly aligned nickel chains through freeze-drying and low-temperature annealing [7]. These materials demonstrate exceptional microwave absorption, low density, and superior mechanical adaptability.

The template-directed method is a fabrication technique that regulates the microstructure of cellulose aerogels by introducing specific templates. The core principle involves using porous templates (e.g., ice crystals, polymer microspheres, or foams) to guide the directional assembly of CNFs, forming biomimetic ordered channels. Zeng et al. prepared aerogels via the Pickering emulsion templating method using TEMPO-oxidized cellulose nanofibers (TOCNFs) and polyvinyltrimethoxysilane (PVTMS) [20]. The resulting aerogels exhibited low density and exceptional toughness, along with outstanding anisotropic thermal insulation performance, achieving a maximum radial thermal conductivity of 115.2 $mW{\cdot}m^{-1}{\cdot}K^{-1}$.

13.2.2 Thermal Insulation Application

The structural design of aerogel fibers can effectively solve problems such as low strength and high brittleness of aerogels, and endow fiber materials with excellent properties of both mechanical and thermal insulation properties. Long et al. developed a PBOZ@CNF composite aerogel with a strength of 41.8 MPa at a density of

0.234 g/cm^3 [21]. The one-dimensional cellulose building blocks confer excellent deformation resistance and fatigue tolerance. The highly porous mesoporous interconnected structure contributes to a low thermal conductivity of 0.02943 W/(m·K) at 25 °C.

Wet-spinning technology has emerged as a primary method for fabricating thermally insulating cellulose aerogel fibers, owing to its exceptional capability for precise regulation of internal porous structures. For instance, Liu et al. developed cellulose aerogel fiber materials for thermal insulation using wet-spinning technology [22]. By precisely regulating the internal pore structure through acid coagulation bath adjustment during wet spinning, they achieved a thermal conductivity of 0.0242 W/(m·K) (As shown in Fig. 13.2). The Ye team developed a novel class of gradient-structured regenerated all-cellulose sponge-aerogel fibers (CGFs) through a microfluidic dynamic dual-crosslinking spinning technique [23]. Using cotton linter pulp dissolved in NaOH/urea aqueous solution and lightly chemically crosslinked with epichlorohydrin as the spinning dope, the researchers engineered fibers with a unique porous sponge outer layer and dense aerogel inner core. This method enabled precise control over sponge layer thickness while endowing the fibers with exceptional toughness (26.20 MJ m^{-3}) and ultralow thermal conductivity (0.023 W m^{-1} K^{-1}). Remarkably, the thermal conductivity surpassed that of stationary air (0.026 W m^{-1} K^{-1}), attributed to Knudsen effect, gaseous conduction suppression, and enhanced infrared scattering. The CGFs demonstrated outstanding wide-temperature-range insulation performance. Also, the material has 40% reduced surface temperature fluctuation range compared to substrate during –20 to 80 °C cycling tests.

Fu et al. developed a microfluidic spinning method to prepare gradient all-nanostructure aramid aerogel fibers (GAFs) [24]. Benefiting from the simultaneous shear alignment and diffusion dilution of a good solvent within the channels, the precursor gel fibers assemble into a structure with a sparse exterior and dense interior, which reverses during supercritical drying to form sheath and core layers with average pore diameters of 150 and 600 nm, respectively. The results show that this gradient nanostructure generates high interfacial thermal resistance at the heat transfer interface, resulting in a radial thermal conductivity as low as 0.0228 W·m^{-1}·K^{-1}, which is much lower than the thermal conductivities of air and wet-spun aerogel fibers.

The use of ionic liquids for wet spinning can effectively regulate the thermal insulation properties of fibers. For instance, Liu et al. developed ultra-tough cellulose aerogel fibers (SAFs) using degreased cotton cellulose dissolved in the ionic liquid 1-allyl-3-methylimidazolium chloride (1–6 wt%) [25]. These fibers were fabricated via wet spinning combined with supercritical CO_2 drying. The SAFs exhibited exceptional mechanical properties, including a tensile strength of 30 MPa, breaking strain of 107%, and toughness of 21.85 MJ/m^3. This toughness surpasses that of human knee ligaments by a factor of 7.4 and exceeds traditional silica aerogel fibers by a remarkable factor of 7280. The SAFs also demonstrate superior thermal insulation. In hot plate tests (124 °C), SAF-woven fabric showed a 24.1 °C surface temperature difference—73% higher than cotton fabric (13.9 °C). Under low-temperature conditions, the SAF fabric maintained a 21 °C difference, significantly outperforming cotton (15.3 °C). This excellent insulation arises from the material's high porosity

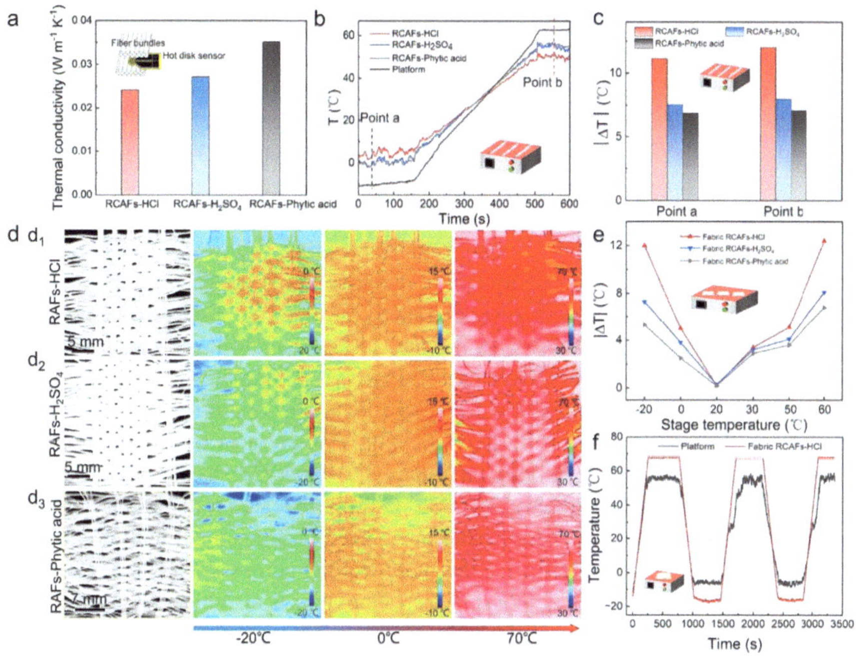

Fig. 13.2 Thermal insulation performance of cellulose aerogel fiber fabrics. Reprinted with the permission from Liu et al. [22], Copyright 2025, Elsevier

(>87%) and nanoporous structure (average pore size: 15–30 nm), making it highly suitable for lightweight thermal textiles.

Some relevant scholars have combined the freeze - spinning technology with the bionic strategy of polar bear hair to develop a CBA fiber material with stable thermal insulation performance. Aerogel fiber applied in textiles is greatly limited by their fragility and poor processability. Bai and co-workers solved these issues by encapsulating the aerogel fiber with a stretchable layer, mimicking the core–shell structure of polar bear hair [26]. The study fabricated these fibers through a two-step process involving freeze-spinning of chitosan aerogel cores followed by thermoplastic polyurethane (TPU) encapsulation, creating a biomimetic hierarchical structure. The resulting fibers exhibited exceptional thermal insulation properties, with an ultralow thermal conductivity of 26.9 ± 1.8 mW·m^{-1}·K^{-1} that outperforms both wool and nylon. When tested as single-layer fabric on a 40 °C hot plate, the material demonstrated an 8.2 °C temperature difference while achieving comparable insulation at just one-fifth the thickness of conventional down feathers. Remarkably, the fibers maintained stable thermal performance across a broad temperature range, sustaining a surface temperature of 3.5 °C in -20 °C environments and retaining 3.3 °C after washing. Despite its high internal porosity over 90%, the fiber was stretchable up to 1000% strain, which was greatly improved compared with that of traditional aerogel fibers (~2% strain). In addition to its washability and dyeability, the prepared fiber was mechanically robust, retaining its stable thermal insulation property after 10,000 stretching cycles (100% strain).

Gui et al. developed biomimetic insulating fibers by mimicking the hollow hair structure of polar bears, employing a freeze-spinning technique with silk fibroin as the primary matrix and chitosan as a mechanical reinforcement [27]. Their innovative freeze-spinning approach integrated directional freezing and solution spinning: the blended solution is extruded through a cold copper ring to guide the aligned growth of ice crystals, forming axially oriented microchannels, followed by freeze-drying to produce fibers with 87% porosity and pore diameters of 20–85 μm. The resulting fibers exhibited an ultralow thermal conductivity of 21.86 ± 1.98 $mW \cdot m^{-1} \cdot K^{-1}$, approaching that of stagnant air. In thermal tests, a single-layer fabric (0.4 mm thick) generated temperature differences (ΔT) of 7.9 °C and 8.2 °C under − 20 and 80 °C heat sources, respectively, while a five-layer stack achieved a ΔT of 20 °C (80 °C environment). When subjected to cyclic temperatures (−22 to 80 °C), the fabric's surface temperature fluctuation range (−10 to 64 °C) was 40% narrower than that of the substrate. Remarkably, infrared imaging revealed a negligible temperature contrast (≤0.5 °C) between the fabric-covered living subject and the ambient background, demonstrating effective thermal cloaking across a broad range (−10 to 40 °C). In addition to the fur of polar bears, camel hair is also one of the sources of heat-insulating fiber design.

Wang et al. developed a micro-extruded physically foamed porous elastic fiber (MEPF) and its double-layer fabric (MEPFT-d) using TPU elastomer via micro-extrusion foaming technology for adaptive thermal management in extreme temperature environments [28]. Inspired by camel hair, the fiber combined radiative cooling (enabled by high solar reflectivity and infrared emissivity), delayed heat dissipation (due to low thermal conductivity), and evaporative cooling (facilitated by unidirectional water transport). In performance tests, the material reduced heat loss in cold conditions (7.2 °C warmer than wool) and provided effective insulation in hot environments (10.2 °C cooler than wool). Additionally, it exhibited a remarkably low radial thermal conductivity (0.048 $W \cdot m^{-1} \cdot K^{-1}$) and a high CLO value (0.545), demonstrating its potential for advanced thermal regulation.

13.3 Fabrication and Thermal Insulation Application of 2D CBAs

Cellulose-based aerogels demonstrate remarkable multifunctionality across the fields of architecture and wearable technology. In architectural insulation, aerogels synthesized through colloidal self-assembly and silanization provide a scalable, hydrophobic solution for glazing systems, effectively reducing thermal transfer while maintaining transparency. For on-skin electronics, composite aerogels with enhanced stretchability enable sensing and thermal healthcare applications, capitalizing on their low thermal conductivity to achieve efficient thermal management in wearable devices.

13.3.1 Fabrication

The preparation of highly transparent silanized cellulose aerogels (SiCellA) for building insulation involves four key steps, as shown in Fig. 13.3a tailored for scalability and glazing compatibility, including (i) Gel Formation: A cellulose-based gel is created, forming the lightweight, porous foundation for the aerogel using dissolved or modified cellulose. (ii) Drying: The gel is dried via supercritical or freeze-drying methods to preserve its porous structure, ensuring thermal insulation and transparency. (iii) Silanization: A hydrophobic layer is added through silanization, enhancing durability and moisture resistance for window applications. (iv) Scalability Design: The process is optimized for roll-to-roll production, enabling large-scale integration into insulating glass units or window retrofits. This streamlined methodology delivers high-performance insulation while meeting industrial demands [8].

The preparation of cellulose aerogel electronics (PAPE) includes 4 steps, as illustrated in Fig. 13.3b [29]. CHP Aerogel Synthesis: A CNF/HPC/PVA gel was prepared by mixing components (6 h stirring), degassed for 7 days, blade-coated (1 mm, −5 °C), and freeze-dried (−50 °C, 10 Pa, 12 h). Plasticization (PCHP): CHP was treated with a DES plasticizer (ChCl/urea/ethanol, 60 °C, 1 h), immersed (2 h), dried (22 h), and rinsed with ethanol to enhance flexibility while retaining porosity. Electrode Fabrication: Conductive Ag ink was screen-printed (300-mesh) onto PCHP and cured (80 °C, 15 min), ensuring uniform conductivity. Device Assembly: The PCHP/Ag construct was encapsulated with porous PU tape, forming a trilayer structure (<180 μm thick) with distinct interfaces.

13.3.2 Thermal Insulation Applications

Buildings account for approximately 40% of global energy consumption, with windows and skylights often being the least energy-efficient components of the building envelope due to the challenge of balancing high transparency with effective thermal insulation. Addressing this, SiCellAs, derived from the abundant biopolymer cellulose, works as an innovative glazing material, as presented by Fig. 13.4a. Fabricated via colloidal self-assembly and roll-to-roll compatible processes, SiCellAs offer scalability and practical application potential. These aerogels exhibit remarkable properties: visible-range light transmission of 97–99% (exceeding traditional glass), haze of ~ 1%, and thermal conductivity lower than still air, positioning them as superior insulators. SiCellAs can be integrated into multi-pane insulating glass

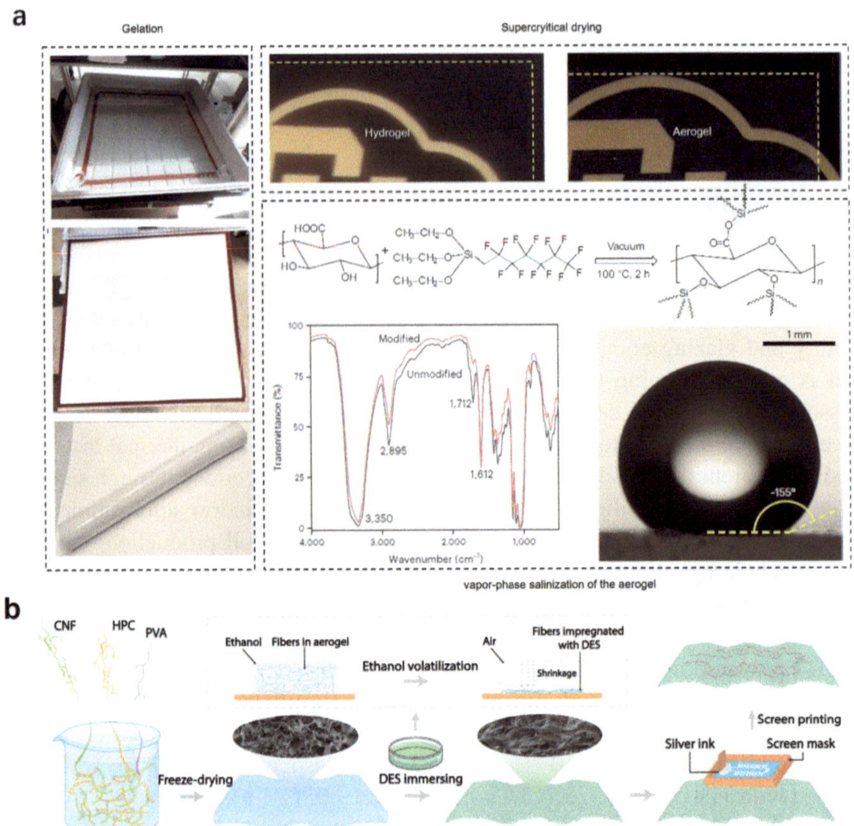

Fig. 13.3 Fabrication of 2D CBA films/membranes: **a** Fabrication of superhydrophobic SiCellAs at window scales with processes of gelation, drying and salinization. Reprinted with the permission from Abraham et al. [8], Copyright 2023, Springer Nature. **b** Schematic diagram for the fabrication of a plasticized aerogel membrane (inset SEM images showing the cross-section of the CNF/HPC/PVA aerogel membrane (CHP) and plasticized CNF/HPC/PVA aerogel membrane (PCHP)). Reprinted with the permission from Liu et al. [29], Copyright 2025, American Chemical Society

units (IGUs) or used to retrofit single-pane windows, significantly enhancing energy efficiency. For example, a retrofitted single-pane window with SiCellA matches or surpasses double-pane window performance. Beyond thermal and optical excellence, SiCellAs are mechanically robust, superhydrophobic, and durable, resisting degradation under environmental tests. The study demonstrates SiCellAs' ability to boost condensation resistance and thermal performance, with potential R-values reaching 5 h ft^2 °F Btu^{-1}. Their lightweight nature—density ~ 1% of glass—and cost-effective production from sustainable sources like wood pulp or industrial waste further enhance their appeal. This work not only tackles a critical challenge in building energy efficiency but also paves the way for advanced glazing technologies. By offering a scalable, high-performance alternative to conventional glazing, SiCellAs could reduce energy consumption and carbon emissions in the built environment, redefining the role of windows in sustainable architecture [8].

The recyclable cellulose aerogel-based electronic system (PCHP/Ag/PU) for on-skin applications, addressing limitations in conventional flexible electronics (Fig. 13.4b). Inspired by human skin, the system integrates a biocompatible CNF/HPC/PVA aerogel hypodermis, a screen-printed silver sensing layer, and a permeable polyurethane epidermis. The aerogel exhibits high tensile strength (1.14 MPa), 43.5% tensile strain, and excellent permeability, enabling long-term skin contact without irritation. The electronics demonstrate superior strain and humidity sensing (gauge factor ~ 238, detection limit 0.1%, response time 18 ms) due to microcrack dynamics in the silver layer, alongside thermal management capabilities. Integrated into an electronic mask, the system supports patient-centered healthcare with power supply, switching, and wireless Bluetooth modules. All components are recyclable in water, and the aerogel is biodegradable with cellulases, enhancing environmental sustainability. This work overcomes challenges of low permeability and biocompatibility in traditional substrates, offering a scalable, eco-friendly solution for wearable electronics. This research highlights significant potential for next-generation on-skin electronics in healthcare monitoring, human–machine interfaces, and smart wearable devices, advancing sustainable and high-performance personal healthcare technologies.[29]

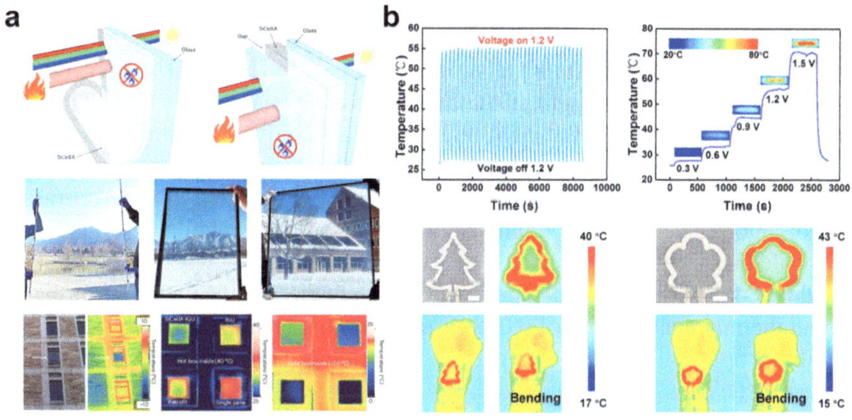

Fig. 13.4 Thermal Insulation and management applications of 2D CBAs **a** SiCellA-based window retrofits and IGUs. Reprinted with the permission from Abraham et al. [8], Copyright 2023, Springer Nature. **b** Permeable, antibacterial, recyclable PCHP/Ag/polyurethane electronics (PAPE) for hyperthermia therapy. Reprinted with the permission from Liu et al. [29], Copyright 2025, American Chemical Society

13.4 Fabrication and Thermal Insulation Application of 3D CBAs

13.4.1 Fabrication

Beyond 1D cellulose aerogel fibers and 2D cellulose aerogel films, the application of 3D cellulose aerogels, foams, and bulk materials in thermal insulation field has also attracted substantial research interest. Currently, the method for fabrication for these 3D cellulose materials are categorized into three types: natural wood chemically treated to remove lignin and hemicellulose (As shown in Fig. 13.5a), the lyophilizing nanocellulose dispersion aqueous (Shown in Fig. 13.5b), and regenerated cellulose-based aerogels (As shown in Fig. 13.5c).

Additionally, to enhance the thermal insulation performance of cellulose aerogels, it is often necessary to incorporate other components. The research group of Professor Liangbing Hu at the University of Maryland focuses on developing functional materials derived from natural wood and has made significant contributions to the thermal insulation field [30, 31]. For instance, Chen et al. removed lignin and hemicellulose from natural wood, obtaining anisotropic cellulose-based porous materials (As shown in Fig. 13.5a) [30]. Using the same method, Li et al. obtained nanowood [31]. Benefiting from the native oriented pores and porous structure of the original wood, the resulting cellulose-based nanowood exhibited distinct anisotropic thermal conductivity, with values of 0.03 W/(m·K) perpendicular to the pore direction and 0.06 W/(m·K) along the pore direction. However, the potential of thermally insulating wood aerogels in energy-saving engineering is limited by their mechanical fragility and insufficient environmental stability. To address these limitations, Liu et al. inspired by diatoms, employed a nanoscale heterogeneous assembly strategy combined with a sol–gel process [32]. The achieved the *in-situ* mineralization of amorphous inorganic oligomers, leading to robust welding of the organic–inorganic interface in the oriented pore, originating from raw woods. The resultant cellulose composite aerogel demonstrated excellent thermal insulation performance (radial thermal conductivity of 0.038 W/(m·K)). This biomimetic material exhibits significant application potential in aerospace, building energy efficiency, and thermal protection fields.

In-situ removal of lignin and hemicellulose, utilizing the abundant cellulose resources in natural wood is employed to prepare fiber-based aerogel materials. In addition to that, the native cellulose raw materials are then nanofibrillated to obtain water-dispersible nanocellulose suspensions, which are used to fabricate nanocellulose-based aerogels and it is a common approach in this field (Fig. 13.5b). Di et al. utilized acid-treated recycled aramid nanofibers (upANFA, derived from Kevlar yarn) and TEMPO-oxidized cellulose nanofibers (CNF, derived from wood) to prepare hybrid upANFA/CNF-based foams via ice templating [9]. When the upANFA content reached 40 wt%, the hybrid foam exhibited excellent thermal stability and extremely low thermal conductivity (18–23 mW m^{-1} K^{-1}, measured perpendicular to the direction of the aligned nanofibers) across a wide relative humidity (RH)

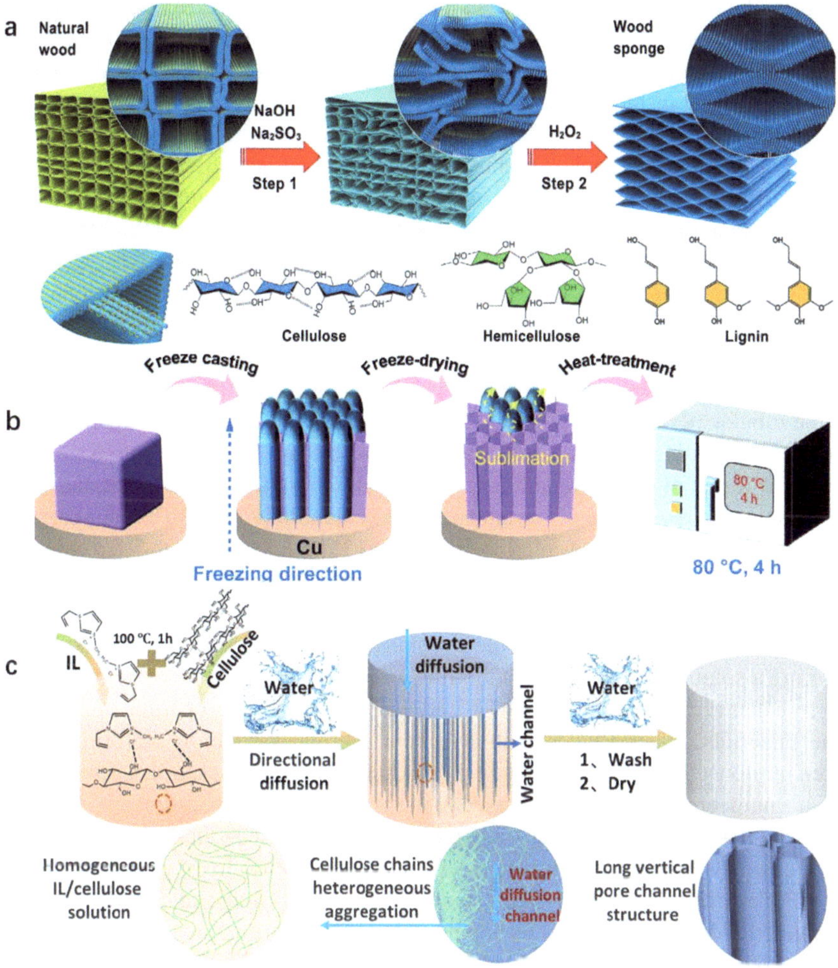

Fig. 13.5 The Preparation method of cellulose aerogels. **a** The raw material is natural wood, which undergoes in-situ removal of lignin and hemicellulose to obtain wood-based cellulose aerogel. Reprinted with the permission from Chen et al. [30], Copyright 2018, American Association for the Advancement of Science. **b** The raw material is nanocellulose. The raw material is nanocellulose, and its aqueous dispersion is freeze-dried to obtain cellulose aerogel. Reprinted with the permission from Sui et al. [15], Copyright 2025, Wiley–VCH. **c** The raw material is cellulose solutions. They undergo regeneration and precipitation in a non-solvent, followed by drying to produce cellulose aerogel. Reprinted with the permission from Liu et al. [40], Copyright 2025, Elsevier

range of 20% to 80%. Ren et al. modified CNF aerogels via in situ supramolecular assembly of melamine (MEL) and phytic acid (PA) [33]. This strategy addressed the flammability issue of CNF while avoiding the environmental concerns associated with conventional flame retardants. The modified aerogel presented a porous honeycomb structure with low density and excellent mechanical properties. After

MEL-PA modification, the aerogel maintained good shape integrity during combustion, demonstrating higher thermal stability and excellent flame-retardant behavior. Heat conduction within the modified aerogel was effectively hindered, showcasing superior thermal insulation performance.

Microstructure design can also effectively enhance the thermal insulation performance of cellulose aerogels. Oriented structures can effectively reduce the thermal conductivity of porous cellulose materials in the radial direction to fabricate porous cellulose materials with low thermal conductivity. For instance, Wicklein et al. utilized directional freeze-casting technology on suspensions of CNFs, graphene oxide (GO), and sepiolite nanorods (SEP) to generate highly anisotropic, strong foam materials, which exhibited super-insulation and flame-retardant properties, outperforming traditional polymer-based insulating materials [34]. The CNF foam achieved a thermal conductivity as low as 18 mW m^{-1} K^{-1} in the radial direction (perpendicular to the tubular pores). Incorporating 10 wt% GO and 10 wt% SEP further reduced the thermal conductivity (λ) of the nanocomposite foam to 15 mW m^{-1} K^{-1}, which is only half that of polystyrene foam. Chen et al. reported a fire-resistant, thermally insulating, and recyclable 3D graphite-cellulose nanofiber (G-CNF) foam prepared from abundant graphite and cellulose resources [35]. This composite foam exhibited low thermal conductivity (0.05 W/(m·K)) without sacrificing recyclability, degradability, or mechanical properties. The demonstrated 3D graphite-cellulose nanofiber foam shows promise as a sustainable alternative to commercial plastic-based foams, offering a solution to the "white pollution" problem.

To enhance the thermal insulation and heat preservation properties of 3D cellulose materials, nanoparticles with excellent thermal insulation performance are often introduced to synergistically improve these characteristics. Silica aerogel (SA) is an engineered material with excellent thermal insulation properties [36, 37]. Further to mitigate the impact of humidity on thermal conductivity, the oriented pore channels which are constructed by CNFs are often hydrophobically modified. For example, Zhao et al. fabricated a lightweight, porous, anisotropic cellulose nanofiber/hydroxylated boron nitride nanosheet/methyltrimethoxysilane composite aerogel (Si-CNF/BNNS) via directional freeze-casting [13]. The resulting cryogel exhibited distinct anisotropic thermal insulation, with an axial thermal conductivity of 0.0621 W m^{-1} K^{-1} and a radial thermal conductivity of 0.0339 W m^{-1} K^{-1}. This high-performance anisotropic Si-CNF/BNNS cryogel holds significant potential for enhancing thermal comfort and environmental safety in electronic packaging and energy efficiency applications. To further improve thermal insulation performance, Sun et al. implemented additional structural designs. Inspired by the "brick–mortar-binder" biomimetic texture, they successfully prepared a layered double hydroxide/carboxylated cellulose nanofiber/Si–O–Si (LCS) composite aerogel via a bottom-up directional freezing followed by drying approach [15]. The introduction of layered double hydroxides significantly enhanced its thermal insulation performance (thermal conductivity = 0.0296 W m^{-1} K^{-1}). These unique properties make the LCS composite aerogel highly promising for thermal management applications under extreme conditions, such as in pipelines. Yan et al. proposed a directional freeze-drying strategy. By combining nanofibrillated cellulose (NFC) with a Si − O −

Si cross-linked network under *p*-toluenesulfonic acid (TsOH)-catalyzed dehydration conditions, they developed a novel anisotropic "strong-yet-soft" NFC aerogel (NFC-Si-T) [38]. The NFC-Si-T aerogel exhibited anisotropic thermal insulation, with an average thermal conductivity of only 0.028 – 0.049 W m^{-1} K^{-1}. Qing et al. proposed a dual ice-templating assembly (DITA) strategy. This involved controlling the assembly of CNF into submicron fibers via cryogenic freezing at -196 °C, followed by further assembly into elastic aerogels composed of interconnected submicron fibers through freezing at -20 °C and freeze-drying [39]. The elastic CNF aerogels could be easily modified via chemical vapor deposition of organosilane, exhibiting superhydrophobicity (water contact angle of 164°), thermal insulation (0.023 W m^{-1} K^{-1}), and infrared shielding properties.

The cellulose could be dissolved to obtain the corresponding solution, followed by regeneration and precipitation in a non-solvent. After drying, CBAs can also be obtained. This method enables manipulation of cellulose molecules and efficient utilization of cellulose molecules to produce functionalized cellulose aerogels. For instance, Liu et al. incorporated SA powder possessing a nanoporous structure into a regenerated nanocellulose network, yielding a composite regenerated cellulose aerogel with superior thermal insulation properties. As shown in Fig. 13.5c, the SA powder was first dispersed in a cellulose/ionic liquid solution. Subsequently, through mechanical stirring foaming combined with a unique self-foaming and novelty bottom-to-top gradual regeneration gelation process, followed by water washing and drying, an isotropic cellulose/silica aerogel composite foam was obtained. Due to the introduction of silica aerogel, this composite exhibited outstanding thermal resistance. The composite foam achieved its lowest thermal conductivity of 0.034 W/(m·K) when incorporating 10 wt% silica aerogel powder. The hydrophilic/hydrophobic properties of materials also significantly impact the insulation performance of cellulose bulk materials. Thermal conductivity typically increases with rising moisture content. Introducing hydrophobic silica aerogel can improve the hydrophobicity of cellulose bulk materials, thereby enhancing their thermal insulation performance [40]. In addition to the aforementioned general methods, 3D printing can also be used to prepare cellulose-based aerogels [41].

13.4.2 Thermal Insulation Applications

In previous discussion, its focus was placed on preparing cellulose-based aerogels using various methods, followed by functional modification or functional treatment to achieve cellulose-based aerogels with excellent thermal insulation properties. Additionally, using cellulose from biomass as thermal insulation materials offers numerous advantages: abundant natural raw materials, reduced toxicity due to biomass characteristics, and minimal environmental harm from the material's natural biodegradability and rapid degradation rate. Based on this, researchers further explored their potential applications. The application fields of cellulose-based aerogel contain military, human thermal management, and architectural fields. To enhance the functional properties of cellulose aerogels, they are often used in combination with other materials. For example, using regeneration technology Han et al. obtained cellulose/lignin composite aerogel [42]. Then, atomic layer deposition was employed to form a TiO_2 coating on the surface of aerogel sheets CLA, resulting in the composite material CLA-TiO_2. Due to its distinctive appearance and favorable characteristics—such as light weight, thermal insulation, and infrared shielding—CLA-TiO_2 is suitable for use as a sheet material in manufacturing cloaks with desert camouflage, heat resistance, and infrared concealment capabilities (Fig. 13.6a). This novel method not only enhances military performance in harsh desert environments but also provides an eco-friendly solution for upcycling discarded textiles into multifunctional advanced materials.

Beyond its potential applications in the military domain, its applications in the field of wearable technology for the human body demonstrate even greater versatility, with a correspondingly broader market prospect. For example, Zhang et al. explored the application of lightweight and superamphiphobic silanized cellulose–silica aerogels (SCSA) on green flexible thermal substrates [43]. They demonstrated that SCSA could be used in wearable electronic devices. As shown in Fig. 13.6b, SCSA can serve as a thermal protection substrate for wearable electronic devices, effectively preventing heat-induced discomfort or skin damage. This demonstrates a practical application of using it as an insulating layer in direct contact with the skin. Within the thermally protective substrate, the encapsulated-void structure formed in the mesopores of the IR-reflective functional phonon barrier—amorphous silica aerogel powder starting material—absorbs a substantial amount of heat generated by the exothermic component. This design ensures that the temperature at the interface between the substrate and the epidermis remains lower than that of the heat source, thereby preventing overheating of the skin (Right in Fig. 13.6b). Furthermore, the morphology of SCSA remained virtually intact without noticeable damage after 1,000 cycles of 360° twisting in both clockwise and counterclockwise directions. This demonstrates the material's outstanding fatigue resistance under repeated torsional stress applied in opposing directions.

The structural design of cellulose-based aerogels enables further reduction of their thermal conductivity, combined with their low density, they can serve as fabric fillers to provide thermal insulation. For example, Wang et al. designed a fuller-dome-structured nanocellulose aerogels (CNF/polyurethane (PU) aerogel) by combination of air bubbles template and ice template. It exhibits an ultralow λ of only 24 mW m^{-1} K^{-1}. The use of air- and ice-templating techniques generates pores of various sizes, forming intricate and irregular channels that introduce numerous solid–gas interfaces. Additionally, the abundant junctions within the network structure led to the formation of more CNF-PU interfaces. These increased solid–gas and solid–solid contact points enhance phonon scattering, thereby effectively reducing heat transfer within the aerogel. Thanks to its low thermal conductivity and lightweight nature, the CNF@PU aerogel has proven effective in practical applications, such as in the production of warming jackets designed to help maintain human thermal comfort (Fig. 13.6c) [44]. To meet the challenge of creating a flexible sensor that performs reliably in outdoor environments, both under intense summer sunlight and in cold

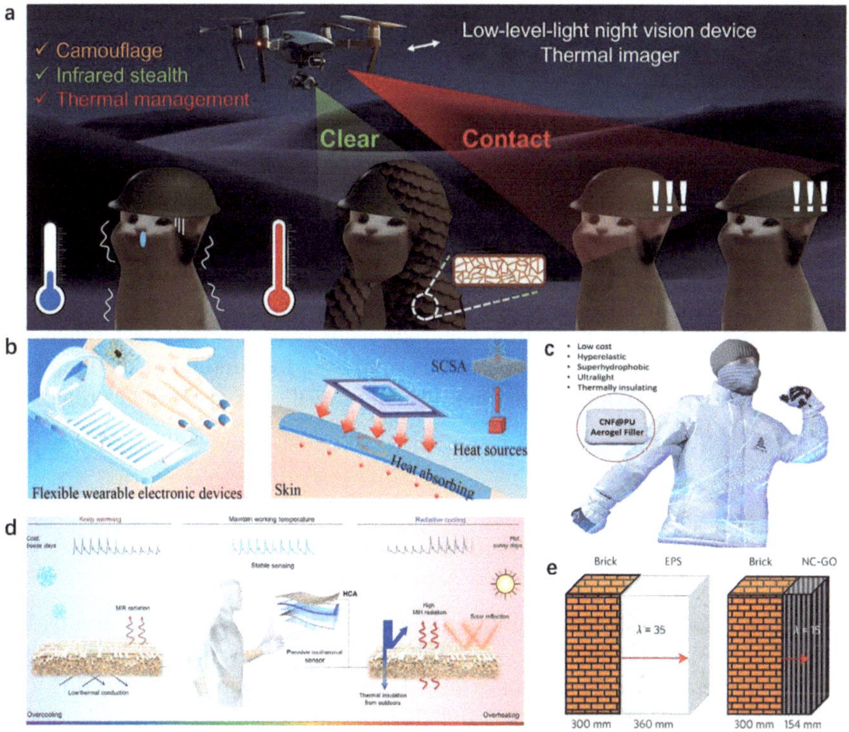

Fig. 13.6 Thermal insulation applications of 3D CBAs: **a** Cellulose aerogel for infrared stealth. Reprinted with the permission from Huang et al. [42], Copyright 2025, Wiley–VCH. **b** Schematic design of thermal superinsulating substrate for wearable electronic devices. Reprinted with the permission from Wang et al. [43], Copyright 2024, Elsevier. **c** Integrated merits of as-prepared CNF@PU aerogels when used as fillers for warming jackets. Reprinted with the permission from Wang et al. [44], Copyright 2024, Wiley–VCH. **d** The HCA results in strong sunlight reflectivity (0.3–2.5 μm), high thermal radiation (8–13 μm), and ultralow thermal conductivity, ensuring that the sensor keeps warm or radiative cooling under dynamic thermal conditions. Reprinted with the permission from Zhong et al. [45], Copyright 2025, Wiley–VCH. **e** Cellulose aerogel for passive house insulation. Reprinted with the permission from Wicklein et al. [34], Copyright 2014, Springer Nature

winter conditions, Zhong et al. presented a passive isothermal sensor utilizing a thermoregulating HCA as the tribonegative layer [45]. In hot weather, elevated solar radiation activates the predominant role of passive daytime radiative cooling (PDRC), ensuring effective heat dissipation. In contrast, during winter's low ambient temperatures, the material's ultralow thermal conductivity becomes essential for retaining heat because of the hollow microfibers in HCA. As a result, the sensor maintains signal stability even in fluctuating environmental conditions (Fig. 13..13.6d).

Building insulation has long been a focal research area, impacting advancements in energy, environmental, and multiple related sectors. There is a growing demand for high-performance thermal insulation materials derived from renewable resources to enhance the energy efficiency of buildings. Conventional insulation options like expanded polystyrene and polyurethane, which originate from fossil fuels, exhibit thermal conductivities that are too high for use in retrofits or in the construction of compact, energy-efficient passive houses. Wicklein et al. demonstrated that freeze-casting a mixture of CNFs, GO, and sepiolite nanorods results in super-insulating, fire-resistant, and mechanically robust anisotropic foams [34]. These foams outperformed traditional polymer-based insulators. They were ultralight, exhibited outstanding flame resistance, and possessed a low thermal conductivity of just 15 mW m^{-1} K^{-1}—approximately half that of expanded polystyrene. These findings highlighted nanoscale engineering as a promising pathway for developing high-performance insulation materials from cellulose and other renewable nanoscale fibers. When this CBA materials were used for building insulation, it not only reduceD the thickness of thermal insulation wall, but also could protect the environment (Fig. 13.6e).

13.5 Conclusion and Outlooks

CBAs, characterized by their lightweight nature, high porosity, renewability, and tunable micro-nano architectures, have emerged as a representative class of next-generation thermal insulation materials. Through dimensional engineering (1D fibers, 2D films, 3D monoliths) and functionalization strategies (e.g., hydrophobic modification, nanocompositing, and polymer incorporation), significant breakthroughs have been achieved in mechanical strength, environmental stability, and thermal insulation performance (thermal conductivity as low as 0.015 W/m·K), enabling applications in smart textiles, building energy efficiency, electronics thermal management, and infrared stealth.

Nevertheless, industrialization faces three challenges. First one is the scalability limitations. Energy-intensive processes (e.g., supercritical/freeze drying) require continuous manufacturing alternatives (e.g., ambient-pressure drying with shape-directed assembly). The second challenge is the cost constraints. Cellulose or nanocellulose sources and functional additives necessitate low-cost biomass precursors (e.g., agroforestry waste) or novelty method for nanocellulose. The third one is the performance trade-offs. Long-term stability under extreme conditions (humidity, high temperature) demands in-situ reinforced interfaces and adaptive protection mechanisms.

Future research could prioritize in several aspects. First, multiscale structural design could be carried out. Biomimetic hierarchical pores to optimize heat-transfer pathways while enhancing mechanical-thermal synergy. Second, sustainable manufacturing is needed. Water-based gel systems and energy-efficient drying to establish full life-cycle eco-efficiency. Third, smart integration is suggested. Coupling with sensors/energy-conversion units for adaptive thermal systems and wearable devices. Addressing these challenges via multidisciplinary collaboration will accelerate the industrial translation of CBAs, positioning them as pivotal green insulation materials in the context of global carbon neutrality goals.

References

1. V. Apostolopoulou-Kalkavoura, P. Munier, L. Bergström, Thermally insulating nanocellulose-based materials. Adv. Mater. **33**, 2001839 (2021)
2. F. Song, Z. Li, P.Y. Jia, C.Y. Bo, M. Zhang, L.H. Hu, Y.H. Zhou, Phosphorus-containing tung oil-based siloxane toughened phenolic foam with good mechanical properties, fire performance and low thermal conductivity. Mater. Des. **192**, 108668 (2020)
3. C.L. Zhang, B. Zhu, D.C. Li, L.J. Lee, Extruded polystyrene foams with bimodal cell morphology. Polymer **53**, 2435–2442 (2012)
4. T.M. Nguyen-Ha, T.B. Nguyen, T.A. Nguyen, L.H. Pham, D.H. Nguyen, D.M. Nguyen, D. Hoang, E. Oh, J. Suhr, Novel high-performance sustainable polyurethane nanocomposite foams: fire resistance, thermal stability, thermal conductivity, and mechanical properties. Chem. Eng. J. **474**, 145585 (2023)
5. Y. Kobayashi, T. Saito, A. Isogai, Aerogels with 3D Ordered nanofiber skeletons of liquid-crystalline nanocellulose derivatives as tough and transparent insulators. Angew. Chem. Int. Ed. **53**, 10394–10397 (2014)
6. B. Wicklein, A. Kocjan, G. Salazar-Alvarez, F. Carosio, G. Camino, M. Antonietti, L. Bergström, Thermally insulating and fire-retardant lightweight anisotropic foams based on nanocellulose and graphene oxide. Nat. Nanotech. **10**, 277–283 (2015)
7. B. Wang, K. Nan, H. Rao, Y. Chen, R. Pei, Y. Wang, Integrated design of multifunctional lightweight magnetic cellulose-ba sed aerogel with 1D/2D/3D hierarchical network for efficient microwave absorption. Compos. Commun **49**, 101987 (2024)
8. E. Abraham, V. Cherpak, B. Senyuk, J.T.T. Hove, T. Lee, Q. Liu, I.J.N.E. Smalyukh, Highly transparent silanized cellulose aerogels for boosting energy efficiency of glazing in buildings. Nat. Energy **8**, 381–396 (2023)
9. A. Di, C. Schiele, S.E. Hadi, L. Bergström, Thermally insulating and moisture-resilient foams based on upcycled aramid nanofibers and nanocellulose. Adv. Mater. **35**, 2305195 (2023)
10. V. Apostolopoulou-Kalkavoura, P. Munier, L.J.A.M. Bergstrm, Thermally insulating nanocellulose based materials. Adv. Mater. **33**, 202001839 (2020)
11. J.Y. Yu, H.B. Wang, D.Q. Wang, X. Cheng, X.S. Du, S. Wang, Z.L. Du, High mechanical strength, flame retardant, and waterproof silanized cellulose nanofiber composite foam for thermal insulation. Cellulose **31**, 9865–9885 (2024)
12. Y.Q. Shi, Y.J. Zhu, S. Liu, L.B. Fu, J.T. Chen, J.W. Liu, L.C. Tang, J.F. Gao, P.G. Song, Multi-hierarchically constructing durable and flame retardant CNF/MXene/PDMS composite aerogels for superhigh electromagnetic shielding performance and ultralow thermal conductivity, Small 2500556 (2025)
13. J. Zhao, J. Zhou, H. Li, A. Xiao, Directional fabricating of flexible and compressible cellulose nanofibril composite cryogel with excellent thermal insulation, flame-retardancy and radiative cooling for efficient thermal management. Cellulose **29**, 9671–9691 (2022)
14. Y.S. Yang, B.K. Dang, C. Wang, Y.P. Chen, K.C. Chen, X.J. Chen, Y.Y. Li, Q.F. Sun, Ultra-strong lightweight nanocellulose-based composite aerogels with robust superhydrophobicity and durable thermal insulation under extremely environment. Carbohydr. Polym. **323**, 121392 (2024)
15. S. Sui, H. Quan, J. Wang, Y. Lu, Y. Yang, Y. Sheng, Z. Sun, Y. Zhang, Brick-Mortar-Binder design toward highly elastic, hydrophobic, and flame-retardant thermal insulator. Adv. Sci. **12**, 2410938 (2025)
16. G. Zhao, L. Shi, G. Yang, X. Zhuang, B. Cheng, 3D fibrous aerogels from 1D polymer nanofibers for energy and environm ental applications. J. Mater. Chem. A **11**, 512–547 (2023)

17. Z.Z. Sheng, Z.W. Liu, Y.L. Hou, H.T. Jiang, Y.Z. Li, G.Y. Li, X.T. Zhang, The rising aerogel fibers: Status, challenges, and opportunities. Adv. Sci. **10**, 2205762 (2023)
18. P. Xu, K. Quan, X. Wei, Y. Li, S. Xu, Vertical porous 1D/2D hybrid aerogels with highly matched charge stora ge performance for aqueous asymmetric supercapacitors. Front. Chem. **13**, 1550285 (2025)
19. X. Long, J. Qin, J. Tang, J. Xue, Y. Wang, L. Zhou, S. Wang, X. Wei, Y. Lin, J. Liao, Hierarchical Design of 1D Building Blocks for Anisotropic Aerogel. Adv. Funct. Mater. **35** (2025)
20. M. Zeng, Q. Xu, J. Yu, Polyvinyltrimethoxysilane-enhanced TEMPO-oxidized cellulose nanofiber aerogels for exceptional anisotropic thermal insulation, flame retarda ncy and oil/water separation. Ind. Crops Prod. **228**, 120862 (2025)
21. X. Long, P. Huang, X. Wei, J. Yu, S. Wang, J. Liao, Ultra-high strength, highly deformable and superhydrophobic polybenzox azine@cellulose nanofiber composite aerogel for thermal insulation. Compos Part A-Appl S. **175**, 107771 (2023)
22. Z. Liu, T. Yang, J. Zhou, J. Xu, K. Ye, F. Chu, C. He, R. Xiao, C. Jiao, S. Han, X. Fu, D. Ye, Customizable structured cellulose aerogel fibers by regulating regener ation rate for enhanced thermal insulation. Int. J. Biol. Macromol. **311**, 143657 (2025)
23. Q. Li, Z. Yuan, C. Zhang, S. Hu, Z. Chen, Y. Wu, P. Chen, H. Qi, D. Ye, Tough, highly oriented, super thermal insulating regenerated all-cellulose sponge-aerogel fibers integrating a graded aligned nanostructure. Nano Lett. **22**, 3516–3524 (2022)
24. X. Fu, L. Si, Z. Zhang, T. Yang, Q. Feng, J. Song, S. Zhu, D. Ye, Gradient all-nanostructured aerogel fibers for enhanced thermal insulation and mechanical properties. Nat. Commun. **16** (2025)
25. Z. Liu, Z. Sheng, Y. Bao, Q. Cheng, P.-X. Wang, Z. Liu, X. Zhang, Ionic liquid directed spinning of cellulose aerogel fibers with superb toughness for weaved thermal insulation and transient impact protection. ACS Nano **17**, 18411–18420 (2023)
26. M. Wu, Z. Shao, N. Zhao, R. Zhang, G. Yuan, L. Tian, Z. Zhang, W. Gao, H. Bai, Biomimetic, knittable aerogel fiber for thermal insulation textile. Sciencc **382**, 1379–1383 (2023)
27. Y. Cui, H. Gong, Y. Wang, D. Li, H. Bai, A thermally insulating textile inspired by polar bear hair. Adv. Mater. **30**, 1706807 (2018)
28. Y. Wang, Z. Wang, H. Huang, Y. Li, W. Zhai, A Camel-Fur-inspired micro-extrusion foaming porous elastic fiber for all-weather dual-mode human thermal regulation. Adv. Sci. **11**, 2407260 (2024)
29. S. Liu, W. Li, X. Wang, L. Lu, Y. Yao, S. Lai, Y. Xu, J. Yang, Z. Hu, X. Gong, Permeable, stretchable, and recyclable cellulose aerogel on-skin electronics for dual-modal sensing and personal healthcare. ACS Nano (2025)
30. C. Chen, J. Song, S. Zhu, Y. Li, L.J.C. Hu, Scalable and sustainable approach toward highly compressible, anisotropic, lamellar carbon sponge. Sci. Adv. **4**, 1–11 (2018)
31. T. Li, J.W. Song, X.P. Zhao, Z. Yang, G. Pastel, S.M. Xu, C. Jia, J.Q. Dai, C.J. Chen, A. Gong, F. Jiang, Y.G. Yao, T.Z. Fan, B. Yang, L. Wågberg, R.G. Yang, L.B. Hu, Anisotropic, lightweight, strong, and super thermally insulating nanowood with naturally aligned nanocellulose. Sci. Adv. **4** eaar3724 (2018)
32. Z. Liu, T. Liu, H. Dong, B. Yang, X. Li, X. Li, Y. Wu, K. Xu, Diatom-inspired nanoscale heterogeneous assembly strategy for constructing thermal insulating wood-based aerogels with exceptional strength, resilience, degradability, and flame retardancy. ACS Nano **19**, 6826–6839 (2025)
33. X. Ren, M. Song, J. Jiang, Z. Yu, Y. Zhang, Y. Zhu, X. Liu, C. Li, H. Oguzlu-Baldelli, F. Jiang, Fire-retardant and thermal-snsulating cellulose nanofibril aerogel modified by in situ supramolecular assembly of melamine and phytic acid. Adv. Eng. Mater. **24**, 2101534 (2022)
34. B. Wicklein, Kocjan, Andraz, Salazar-Alvarez, German, Carosio, Federico, Camino, Giovanni, Thermally insulating and fire-retardant lightweight anisotropic foams based on nanocellulose and graphene oxide. Nat. Nanotech. **10**, 277 (2015)
35. C.J. Chen, Y.B. Zhou, W.Q. Xie, T.T. Meng, X.P. Zhao, Z.Q. Pang, Q.Y. Chen, D.P. Liu, R.L. Wang, V.A. Yang, H.L. Zhang, H. Xie, U.H. Leiste, W.L. Fourney, S.M. He, Z.Y. Cai, Z.Q. Ma, T. Li, L.B. Hu, Lightweight, thermally insulating, fire-Proof graphite-cellulose foam. Adv. Funct. Mater. **33** (2023)

36. S.Y. Zhao, G. Siqueira, S. Drdova, D. Norris, C. Ubert, A. Bonnin, S. Galmarini, M. Ganobjak, Z. Pan, S. Brunner, G. Nyström, J. Wang, M.M. Koebel, W.J. Malfait, Additive manufacturing of silica aerogels. Nature **584**, 387 (2020)
37. G.G. Kaya, H. Deveci, Synergistic effects of silica aerogels/xerogels on properties of polymer composites: a review. J. Ind. Eng. Chem. **89**, 13–27 (2020)
38. M.Y. Yan, Y.L. Pan, X.D. Cheng, Z.X. Zhang, Y.R. Deng, Z.Y. Lun, L.L. Gong, M.Y. Gao, H.P. Zhang, Robust-soft anisotropic nanofibrillated cellulose aerogels with superior mechanical, flame-Retardant, and thermal insulating properties. ACS Appl. Mater. Interfaces **13**, 27458–27470 (2021)
39. H.F. Qin, Y.F. Zhang, J.G. Jiang, L.L. Wang, M.Y. Song, R. Bi, P.H. Zhu, F. Jiang, Multifunctional superelastic cellulose nanofibrils aerogel by dual ice-templating assembly. Adv. Funct. Mater. **31**, 2106269 (2021)
40. X.R. Liu, T.P. Zhang, H.Z. Xu, B.X. Li, G. Wei, X.F. Zhang, J.M. Zhang, Ionic liquid-mediated facile synthesis of porous cellulose/silica aerogel composites with improved thermal insulation. Int. J. Biol. Macromol. **307**, 142418 (2025)
41. T.Y. Zhu, Z.Y. Ren, D.B. Wang, S.D. Zhao, X. Liu, W. Fan, Y.E. Miao, C. Zhang, T.X. Liu, Reactive 3D printed silanized cellulose nanofiber aerogels for solar-thermal regulatory cooling. Compos. Part A-Appl. S. **192**, 108761 (2025)
42. Z. Huang, A. Tong, T. Xing, A. He, Y. Luo, Y. Zhang, M. Wang, S. Qiao, Z. Shi, F.J.A.F.M. Chen, Regenerated cellulose/Lignin composite aerogel with unique toast-like structure and their potential applications in thermal camouflage. **35,** 2414696 (2025)
43. K.K. Zhang, L. Liu, X. Wang, J.J. Shi, N. Li, Y.X. Huang, C.W. Yang, S.S. Yang, C.Z. Li, H. Lu, H.J. Liu, J. Li, J. Qiu, Y.S. Yang, Lightweight and superamphiphobic silanized cellulose-silica aerogels on green flexible thermal substrates. Chem. Eng. J. **502**, 158238 (2024)
44. G. Wang, J.B. Feng, Z.Z. Zhou, Z. Liu, J.P. Wu, J.C. Li, Q. Gao, M. Lynch, J.Z. Li, P.G. Song, Low-cost hyperelastic fuller-dome-structured nanocellulose aerogels by dual templates for personal thermal management. Adv. Mater. 2414896 (2024)
45. S.J. Zhong, B.H. Lu, D.C. Wang, B. Arianpour, S.L. Wang, H.Y. Han, J.Y. Yin, H. Bao, Y.A. Liu, Z. Wen, Y.L. Zhou, Passive isothermal flexible sensor enabled by smart thermal-regulating aerogels. Adv. Mater. **37** (2025)

Chapter 14
Cellulose-Based Aerogels for Adsorption and Catalysis

Shuxian Tang and Yan Wang

Abstract Cellulose-based aerogels (CBAs) have emerged as highly versatile materials for environmental remediation and catalysis, thanks to their unique combination of high porosity, low density, biodegradability, and tunable surface functionality. This chapter provides a comprehensive overview of the latest advancements in the synthesis, modification, and application of CBAs in these fields. The focus is on their performance in key areas such as heavy metal ion adsorption, organic pollutant removal, oil–water separation, gas adsorption and separation, photocatalytic degradation, and catalytic organic transformations. The inherent hydroxyl-rich surface and three-dimensional network architecture of CBAs offer abundant active sites and enhance mass transport, making them ideal for both adsorption and catalysis. Additionally, CBAs have shown strong potential in enzyme immobilization and biocatalysis due to their structural compatibility and chemical modifiability. Continued innovations in material engineering and sustainable chemistry are expected to solidify CBAs as crucial components in next-generation environmental and catalytic technologies.

Keywords Cellulose · Aerogels · Adsorption · Catalysis · Environmental remediation

14.1 Introduction

With the rapid advancement of modern industry, the excessive consumption of petroleum-derived polymers has triggered a dual crisis of non-renewable resource scarcity and environmental pollution. In response, natural polymers have emerged as promising alternatives in the development of advanced functional materials, owing

S. Tang · Y. Wang (✉)
College of Chemistry and Chemical Engineering, Qingdao University, Qingdao, China
e-mail: yanwang@qdu.edu.cn

G. Wei and J. Zhang (eds.), *Cellulose-Based Hydrogels and Aerogels*, Lecture Notes in Nanoscale Science and Technology 41,
https://doi.org/10.1007/978-3-032-03008-5_14

to their intrinsic renewability, environmental compatibility, and favorable physicochemical properties. Traditional inorganic aerogels, such as silica and titania aerogels, although notable for their high porosity and superior thermal insulation, are inherently limited by their brittleness and poor mechanical flexibility [1, 2]. Consequently, there has been a growing research interest in organic aerogels derived from biopolymers, which offer advantages including enhanced biodegradability, biocompatibility, non-toxicity, and sustainable sourcing. Compared to petroleum-based counterparts, biopolymer-based materials exhibit faster degradation rates and lower environmental impact during synthesis and disposal, thereby aligning with the principles of green chemistry and sustainable development [3]. Among various biopolymers, cellulose is particularly attractive for the fabrication of aerogels due to its abundance, low cost, excellent biocompatibility, and high biodegradability. The intrinsic three-dimensional (3D) porous architecture of cellulose-based aerogels (CBAs) imparts them with a high specific surface area, low density, strong adsorption capacity, and favorable mechanical strength, positioning them as versatile candidates for the applications in adsorption, catalysis, and related fields [4].

CBAs possess exceptional adsorption and catalytic properties, primarily attributed to their high specific surface area, hierarchical porous architecture, renewability, and intrinsic biocompatibility. A key feature contributing to their functional performance is the abundance of surface hydroxyl groups (–OH), which serve as active sites for the adsorption of various pollutants. Nevertheless, the intrinsic hydrophilicity of cellulose imposes certain limitations on the applicability of CBAs in environments requiring hydrophobic or selective interactions [5]. To overcome these problems, extensive efforts have been directed toward enhancing their functional properties through chemical and physical modifications. Strategies such as surface functionalization [6] and the incorporation of functional nanomaterials including metal oxides and graphene derivatives [7], have been employed to improve the adsorption capacity, catalytic activity, hydrophobicity, electrical conductivity, and thermal stability of CBAs.

In recent years, CBAs have exhibited significant potential across a wide spectrum of applications, including the adsorption of heavy metal ions, removal of organic contaminants, gas capture and separation, oil–water separation, photocatalytic degradation, heterogeneous catalysis of organic transformations, and biocatalysis [8–13]. Drawing upon the recent advancements in this domain, the following section provides a comprehensive review of the progress in CBAs for both adsorption and catalytic applications, elucidates their performance advantages, and outlines prospective research directions aimed at further optimizing their multifunctionality and application scope.

14.2 CBAs for the Removal of Heavy Metal Ions

Heavy metal ions represent a particularly hazardous class of waterborne pollutants, posing significant threats to both human health and ecological systems. Common species encountered in contaminated aqueous environments include Pb^{2+}, Cd^{2+}, Cr^{6+}, Hg^{2+}, and Cu^{2+}, all of which are characterized by high toxicity, strong bioaccumulative tendencies, and persistence due to their resistance to natural degradation processes. Upon entering aquatic ecosystems, these ions can disrupt the physiological and metabolic functions of living organisms over extended periods. Owing to their high specific surface area and well-defined porous architecture, porous materials have demonstrated considerable potential for the efficient removal of heavy metal ions from aqueous media [14]. Among them, CBAs have garnered increasing interest as sustainable, environmentally benign adsorbents. The abundance of surface functional groups (e.g., hydroxyl and carboxyl groups) in CBAs facilitates the formation of stable coordination complexes with heavy metal ions, thereby enabling efficient chemical adsorption. Simultaneously, their internal porous networks offer numerous physical adsorption sites, further enhancing overall adsorption capacity. This synergistic effect of chemical interaction and physical entrapment imparts CBAs with excellent performance for the remediation of heavy metal-contaminated water [15].

In recent years, significant progress has been made in the research on CBAs for heavy metal ion removal. Functional modification strategies, including compositing cellulose with synthetic polymers or inorganic nanomaterials, have been shown to significantly improve the adsorption efficiency and mechanical robustness of the aerogels [16, 17]. For instance, Mo et al. developed a 3D layered porous composite aerogel based on cellulose nanofibers (CNFs) and polyacrylamide (TOCNF-TMPTAP-APAM aerogels) via an in-situ dual cross-linking physical and chemical strategy, followed by freeze-drying. As illustrated in Fig. 14.1a, the resulting composite exhibited rapid and efficient adsorption of Cu^{2+}, driven by the mechanisms including electrostatic interactions, chelation, and complexation (Fig. 14.1c). The maximum adsorption capacity reached 240 $mg{\cdot}g^{-1}$, and the material retained over 80% removal efficiency after 10 adsorption–desorption cycles, underscoring its excellent reusability (Fig. 14.1b and d) [18]. This study highlights the practical potential of such composite aerogels in the treatment of heavy metal-laden wastewater. Similarly, Guo et al. prepared a porous cellulose-polymer-graphene aerogel (CPGA) with excellent mechanical properties by using polyvinyl alcohol (PVA), reduced graphene oxide (rGO), and CNFs as precursors. The covalently cross-linked structure of CPGA exhibited high adsorption efficiency for a broad range of heavy metal ions, including Cu^{2+}, Cd^{2+}, Pb^{2+}, Cr^{3+}, and Co^{2+} [19]. These findings underscore the significant enhancements in adsorption performance achieved through the integration of CBAs with functional materials, pointing toward promising directions for the design of next-generation water purification systems.

Moreover, the selective adsorption capacity of CBAs toward specific heavy metal ions can be significantly enhanced through the introduction of functional groups such as thiol ($-SH$) and amine ($-NH_2$) via surface chemical modification [20]. In a typical

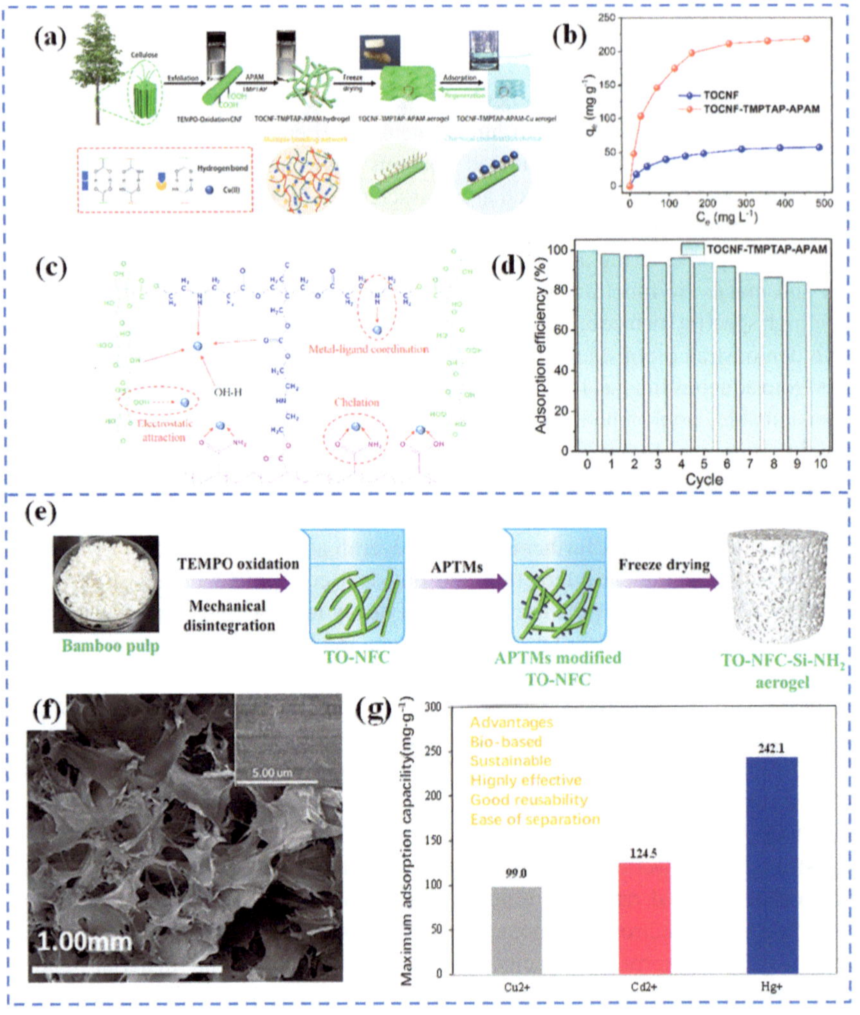

Fig. 14.1 **a** Schematic diagram of TOCNF-TMPTAP-APAM aerogel; **b** Adsorption isotherms; **c** Adsorption mechanism of TOCNF-TMPTAP-APAM aerogel and **d** absorption–desorption cycle diagram. Reproduced with the permission from Reference [18]. Copyright 2022, Elsevier. **e** Schematic diagram of TO-NFC-Si-NH_2 preparation; **f** SEM images of TO-NFC-Si-NH_2 aerogel and **g** Adsorption capacity of TO-NFC-Si-NH_2 aerogel for different ions. Reproduced with the permission from Reference [21]. Copyright 2021, Elsevier

study, Geng et al. ingeniously introduced -NH_2 into CBAs by using the TEMPO-oxidized nanofibrillated cellulose (TO-NFC) derived from bamboo fibers and amino-propyltrimethoxysilane (APTMs) as raw materials, and successfully prepared a 3D macroscopic aminosilylated nanocellulose aerogel (TO-NFC-Si-NH_2) for efficient removal of heavy metal ions from water (Fig. 14.1e). The resultant aerogel exhibited a roughened surface morphology and a highly porous microstructure, as evidenced by SEM imaging (Fig. 14.1f), leading to a substantial increase in specific surface area. These structural enhancements, combined with the introduction of amine functional groups, enabled the aerogel to achieve outstanding adsorption capacities for Cu^{2+} (99.0 mg·g^{-1}), Cd^{2+} (124.5 mg·g^{-1}), and Hg^{2+} (242.1 mg·g^{-1}), as shown in Fig. 14.1g [21].

In addition to chemical functionalization, the incorporation of magnetic nanofillers has emerged as an effective strategy to further enhance the practical utility of CBAs. Magnetic hybrid aerogels, fabricated by integrating nanocellulose with iron oxide (Fe_3O_4) nanoparticles, exhibited several advantageous properties including high specific surface area, strong affinity towards metal ions, facile chemical functionalization, as well as easy separation and recovery via external magnetic fields. Recent study has demonstrated that such magnetic aerogels possess high adsorption capacities for heavy metal ions such as Cr^{6+} and Cu^{2+}, while retaining the structural and compositional integrity of the embedded Fe_3O_4 nanoparticles through physical adsorption [22]. Notably, these hybrid aerogels also maintain robust ferromagnetic properties, enabling efficient and controllable magnetic recovery, thereby improving their operational efficiency and cost-effectiveness for real-world applications.

CBAs exhibit remarkable potential for the adsorption of heavy metal ions due to their tunable surface chemistry and hierarchical porosity. However, several challenges remain, particularly concerning their structural stability under harsh environmental conditions, scalability of production, and cost-effectiveness. Future research efforts could be focused on the rational design of aerogel architectures, development of advanced functionalization techniques, and integration of CBAs into scalable and economically viable treatment systems to facilitate their broader application in heavy metal pollution remediation.

14.3 CBAs for the Adsorption of Dyes and Organic Pollutants

With the rapid development of the textile and dyeing industries, effluents discharged from dyeing processes have emerged as a major source of environmental contamination. Dye-containing wastewater is particularly concerning due to its high chemical stability, toxicity, and substantial organic load, which collectively pose serious risks to human health and aquatic ecosystems. The persistence of dyes in water bodies not only impairs light penetration and aquatic photosynthesis but also reduces the self-purification and oxidative capacity of receiving water systems, thereby complicating wastewater treatment processes. Consequently, the efficient removal of dyes from industrial effluents prior to discharge is essential to address pressing ecological, public health, and industrial challenges [23].

Among existing treatment methods, adsorption technology is widely regarded as one of the most efficient and practical approaches for dye removal from wastewater. Its advantages include operational simplicity, high removal efficiency, cost-effectiveness, and wide adaptability to various pollutants, including both organic and inorganic contaminants. Additionally, adsorption processes are typically unaffected by the toxicity or complexity of the wastewater matrix [24]. CBAs, characterized by their high porosity, large specific surface area, and abundance of surface hydroxyl groups, have emerged as highly promising adsorbents for dye removal. Recent advances in surface engineering and material modification have further improved the adsorption performance of CBAs, enabling enhanced interaction with dye molecules and expanding their practical applications.

Chemical functionalization is a widely employed strategy to tailor the surface properties of cellulose aerogels and improve their dye adsorption capacities, which can be achieved by introducing specific functional groups onto the cellulose framework to promote selective interactions with dye molecules. For instance, Zhang et al. synthesized a biomass aerogel composed of TEMPO-oxidized cellulose (TOC) and nanofibrillated cellulose (NFC) via freeze-drying. The introduction of negatively charged carboxyl groups (–COOH) through TEMPO-mediated oxidation significantly improved the aerogel's affinity for cationic dyes via electrostatic interactions (Fig. 14.2a). SEM analysis confirmed the formation of a regular porous network and enhanced mechanical integrity (Fig. 14.2b). The TOC/NFC composite aerogel exhibited excellent adsorption performance, achieving maximum adsorption capacities of 248.27 $mg{\cdot}g^{-1}$ for methylene blue (MB) and 227.30 $mg{\cdot}g^{-1}$ for malachite green (MG), as illustrated in Fig. 14.2c and d [25]. In another study, Shi et al. functionalized cellulose with sulfamic acid (SA) to introduce sulfonic acid groups ($-SO_3H$), yielding sulfamic acid-modified cellulose (SAMC), which was subsequently fabricated into aerogel spheres for dye removal applications. The strong electrostatic attraction between the sulfonic acid groups and cationic dye molecules, such as methylene blue, significantly enhanced adsorption efficiency. The SAMC aerogel spheres exhibited a maximum adsorption capacity of 14.53 $mg{\cdot}g^{-1}$ for MB, demonstrating their potential as a novel and effective adsorbent for dye-laden wastewater remediation [26]. The above findings collectively highlight the effectiveness of chemical modification strategies in improving the dye adsorption performance of CBAs, offering valuable insights for the development of advanced bio-based materials for environmental remediation.

The functional performance of CBAs can be substantially enhanced through physical integration with other functional materials, enabling the development of advanced composites with tailored physicochemical properties. For example, Zheng et al. successfully fabricated a multifunctional CNF/MXene/PVA composite aerogel (C-P-MX) via a bidirectional freezing technique, followed by surface modification through chemical vapor deposition of polysiloxane. The resulting aerogel exhibited a highly ordered hierarchical structure attributed to the synergistic interaction among CNFs, PVA, and MXene, reinforced through chemical cross-linking and hydrogen bonding (Fig. 14.2e). This composite aerogel demonstrated ultralow density, remarkable compressibility, and superior recyclability. Furthermore, the

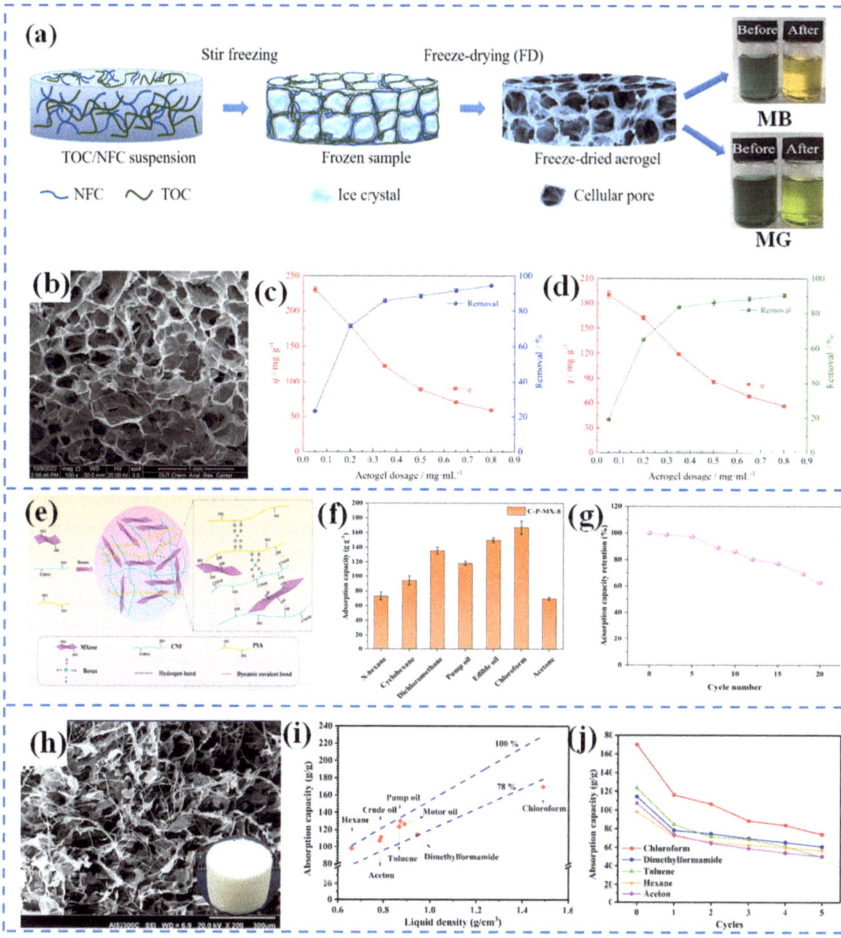

Fig. 14.2 **a** Preparation process of TOC/NFC biomass aerogel; **b** SEM images of TOC/NFC biomass aerogel; Adsorption capacity of **c** MB and **d** MG. Reproduced with the permission from Reference [25]. Copyright 2023, Elsevier. **e** Schematic representation of interactions in C-P-MX composite aerogels; **f** absorption capacity of C-P-MX aerogel and **g** cycling properties of C-P-MX composite aerogels. Reproduced with the permission from Reference [27]. Copyright 2025, Elsevier. **h** SEM images of MNCA; **i** mass adsorption capacity of s-MNCA and **j** reusability of S-MNCA. Reproduced with the permission from Reference [30]. Copyright 2021, Elsevier

highly porous framework and hydrophobic surface coating endowed the material with excellent oil and organic solvent adsorption capacities, achieving an uptake of up to 167.00 $g{\cdot}g^{-1}$ (Fig. 14.2f). Notably, the aerogel retained its structural integrity and adsorption efficiency under varying humidity conditions, exhibiting robust environmental stability and reusability (Fig. 14.2g) [27]. In another study, Ruan et al. developed a novel cellulose nanocrystal (CNC)/SiO_2 composite aerogel via freeze-drying. This self-assembled porous material exhibited outstanding adsorption capacity for

MB, reaching a maximum uptake of 190.85 mg·g^{-1}. The adsorption mechanism was primarily governed by hydrogen bonding, electrostatic interactions, and van der Waals forces. The incorporation of SiO_2 not only enhanced the mechanical stability of the aerogel but also introduced additional adsorption sites, collectively improving its efficiency and durability. These attributes highlight the potential of CNC/SiO_2 composite aerogels for effective dye removal in wastewater treatment applications [28].

Hydrophobic modification of CBAs has also emerged as a promising strategy for the selective removal of hydrophobic organic pollutants [29]. For instance, Dilamian et al. synthesized a superhydrophobic CBA using rice straw-derived native cellulose. A highly porous, cross-linked aerogel (MNCA) was first prepared via freeze-drying, exhibiting a specific surface area of 178.8 m^2·g^{-1} and a mesopore volume of 0.8 cm^3·g^{-1} (Fig. 14.2h). Subsequent hydrophobic surface treatment yielded a superhydrophobic variant (s-MNCA), capable of efficiently and selectively adsorbing oils and nonpolar hydrocarbons. Depending on the density of the organic liquid, the s-MNCA achieved adsorption capacities ranging from 98 to 170 g·g^{-1} (Fig. 14.2i). Importantly, the adsorbed substances could be easily recovered through mechanical compression, and the aerogel maintained its performance over multiple cycles, demonstrating excellent reusability (Fig. 14.2j). These properties make s-MNCA an economically viable and highly effective candidate for the remediation of oil-contaminated wastewater [30].

The above introduced studies underscore the versatility of composite and modified CBAs in addressing diverse environmental challenges. Through material engineering strategies such as composite formation, surface functionalization, and structural optimization, CBAs can be tailored for specific applications, particularly in the selective adsorption of dyes, oils, and organic solvents from polluted aqueous systems.

14.4 CBAs for Gas Adsorption and Separation

With the acceleration of industrialization, atmospheric pollution, particularly the emission of greenhouse gases and the leakage of industrial exhaust, has become increasingly critical, posing serious threats to both environmental sustainability and human health. Conventional adsorbents such as activated carbon and zeolite molecular sieves, although widely used, are limited by disadvantages including high energy consumption, environmental impact, and restricted functionality [31, 32]. In contrast, CBAs, owing to their distinctive physicochemical properties such as high specific surface area, hierarchical porosity, lightweight structure, and excellent regeneration ability, have emerged as promising alternatives for gas adsorption applications, including carbon dioxide (CO_2) capture [33] and the removal of volatile organic compounds (VOCs) [34].

Wei and colleagues employed a freeze-casting technique to fabricate ultralight, flexible CNF-based aerogels with an interconnected 3D porous architecture. By impregnating these aerogels with cellulose acetate or acetylated cellulose nanocrystals (a-CNCs), they successfully introduced acetate functionalities to the aerogel surface. The incorporation of acetylated carbon nanotubes contributed to the formation of electron-rich sites on the aerogel framework, significantly enhancing CO_2 affinity. The modified aerogels demonstrated a CO_2 adsorption capacity of 1.14 $mmol \cdot g^{-1}$ under ambient pressure (101 kPa) and low-temperature conditions (273 K), while maintaining mechanical robustness, underscoring their potential in environmental remediation applications [35]. To further improve adsorption performance and mechanical stability, recent research has focused on developing hybrid CBA-based composites. For example, Li et al. synthesized a composite material by immobilizing the metal–organic framework UiO-66(Zr)-$(OH)_2$ onto highly porous CNF aerogels via a freeze-drying process (Fig. 14.3a). The resulting material, termed MOF-CNFs, exhibited a flake-like morphology with uniformly dispersed octahedral MOF particles adhered to the CNF matrix (Fig. 14.3b and c), which significantly enhanced the interfacial area for CO_2 interaction. Under conditions of 298 K and 0.15 bar, the composite achieved a CO_2 adsorption capacity of 1.8 $mmol \cdot g^{-1}$ and demonstrated excellent selectivity ($CO_2/N_2 = 66$) for binary gas mixtures (15:85 molar ratio) (Fig. 14.3d and e) [36]. In another study, researchers developed CNFs/PVA/GO aerogel composites by inducing polymer chain interactions between PVA and a CNFs–graphene oxide (GO) dispersion through controlled heating, followed by freeze-drying (Fig. 14.3f). The resulting aerogels displayed a layered, porous architecture, and exhibited enhanced CO_2 adsorption performance with increasing GO content and pressure, reaching up to 432.76 $cm^3 \cdot g^{-1}$ at 273 K (Fig. 14.3g) [37].

Surface functionalization via amination has also proven effective in modulating the gas adsorption characteristics of nanocellulose aerogels. Zhu et al. fabricated amino-functionalized nanocellulose (a-NC) aerogels using CNFs and 3-aminopropyltriethoxysilane (AEAPDMS) through a sol–gel process coupled with freeze-drying (Fig. 14.3h). The addition of AEAPDMS increased the solvent contact angle to 118.6°, indicating enhanced hydrophobicity, while also modulating capillary forces and yielding a well-organized polygonal porous structure with abundant active sites. The modified aerogels exhibited a CO_2 adsorption capacity of 2.4 $mmol \cdot L^{-1}$, highlighting the effectiveness of amino group incorporation in enhancing gas uptake (Fig. 14.3i) [38].

Similarly, another recent study employed N-[3-(trimethoxysilyl)propyl]ethylenediamine (DAMO) to silanize nanocellulose aerogels, thereby increasing amino group density on the surface. The CO_2 adsorption performance of these DAMO-functionalized aerogels increased proportionally with the amino group loading, achieving a maximum uptake of 7.86 $mmol \cdot g^{-1}$ under conditions of 298 K and 4 bar, at a DAMO loading level of 13.87 mmol [39].

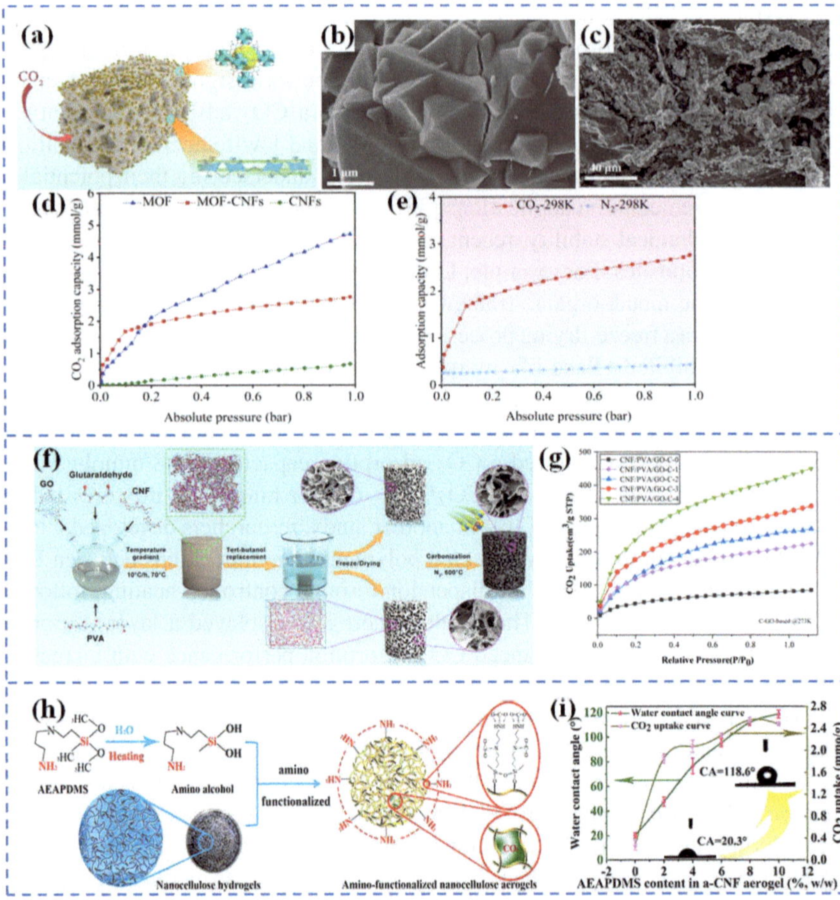

Fig. 14.3 **a** Structural schematic of MOF-CNFs aerogel; **b** and **c** SEM images of MOF-CNFs aerogel; **d** CO_2 adsorption capacity of CNFs, MOF, and MOF-CNFs and **e** CO_2 and N_2 adsorption capacity of MOF-CNFs. Reproduced with the permission from Reference [36]. Copyright 2025, Elsevier. **f** Preparation schematic and **g** CO_2 adsorption capacity of CNFs/PVA/GO aerogel. Reproduced with the permission from Reference [37]. Copyright 2022, Elsevier. **h** Mechanism diagram of CO_2 capture by a-NC aerogel and **i** Hydrophobicity and CO_2 adsorption properties of a-CNF aerogels. Reproduced with the permission from Reference [38]. Copyright 2024, Elsevier

14.5 CBAs for Oil–Water Separation

With the rapid advancement of global industrialization and urbanization, environmental issues such as oil spills, unauthorized industrial wastewater discharge, and the increasing burden of domestic sewage have become increasingly severe. These environmental challenges have rendered oil–water separation technologies a focal point for the development of innovative pollution mitigation strategies. Conventional separation methods such as mechanical separation, adsorption, and filtration often suffer from limitations including low efficiency, complex operational procedures, and the risk of secondary pollution [40–42]. CBAs, derived from renewable biomass, combine the inherent advantages of cellulose (biocompatibility and sustainability) with structural features such as ultralow density, high porosity, and large specific surface area, making them highly attractive candidates for oil–water separation. However, their intrinsic hydrophilicity limits their ability to selectively adsorb oils from aqueous environments, thus necessitating surface modification strategies to impart hydrophobicity. Chemical functionalization or incorporation of hydrophobic components can significantly enhance the oil–water separation performance of CBAs [43].

Recent research efforts have focused on modifying the hydrophilic surface of CBAs to confer hydrophobic or even superhydrophobic characteristics, enabling selective oil uptake in aqueous media [44, 45]. A common approach involves the use of silane coupling agents or hydrophobic metal oxides. For instance, Fan et al. engineered a multifunctional aerogel by constructing a physically and chemically cross-linked oxidized nanocellulose (ONC) network with polyethyleneimine (PEI) and ethylene glycol diglycidyl ether (EGDE) (Fig. 14.4a). The hybrid network was stabilized via hydrogen bonding, electrostatic interactions, and covalent crosslinking. Subsequent in situ grafting of polymethyltrimethoxysilane (PMTS) rendered the composite (C-g-PEI-PMTS) hydrophobic. This aerogel exhibited a hierarchical porous structure and rough surface morphology, enabling oil and organic solvent adsorption capacities ranging from 11.68 to 28.03 $g \cdot g^{-1}$, while maintaining excellent separation efficiency and cyclic stability (Fig. 14.4b and c) [46].

Li et al. reported the fabrication of a three-dimensional superhydrophobic aerogel composed of sodium alginate (SA) and carboxymethyl cellulose (CMC), modified with SiO_2 nanoparticles and methyltrimethoxysilane (MTMS) via chemical vapor deposition. This SA/CMC aerogel exhibited superior water resistance, mechanical robustness, and effective oil adsorption capabilities, demonstrating high separation efficiency for oil–water mixtures [44]. In a related study, Hu et al. employed freeze-drying to incorporate MTMS into a sulfonated nanocellulose (SNBC) matrix, yielding a SNBC/MTMS aerogel with a continuous 3D network. This material demonstrated excellent hydrophobicity, with a water contact angle of 152.4°, and exhibited selective adsorption capacities for various oils and organic solvents in the range of 42.14–85.37 $g \cdot g^{-1}$. Notably, the aerogel achieved a high separation efficiency of 98.48% during continuous oil–water separation cycles [49].

An alternative and effective strategy for achieving hydrophobicity in cellulose aerogels is carbonization. Pyrolytic treatment removes polar hydrophilic groups and generates a carbon-rich, nonpolar framework that significantly enhances hydrophobicity and oil sorption capacity [50]. For example, Long et al. employed a deep eutectic solvent (DES) composed of choline chloride and glucose to modify CNFs, followed by carbonization at 350 °C to prepare a hydrophobic carbon aerogel (DCCA-350). The resulting material exhibited a rough surface morphology and a pronounced reduction in polar functional groups, resulting in high oil absorption efficiency (volumetric uptake of 74.1%–95.2%) and excellent reusability, retaining over 90% of its performance after 100 cycles [51]. In addition, carbonization can also

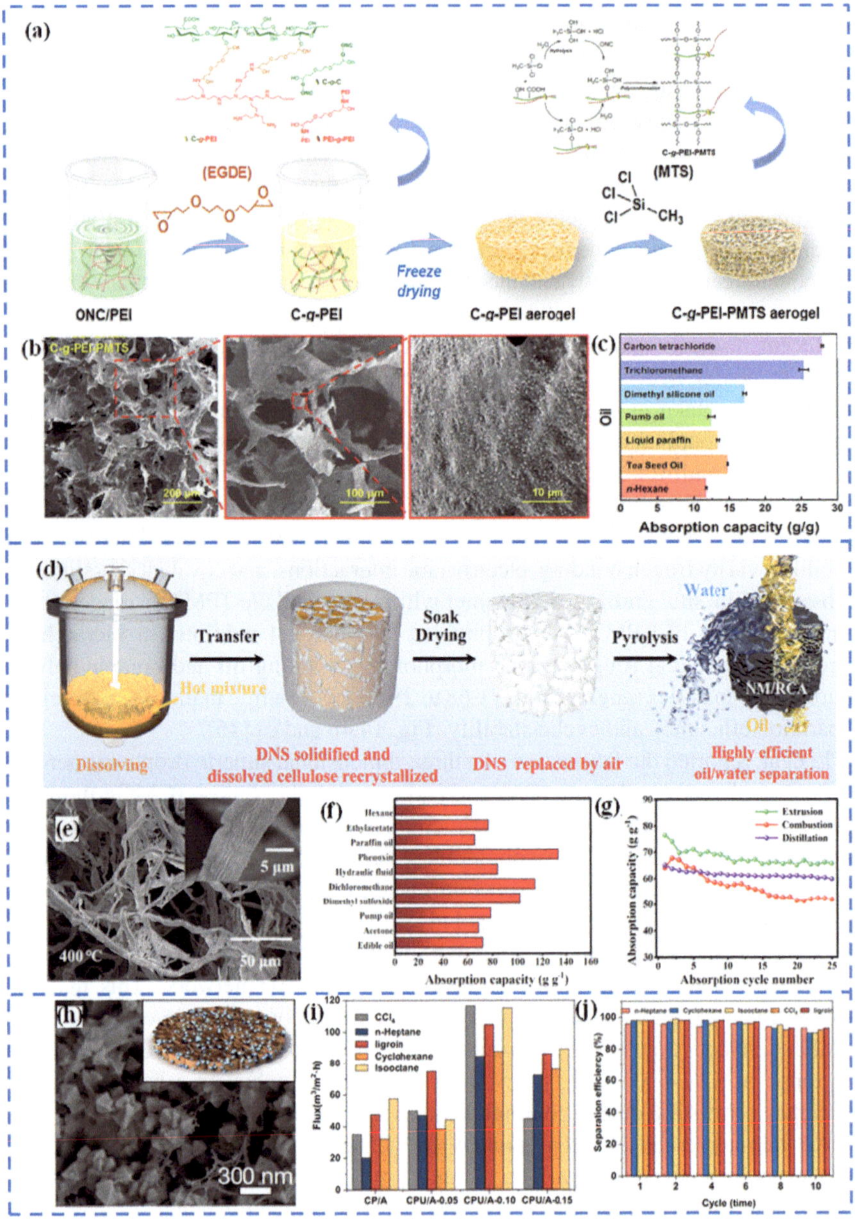
(a)
(EGDE)
(MTS)
Freeze drying
ONC/PEI
C-g-PEI
C-g-PEI aerogel
C-g-PEI-PMTS aerogel
(b)
C-g-PEI-PMTS
200 μm
100 μm
10 μm
(c)
Carbon tetrachloride
Trichloromethane
Dimethyl silicone oil
Pumb oil
Liquid paraffin
Tea Seed Oil
n-Hexane
Oil
Absorption capacity (g/g)
(d)
Transfer
Soak Drying
Pyrolysis
Water
Hot mixture
NM/RCA
Oil
Dissolving
DNS solidified and dissolved cellulose recrystallized
DNS replaced by air
Highly efficient oil/water separation
(e)
5 μm
400 °C
50 μm
(f)
Hexane
Ethylacetate
Paraffin oil
Phenoxin
Hydraulic fluid
Dichloromethane
Dimethyl sulfoxide
Pump oil
Acetone
Edible oil
Absorption capacity (g g⁻¹)
(g)
Extrusion
Combustion
Distillation
Absorption capacity (g g⁻¹)
Absorption cycle number
(h)
300 nm
(i)
CCl4
n-Heptane
ligroin
Cyclohexane
Isooctane
Flux(m³/m²·h)
CP/A
CPU/A-0.05
CPU/A-0.10
CPU/A-0.15
(j)
n-Heptane
Cyclohexane
Isooctane
CCl4
ligroin
Separation efficiency (%)
Cycle (time)

◂**Fig. 14.4** **a** Schematic diagram of the preparation of C-g-PEI-PMTS composite aerogel; **b** SEM images of C-g-PEI-PMTS composite aerogel and **c** Adsorption capacity of C-g-PEI-PMTS composite aerogel for different oils. Reproduced with the permission from Reference [46]. Copyright 2023, Elsevier. **d** Schematic diagram of the preparation of NM/RCA; **e** SEM images of NM/RCA; **f** Adsorption capacity of NM/RCA composite aerogel for different oils and **g** Recyclable oil absorption performance of NM/RCA. Reproduced with the permission from Reference [47]. Copyright 2023, Elsevier. **h** SEM images of CPU/A; **i** Separation flux of CP/A, CPU/A$_{-0.05}$, CPU/A$_{-0.10}$, and CPU/A$_{-0.15}$ aerogels and **j** Separation efficiency of CPU/A$_{-0.10}$ in cyclic oil–water mixture tests. Reproduced with the permission from Reference [48]. Copyright 2023, Elsevier

preserve the original porous structure of CBAs, enabling rapid oil absorption. For instance, Ma et al. synthesized a carbon aerogel (NM/RCA) via the carbonization of natural microfibrils (NM) and regenerated cellulose (RCA) at 400 °C (Fig. 14.4d). This aerogel retained a 3D network architecture with a high specific surface area (768.89 $m^2 \cdot g^{-1}$) and superhydrophobicity (contact angle: 151.5°) (Fig. 14.4e). It demonstrated exceptional oil–water separation performance, with adsorption capacities for oils and organic solvents ranging from 65 to 133 $g \cdot g^{-1}$ and a retention rate of 91.9% after 25 reuse cycles (Fig. 14.4f and g) [47].

Further enhancement of CBA performance can be achieved through the incorporation of additional functional components. For example, a composite aerogel (CPU/A) was fabricated by integrating UiO-66-NH_2 metal–organic framework particles into a PVA/CNF (PVA/CN) matrix, followed by hydrophobic treatment with octadecyltrichlorosilane (OTS) under thermal conditions (Fig. 14.4h). The presence of UiO-66-NH_2 significantly decreased pore size, increased specific surface area and pore volume, and improved mechanical properties. Among various formulations, CPU/A-0.10 (10% UiO-66-NH_2 content) exhibited the most effective separation performance, maintaining over 98% of its original flux after 10 reuse cycles (Fig. 14.4i and j) [48]. Miao et al. developed a multifunctional, magnetic, and superhydrophobic CNF/Fe_3O_4@BAC aerogel, using CNFs as the matrix and bamboo-activated carbon (BAC) and Fe_3O_4 nanoparticles as functional fillers. Following hydrophobic surface modification, the resulting aerogel exhibited ultra-low density (30.12 $mg \cdot cm^{-3}$), reasonable specific surface area (25.38 $m^2 \cdot g^{-1}$), magnetization (9.44 $emu \cdot g^{-1}$), and excellent superhydrophobicity. Owing to its layered porous architecture, the material achieved high oil adsorption capacities ranging from 16.79 to 51.93 $g \cdot g^{-1}$ for various oils [52].

14.6 CBAs for Photocatalytic Degradation of Organic Pollutants

Organic pollutants present in such effluents pose severe threats to aquatic ecosystems and public health, thereby necessitating the development of efficient wastewater treatment strategies. Conventional treatment techniques, including physical adsorption, biodegradation, and electrochemical oxidation, often suffer from limitations such as high energy consumption, low degradation efficiency, and the potential for secondary pollution [53, 54]. Photocatalytic degradation has emerged as a promising alternative due to its advantages of mild operating conditions, environmental compatibility, high efficiency, and low operational cost [55]. However, traditional photocatalysts often exhibit drawbacks such as aggregation, poor dispersion, and challenging recyclability, which significantly hinder their practical application [56, 57]. CBAs facilitate pollutant adsorption and provide accessible sites for photocatalytic reactions, thereby improving degradation efficiency [58].

Surface structure optimization and pore architecture modulation in CBAs can significantly enhance their capacity to support photocatalytic materials and adsorb contaminants. For instance, Zhu et al. fabricated a multilayer composite aerogel (BC@SiO_2-TiO_2) using a bacterial cellulose (BC) scaffold coated with silica (SiO_2), followed by TiO_2 loading. The incorporation of SiO_2 not only improved the interfacial adhesion between TiO_2 and the cellulose matrix but also increased the specific surface area and pore openness of the aerogel (Fig. 14.5a and b). Under UV irradiation, this composite completely degraded methylene blue (MB) within 60 min, achieving a 12-fold enhancement in degradation efficiency, alongside robust structural integrity and photocatalytic performance (Fig. 14.5c and d) [59]. Beyond structural optimization, the incorporation of multiple functional photocatalytic components to leverage synergistic effects has gained prominence. Jing et al. synthesized a ternary β-FeOOH/TiO_2/cellulose composite aerogel that exploits the photocatalytic activity of TiO_2 and the Fenton-like catalytic behavior of β-FeOOH. Under weak UV light, this composite achieved a MB degradation efficiency of 97.2% and demonstrated stable recyclability over five consecutive cycles [60]. Wang et al. reported a 3D composite aerogel catalyst (CLGM-A), developed via freeze-drying, which integrates two-dimensional cobalt–iron metal–organic frameworks (2D CoFe-MOF) and nitrogen-doped graphene (2D N-GE) on a CNF matrix. The intimate 2D/2D heterojunction and abundant active sites promote efficient charge separation and transfer. Additionally, nitrogen doping enhances the generation of unsaturated metal–oxygen bonds, further accelerating reactive oxygen species formation. This catalyst achieved a toluene conversion rate of 95.4% under visible light and retained 90% efficiency after five cycles, underscoring its excellent photocatalytic stability and reusability (Fig. 14.5e and f) [61].

In addition, carbon-based composite systems have also demonstrated enhanced visible-light responsiveness and photodegradation performance. For example, Ma et al. developed a recyclable 3D porous g-C_3N_4@cellulose aerogel (CN@CA) via chemical cross-linking. The amino groups in g-C_3N_4 form strong hydrogen bonds with hydroxyl groups on the cellulose backbone, ensuring firm anchorage and improved structural stability. The aerogel exhibits high specific surface area, effective charge separation, and superior photocatalytic performance, with a reported dye degradation efficiency of 99.0% [63]. To further enhance electron transfer dynamics, Susana and co-workers designed a TiO_2-carbon nanotube-cellulose composite aerogel (TiO_2-CNTs-CA). Scanning electron microscopy revealed a highly porous 3D network structure conducive to pollutant diffusion and effective exposure of TiO_2 active sites. The incorporation of CNTs significantly improved the separation

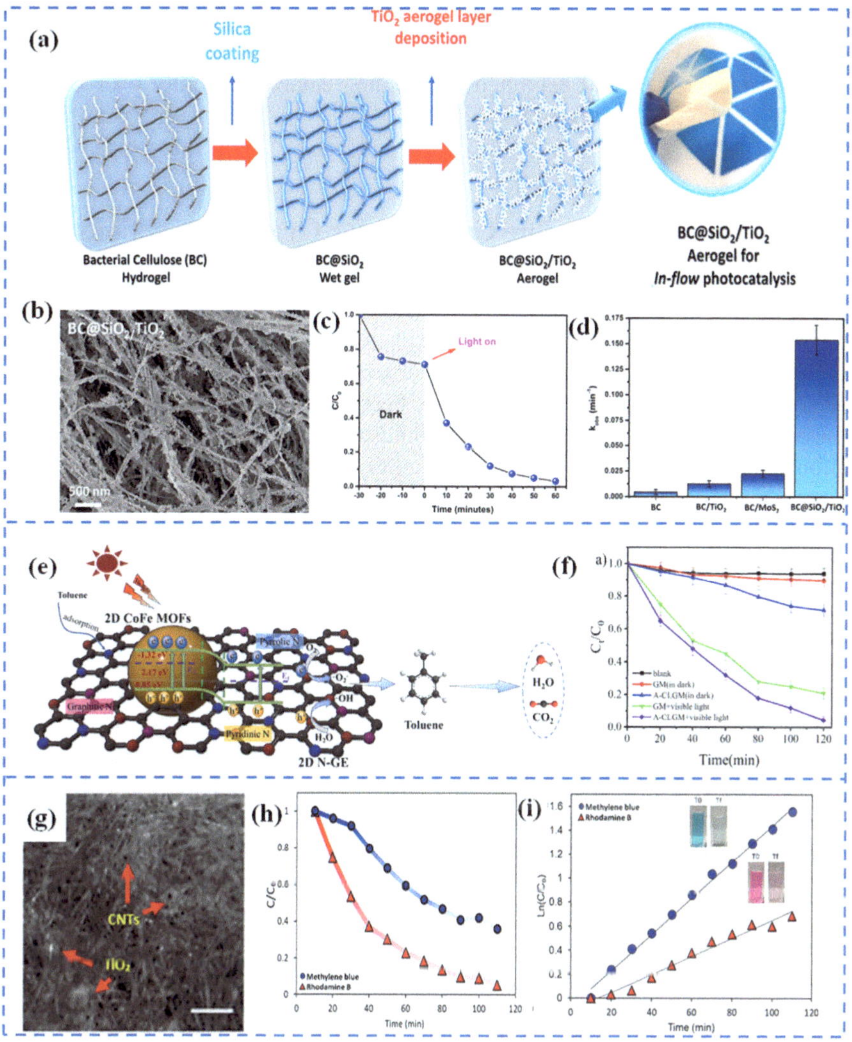

Fig. 14.5 **a** Schematic diagram of the preparation of BC@SiO_2/TiO_2 aerogel; **b** SEM images of BC@SiO_2/TiO_2 aerogel; **c** Comparison of the performance of aerogels for MB removal under dark (dark adsorption) and light (photocatalytic) conditions and **d** comparison of the kinetic constants of MB photocatalytic degradation for different samples. Reproduced with the permission from Reference [59]. Copyright 2023, American Chemical Society. **e** Mechanism diagram of CLGM-A photocatalytic degradation of toluene and **f** Photocatalytic degradation conversion of toluene under different reaction systems. Reproduced with the permission from Reference [61]. Copyright 2024, Elsevier. **g** SEM images of TiO_2-CNTs-CA; **h** and **i** Kinetic modeling of TiO_2-CNTs-CA degradation of Rhodamine B (RB) and MB under UV irradiation. Reproduced with the permission from Reference [62]. Copyright 2023, Elsevier

efficiency of photogenerated electron–hole pairs, thereby enhancing photocatalytic activity. Under UV irradiation, the composite achieved a degradation rate exceeding 97% for organic dyes and exhibited outstanding catalytic stability over multiple cycles (Fig. 14.5 g, h, and i) [62].

14.7 CBAs as Sustainable Catalysis for Organic Reactions

CBAs have been successfully employed in a wide range of prototypical organic transformations, such as hydrogenation, oxidation, and esterification reactions [64]. To impart catalytic functionality to cellulose aerogels, a variety of functionalization and active site incorporation strategies have been developed. For example, Fauziyah et al. employed pyrolysis of coir-derived cellulose followed by sulfonation to generate sulfonated carbon aerogels (SCAs) bearing $-SO_3H$ groups. As depicted in Fig. 14.6a and b, nitrogen adsorption–desorption analyses revealed that the SCAs exhibited a high specific surface area approaching 1000 m^2/g, indicative of a well-developed porous framework conducive to efficient mass transport. In the model esterification of acetic acid and ethanol at 80 °C over 4 h, the catalyst achieved a conversion exceeding 95% (Fig. 14.6c). Furthermore, the SCA retained more than 80% of its catalytic activity after five successive reaction cycles, demonstrating excellent operational stability and reusability [65]. Building upon this work, Suryandari and co-workers introduced hydrophobic surface modifications by grafting 4-tert-butylaniline (TBA) onto the SCAs. This surface engineering approach markedly enhanced the hydrophobicity of the catalyst, mitigating deactivation phenomena during reaction. The modified catalysts achieved acetic acid–ethanol esterification conversions in the range of 65–74% under shorter reaction durations [66].

In the domain of hydrogenation reactions, Ho et al. utilized cellulose aerogels derived from water hyacinth as support matrices and employed plant extracts for the in-situ reduction of Ag^+ ions. This green synthesis strategy enabled the uniform immobilization of silver nanoparticles within the aerogel's three-dimensional porous architecture. The resultant Ag/CA composite catalyst exhibited a high specific surface area and demonstrated remarkable catalytic activity in the hydrogenation of *p*-nitrophenol (PNP), achieving 99% conversion within 10 min at ambient temperature. The apparent rate constant (k) was calculated to be 0.34 min^{-1}, significantly outperforming conventional catalytic systems. Moreover, the Ag/CA catalyst retained over 90% of its catalytic performance across five consecutive reaction cycles, highlighting its outstanding recyclability and operational stability (Fig. 14.6d, e, and f) [67]. In another study, Zhou et al. synthesized MnOx/N-doped carbon aerogels (Mn-NCA) via an in situ approach. The resulting composite displayed a three-dimensional porous morphology with homogeneously dispersed MnOx nanoparticles, contributing to high catalytic efficiency, favorable dispersibility, and facile separation. In the transfer hydrogenation of biomass-derived aldehydes, the Mn-NCA catalyst achieved conversion rates of 90–100%, with product selectivity ranging from 64–100% [68].

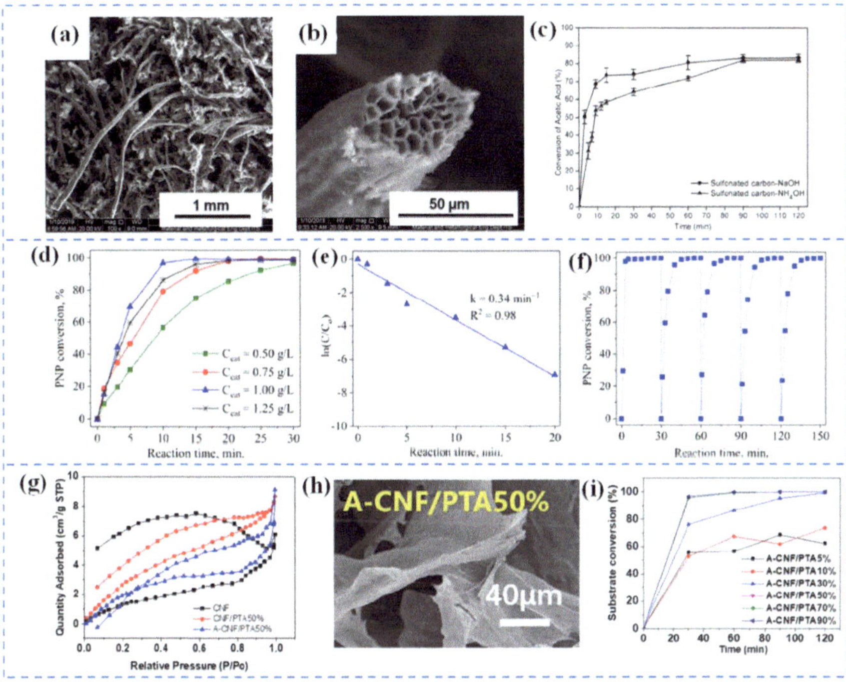

Fig. 14.6 **a** and **b** SEM images of SCA and **c** Conversion of acetic acid using SCA-NaOH and SCA-NH4OH catalysts. Reproduced with the permission from Reference [65]. Copyright 2020, Elsevier. **d** Conversion of PNP using Ag/CA catalysts; **e** Plot of L_n (C_t/C_0) versus reaction time and **f** Cyclic Stability of Ag/CA Catalysts in PNP hydrogenation. Reproduced with the permission from Reference [67]. Copyright 2023, Elsevier. **g** Adsorption and desorption profiles of CNF, CNF/PTA$_{50\%}$ and A-CNF/PTA$_{50\%}$ aerogels; **h** SEM images of A-CNF/PTA aerogel and **i** Catalytic properties of A-CNF/PTA aerogels. Reproduced with the permission from Reference [70]. Copyright 2022, MDPI

Regarding oxidative transformations, a nanocomposite catalyst designated as STGP@CA was developed for the aerobic oxidation of alcohols, ethylbenzene, and alkenes. At ambient temperature, the catalyst facilitated the oxidation of 1-pentanol with 84% conversion and 98.1% selectivity, delivering a turnover number (TON) of 214. For benzyl alcohol oxidation, the TON reached 314. The catalyst exhibited notable reusability and chemical stability, with no detectable leaching of palladium during the reactions [69]. In a complementary study, Song et al. modified CNFs with a minimal amount of (3-aminopropyl)trimethoxysilane (APTS) to introduce surface positive charges, followed by the immobilization of phosphotungstic acid (PTA) to construct A-CNF/PTA aerogels for oxidative desulfurization (ODS) processes. BET surface area measurements and SEM analysis revealed a significantly increased surface area and a uniform porous network (Fig. 14.6 g and h). The A-CNF/PTA50% aerogel demonstrated exceptional catalytic performance, achieving complete substrate conversion within 120 min at room temperature. Even after five cycles under dynamic catalytic conditions, the catalyst retained over 90% of its activity, underscoring its excellent recyclability and structural robustness (Fig. 14.6i) [70].

14.8 Conclusions and Perspectives

In summary, CBAs possess a combination of hierarchical porous architectures, intrinsic biodegradability, favorable mechanical resilience, and tunable surface chemistry, positioning them as highly promising materials for a wide range of adsorption and catalytic applications. This review has comprehensively summarized recent advancements in the use of CBAs for removing heavy metal ions, adsorbing organic pollutants, oil–water separation, gas capture and separation, photocatalytic degradation, and catalytic organic transformations. Thanks to their high surface area and abundant functional groups, CBAs excel at pollutant capture and show great promise in the design of green, efficient catalytic platforms. Their 3D network structures, coupled with modifiable surface functionalities, make them particularly advantageous for interfacial engineering and enhancing charge transfer in photocatalytic and sustainable organic reaction systems.

Despite the significant advancements made in the development of CBAs, several critical challenges remain that need to be addressed for their successful application in real-world scenarios. These challenges include limited structural and functional stability under complex environmental conditions, difficulties in achieving scalable and economically viable fabrication methods, and reliance on non-sustainable or chemically intensive surface modification techniques. To fully realize the potential of CBAs and advance their practical deployment, future research efforts should focus on the following strategic areas. Firstly, the rational, synergistic structure–function integration aimed at enhancing selectivity and catalytic/adsorptive performance; Secondly, the development of multifunctional composite systems through the incorporation of CBAs with complementary materials such as semiconductor photocatalysts, MOFs, or magnetic nanoparticles to enable synergistic functionalities; Thirdly, the innovation in green and energy-efficient synthesis methodologies, including aqueous-phase crosslinking techniques and enzyme-assisted surface modifications; Fourthly, the exploration of CBAs in emerging domains such as biosensing, electrochemical energy storage, and biomedical engineering; Finally, the enhancement of operational durability, reusability, and performance retention under real-world conditions. Overall, with continued innovation in material design and green processing, CBAs are expected to play a key role in next-generation environmental and catalytic technologies.

Acknowledgements The authors acknowledge financial support from the Natural Science Foundation of Shandong Province (ZR2023MB131).

References

1. Z.L. Yang, Q. Hu, L. Wang, J. Cao, J.R. Song, L.J. Song, Y.J. Zhang, Recent advances in the synthesis and application of graphene aerogel and silica aerogel for environment and energy storage: a review. J. Environ. Manag. **377**, 124668 (2025)
2. N.T.H. Giang, N.N. Tan, L. Huong, N.D. Hai, N.T. Thinh, N.T. Phuc, N.M. Dat, M.T. Phong, N.H. Hieu, Photocatalytic degradation of crystal violet on titanium dioxide/graphene aerogel doped sulfur. J. Mol. Struct. **1271**, 134013 (2023)
3. G.Y. Zheng, X.L. Kang, H.R. Ye, W. Fan, C. Sonne, S.S. Lam, R.K. Liew, C.L. Xia, Y. Shi, S.B. Ge, Recent advances in functional utilisation of environmentally friendly and recyclable high-performance green biocomposites: a review. Chinese. Chem. Lett. **35**, 108817 (2024)
4. Q.Q. Wang, Q. Yao, J. Liu, J.Z. Sun, Q.Q. Zhu, H.L. Chen, Processing nanocellulose to bulk materials: a review. Cellulose **26**, 7585–7617 (2019)
5. N.C. Ellebracht, C.W. Jones, Functionalized cellulose nanofibril aerogels as cooperative acid-base organocatalysts for liquid flow reactions. Carbohyd. Polym. **233**, 115825 (2020)
6. K.H. Fan, T.J. Zhang, S.Y. Xiao, H. He, J.S. Yang, Z.Y. Qin, Preparation and adsorption performance of functionalization cellulose-based composite aerogel. Int. J. Biol. Macromol. **211**, 1–14 (2022)
7. W.Y. Wei, S.H. Song, C. Meng, R.Z. Li, Y.Y. Feng, X.G. Chen, J.L. Chang, B. Fei, W.Q. Yang, J.S. Li, Bacterial cellulose/polyethylene glycol composite aerogel with incorporated graphene and metal oxides for VOCs detection. Chem. Eng. J. **499**, 156510 (2024)
8. L.T. Mo, H.W. Pang, Y. Tan, S.F. Zhang, J.Z. Li, 3D multi-wall perforated nanocellulose-based polyethylenimine aerogels for ultrahigh efficient and reversible removal of Cu(II) ions from water. Chem. Eng. J. **378**, 122157 (2019)
9. A. Ashori, E. Chiani, S. Shokrollahzadeh, M. Madadi, F.B. Sun, X.M. Zhang, Cellulose-based aerogels for sustainable dye removal: advances and prospects. J. Polym. Environ. **32**, 6149–6181 (2024)
10. F. Baraka, J. Labidi, The emergence of nanocellulose aerogels in CO_2 adsorption. Sci. Total. Environ. **912**, 169093 (2024)
11. Q.Q. Shang, J.Q. Chen, Y. Hu, X.H. Yang, L.H. Hu, C.G. Liu, X.L. Ren, Y.H. Zhou, Facile fabrication of superhydrophobic cross-linked nanocellulose aerogels for oil-water separation. Polymers-Basel **13**, 625 (2021)
12. X.Y. Zou, L.Y. Yao, S.Q. Zhou, G.N. Chen, S.F. Wang, X.Y. Liu, Y. Jiang, Sulfated ligno-cellulose nanofibril based composite aerogel towards adsorption-photocatalytic removal of tetracycline. Carbohyd. Polym. **296**, 119970 (2022)
13. Q. Luan, H. Zhang, Y.J. Lei, Y.X. Cai, Y.P. Bao, Y. Li, H. Tang, X.T. Li, Microporous regenerated cellulose-based macrogels for covalent immobilization of enzymes. Cellulose **28**, 5735–5744 (2021)
14. L.A. Malik, A. Bashir, A. Qureashi, A.H. Pandith, Detection and removal of heavy metal ions: a review. Environ. Chem. Lett. **17**, 1495–1521 (2019)
15. K.J. Yu, L.N. Yang, S.Y. Zhang, N. Zhang, Nanocellulose-based aerogels for the adsorption and removal of heavy-metal ions from wastewater: a review. Mater. Today. Commun. **43**, 111744 (2025)
16. Y.P. Guo, Q.M. Hasi, J.L. Yu, Y.Y. Guo, L.Y. Song, S. Wu, X.P. Luo, L.H. Chen, Carboxymethyl cellulose/sulfonated conjugated microporous polymer composite aerogel for efficient pollution removal and water evaporation. Sep. Purif. Technol. **324**, 124518 (2023)
17. S. Perumal, H. Lee, S. Jeon, D.H. Yoon, I.W. Cheong, Synthetization of hybrid nanocellulose aerogels for the removal of heavy metal ions. J. Polym. Res. **28**, 325 (2021)
18. L.T. Mo, S.F. Zhang, F. Qi, A.M. Huang, Highly stable cellulose nanofiber/polyacrylamide aerogel via in-situ physical/chemical double crosslinking for highly efficient Cu(II) ions removal. Int. J. Biol. Macromol. **209**, 1922–1932 (2022)
19. P.P. Sun, M. Wang, T.T. Wu, L.S. Guo, W.J. Han, Covalent crosslinking cellulose/graphene aerogels with high elasticity and adsorbability for heavy metal ions adsorption. Polymers-Basel **15**, 2434 (2023)

20. K. Singh, J.K. Arora, T.J.M. Sinha, S. Srivastava, Functionalization of nanocrystalline cellulose for decontamination of Cr(III) and Cr(VI) from aqueous system: computational modeling approach. Clean. Technol. Envir. **16**, 1179–1191 (2014)
21. B.Y. Geng, Z.Y. Xu, P. Liang, J. Zhang, P. Christie, H.Z. Liu, S.C. Wu, X.H. Liu, Three-dimensional macroscopic aminosilylated nanocellulose aerogels as sustainable bio-adsorbents for the effective removal of heavy metal ions. Int. J. Biol. Macromol. **190**, 170–177 (2021)
22. J. Wei, Z.X. Yang, Y. Sun, C.K. Wang, J.L. Fan, G.Y. Kang, R. Zhang, X.Y. Dong, Y.F. Li, Nanocellulose-based magnetic hybrid aerogel for adsorption of heavy metal ions from water. J. Mater. Sci. **54**, 6709–6718 (2019)
23. V. Katheresan, J. Kansedo, S.Y. Lau, Efficiency of various recent wastewater dye removal methods: a review. J. Environ. Chem. Eng. **6**, 4676–4697 (2018)
24. K.A. Adegoke, S.O. Akinnawo, T.A. Adebusuyi, O.A. Ajala, R.O. Adegoke, N.W. Maxakato, O.S. Bello, Modified biomass adsorbents for removal of organic pollutants: a review of batch and optimization studies. Int. J. Environ. Sci. Technol. **20**, 11615–11644 (2023)
25. S. Zhang, Y.Q. Pan, W. Wang, R.Z. Lin, X.W. Liu, Carboxylated cellulose-based aerogel with cellular pores prepared by stir freezing for cationic dye adsorption. Process. Saf. Environ. **177**, 807–817 (2023)
26. X.H. Shi, H.R. Ji, X.X. Ji, Z.J. Tian, J.C. Chen, Cellulose modification using sulfamic acid for adsorption of methylene blue and lead ions. Cellulose **31**, 4381–4393 (2024)
27. L.L. Zheng, Y. Yang, H.Y. Zhang, T. Liao, Y.Y. Han, M. Yu, Mxene-assisted construction of versatile cellulose-based aerogels with multilayered structure for piezoresistive sensing and contaminant adsorption. Int. J. Biol. Macromol. **308**, 142382 (2025)
28. C. Ruan, Y. Ma, G. Shi, C. He, C. Du, X. Jin, X. Liu, S. He, Y. Huang, Self-assembly cellulose nanocrystals/SiO_2 composite aerogel under freeze-drying: adsorption towards dye contaminant. Appl. Surf. Sci. **592**, 153280 (2022)
29. M. Hasan, D.A. Gopakumar, V. Arumughan, Y.B. Pottathara, K.S. Sisanth, D. Pasquini, M. Bracic, B. Seantier, A. Nzihou, S. Thomas, S. Rizal, H.P.S.A. Khalil, Robust superhydrophobic cellulose nanofiber aerogel for multifunctional environmental applications. Polymers-Basel **11**, 495 (2019)
30. M. Dilamian, B. Noroozi, Rice straw agri-waste for water pollutant adsorption: relevant mesoporous super hydrophobic cellulose aerogel. Carbohyd. Polym. **251**, 117016 (2021)
31. N. Kostoglou, C. Koczwara, C. Prehal, V. Terziyska, B. Babic, B. Matovic, G. Constantinides, C. Tampaxis, G. Charalambopoulou, T. Steriotis, S. Hinder, M. Baker, K. Polychronopoulou, C. Doumanidis, O. Paris, C. Mitterer, C. Rebholz, Nanoporous activated carbon cloth as a versatile material for hydrogen adsorption, selective gas separation and electrochemical energy storage. Nano Energy **40**, 49–64 (2017)
32. V. Sacchetto, D.F.O. Olivera, G. Paul, G. Gatti, I. Braschi, L. Marchese, C. Bisio, On the adsorption of gaseous mixtures of hydrocarbons on high silica zeolites. J. Phys. Chem. C **121**, 6081–6089 (2017)
33. J.J. Yuan, X.Y. Song, X.Y. Yang, C. Yang, Y.X. Wang, G.F. Deng, Z.C. Wang, J.B. Gao, Indoor carbon dioxide capture technologies: a review. Environ. Chem. Lett. **21**, 2559–2581 (2023)
34. Z. Wen, J. Lyu, Y. Ding, B. Liu, X. Zhang, Aerogel fibers made via generic sol-gel centrifugal spinning strategy enable dynamic removal of volatile organic compounds from high-flux gas. Adv. Funct. Mater. **34**, 2407221 (2024)
35. J.Y. Wei, S.Y. Geng, J. Hedlund, K. Oksman, Lightweight, flexible, and multifunctional anisotropic nanocellulose-based aerogels for CO_2 adsorption. Cellulose **27**, 2695–2707 (2020)
36. L.L. Li, Z.Y. Zhou, Y.D. Yang, T.T. Liu, C.C. Wan, Enhanced capacity in cellulose aerogel for carbon dioxide capture through modified by metal-organic framework. Int. J. Biol. Macromol. **303**, 140423 (2025)
37. Q. Zhang, W. Lu, M.Y. Wu, G.S. Qi, Y. Yuan, J. Li, H. Su, H.L. Zhang, Preparation and properties of cellulosenanofiber (CNF)/polyvinyl alcohol (PVA)/graphene oxide (GO): application of CO_2 absorption capacity and molecular dynamics simulation. J. Environ. Manage. **302**, 114044 (2022)

38. W.K. Zhu, M.L. Chen, J. Jang, M. Han, Y. Moon, J. Kim, J. You, S. Li, T. Park, J. Kim, Amino-functionalized nanocellulose aerogels for the superior adsorption of CO_2 and separation of CO_2/CH_4 mixture gas. Carbohyd. Polym. **323**, 121393 (2024)
39. F. Valdebenito, C. Albornoz, V. Rivera, E. Elgueta, M. Nisar, S. Lira, O. Valerio, A. Narváez, C. Quezada, R. Muñoz, L. Azócar, F. Sandoval, Stable reusability of nanocellulose aerogels with amino group modification in adsorption/desorption cycles for CO_2 capture. Materials **18**, 243 (2025)
40. Y.Y. Liu, X. Wang, S.Y. Feng, Nonflammable and magnetic sponge decorated with polydimethylsiloxane brush for multitasking and highly efficient oil-water separation. Adv. Funct. Mater. **29**, 1902488 (2019)
41. K. Yang, J. Ren, Y. Cui, T. Shah, Q. Zhang, B. Zhang, Length controllable tubular carbon nanofibers: surface adjustment and oil adsorption performances. Colloid. Surface. A **615**, 126272 (2021)
42. M. Yin, H. Huang, T.B. Oldenburg, An application of exploratory factor analysis in the deconvolution of heavy oil biodegradation, charging and mixing history in southeastern Mexico. Org. Geochem. **151**, 104161 (2021)
43. J.K. Gao, J.Q. Wang, M.M. Cai, Q.Y. Xu, J.W. Zhang, X. Cao, J.S. Zhang, Y. Chen, Advanced superhydrophobic and multifunctional nanocellulose aerogels for oil/water separation: a review. Carbohyd. Polym. **300**, 120242 (2023)
44. H.M. Li, J.Y. Huang, S. Shen, C.R. Meng, H.B. Wang, J.J. Fu, Superhydrophobic sodium alginate/cellulose aerogel for dye adsorption and oil-water separation. Cellulose **30**, 7157–7175 (2023)
45. Y. Sun, Y.L. Chu, W.B. Wu, H.N. Xiao, Nanocellulose-based lightweight porous materials: a review. Carbohyd. Polym. **255**, 117489 (2021)
46. B.J. Fan, L.L. Wu, A.X. Ming, Y. Liu, Y.Y. Yu, L. Cui, M. Zhou, Q. Wang, P. Wang, Highly compressible and hydrophobic nanofibrillated cellulose aerogels for cyclic oil/water separation. Int. J. Biol. Macromol. **242**, 125066 (2023)
47. X. Ma, S. Zhou, J.T. Li, F. Xie, H. Yang, C. Wang, B.D. Fahlman, W.J. Li, Natural microfibrils/regenerated cellulose-based carbon aerogel for highly efficient oil/water separation. J. Hazard. Mater. **454**, 131397 (2023)
48. W.J. Qu, Z.L. Wang, M.Y. Qin, X.J. Yang, F.S. Zhang, Z. Wang, D.D. Ji, D.H. Yu, Synthesis and characterization of UiO-66-NH_2 incorporated PVA/cellulose nanofibers composite aerogel for enhanced oil-water separation and formaldehyde adsorption. Sep. Purif. Technol. **325**, 124673 (2023)
49. C. Hu, J.S. Zhuang, Y.Y. Xia, J.C. Zhang, X.J. Zhang, Facile fabrication of sulfonated bacterial cellulose-silane composite aerogels via in suit polymerization for enhanced oil/water separation performance. Int. J. Biol. Macromol. **283**, 137650 (2024)
50. Y. Zhang, E.K. Sam, J. Liu, X. Lv, Biomass-based/derived value-added porous absorbents for oil/water separation. Waste Biomass Valori. **14**, 3147–3168 (2023)
51. S.S. Long, Y.C. Feng, Y.Z. Liu, L.L. Zheng, L.H. Gan, J. Liu, X.H. Zeng, M.N. Long, Renewable and robust biomass carbon aerogel derived from deep eutectic solvents modified cellulose nanofiber under a low carbonization temperature for oil-water separation. Sep. Purif. Technol. **254**, 117577 (2021)
52. Y. Miao, H. Xu, E. Wang, Y. Liang, Z. Wang, K. Sheng, C. Dai, W. Zhang, J. Huang, Ultralight, elastic, magnetic and superhydrophobic cellulose nanofibril based aerogel with layer-support structure designed for both excellent oil-water separation and efficient PM2. 5 removal. Sep. Purif. Technol. **327**, 124986 (2023)
53. S. Kim, S.-N. Nam, A. Jang, M. Jang, C.M. Park, A. Son, N. Her, J. Heo, Y. Yoon, Review of adsorption-membrane hybrid systems for water and wastewater treatment. Chemosphere **286**, 131916 (2022)
54. Y. He, Z. Wang, T. Li, X. Peng, Y. Tang, X. Jia, Biodegradation of phenol by candida tropicalis sp.: kinetics, identification of putative genes and reconstruction of catabolic pathways by genomic and transcriptomic characteristics. Chemosphere **308**, 136443 (2022)

55. F. Deng, H. Shi, Y. Guo, X. Luo, J. Zhou, Engineering paths of sustainable and green photocatalytic degradation technology for pharmaceuticals and organic contaminants of emerging concern. Curr. Opin. Green Sustain. Chem. **29**, 100465 (2021)
56. H. Dong, G. Zeng, L. Tang, C. Fan, C. Zhang, X. He, Y. He, An overview on limitations of TiO_2-based particles for photocatalytic degradation of organic pollutants and the corresponding countermeasures. Water Res. **79**, 128–146 (2015)
57. S.R.D. Gamelas, J.P.C. Tomé, A.C. Tomé, L.M.O. Lourenço, Advances in photocatalytic degradation of organic pollutants in wastewaters: harnessing the power of phthalocyanines and phthalocyanine-containing materials. RSC Adv. **13**, 33957–33993 (2023)
58. K. Milosevic, D. Loncarevic, M.K. Krusic, M. Hadnadev-Kostic, J. Dostanic, Eco-friendly g-C_3N_4/carboxymethyl cellulose/alginate composite hydrogels for simultaneous photocatalytic degradation of organic dye pollutants. Int. J. Mol. Sci. **25**, 7896 (2024)
59. T.C.A. da Silva, L. Marchiori, B.O. Mattos, S. Ullah, H.D. Barud, R.R. Domeneguetti, H.D. Rojas-Mantilla, M.V.B. Zanoni, U.P. Rodrigues, E.P. Ferreira-Neto, S.J.L. Ribeiro, Designing highly photoactive hybrid aerogels for in-flow photocatalytic contaminant removal using silica-coated bacterial nanocellulose supports. ACS Appl. Mater. Inter. **15**, 23146–23159 (2023)
60. J. Jing, Y.C. Feng, S.Y. Wu, Z.J. Ye, L. Yang, J.L. Li, Y.Y. Chen, F. Yang, B-feooh/TiO_2/cellulose nanocomposite aerogel as a novel heterogeneous photocatalyst for highly efficient photo-fenton degradation. RSC Adv. **13**, 14190–14197 (2023)
61. Y. Wang, F. Xu, L.N. Zhou, H. Li, Q.J. Meng, L.M. Jing, Z.H. Tian, C. Hou, 2D N-doped graphene/CoFe mofs heterostructure functionalized cnf aerogels impart highly efficient photocatalytic oxidation of gaseous vocs. J. Environ. Chem. Eng. **12**, 112225 (2024)
62. S.D. Gallegos-Cerda, J.D. Hernández-Varela, J.J. Chanona Pérez, C.A. Huerta-Aguilar, L. González Victoriano, B. Arredondo-Tamayo, O. Reséndiz Hernández, Development of a low-cost photocatalytic aerogel based on cellulose, carbon nanotubes, and TiO_2 nanoparticles for the degradation of organic dyes. Carbohyd. Polym. **324**, 121476 (2024)
63. Z. Ma, P. Zhou, L. Zhang, Y. Zhong, X. Sui, B. Wang, Y. Ma, X. Feng, H. Xu, Z. Mao, A recyclable 3D g-C_3N_4 based nanocellulose aerogel composite for photodegradation of organic pollutants. Cellulose **28**, 3531–3547 (2021)
64. P. Sudarsanam, E. Peeters, E.V. Makshina, V.I. Parvulescu, B.F. Sels, Advances in porous and nanoscale catalysts for viable biomass conversion. Chem. Soc. Rev. **48**, 2366–2421 (2019)
65. M.a. Fauziyah, W. Widiyastuti, H. Setyawan, Sulfonated carbon aerogel derived from coir fiber as high performance solid acid catalyst for esterification. Adv. Powder Technol. **31**, 1412–1419 (2020)
66. A.S. Suryandari, T. Nurtono, W. Widiyastuti, H. Setyawan, Hydrophobic modification of sulfonated carbon aerogels from coir fibers to enhance their catalytic performance for esterification. ACS Omega **8**, 27139–27145 (2023)
67. T. Gia-Thien Ho, D.P. Thao Truong, H.B. Nguyen, B. Long Do, T.A. Dinh, P. Ton-That, T.T. Van Nguyen, T.B. Ta Truong, K.P. Ha Huynh, N. Tri, Green synthesized nano-silver/cellulose aerogel as a robust active and recyclable catalyst towards nitrophenol hydrogenation. Chem. Eng. J. **471**, 144604 (2023)
68. S. Zhou, G. Chen, X. Feng, M. Wang, T. Song, D. Liu, F. Lu, H. Qi, In situ MnO_x/N-doped carbon aerogels from cellulose as monolithic and highly efficient catalysts for the upgrading of bioderived aldehydes. Green Chem. **20**, 3593–3603 (2018)
69. S. Keshipour, M. Khezerloo, Nanocomposite of hydrophobic cellulose aerogel/graphene quantum dot/Pd: synthesis, characterization, and catalytic application. RSC Adv. **9**, 17129–17136 (2019)
70. R. Song, X.Q. Zhang, H.H. Wang, C.F. Liu, Polyoxometalate/cellulose nanofibrils aerogels for highly efficient oxidative desulfurization. Molecules **27**, 2782 (2022)

Chapter 15
Cellulose-Based Aerogels for Solar Steam Evaporation

Jun Zhao, Jingjing Li, Yuke Sun, and Xiaofeng Xu

Abstract Solar-driven interfacial water evaporation technology has garnered significant attention due to its considerable potential in seawater desalination and wastewater treatment. Cellulose-based aerogel (CBA) has become a highly promising photothermal evaporation material due to its excellent hydrophilicity, controllable porous structure, low thermal conductivity, abundant sources, and biodegradability. This chapter systematically discusses the research progress and related applications of CBAs in solar steam evaporation. The core design principles of efficient photothermal evaporators were introduced, encompassing three key aspects: light absorption and photothermal conversion, thermal management optimization, and synergy between water and materials. In addition, various preparation methods for cellulose aerogel (such as freeze-drying and chemical cross-linking) were introduced, and the excellent performance of cellulose aerogel in a photothermal evaporation system was demonstrated through specific examples. Finally, the multifunctional and integrated applications of CBA photothermal evaporation systems beyond sustainable water production are discussed, and the future development of CBAs is prospected. This chapter provides a theoretical reference and basis for the development of high-performance, multi-functional, and environmentally friendly cellulose-based solar water evaporation materials.

Keywords Clean water production · Photothermal conversion · Solar interfacial evaporation · Solar desalination · Water purification

15.1 Introduction

Due to the explosive growth of the population, rapid economic and social development, and large-scale water pollution, the shortage of freshwater resources has become increasingly prominent, and the safety of drinking water has also received

J. Zhao · J. Li · Y. Sun · X. Xu (✉)
College of Materials Science and Engineerinng, Ocean University of China, Qingdao, China
e-mail: xuxiaofeng@ouc.edu.cn

G. Wei and J. Zhang (eds.), *Cellulose-Based Hydrogels and Aerogels*, Lecture Notes in Nanoscale Science and Technology 41,
https://doi.org/10.1007/978-3-032-03008-5_15

growing attention [1–3]. There are about 1.4 billion cubic kilometers of water on Earth; however, 97% of it is seawater, and only 0.06% of the remaining fresh water is available directly for human use [4, 5]. The latest report projects that some 4.4 billion people in low and middle income countries will lack access to safe drinking water [6, 7]. Access to safely managed drinking water (SMDW) has become an urgent need for people worldwide and is included in the framework of Sustainable Development Goal 6.1 [4, 8]. Due to surface water and groundwater reserves being small, desalination to obtain fresh water has become the primary way to solve the problem of water resources shortage. Traditional desalination technologies mainly include thermal separation and membrane separation processes. Although these technologies have proven to have mature freshwater production processes and reliable freshwater supply capacity, their high operating costs and high demand for fossil fuels require large investments, exacerbating the inequalities in global water supply [9–12]. The dual requirements of "affordability" and "decarbonization" are difficult to meet with traditional technologies, so universal and equitable access to affordable and safely managed drinking water is an important challenge that needs to be urgently addressed (Fig. 15.1a) [13].

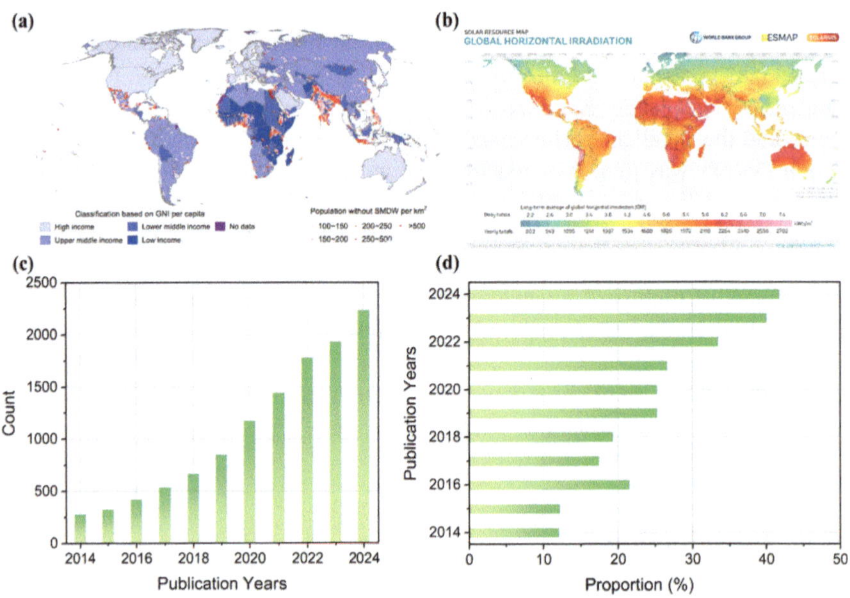

Fig. 15.1 **a** Geography of the global water-economy nexus. Reprinted from Reference [13], Copyright 2024, Springer Nature. **b** Global horizontal solar irradiation (https://globalsolaratlas.info.). **c** Number of articles published per year about Solar steam generation from 2014 and **d** The change in the proportion of articles related to solar-driven water evaporation based on cellulose materials since 2014 (*Data source* WOS web of knowledge)

Compared with traditional desalination methods that rely on various secondary energy sources for clean water production, solar steam evaporation driven by abundant and clean natural sunlight has attracted more and more attention in recent years due to its advantages of low cost and being environmentally friendly (Fig. 15.1b). Since 2014, more than 12,000 relevant articles have been published, and the number is increasing year by year (Fig. 15.1c). The solar steam evaporation system is composed of photothermal materials, water transmission channels and insulation layers, and converts solar energy directly into clean water production capacity through the synergistic mechanism of "light capture-heat location-water transport-steam recovery" [14–16]. In addition, based on the principle of phase transformation and diffusion, this technology is suitable for seawater desalination and complex water treatment containing organic/biological pollutants [17]. It is an efficient and economical water purification method, which is in line with decarbonization and sustainable development goals. Although synthetic polymers such as polyurethane (PU) [18], polyvinyl alcohol (PVA) [19], and polystyrene (PS) [20] are often used as substrates for interfacial photothermal evaporation systems due to their low thermal conductivity, tunable pore structure, and low-cost processing characteristics. However, synthetic polymers are not biodegradable and difficult to recycle, which conflicts with the global carbon neutrality goal and the principles of circular economy [21]. Therefore, the development of degradable alternative materials has become an inevitable demand for green and sustainable development in this field. Cellulose, as the most abundant natural polysaccharide, has a high density of hydroxyl groups on its surface, which provides an active site for photothermal material modification and can be functionalized through esterification/etherification reactions [22]. The 3D network formed by chemical cross-linking has structural stability, multi-stage pores, high crystallinity, and low thermal conductivity, which gives the material excellent mechanical strength and thermal insulation performance [23, 24]. Combined with abundant raw materials, low cost and complete biodegradation, cellulose-based materials have significant advantages in the field of solar interface evaporation. The number and proportion of research articles related to photothermal seawater desalination based on cellulose aerogel have also been increasing in the last five years (Fig. 15.1d).

In the following chapters, we will review the design principles of the interface evaporation system, the design methods and performance improvement strategies of the cellulose aerogel-based evaporator, and the current application and multifunctional integration research status. Based on the above summary, the future development of cellulose-based aerogels (CBAs) is prospected and discussed, which provides a new perspective and theoretical guidance for the design of high-performance cellulosic materials and the development of interfacial evaporation technology.

15.2 Design Principles of CBA Evaporators

15.2.1 Key Points of Solar Steam Evaporation Design

In recent years, the related materials and structures of solar-driven photothermal evaporation systems have been developed rapidly. Solar-driven evaporative systems can be divided into the two-dimensional (2D) evaporator and the three-dimensional (3D) evaporator according to the spatial dimension, and can be divided into rigid/flexible interfacial solar-driven structural evaporators (ISSEs) from the perspective of deformation [25]. However, regardless of the type of evaporator, to achieve efficient clean water production capacity, it is necessary to control the efficiency of the evaporator from the material and structural design, including light absorption, photothermal conversion, water transport, vapor-water phase transition, thermal management, and other aspects [26].

As a key component, the light-absorbing layer converts solar energy into thermal energy through efficient photothermal conversion, and its conversion efficiency directly determines the evaporation rate. The carrier material should have low thermal conductivity properties to reduce the conduction loss of heat to the water body. The substrate structure should form a stable water delivery channel to ensure that water is continuously transported to the evaporation interface. Besides, the chemical structure design of polymer backbones can optimize the phase transition process of water by regulating the interaction between water molecules. The systematic integration of these elements enables efficient solar evaporation [27]. In order to achieve more efficient evaporation performance, CBA evaporator also needs to be regulated from three aspects: solar absorption and utilization, heat management and application, and water transfer and evaporation.

15.2.2 Improvement of Solar Light Absorption

The rapid development of solar steam evaporation technology is due to the synergistic optimization design of photothermal conversion materials and evaporator structure. As the most abundant renewable energy on the Earth's surface, solar radiation has an annual total of 3.4 million EJ. Its spectral energy is mainly distributed in the 280–2500 nm band, of which the visible light (400–700 nm) and near-infrared (700–2500 nm) region contribute 97% of the energy, constituting the core energy source of photothermal conversion (Fig. 15.2a) [16].

As shown in Fig. 15.2b, when an incident photon interacts with a material surface, the energy distribution depends on the reflection, transmission, and absorption properties of the material. Therefore, the ideal photothermal material should have a wide spectrum of light absorption ability in the 400–2500 nm band, while reducing the light reflectance and thermal radiation emissivity of the material, in order to maximize the capture of energy and realize the transition from light to heat. The coupling

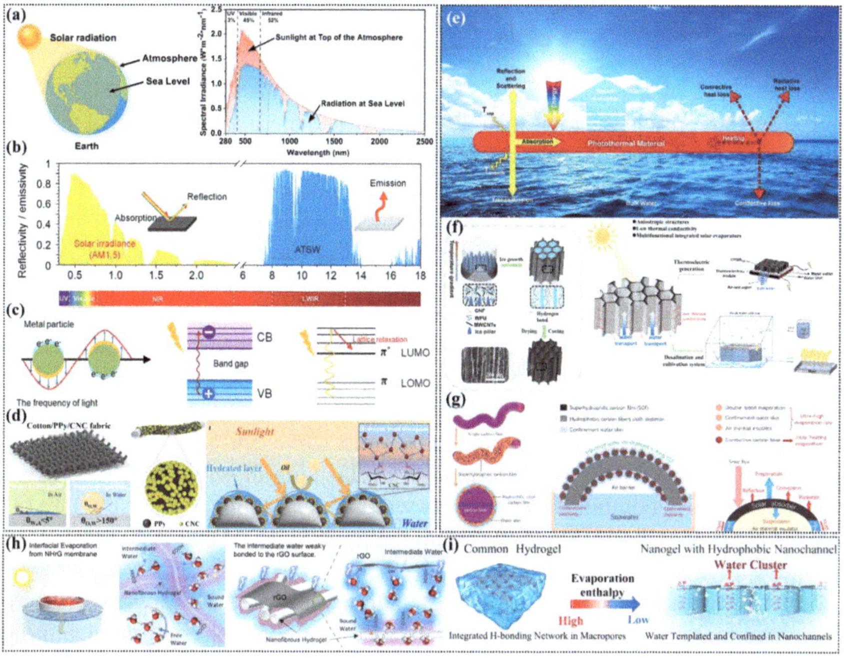

Fig. 15.2 **a** Schematic diagram of solar radiation and solar spectrum (AM1.5). Reprinted from Reference [16], Copyright 2024, Elsevier. **b** Mechanisms of light absorptions and photothermal conversion, **c** from left to right are respectively localized heating via plasmonic effects of metal nanoparticles, non-radiative relaxation in semiconductors, and molecular thermal vibration. Reprinted from Reference [28], Copyright 2025, Elsevier. **d** Fabric contact Angle, light capture effect and function of fabric hydration layer. Reprinted from Reference [29], Copyright 2025, Elsevier. **e** Schematic diagram of heat loss. Reprinted from Reference [15], Copyright 2024, Wiley. **f** Manufacturing process of anisotropic low thermal conductivity CPMA aerogel-based evaporator. Reprinted from Reference [30], Copyright 2025, Elsevier. **g** Arched double-sided evaporation promoting all-weather desalination. Reprinted from Reference [31], Copyright 2023, Wiley. **h** Schematic diagram of the state of water in the NHrG membrane. Reprinted from Reference [32], Copyright 2021, Elsevier. **i** The schematic of water-cluster evaporation. Reprinted from Reference [33], Copyright 2025, Wiley

design of such materials and porous evaporators can further localize thermal energy and promote the practical process of solar-driven water purification and seawater desalination.

In the process of solar photothermal conversion, the controllable selective light absorption layer is a way to improve the operating temperature and efficiency, and its behavior is determined by the solar absorption rate (Eq. (15.1)) and the thermal emission rate (Eq. (15.2)) [25, 34, 35]:

$$\overline{\alpha} = \frac{\int_0^{\infty} d\lambda \alpha(\lambda) E_{\lambda,\,\mathrm{Solar}}(\lambda)}{q_{\mathrm{Solar}}} \tag{15.1}$$

$$\bar{\varepsilon} = \frac{\int_0^{\infty} d\lambda \varepsilon(\lambda) E_{\lambda,B}(\lambda, T_A)}{\sigma {T_A}^4} \tag{15.2}$$

Here,$\bar{\alpha}$ and $\bar{\varepsilon}$ represent the total solar absorptivity and thermal emissivity of the solar absorber, respectively. The terms $\alpha(\lambda)$ and $\varepsilon(\lambda)$ denote the spectral absorptivity and spectral emissivity at a specific wavelength λ. $E_{\lambda,Solar}(\lambda)$ corresponds to the solar spectral irradiance (AM 1.5G spectrum), while $E_{\lambda,B}(\lambda, T_A)$ represents the blackbody spectral emissive power at absorber temperature T_A. The parameter q_{Solar} is the total solar irradiance (1 kW m^{-2} under AM 1.5G conditions), σ is the Stefan-Boltzmann constant, and T_A is the absorber temperature.

As shown in Fig. 15.2c, according to the interaction mechanism of materials with electromagnetic radiation and their electronic structure properties, the photothermal conversion mechanisms of materials can be divided into three different categories: (1) Plasmonic resonance heating, caused by collective electronic oscillations in the metal nanostructures [36]. (2) Non-radiative relaxation in semiconductors, driven by electron–hole recombination processes within the custom bandgap structure [37, 38]. (3) Molecular thermal motion, promoted by vibrational and rotational energy transitions in organic or polymer systems [39, 40].

To maximize solar energy harvesting, materials must be codesigned at macroscopic and microscopic structural levels. The structure and physicochemical properties of cellulose-based gels determine their evaporation efficiency. Based on the above discussion, in order to further improve the light absorption and photothermal conversion efficiency of fiber-based aerogel evaporator, the following two strategies can be adopted: (1) Incorporation of photothermal materials: different incorporation strategies of light-absorbing materials are used to ensure good dispersion of light-absorbing materials, so as to improve the ability of light absorption and photothermal conversion efficiency [41]. (2) By controlling the surface structure and interface structure of the material, the surface area of the evaporator can be increased, thereby increasing the absorption area, while reducing the surface wettability, and further improving the water transmission capacity (Fig. 15.2d) [29, 42].

15.2.3 Optimization of Energy Application

According to the second law of thermodynamics, heat will spontaneously flow from the warmer part to the cooler part. As shown in Fig. 15.2e, an ideal solar-driven evaporation system needs to concentrate the heat generated by photothermal conversion at the evaporative interface, in practice, if the evaporative surface temperature is higher than the ambient temperature, heat loss will occur through radiation and convection, so the influence of environmental factors such as temperature, humidity, and airflow on the energy distribution should be comprehensively considered. There are three main types of heat losses in solar-driven evaporation system: (1) Conduction loss ($E_{\text{conduction}}$): heat is transmitted down to the lower water through the absorber.

(2) Radiation loss ($E_{radiation}$): the surface of the high temperature absorber releases infrared radiation heat to the environment. (3) Convective loss ($E_{convection}$): heat exchange between surface and airflow [17].

For the 2D evaporator, the static energy of the system can be expressed by Eq. (15.3). In the formula, α is the light absorption coefficient.

$$E_{\text{evaporation}} = \alpha E_{\text{in}} - E_{\text{conduction}} - E_{\text{convection}} - E_{\text{radiation}} \quad (15.3)$$

Compared with the traditional 2D evaporator, the 3D evaporator not only has a top surface that directly absorbs light energy, but also introduces a side surface that does not directly absorb light radiation. This side surface exhibits a unique energy transfer property during evaporation: due to the evaporative cooling effect, the side surface temperature can be dropped below the ambient temperature, and additional energy can be actively absorbed from the surrounding environment through the heat radiation and convection heat transfer mechanism. In order to accurately describe the energy relationship of the system, a modified energy balance equation should be established. The core of this equation is to quantify the environmental energy capture contribution of the cold evaporation surface, and it should be coupled with the energy dissipation of the traditional top evaporation process. The energy equation should be updated as shown in Eq. (15.4) [43].

$$E_{\text{evaporation}} = \alpha E_{\text{in}} - E_{\text{conduction}} - E_{\text{convection,A1}} - E_{\text{radiation,A1}} + E_{\text{convection,A2}} + E_{\text{radiation,A2}} \quad (15.4)$$

In this context, $E_{convection,A1}$ and $E_{radiation,A1}$ quantify the convective and radiative thermal losses from photothermal evaporation surface *A1*, respectively. Correspondingly, $E_{convection,A2}$ and $E_{radiation,A2}$ characterize the convective and radiative energy gains occurring at the cold evaporation surface *A2*. The detailed explanation of the above formula will be introduced in detail in Sect. 15.3.2 of the chapter.

To mitigate environmental heat loss and enhance photothermal evaporation efficiency in CBAs for seawater desalination, the following optimization strategies are proposed based on thermal management principles: (1) Reduce the overall thermal conductivity of the material by adjusting the porosity of the aerogel. High porosity (>90%) can effectively reduce the solid phase heat conduction path and form a "thermal barrier" effect, thus controlling the heat conduction loss (Fig. 15.2f) [30, 44]. (2) To reduce the radiation loss by constructing microstructures on the aerogel surface or selecting coatings with low mid-infrared emissivity [45]. (3) Thermal localization is achieved through hydrophilic and hydrophobic regulation. (4) The 3D structure of the aerogel was designed to improve environmental energy capture and reuse through the cold evaporation surface (Fig. 15.2g) [31, 46].

15.2.4 Water-Material Synergistic Regulation

To achieve efficient operation of the solar-driven evaporation system, it is necessary to balance the energy relationship between water transport and evaporation process. Excessive water supply leads to heat loss, and insufficient water supply leads to interface drying or salt crystallization. To solve the above problems, current research focuses on two core control strategies: (1) Construct multi-level water transport channels to enhance capillary water supply, thereby precisely regulating the heat distribution at the evaporation interface and improving the continuous stability of the evaporation system [47]. (2) By adjusting the interface wettability and reconstructing the hydrogen bond network, the evaporation rate was improved by reducing the evaporative phase change enthalpy per unit mass of water while maintaining the photothermal conversion efficiency [48].

Ensuring the continuous delivery of liquid to the hot local area is an important criterion for guaranteeing the high evaporation rate and efficiency of solar-driven interface evaporation. For this reason, most studies on interfacial evaporation have employed porous materials to facilitate the transport of fluids to the liquid–gas interface through capillary forces. The force required for the liquid to flow through a capillary tube is given by Young's relation (Eq. (15.5)) and Young–Laplace equation (Eq. (15.6)).

$$\gamma_{SV} - \gamma_{SL} = \gamma_{LV} \cos\theta \tag{15.5}$$

Young's equation is used to describe the surface energy relationship at solid-liquid-gas equilibrium. Here, γ_{SV} and γ_{SL} are the solid–gas and solid–liquid interface energies, respectively, γ_{LV} is the liquid–gas surface tension, and θ is the contact Angle.

$$\Delta P = \frac{2\gamma \cos\theta}{r} \tag{15.6}$$

The Young–Laplace equation can be used to calculate the pressure difference at the liquid–gas interface in the capillary. Where r is the capillary radius (or pore characteristic size), γ is the water gas–liquid surface tension and the pressure difference ΔP drives the liquid up in the capillary [49].

When the pore radius r of a porous material is much smaller than the capillary length, the capillary force is much larger than gravity, the effect of gravity is negligible in small-scale systems, and the dynamic balance between the capillary force ΔP (driven by surface tension) and viscous resistance dominates the flow behavior. At this point, the relationship between the penetration depth z and the time t in the capillary or porous medium can be described by the Lucas-Washburn equation (Eq. (15.7)) [50].

$$Z = \sqrt{\frac{\gamma r \cos\theta}{2\mu} t} \tag{15.7}$$

where Z is the penetration depth of water, μ is the liquid viscosity, and t is the time required to penetrate a certain depth. Therefore, the water transport rate is balanced against the evaporation demand by adjusting the wettability of the material and optimizing the pore structure such as gradient pore size design.

In addition to ensuring an effective water supply strategy, from the perspective of energy utilization, reducing the energy required for water evaporation, that is, evaporation enthalpy, can also obtain a faster evaporation rate under the same energy input. The enthalpy of water evaporation is a core thermodynamic parameter for measuring the energy required for liquid water to transform into gas, and its value is jointly determined by intermolecular interactions (hydrogen bonds) and external conditions (temperature, pressure). According to the thermodynamic definition, the latent heat of water evaporation (ΔH_{vap}) is the enthalpy difference between the gas and liquid phases, the calculation formula for the enthalpy of water evaporation is as Eq. (15.8) [51].

$$\Delta H_{\text{vap}} = H_g - H_l = E_g + p_g V_g - E_l - p_l V_l \tag{15.8}$$

The internal energy E can be decomposed into molecular kinetic energy K and potential energy U. For liquid water $E_l = K_l + U_l$, $U_l = U_l^{ie} + U_l^{ia}$, where U_l^{ie} is the intermolecular interaction potential energy (such as hydrogen bonds) and U_l^{ia} is the intramolecular potential. For ideal gas the intermolecular potential energy is 0, so $E_g \approx K_g$. According to the ideal gas equation $p_g V_g = NK_BT$, the evaporation enthalpy formula can be approximately simplified as Eq. (15.9) [43].

$$\Delta H_{\text{vap}} = -U_l^{ie} + Nk_BT \tag{15.9}$$

As shown in Fig. 15.2 h, compared with the bound water (BW) that is tightly bound to the material surface through strong interactions (such as hydrogen bonds and electrostatic adsorption) and the free water (FW) that needs to overcome the complete hydrogen bond network in the bulk phase, the intermediate water (IW) is located in the interface or nano-constrained environment, with the hydrogen bond network partially broken and the intermolecular forces significantly weakened [32, 52]. Meanwhile, small cluster water has a significantly lower number of hydrogen bonds than bulk water and a higher specific surface area at the same time (Fig. 15.2i) [33]. Therefore, current research often weakens the non-covalent interactions between water molecules by increasing the effective evaporation surface area of the material, raising the proportion of intermediate water, and constructing clustered water or forming capillary water, etc., in order to reduce the evaporation enthalpy and improve the evaporation efficiency [53].

For CBAs, the following two strategies can be considered for water supply optimization: (1) Adjust the internal structure of the aerogel to improve the pore distribution inside the evaporator, thereby increasing the water delivery capacity [54]. (2) Introduce hydrophilic groups such as amino–NH_2), carboxylate (−COOH), and sulfonic acid (−SO_3H) to regulate the interaction between polymerization and water, reduce the enthalpy of evaporation, and improve the evaporation performance [55, 56].

15.3 Structural Design, Preparation, and Performance Enhancement of CBA Evaporators

In contrast to conventional desalination technologies mentioned earlier, solar-driven desalination systems leverage direct solar-thermal energy conversion to locally heat brine, enabling freshwater generation via evaporation–condensation processes followed by targeted collection. The development of solar-driven desalination technology to date can be delineated into three phases: Phase I: Bottom heating evaporation, where thermal energy is applied to the bulk solution, inadvertently heating non-evaporative water layers and incurring parasitic heat loss. Phase II: Bulk heating evaporation, which disperses photothermal materials uniformly within the brine, yet still dissipates heat into the entire water column, leading to unavoidable energy waste. Phase III: Interfacial heating evaporation, which localizes solar-thermal conversion and heating precisely at the air–liquid interface (Fig. 15.3). This approach selectively activates the evaporative surface layer rather than the entire bulk water, achieving thermal localization to minimize conductive/convective losses. The design and fabrication of ISSEs have significantly enhanced the performance of solar desalination technologies [43].

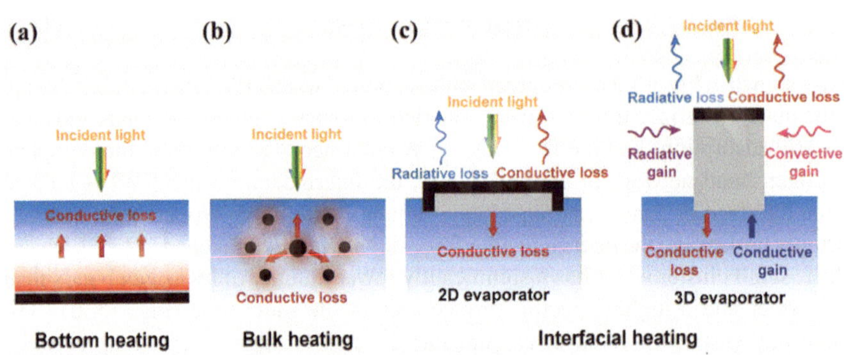

Fig. 15.3 Energy nexus during different solar steam generation processes

15.3.1 *Structural Design and Preparation*

ISSEs can be spatially categorized into 2D and 3D architectures. 2D ISSEs predominantly adopt composite film-based evaporator, while most 3D ISSEs are typically engineered as cylindrical, cuboidal, bridge-arch, spherical, T-shaped disc, or conical structures, composed of biomass-derived carbon materials or hybrid systems integrating support scaffolds and photothermal absorbers. Primary fabrication methods for ISSEs include dip-coating, electrospinning, bio-carbonization, gas-foaming, and freeze-drying techniques.

15.3.1.1 2D Evaporators

2D ISSEs predominantly utilize low-cost, flexible substrates such as cellulose-based paper and porous aerogel membranes as carriers for photothermal materials. Cellulose-based paper has emerged as a promising supporting substrate for hosting photothermal nanomaterials, leveraging its inherent microporous architecture, hydrophilicity, superior mechanical robustness, flexibility, ultralow thermal conductivity (~0.02 W m^{-1} K^{-1}), and cost-effectiveness. A representative application involves the integration of cellulose paper with plasmonic Au nanoparticles [57] and RGO [58] for solar-driven interfacial evaporation systems. Notably, Wang et al. pioneered a bilayered evaporator architecture by depositing Ti_2O_3 nanoparticles onto cellulose membranes [59]. The hierarchical microporous network of cellulose facilitates efficient water infiltration via capillary action while enabling rapid interfacial water replenishment post-evaporation. This substrate's intrinsic hydrophilicity further enhances hydraulic connectivity through hydrogen-bonding interactions, promoting unidirectional water transport to the evaporation interface. Moreover, cellulose membranes significantly mitigate parasitic thermal dissipation from the Ti_2O_3 photothermal layer to the bulk water phase, attributed to their exceptionally low thermal conductivity—nearly two orders of magnitude below that of liquid water [60].

15.3.1.2 3D Evaporators

2D ISSEs drive the water evaporation directly at the water–air interfaces, resulting in the heat conduction loss to the bulk water. A thermal insulation layer is normally incorporated into 3D ISSEs as the supporting substrate to localize the generated heat on the photothermal surface to facilitate the quick vaporization of water. Therefore, designing different structures of 3D ISSEs can effectively control factors such as light absorption area and steam flow resistance, improving the solar-to-vapor conversion efficiency.

Diverse fabrication strategies and structural designs have been developed for 3D cellulose aerogels. Freeze-drying, ambient pressure drying, and supercritical drying are conventional drying methods for fabricating cellulose aerogels. These approaches are typically combined with auxiliary techniques such as chemical reaction, template-assisted strategies, or 3D printing to construct cellulose aerogel-based evaporators with diverse structures.

Xu et al. developed hybrid fibrous aerogels with tunable wettability by controlling the assembly pathways of vinyltrimethoxysilane (VTMS) on bacterial cellulose (BC) networks, and creating superhydrophobic (VNFs) and superhydrophilic (PNFs) aerogels through surface-exposed siloxane or carbon groups via freeze-drying technology. The integrated double-layered evaporator, combining VNFs (thermal insulation) and PPy-doped PNFs (solar absorption), achieved exceptional evaporation rates of 1.91 (lab, 1 sun) and 4.20 kg m^{-2} h^{-1} (outdoor with wind), while demonstrating mechanical robustness (1,000 compression cycles), salt rejection (stable in 20 wt.% brine), and thermal stability (29.5 mW m^{-1} K^{-1} conductivity). This lightweight (6–8 mg cm^{-3}), hierarchically porous aerogel system highlights a sustainable approach for efficient solar desalination with scalable fabrication and extreme-condition durability (Fig. 15.4a) [61].

Bin et al. prepared a novel superhydrophobic/super-oleophilic methyltrichlorosilane-modified polyvinyl alcohol methyltrichlorosilane-modified polyvinyl alcohol/cellulose aerogel (MPAC) with asymmetric structure via freeze-drying, chemical crosslinking and chemical vapor deposition (CVD) methods. The MPCA demonstrated a notable evaporation rate of 1.39 kg m^{-2} h^{-1} and energy conversion efficiency of 89.7%. Furthermore, its novel asymmetric architecture conferred enhanced salt rejection capability to the evaporator, enabling autonomous dissolution of accumulated salt deposits within 20 min under continuous operation (Fig. 15.4b) [62].

In addition, Li et al. fabricated bacterial cellulose-based cylindrical aerogels via template-assisted modulation and freeze-drying, creating a 3D steam generator with well-defined hybrid structures and several self-contained properties of lightweight, efficient evaporation (1.82 and 4.32 kg m^{-2} h^{-1} under calm and light air conditions under 1 sun, respectively.), and good durability (Fig. 15.4c) [63].

However, the preparation of aerogels is currently limited by high-cost strategies such as freeze drying or supercritical drying. Here, Zong et al. presented a foam-template-assisted (FTA) strategy to fabricate ultralight (9 mg cm^{-3}), highly porous (porosity: >99%) CNF aerogels via ambient-pressure drying, with tannic acid polymerization yielding tannic-coated aerogels (TCA) exhibiting superior photothermal conversion. Under 1 sun, the TCA achieved an evaporation rate of 3.58 kg m^{-2} h^{-1} and energy efficiency of 97.2% (Fig. 15.4d) [64].

Zong et al. developed a foam skeleton template (FST) strategy to synthesize cellulose/graphite aerogels (CGAs) via ambient-pressure drying without organic solvent displacement, achieving a lightweight (36 mg $cm^{-3)}$ and highly porous (>97%) architecture. The CGA's 3D interconnected porous framework, coupled with its superhydrophilicity and ionic cross-linking between TEMPO-oxidized cellulose nanofibers (CNFs), flake graphite, and in situ-generated Ca^{2+}, facilitated efficient thermal localization, rapid capillary-driven water transport, and reduced evaporation enthalpy (1246 J g^{-1}), yielding a evaporation rate of 3.8 kg m^{-2} h^{-1} and energy efficiency of

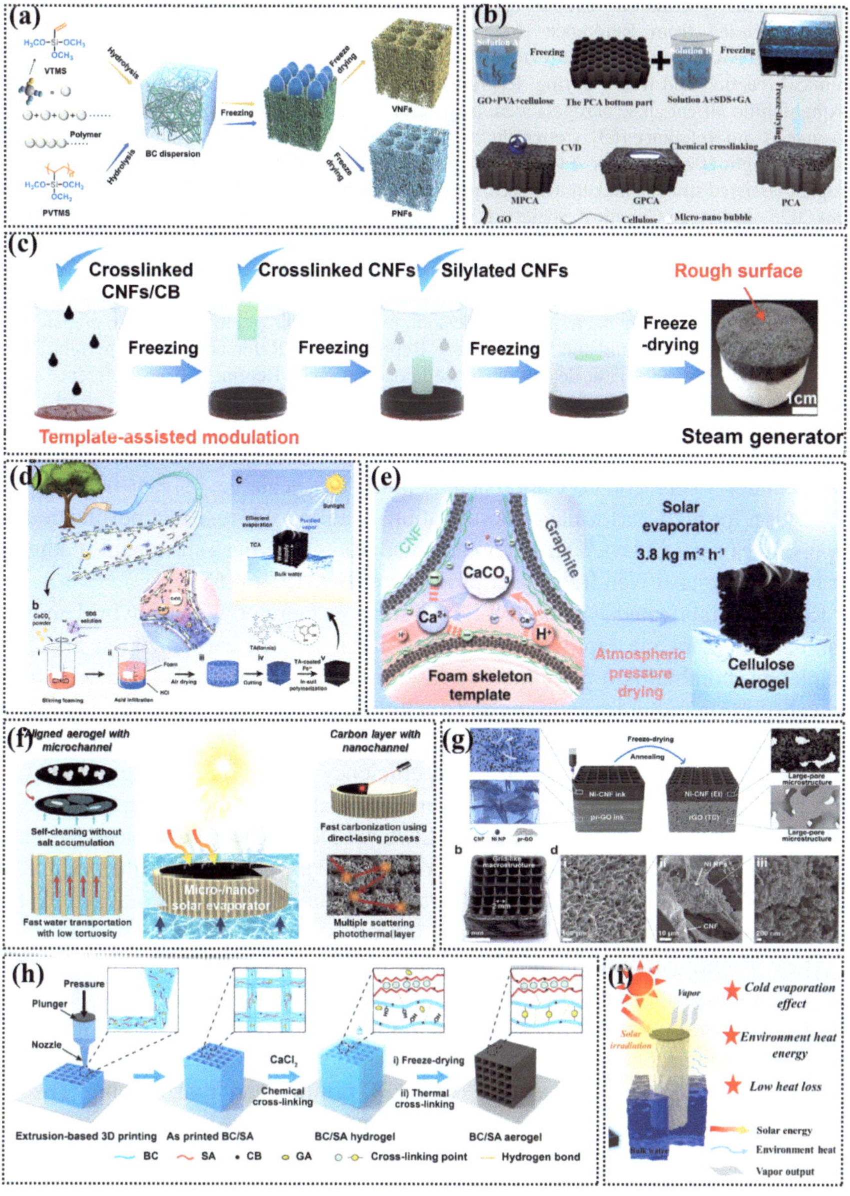
(a)
VTMS
PVTMS
Hydrolysis
BC dispersion
Freezing
Freeze drying
VNFs
PNFs
(b)
Freezing
GO+PVA+cellulose
The PCA bottom part
Solution A+SDS+GA
Freeze-drying
CVD
Chemical crosslinking
MPCA
GPCA
PCA
GO
Cellulose
Micro-nano bubble
(c)
Crosslinked CNFs/CB
Crosslinked CNFs
Silylated CNFs
Rough surface
Freezing
Freeze -drying
1cm
Template-assisted modulation
Steam generator
(d)
(e)
CNF
Graphite
Solar evaporator
3.8 kg m-2 h-1
Atmospheric pressure drying
Foam skeleton template
Cellulose Aerogel
(f)
Aligned aerogel with microchannel
Carbon layer with nanochannel
Self-cleaning without salt accumulation
Fast carbonization using direct-lasing process
Micro-/nano- solar evaporator
Fast water transportation with low tortuosity
Multiple scattering photothermal layer
(g)
Freeze-drying
Annealing
(h)
Pressure
Plunger
Nozzle
Chemical cross-linking
i) Freeze-drying
ii) Thermal cross-linking
Extrusion-based 3D printing
As printed BC/SA
BC/SA hydrogel
BC/SA aerogel
BC
SA
CB
GA
Cross-linking point
Hydrogen bond
(i)
Vapor
Solar irradiation
Cold evaporation effect
Environment heat energy
Low heat loss
Solar energy
Environment heat
Vapor output

◀**Fig. 15.4** The preparation method of 3D CBAs. **a** Hybrid aerogels with tunable wettability. Reprinted from Reference [61], Copyright 2023, Springer Nature. **b** Asymmetric methyltrichlorosilane-modified polyvinyl alcohol/cellulose aerogels were prepared by integrating chemical reaction and freeze-drying. Reprinted from Reference [62], Copyright 2024, Elsevier. **c** Amphiphilic all-cellulose aerogels were prepared by integrating templating with freeze-drying. Reprinted from Reference [63], Copyright 2021,Wiley. **d** CNF aerogels. Reprinted from Reference [64], Copyright 2024, Elsevier. **e** Cellulose/graphite aerogels were prepared by integrating a foam template-assisted strategy and atmospheric pressure drying. Reprinted from Reference [65], Copyright 2025, American Chemical Society. **f** A cellulose aerogel featuring nanopores and unidirectional microchannels was prepared by integrating rapid carbon-dioxide-laser-induced carbonization and ice-templating. Reprinted from Reference [66], Copyright 2025, Wiley. **g** A CBA with nano-micro–macro hierarchical structure. Reprinted from Reference [67], Copyright 2023, Wiley. **h** Monolithic, self-standing bacterial cellulose/sodium alginate aerogel matrices were prepared by direct ink writing 3D printing and freeze-drying. Reprinted from Reference [68], Copyright 2023, Wiley. **i** Thermal management design of 3D cellulose aerogels. Reprinted from Reference [69], Copyright 2024, Elsevier

98.4% under 1 sun. The FST approach demonstrates scalable, solvent-free aerogel fabrication for high-performance desalination, enabling salt-resistant purification of hypersaline brines (25 wt.%) and contaminated water, offering a sustainable pathway for low-cost solar-driven freshwater production (Fig. 15.4e) [65].

Ko et al. fabricated a cellulose-based solar evaporator with a hybrid micro-/nano-architecture via rapid carbon-dioxide-laser-induced carbonization and ice-templating, combining a photothermal laser-induced carbon (LC) layer with high light absorption/multiple scattering and a cellulose aerogel (CA) featuring nanopores and unidirectional microchannels for capillary-driven water transport, achieving an evaporation rate of 1.62 kg m^{-2} h^{-1} and efficiency of 66.6%. Coating the LC/CA with polydimethylsiloxane (PDMS) enhanced floatability, elevating performance to 1.9 kg m^{-2} h^{-1} and 83.8%, while the LC/CA/PDMS system demonstrated salt-resistant self-cleaning in seawater (1.68 kg m^{-2} h^{-1}). The design further exhibited scalable fabrication and inherent biodegradability, underscoring its practicality for sustainable solar desalination (Fig. 15.4f) [66].

The conventional 3D structures and related fabrication technologies intrinsically fail to maximize important performance-enhancing strategies and this technology still faces several performance roadblocks. Yang et al. prepared a CBA with nano-micro–macro hierarchical structure by using Ni nanoparticles and natural hydrophilic CNF as ink through direct ink writing (DIW) 3D printing and freeze-drying. The 3D plasmonic evaporator consisting of a Ni-CNF and a porous reduced graphene oxide (rGO) transport channel (TC) achieves an ultrafast freshwater flux of 29.1 L m^{-2} h^{-1} (Fig. 15.4g) [67].

Li et al. developed monolithic, self-standing bacterial cellulose/sodium alginate aerogel matrices via composite photothermal ink-based 3D extrusion printing, creating integrated interfacial steam generators (SGs) with multiscale hierarchical structures that synergistically reduce evaporation energy demand, expand active evaporation surfaces, and enhance environmental energy utilization for accelerated mass transport. The optimized aerogel achieves exceptional evaporation rates of

3.74 kg m^{-2} h^{-1} in calm air and 25.3 kg m^{-2} h^{-1} with 2 m s^{-1} airflow under 1 sun (Fig. 15.4h) [68].

15.3.2 Thermal Management Design OfCBAs

During interfacial evaporation processes, the water path not only facilitates the migration of water to the evaporation interface, but also induces thermal dissipation from the evaporation surface to the bulk water. 2D evaporators were developed for solar-driven interfacial evaporation and typically exhibited in direct contact with water.

As illustrated in Fig. 15.3, the incident solar energy (E_{in}) is predominantly absorbed by photothermal materials and transduced into thermal energy to drive interfacial water vaporization ($E_{\text{evaporation}}$). However, a fraction of the incident radiation is inevitably dissipated as reflection losses ($E_{\text{reflection}}$) to environment through scattering and reflection. The thermal localization effect induces a thermal gradient at the evaporation surface and surroundings, where the surface temperature exceeds that of both the underlying bulk water and surrounding air. This gradient drives energy dissipation through three principal pathways: conductive heat transfer to bulk water ($E_{\text{conduction}}$), convective losses to air ($E_{\text{convection}}$), and radiative emission ($E_{\text{radiation}}$). Consequently, the net energy available for interfacial solar steam generation (ISSG) in conventional 2D evaporation systems can be described via Eq. (15.3).

The energy loss of 2D evaporators is inevitable, 3D evaporators with enlarged surface evaporation area were designed to suppress and compensate for the radiation and convection losses. In addition to the top evaporation surface that primarily light-absorbing, 3D evaporators can incorporate evaporative cooling on the side surfaces without light absorption. The cold evaporative surface can harvest extra energy from the ambient environment through radiation and convection mechanisms to enhance evaporation performance (Fig. 15.3d). By strategically tuning the area ratio between the top surface (which dissipates energy to the surroundings) and the side surfaces (which actively harvest energy from the environment), it becomes feasible to achieve energy gain from the surroundings, resulting in evaporation efficiencies that exceed conventional theoretical limits. Under such conditions, the energy balance relationship of 3D interfacial evaporation systems with cold evaporative surfaces fundamentally differs from traditional 2D evaporative configurations. Consequently, the governing energy equation requires revision to account for these additional energy harvesting pathways, as Eq. (15.4).

The conduction, convection, and radiation losses can be calculated via Eqs. (15.10–15.12) listed below:

$$E_{\text{conduction}} = C\text{m}\Delta T \tag{15.10}$$

where C is the specific heat capacity of water (4.2 J g^{-1} K^{-1}), m is the mass weight of water, and ΔT is the temperature change of bulk water during the evaporation period.

$$E_{\text{convection}} = Ah(T_1 - T_0) \tag{15.11}$$

where A represents the evaporation surface area (m^2), h is the convection heat transfer coefficient, T_1 is the average temperature of the photothermal evaporation surface, and T_0 is the environment temperature during evaporation period.

$$E_{\text{radiation}} = \varepsilon A\sigma\left({T_1}^4 - {T_0}^4\right) \tag{15.12}$$

ε is the emissivity, σ is Stefan-Boltzmann constant (5.67×10^{-8} Wm^{-2} K^{-4}).

Wei et al. developed a hydrophilic 3D cellulose aerogel evaporator (ACP-7LM) inspired by tree transpiration, featuring Schiff base crosslinking for enhanced structural stability in water. Moreover, the exposure height from 0 to 4 cm of ACP-7LM aerogel was adjusted to increase the cold evaporation surface, maintain a lower temperature on the side surface, and harvest heat energy from the environment during evaporation. This augmented environmental heat absorption elevated the evaporation rate of ACP-7LM from 1.60 to 3.76 kg m^{-2} h^{-1}, representing a 2.35 fold enhancement relative over the ACP-7LM (0 cm) (Fig. 15.4i) [69].

To achieve heat management, in addition to increasing the height of the evaporator, it can also be realized from a material perspective. Liu et al. fabricated CNF/BNNS aerogels by introducing boron nitride nanosheets (BNNS) with excellent thermal conductivity. CNF/BNNS aerogel exhibited a thermal conductivity of 0.57 W m^{-1} K^{-1}, which is an enhancement of 443% compared to 0.105 W m^{-1} K^{-1} of CNF aerogel. The surface temperature changes of the aerogels were confirmed a strong thermal management capacity [70].

15.4 Applications and Multifunctional Integration of CBA Evaporators

15.4.1 Seawater Desalination

Due to the rapid urbanization, socioeconomic development, and climate change, the world is experiencing exacerbated water scarcity. Cellulose-based evaporator combined solar energy with renewable materials, which provide potential strategy for water treatment. However, the formation of salt deposits would impair solar evaporation performance. Within the solar-driven evaporation proceeds, water molecules gradually evaporate, leaving behind concentrated saline residues that finally deposit on the evaporator surface. The diffusion–advection-osmotic, acting in opposition to solar evaporation, controls the salt accumulation by isolating the evaporation surface, introducing excess water, and controlling the salt concentration pathway (Fig. 15.5a) [71]. As a result, salt accumulation can be derived from solar-driven net mass flux of solutes (m_{excess}) to the evaporative interface and the total salt rejecting flux (J). m_{excess} can be calculated by multiplying the solar evaporation rate by the concentration of seawater (Eq. (15.13)).

$$\mathrm{m_{excess}} = \mathrm{m}\frac{cwt\%}{(100-c)wt\%} = \frac{\eta_{evap}A_{\mathrm{evap}}E_{\mathrm{solar}}}{h_{LV}} \times \frac{cwt\%}{(100-c)wt\%} \quad (15.13)$$

where c is the initial concentration of seawater, η_{evap} is the solar-to-vapor efficiency, A_{evap} is the surface area of the solar evaporator (m^{-2}), E_{solar} is the density of solar energy (kW m^{-2} h^{-1}), and h_{LV} represents the total enthalpy of liquid–vapor phase change.

The condensed salt ions can be rejected via three mechanisms: (i) diffusion, (ii) advection, and (iii) osmotic. The cumulative salt dilution capacity provided by these three mechanisms is represented as the total salt rejecting flux (J) (Eq. (15.14)).

$$J = J_{diffusion} + J_{advection} + J_{osmotic} \quad (15.14)$$

where $J_{diffusion}$, $J_{advection}$, and $J_{osmotic}$ represent the salt rejection flux via diffusion, advection, and osmotic, respectively.

Below are selected design examples of cellulose-based ISSEs for seawater desalination. Wu et al. presented a flexible single-layer salt-rejecting monolithic porous evaporator (SRMPE) (Fig. 15.5b). During desalination, increasing surface salinity creates a surface tension gradient, inducing Marangoni convection that transports salt away from the surface and prevents crystallization. The SRMPE demonstrates rapid self-cleaning, removing NaCl powder within 1.25 h under 1 sun irradiation [72]. Wang et al. designed a CNF-based evaporator integrating a lotus-inspired structure and Janus wettability (super hydrophilic surface) (Fig. 15.5c). Its Janus dual-porous structure enables floating and directional salt ion backflow, mitigating salt accumulation and achieving a high evaporation rate of 3.61 kg m^{-2} h^{-1} under 1 sun [73]. To further address salt accumulation, Liao et al. developed a floating 3D solar evaporator by encapsulating lightweight foam spheres within porous cellulose hydrogel (Fig. 15.5d). Achieving an evaporation rate of ~2.01 kg m^{-2} h^{-1} under 1 sun, it accomplishes self-cleaning through spontaneous rotation driven by salt accumulation, offering a novel, cost-effective solution [74].

Evaporators fabricated via 3D printing technology exhibit exceptional characteristics, including rapid water transport, salt resistance, enhanced solar absorption, and expanded evaporation surface area. Yuan et al. pioneered 3D printed cellulose/alginate/carbon black hierarchical porous hydrogel evaporators. The hydrogel shows super-hydrophilic wettability and excellent light absorption, achieving an evaporation rate of 1.33 kg m^{-2} h^{-1} under 1 sun and superior salt resistance, with collected water meeting WHO drinking standards after purification [75]. Li et al. demonstrated that silylation of the CNFs rendered the 3D-printed surface highly hydrophobic, enabling a novel icephobic functionality. Under 1 sun illumination, interfacial heating rapidly melted ice droplets, while the oriented microstructures facilitated self-propelled meltwater removal within 120 s (Fig. 15.5e) [68]. Gao et al.

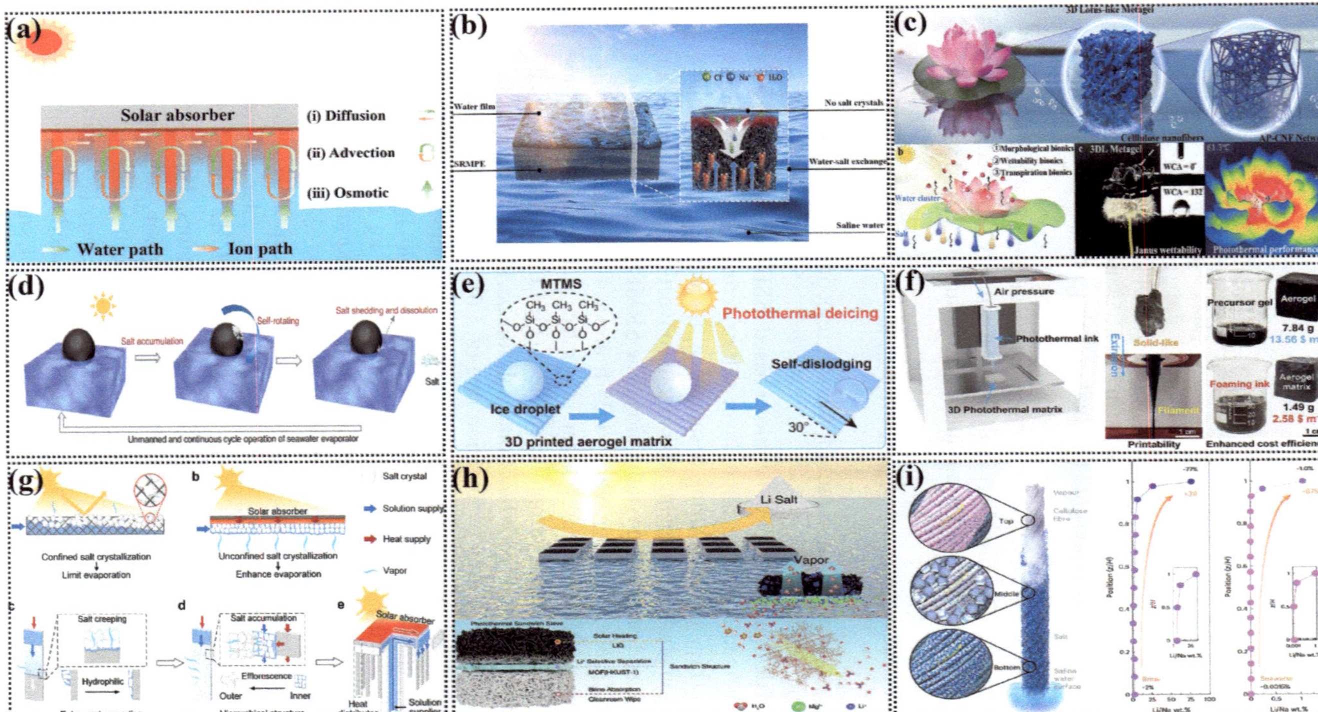

Fig. 15.5 Applications in seawater desalination and solute separation: **a** Illustration of the mass transfer during the solar desalination process. Reprinted from Reference [71], Copyright 2024, American Chemical Society. **b** Multilayer micro-/nano-architected aerogel via CO_2-laser carbonization and ice-templating. Reprinted from Reference [72], Copyright 2024, Wiley. **c** Lotus-inspired Janus evaporator with dual-porous structure and directional salt backflow. Reprinted from Reference [73], Copyright 2025, Springer Nature. **d** Rotational 3D evaporator with foam-core/hydrogel-shell design. Reprinted from Reference [74], Copyright 2024, Wiley. **e** Multiscale structured evaporator with silylated CNF surface. Reprinted from Reference [68], Copyright 2023, Wiley. **f** Hierarchical porous foam fabricated via "foaming" photothermal inks. Reprinted from Reference [76], Copyright 2024, Royal Society of Chemistry. **g** Ethylcellulose-based salt-crystallized self-assembled porous evaporator. Reprinted from Reference [77], Copyright 2025, Royal Society of Chemistry. **h** Paper-based solar sieve structure with HKUST-1 sub-nano-channels for selective lithium extraction. Reprinted from Reference [78], Copyright 2025, Wiley. **i** Twisted cellulose fiber crystallizer with capillary evaporation flow-driven ion-graded crystallization. Reprinted from Reference [79], Copyright 2023, Springer Nature

implemented "foaming" photothermal inks to enable the direct fabrication of hierarchically porous structures (Fig. 15.5f). This approach boosts solar evaporation while concurrently minimizing material consumption, achieving a high water evaporation rate of 2.94 kg m^{-2} h^{-1} under 1 sun [76].

15.4.2 Solute Separation

An alternative strategy for achieving salt-rejecting desalination involves controlled salt crystallization. This technique enables the co-generation of salt and freshwater while preventing salt ion reflux and salinity elevation. Salt crystallization is directed to occur at evaporator edges or peaks, preserving a clean evaporation surface for continuous operation. The capability to process highly concentrated brine facilitates salt recovery with zero liquid discharge (ZLD). Xu et al. employed vinyltrimethoxysilane hybridized with bacterial cellulose to create aerogels, achieving a water evaporation rate of 1.91 kg m^{-2} h^{-1} under 1 sun. This structure enabled specific salt accumulation and crystallization at the evaporator edge [61]. Yu et al. developed a hierarchical porous salt evaporator (HPSE) based on ethylcellulose, leveraging self-amplifying salt creeping and efflorescence effects. The HPSE achieved stable evaporation (18.78 kg m^{-2} h^{-1} under 1 sun) and crystallization when treating hypersaline brine (27.3 wt.% salinity), enabling passive liquid supply [77].

Lithium is a pivotal raw material for renewable energy. Extracting lithium from brine, which often has high Mg^{2+}/Li^{+} ratios, is challenging due to interfering cations like Mg^{2+}. Photothermal evaporation, powered by renewable energy sources and with low energy consumption, shows potential for reducing the overall cost of lithium extraction. By integrating photothermal materials with specific materials or structural designs, the extraction efficiency of lithium ions can be significantly enhanced. Zhong et al. utilized dust-free paper as a dual-functional substrate for water transport and photothermal conversion. They implemented a solar-thermal sandwich sieve structure with HKUST-1 providing sub-nanochannels, enabling selective permeation of hydrated Li^{+} through the MOF layer. This achieved a maximum lithium extraction capacity of 1467 mg m^{-2} per cycle, demonstrating high Li^{+} selectivity under elevated Mg^{2+}/Li^{+} ratios under 1 sun irradiation [78]. Chen et al. reported a passive method for rapid selective lithium extraction from saline water using twisted cellulose fiber crystallizers. The 3D porous architecture mediated sequential and separable crystallization of cations through synergistic capillary and evaporative flows, achieving a 9.8 kg m^{-2} h^{-1} evaporation rate and order-of-magnitude lithium concentration enhancement [79].

15.4.3 Wastewater Treatment

With accelerating industrialization, modern society generates vast quantities of wastewater that require complex treatment processes for resource recovery due to limited direct usability. Solar interfacial evaporation technology provides a sustainable approach for treating domestic sewage and industrial wastewater. However, abundant microorganisms (e.g., bacteria and fungi) proliferate within evaporators, causing pore clogging and compromising long-term operational stability. Consequently, efficient removal of heavy metal ions, organic pollutants, and microorganisms is critical during treatment.

Current strategies employ photocatalytic degradation, adsorption, and high-temperature sterilization for synergistic pollutant removal. Zhao et al. designed a biomass-based Janus evaporator with an oriented nanostructure (Fig. 15.6a). Utilizing the nanostructure's efficient water transport and sodium alginate's coordination capacity for heavy-metal ion enrichment, metal-ion concentrations were reduced by 2–4 orders of magnitude [80]. Wang et al. developed a bionic evaporator inspired by plant transpiration (Fig. 15.6b). Electrostatic adsorption between polypyrrole (PPy) and bacteria, combined with PPy's photothermal conversion, confers excellent antibacterial properties against Escherichia coli and Staphylococcus aureus [81].

Li et al. fabricated a dual-layer Janus-structured nanofilament cellulose (JNC) aerogel photothermal evaporator featuring a vertically ordered cellular network with multi-scale crosslinking (Fig. 15.6c). For emulsion separation, the aerogel's underwater superoleophobicity enables efficient interception of nano/submicrometer-sized oil droplets. The evaporator maintains stable evaporation rates during purification, achieving >99.7% removal efficiency for diverse emulsions [82]. Lu et al. integrated cationic cellulose nanofibrils (CCNF) with a dynamic covalent tannin/borate (T/B) complex to fabricate 3D continuous foams. These foams serve as customizable platforms for metal ions (Fe^{3+}, Cu^{2+}, Ag^{+}). pH-responsive coordination bonds and electrostatic attraction between T/B complexes and metal ions enable controlled ion release in response to environmental pH changes, ensuring long-term antimicrobial efficacy [83]. Tian et al. demonstrated a Janus solar evaporator (HZC-CNF) based on TEMPO-oxidized cellulose nanofibril (TOCNF) aerogel (Fig. 15.6d). Well-dispersed ZIF-67 catalysts within the matrix achieve >90% removal efficiency for pollutants including ciprofloxacin [84].

15.4.4 Water-Electricity Cogeneration

In cellulose-based ISSEs, thermal energy transfer and conversion occur concurrently during the evaporation process. The substantial low-grade thermal energy involved holds significant potential for further utilization in electricity generation. By incorporating multi-physics coupling mechanisms—such as water and ion flows, thermoelectric effects, salinity gradients, humidity variations, photovoltaics, thermos-galvanic

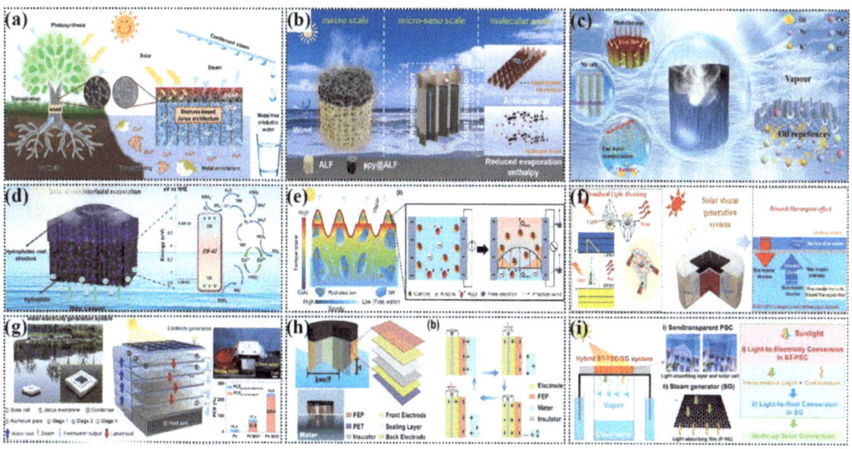

Fig. 15.6 Applications in wastewater treatment water-electricity cogeneration: **a** Janus evaporator with oriented nanostructured grid coordination for the treatment of heavy metal ions. Reprinted from Reference [80], Copyright 2022, Wiley. **b** Bionic evaporator with integrated PPy achieves antimicrobial activity against E. coli/Staphylococcus aureus. Reprinted from Reference [81], Copyright 2023, Elsevier. **c** Janus nanofibrillar cellulose aerogels with cellular networks for nanoemulsion separation. Reprinted from Reference [82], Copyright 2025, Wiley. (d) ZIF-67/TOCNF Janus Evaporator removes ciprofloxacin contaminants. Reprinted from Reference [84], Copyright 2025, Elsevier. **e** Evaporation-driven nanogenerators achieve 11.8 μW cm^{-2} continuous power output. Reprinted from Reference [85], Copyright 2022, Springer Nature. (f) CNC/EGaIn-PANI hydrogel thermoelectric generator utilizes temperature difference to generate electricity. Reprinted from Reference [86], Copyright 2022, Wiley. **g** Floating solar panel-integrated membrane distillation unit realizes hydropower cogeneration. Reprinted from Reference [87], Copyright 2024, Elsevier. **h** BC-CNF evaporator with wrapped TENG harvests wave energy for self-powered water monitoring. Reprinted from Reference [63], Copyright 2021, Wiley. **i** Semitransparent PV-interfacial steam generator enable hydroelectric co-generation. Reprinted from Reference [88], Copyright 2021, Royal Society of Chemistry

cells, piezoelectric responses, triboelectricity from water electrification, and wave energy-water-electricity cogeneration can be achieved. Particular focus should be placed on electricity generation driven by evaporation-induced water and ion flows, thermoelectric power generation based on the Seebeck effect, and electricity generation from salinity gradients. Furthermore, the electrical energy produced can enable electrothermal-driven processes during nighttime, thereby ensuring all-weather evaporation. These low-carbon approaches serve as continuous, clean, and renewable sources for energy production, thus maintaining freshwater production efficiency.

Sun et al. developed a simple yet efficient interfacial evaporation-driven nanogenerator (IENG), which achieved layered functionalization and incorporated a biomimetic light-trapping structure to enhance both photothermal and electrical energy conversion (Fig. 15.6e). Their results demonstrated a continuous power output as high as 11.8 μW cm^{-2} under optimal conditions—more than 6.8 times higher than the currently reported average values. The escape of water molecules during evaporation leads to the formation of a localized high-salt concentration zone near the interface, establishing a salinity gradient between the brine beneath the evaporator and the surrounding bulk water. This gradient can be harnessed for osmotic energy harvesting (salinity gradient power generation). Additionally, thermoelectric generators (TEGs) can utilize the temperature difference (ΔT) between the evaporator surface and the underlying water to enable efficient thermal-to-electrical energy conversion [85]. Wei et al. investigated a novel eutectic gallium-indium (EGaIn)/polyaniline (PANI) complex (EP) combined with cellulose nanocrystals (CNCs) (Fig. 15.6f). The resulting hydrogel exhibited excellent photothermal conversion properties, absorbing solar energy to create a temperature gradient between the upper membrane surface and the water interface. This temperature difference was exploited for electricity generation using a thermoelectric heating sheet, based on the Seebeck effect [86].

Furthermore, cellulose-based ISSEs can be integrated with other energy conversion systems to broaden their functionality. Leveraging the inherent hydrophilicity of cellulose, Liu et al. combined a floating solar panel with a five-stage membrane distillation (MD) unit to simultaneously achieve clean water production and electricity generation on water surfaces (Fig. 15.6g). The integrated system achieved a notable water production rate of 4.14 kg m^{-2} h^{-1} and an electricity generation efficiency of 16.4% under 1 sun. Triboelectric nanogenerators (TENGs) can harvest wave energy via contact electrification with water [87]. Li et al. designed an all-cellulose-based, 3D monolithic steam generator constructed from BC-based CNFs (Fig. 15.6h). A thin, lightweight, and flexible TENG was fabricated and appropriately wrapped around the evaporator's outer surface. This integrated system demonstrated efficient water wave sensing in a self-powered manner [63]. Ji et al. fabricated a solar-powered water-electricity generator by integrating a semitransparent photovoltaic (ST-PV) cell with an interfacial steam generator (SG) (Fig. 15.6i). The ST-PV/SG hybrid system achieved a high electrical output of 122 W m^{-2} and a stable water evaporation rate of 1.30 kg m^{-2} h^{-1} under 1 sun, maximizing total solar energy utilization with a combined efficiency of 88.8% [88].

15.5 Conclusion and Outlooks

Here, we systematically summarize the design points and the latest research progress of cellulosic nanomaterials as support substrates for solar-powered evaporation systems. At present, a variety of cellulose-based functional substrate materials have been developed, including nanofiber aerogel, aerogel paper, etc., and efficient solar evaporation systems have been constructed through structural design and material composite strategies. These cellulose-based materials are attractive for efficient solar-driven evaporation to alleviate fresh water scarcity due to their rapid water transport properties, stable binding to photothermal materials due to their rich surface functional groups, as well as their natural biocompatibility and renewability. However, although laboratory studies have fully verified their performance advantages, cellulose-based evaporators still face multiple challenges to move toward practical applications.

The typical preparation method of aerogel is the process of removing solvent from wet gel. In order to minimize the effect of solvent surface tension during drying, most aerogels are currently freeze-dried or supercritical drying. However, this method is not cost-effective, and the drying process will consume a lot of energy and increase the manufacturing cost [89, 90]. This makes the scalability of the solar evaporator limited. Therefore, the scalability and cost-effectiveness of cellulose aerogels in solar evaporator applications should be further explored.

Solar-driven evaporators are often deployed in complex water environments, which put forward strict requirements for the durability and environmental stability of materials. The practical application of cellulosic aerogels faces two challenges: first, its excellent biocompatibility is easy to induce the attachment of microorganisms to form biofilms. Cellulase-producing microorganisms (such as fungi and bacteria) in natural water will catalyse the degradation of materials and damage the structural integrity and evaporation performance. Second, the accumulation of salt crystals at the interface hinders light absorption and water transport, leading to a continuous decline in evaporation efficiency. At present, cellulose-based evaporators with salt removal capacity have been developed, but there is a lack of systematic evaluation of their pollution. Future research should focus on developing high performance evaporators resistant to pollution, especially enzymatic hydrolysis, and systematically evaluating their operational degradation and stability.

In addition, the current research on solar evaporator focuses too much on the optimization of photothermal evaporation performance of materials, but ignores the bottleneck of practical application. For instance, in practical applications, the demand for the all-weather operation of evaporation devices and the difficulty of continuous water collection in closed fresh water collection devices. The design of evaporation devices is mostly limited to small area components in laboratories, and breakthroughs are still needed for large-scale application.

Although there are still some challenges in transitioning cellulose-based solar evaporators from laboratory prototypes to actual implementation, their inherent advantages, including environmental friendliness, abundant material reserves, and performance controllability, make cellulose-based materials an excellent choice for solar-driven evaporation. In order to bridge the gap between theoretical potential and practical application, it is still necessary to synergistically optimize material modification (such as chemical modification), structural regulation (such as pore design) and system integration (such as water-energy coupling), and further promote the leap of this technology from experimental verification to industrial application.

References

1. A. Mukherji, Climate change: put water at the heart of solutions. Nature **605**, 195 (2022)

2. M. Rodell, J.S. Famiglietti, D.N. Wiese, J.T. Reager, H.K. Beaudoing, F.W. Landerer, M.H. Lo, Emerging trends in global freshwater availability. Nature **557**, 651–659 (2018)
3. M. Wang, B.L. Bodirsky, R. Rijneveld, F. Beier, M.P. Bak, M. Batool, B. Droppers, A. Popp, M.T.H. Van Vliet, M. Strokal, A triple increase in global river basins with water scarcity due to future pollution. Nat. Commun. **15**, 880 (2024)
4. F. Silveira, A.L. Martins, P. Gadelha, Paes-Sousa, quantifying convergence on health-related indicators of the 2030 agenda for sustainable development. Bull. World Health Organ. **99**, 228 (2021)
5. K. Brown, Water scarcity: forecasting the future with spotty data. Science **297**, 926–927 (2002)
6. R. Hope, Four billion people lack safe water. Science **385**, 708–709 (2024)
7. E.E. Greenwood, T. Lauber, J. Van Den Hoogen, A. Donmez, R.E. Bain, R. Johnston, T.W. Crowther, T.R. Julian, Mapping safe drinking water use in low-and middle-income countries. Science **385**, 784–790 (2024)
8. U J D S Statistics, Youth, Global indicator framework for the sustainable development goals and targets of the 2030 agenda for sustainable development, in *Developmental science sustainable development goals for children*, 439 (2019)
9. H. Nassrullah, S.F. Anis, R. Hashaikeh, N. Hilal, Energy for desalination: a state-of-the-art review. Desalination **491**, 114569 (2020)
10. K. Elsaid, M. Kamil, E.T. Sayed, M.A. Abdelkareem, T. Wilberforce, A. Olabi, Environmental impact of desalination technologies: a review. Sci. Total. Environ. **748**, 141528 (2020)
11. Y. Cai, J. Wu, S.Q. Shi, J. Li, K.-H. Kim, Advances in desalination technology and its environmental and economic assessment. J. Cleaner Prod. **397**, 136498 (2023)
12. S.M. Alawad, R.B. Mansour, F.A. Al-Sulaiman, S. Rehman, Renewable energy systems for water desalination applications: a comprehensive review. Energy Convers. Manag. **286**, 117035 (2023)
13. W. Zhang, Y. Chen, Q. Ji, Y. Fan, G. Zhang, X. Lu, C. Hu, H. Liu, J. Qu, Assessing global drinking water potential from electricity-free solar water evaporation device. Nat. Commun. **15**, 6784 (2024)
14. X. Li, W. Xie, J. Zhu, Interfacial solar steam/vapor generation for heating and cooling. Adv. Sci. **9**, e2104181 (2022)
15. F. Nawaz, Y. Yang, Q. Zhao, Y. Mo, Z. Jiang, J. Wu, Y. Liu, B. Liu, B. Gao, W. Que, Can the interfacial solar vapor generation performance be really "beyond" theoretical limit? Adv. Energy Mater. **14**, 2400135 (2024)
16. M. Zhou, L. Zhang, S. Tao, R. Li, Y. Wang, Recent research advances in efficient solar-driven interfacial evaporation. Chem. Eng. J. **489**, 151157 (2024)
17. Y. Li, Y. Shi, H. Wang, T. Liu, X. Zheng, S. Gao, J. Lu, Recent advances in carbon-based materials for solar-driven interfacial photothermal conversion water evaporation: assemblies, structures, applications, and prospective. Carbon Energy **5**, e331 (2023)
18. R. Li, M. Wu, H. Ma, Y. Zhu, H. Zhang, Q. Chen, C. Zhang, Y.J.M. Wei, A single component, single layer flexile foam evaporator with the higher efficiency for water generation. Adv. Mater. **36**, 2402016 (2024)
19. H.-T. Ren, H.-B. He, Q.-K. Zhu, T.-T. Li, C.-W. Lou, J.-H. Lin, A Janus evaporator based on Donnan effect for efficient solar thermal desalination. Desalination **586**, 117814 (2024)
20. L. Noureen, S. Zaman, W. Ali Shah, Q. Wang, M. Humayun, Q. Xu, X. Wang, Bifunctional photothermal membrane for high-temperature interfacial solar steam generation and off-grid sterilization. Chem. Eng. J. **473**, 145122 (2023)
21. S. Cao, P. Rathi, X. Wu, D. Ghim, Y.S. Jun, S.J.M. Singamaneni, Cellulose nanomaterials in interfacial evaporators for desalination: a "natural" choice. Adv. Mater. **33**, 2000922 (2021)
22. R.J. Moon, A. Martini, J. Nairn, J. Simonsen, J. Youngblood, Cellulose nanomaterials review: structure, properties and nanocomposites. Chem. Soc. Rev. **40**, 3941–3994 (2011)
23. W. Liu, K. Liu, H. Du, T. Zheng, N. Zhang, T. Xu, B. Pang, X. Zhang, C. Si, K. Zhang, Cellulose nanopaper: fabrication, functionalization, and applications. Nano-Micro Lett. **14**, 104 (2022)
24. H. Seddiqi, E. Oliaei, H. Honarkar, J. Jin, L.C. Geonzon, R.G. Bacabac, J. Klein-Nulend, Cellulose and its derivatives: towards biomedical applications. Cellulose **28**, 1893–1931 (2021)

25. J. Wang, Y. Kong, Z. Liu, H. Wang, Solar-driven interfacial evaporation: design and application progress of structural evaporators and functional distillers. Nano Energy **108**, 108115 (2023)
26. G. Maqsood, M.S. Irshad, N. Arshad, M.S. Asghar, M.A. Ali, T. Mei, X. Wang, Can the high throughput yield of solar thermal interfacial evaporation systems be beyond theoretical efficiency? J. Mater. Chem. A **13**, 8315–8336 (2025)
27. J. Chen, X. Wang, B. Wang, T. Wu, L. Zhang, K. Zhang, G. Fang, Y. Wang, Y. Zhao, G. Yang, Recent advances of bio-based hydrogel derived interfacial evaporator for sustainable water and collaborative energy storage applications. Small **20**, 2403221 (2024)
28. C. Wen, X. Li, S. Yan, J. Wen, R. Zheng, X. Wang, H. Zhao, J. Zhou, B. Sa, Z. Sun, Progress in MXene-based photothermal materials for solar-driven water evaporation and desalination. Chem. Eng. J. **510**, 161873 (2025)
29. M. Wang, J. Hu, M. Li, L. Zhang, M. Salimi, M. Amidpour, Z. Chen, Bioinspired design of photothermal anti-fouling fabrics for solar-driven sustainable seawater desalination. Nano Energy **136**, 110726 (2025)
30. M. Sun, X. Ji, M. Li, Y. Yu, K. Zhang, C. Wang, H. Yang, Anisotropic solar evaporator with low thermal conductivity for desalination, thermoelectric generation and cultivation. Sep. Purif. Technol. **369**, 133124 (2025)
31. C. Wang, K. Xu, G. Shi, D. Wei, Water skin effect and arched double-sided evaporation for boosting all-weather high salinity desalination. Adv. Energy Mater. **13**, 2300134 (2023)
32. L. Zang, L. Sun, S. Zhang, C. Finnerty, A. Kim, J. Ma, B. Mi, Nanofibrous hydrogel-reduced graphene oxide membranes for effective solar-driven interfacial evaporation and desalination. Chem. Eng. J. **422**, 129998 (2021)
33. H. Zhu, J. Yang, C. Li, Y. Zhong, X. Tian, M. Zhang, W. Huang, Mesoporous nanogel sprays as universal evaporation interface modifiers for boosting water-cluster evaporation. Adv. Mater. **37**, 2419243 (2025)
34. L. Zhou, X. Li, G.W. Ni, S. Zhu, J. Zhu, The revival of thermal utilization from the Sun: interfacial solar vapor generation. Natl. Sci. Rev. **6**, 562–578 (2019)
35. X. Li, T. Cooper, W. Xie, P.-C. Hsu, Design and utilization of infrared light for interfacial solar water purification. ACS Energy Lett. **6**, 2645–2657 (2021)
36. G. Liu, J. Xu, T. Chen, K. Wang, Progress in thermoplasmonics for solar energy applications. Phys. Rep. **981**, 1–50 (2022)
37. J. Hu, X. Yao, K. Han, Y. Ge, S. Xu, G. Bai, Boosted near-infrared photothermal conversion in rare earth ions-doped 2D SnSe nanosheets for solar-powered water evaporation systems. Small **20**, 2405742 (2024)
38. W.-Q. Zhao, Y. Liu, Z.-X. Zheng, L. Ma, K.-W. Xiong, X.-B. Chen, Q.-Q. Wang, Broad-spectrum light absorption and efficient charge transfer in plasmonic PbS/Cu_2-xS/CdS polyhedral heterojunction for improved photothermal/photocatalytic performance. Chem. Eng. J. **475**, 146136 (2023)
39. M. Kim, K. Yang, Y.S. Kim, J.C. Won, P. Kang, Y.H. Kim, B.G. Kim, Laser-induced photothermal generation of flexible and salt-resistant monolithic bilayer membranes for efficient solar desalination. Carbon **164**, 349–356 (2020)
40. J. Wang, Y. Lian, J. Shao, S. Cheng, H. Zhao, C. Zhou, G. Yu, Y. Dai, H. Zhang, A review: solar-driven water evaporation based on biomass carbon materials. Renew. Sustain. Energy Rev. **217**, 115755 (2025)
41. J. Gu, Z. Luan, X. Zhang, H. Wang, X. Cai, W. Zhan, X. Ji, J. Liang, Highly salt-resistant and efficient dynamic Janus absorber based on thermo-responsive hydroxypropyl cellulose. Mater. Horiz. **12**, 4248–4259 (2025)
42. G. Li, J. Xu, C. Xu, X. Xie, W. Chen, Y. Liu, Equilibrium and optimization of water-heat transfer in natural cellulose-based hybrid hydrogel for efficient solar desalination. Chem. Eng. J. **486**, 150398 (2024)
43. X. Wu, Y. Lu, X. Ren, P. Wu, D. Chu, X. Yang, H. Xu, Interfacial solar evaporation: from fundamental research to applications. Adv. Mater. **36**, 2313090 (2024)
44. C. Chen, M. Wang, X. Chen, X. Chen, Q. Fu, H. Deng, Recent progress in solar photothermal steam technology for water purification and energy utilization. Chem. Eng. J. **448**, 137603 (2022)

45. D. Ding, H. Wu, X. He, F. Yang, C. Gao, Y. Yin, S. Ding, A metal nanoparticle assembly with broadband absorption and suppressed thermal radiation for enhanced solar steam generation. J. Mater. Chem. A **9**, 11241–11247 (2021)
46. Y. Gao, Q. Sun, Y. Chen, X. Zhou, C. Wei, L. Lyu, A highly efficient bio-inspired 3D solar-driven evaporator with advanced heat management and salt fouling resistance design. Chem. Eng. J. **455**, 140500 (2023)
47. Y. Wang, X. Wu, X. Yang, G. Owens, H. Xu, Reversing heat conduction loss: extracting energy from bulk water to enhance solar steam generation. Nano Energy **78**, 105269 (2020)
48. J. Sun, S. Zhao, X. Wang, W. Kong, W. Li, S. Wang, S. Liu, S. Nie, Minimizing enthalpy of evaporation in solar steam generation: an emerging strategy beyond theoretical evaporation limitation. Mater. Today **80**, 619–647 (2024)
49. M. Xia, D. Cai, J. Feng, P. Zhao, J. Li, R. Lv, G. Li, L. Yan, W. Huang, Y. Li, Z. Sui, M. Li, H. Wu, Y. Shen, J. Xiao, D. Wang, Q. Chen, Biomimetic hygroscopic fibrous membrane with hierarchically porous structure for rapid atmospheric water harvesting. Adv. Funct. Mater. **33**, 2214813 (2023)
50. V. Kashyap, H. Ghasemi, Solar heat localization: concept and emerging applications. J. Mater. Chem. A **8**, 7035–7065 (2020)
51. D. Wei, C. Wang, J. Zhang, H. Zhao, Y. Asakura, M. Eguchi, X. Xu, Y. Yamauchi, Water activation in solar-powered vapor generation. Adv. Mater. **35**, 2212100 (2023)
52. M. Wang, G. Xu, Y. Wu, R.R. Gonzales, K. Xu, H. Zhao, F. Wang, Engineering hydrogels towards next-generation multifunctional interfacial solar evaporators beyond seawater desalination. Chem. Eng. J. **502**, 157988 (2024)
53. C. Gao, Y. Li, L. Lan, Q. Wang, B. Zhou, Y. Chen, J. Li, J. Guo, J. Mao, Bioinspired asymmetric polypyrrole membranes with enhanced photothermal conversion for highly efficient solar evaporation. Adv. Sci. **11**, 2306833 (2024)
54. B. Jin, T. Cui, Y. He, M. Zhang, Y. Qi, C. Ye, 3D cellulose-based solar evaporator with tunable porous structures for high steam generation. Adv. Sustain. Syst. **8**, 2300670 (2024)
55. J. Chen, Y. Xie, M. Tian, J. Tian, Z. Liang, H. Xie, M. Lv, Cellulose-derived hydrogel with low vaporization enthalpy enables highly efficient solar steam generation. Carbohydr. Polym. **356**, 123385 (2025)
56. W. Zhong, F. Gao, J. Qu, Y. Zang, Z. Jiao, Quantitative tailoring of water state of sulfonated corncob for boosting solar interfacial evaporation and salt resistance. Carbon **224**, 119087 (2024)
57. X. Wang, Y. He, X. Liu, G. Cheng, J. Zhu, Solar steam generation through bio-inspired interface heating of broadband-absorbing plasmonic membranes. Appl. Energy **195**, 414–425 (2017)
58. G. Wang, Y. Fu, X. Ma, W. Pi, D. Liu, X. Wang, Reusable reduced graphene oxide based double-layer system modified by polyethylenimine for solar steam generation. Carbon **114**, 117–124 (2017)
59. J. Wang, Y. Li, L. Deng, N. Wei, Y. Weng, S. Dong, D. Qi, J. Qiu, X. Chen, T. Wu, High-performance photothermal conversion of narrow-bandgap Ti_2O_3 nanoparticles. Adv. Mater. **29**, 1603730 (2017)
60. A. Alkhudhiri, N. Darwish, N. Hilal, Membrane distillation: a comprehensive review. Desalination **287**, 2–18 (2012)
61. C. Xu, M. Gao, X. Yu, J. Zhang, Y. Cheng, M. Zhu, Fibrous aerogels with tunable superwettability for high-performance solar-driven interfacial evaporation. Nano-Micro Lett. **15**, 64 (2023)
62. Z. Bin, Z. Chen, W. Zhou, X. Li, G. Wang, Y. Liu, Superwetting PVA/cellulose aerogel with asymmetric structure for oil/water separation and solar-driven seawater desalination. J. Hazard. Mater. **476**, 135131 (2024)
63. N. Li, L. Qiao, J. He, S. Wang, L. Yu, P. Murto, X. Li, X. Xu, Solar-driven interfacial evaporation and self-powered water wave detection based on an all-cellulose monolithic design. Adv. Funct. Mater. **31**, 2008681 (2021)
64. S. Zong, L. Zhu, F. Lei, J. Jiang, J. Duan, Low cost non freeze-drying construction strategy of cellulose aerogels for efficient solar desalination. Chem. Eng. J. **500**, 156859 (2024)

65. S. Zong, C. Feng, F. Lei, L. Zhu, J. Jiang, J. Duan, Construction of nanocellulose aerogels with environmental drying strategy without organic solvent displacement for high-efficiency solar steam generation. ACS Nano **19**, 5305–5315 (2025)
66. Y. Ko, S. Lee, J. Jang, G. Kwon, K. Lee, Y. Jeon, A. Lee, T. Park, J. Kim, J. You, Nanocellulose-based interfacial solar evaporator: integrating sustainable materials and micro-/nano-architectures for solar desalination. Adv. Funct. Mater. **35**, 2414576 (2025)
67. H. Yang, D. Li, X. Zheng, J. Zuo, B. Zhao, D. Li, J. Zhang, Z. Liang, J. Jin, S. Ju, M. Peng, Y. Sun, L. Jiang, High freshwater flux solar desalination via a 3D plasmonic evaporator with an efficient heat-mass evaporation interface. Adv. Mater. **35**, 2304699 (2023)
68. N. Li, K. Shao, J. He, S. Wang, S. Li, X. Wu, J. Li, C. Guo, L. Yu, P. Murto, J. Chen, X. Xu, Solar-powered interfacial evaporation and deicing based on a 3d-printed multiscale hierarchical design. Small **19**, 2301474 (2023)
69. Z. Wei, W. Cai, C. Cai, Y. Fu, Tree transpiration-inspired cellulose aerogel with engineered cold-evaporated surface for promoting structural stability and minimizing energy loss. Carbohydr. Polym. **328**, 121729 (2024)
70. Y. Liu, Y. Zhang, T. Liao, L. Gao, M. Wang, X. Xu, X. Yang, H. Liu, Boron nitride–nanosheet enhanced cellulose nanofiber aerogel with excellent thermal management properties. Carbohydr. Polym. **241**, 116425 (2020)
71. Z. Yang, D. Li, Y. Zhu, X. Zhu, W. Yu, K. Yang, B. Chen, Developing salt-rejecting evaporators for solar desalination: a critical review. Environ. Sci. Technol. **58**, 8610–8630 (2024)
72. M. Wu, Y. Wei, Y. Zhu, Y. Bai, Y. Wang, X. Wang, S.-H. Ho, W. Wang, R. Li, Hydrophilic polymer foam as a monolithic interfacial solar evaporator with rapid self-cleaning, high evaporation efficiency, and salt resistance. Adv. Funct. Mater. **34**, 2410729 (2024)
73. D. Wang, R. Zhu, X. Tang, J.Y. Tan, H. Li, Y. Chen, Z. Lin, X. Xia, S. Fu, Multi-bionic strategies integration in cellulose nanofiber-based metagels with strong hydrogen-bonded network for solar-driven water evaporation. Adv. Fiber Mater. **7**, 748–761 (2025)
74. Y. Liao, C. Wang, Y. Dong, Z. Miao, H.-Y. Yu, G. Chen, J. Yao, Y. Zhou, Y. Liu, Spherical design-driven scalable solar-powered water treatment with salt self-cleaning and light self-adaptivity. Adv. Funct. Mater. **34**, 2409813 (2024)
75. J. Yuan, X. Lei, C. Yi, H. Jiang, F. Liu, G.J. Cheng, 3D-printed hierarchical porous cellulose/alginate/carbon black hydrogel for high-efficiency solar steam generation. Chem. Eng. J. **430**, 132765 (2022)
76. J. Gao, K. Shao, J. Li, N. Li, S. Wang, X. Wu, P. Murto, Z. Wang, Y. Zhou, X. Xu, Foaming photothermal inks for direct-ink writing: hierarchical design and enhanced solar-powered interfacial evaporation. J. Mater. Chem. A **12**, 6592–6609 (2024)
77. J. Yu, L. Zhang, J. Gao, W. Han, R. Wang, Z. Xu, Self-assembled porous salt crystals for solar-powered crystallization. Energy Environ. Sci. **18**, 454–467 (2025)
78. H. Zhong, Q. Zhao, Z. Lai, D. Xu, Y. Ji, S. Wang, Q. Qin, Efficient and selective lithium extraction from brine water via a photothermal sandwich sieve structure. Adv. Funct. Mater. **35**, 2418358 (2025)
79. X. Chen, M. Yang, S. Zheng, F. Temprano-Coleto, Q. Dong, G. Cheng, N. Yao, H.A. Stone, L. Hu, Z.J. Ren, Spatially separated crystallization for selective lithium extraction from saline water. Nat. Water **1**, 808–817 (2023)
80. H.-Y. Zhao, W.-X. Dong, Y. Deng, L.-F. Chen, C.-F. Zhao, C.-L. Zhang, J. Zhou, Y.-F. Qu, Y.-S. Li, D.-J. Li, S.-H. Yu, Biomass-based biomimetic-oriented Janus nanoarchitecture for efficient heavy-metal enrichment and interfacial solar water sanitation. Interdiscip. Mater. **1**, 537–547 (2022)
81. J. Wang, Z. Chen, L. Feng, F. Yu, C. Ran, N. Xu, Z. Jia, C. Li, Y. Zheng, W. Shi, M. Li, Plants transpiration-inspired antibacterial evaporator with multiscale structure and low vaporization enthalpy for solar steam generation. Nano Energy **114**, 108631 (2023)
82. B. Li, Z. Bian, Z. Wang, F. Zhang, Y. Liu, W. Zhang, M. Bao, Y. Li, Advanced janus aerogels evaporator featuring with multiscale architectures and dual-layer design for efficient oily seawater purification. Adv. Funct. Mater. 2504591 (2025)

83. Z. Lu, H. Zhang, Q. Wang, M. Toivakka, C. Xu, X. Wang, Hierarchical assembly of cellulose fibrils and tannin in biocomposite foam: scalable production via oven drying and customizable metal ions release for antimicrobial activity. Small **21**, 2501776 (2025)
84. G. Tian, C. Duan, B. Zhao, K. Yang, J. Guo, Y. Wen, B. Zhou, J. Wang, Y. Ni, Janus structured cellulose-based aerogel with vertical channels and conical roof for efficient solar-driven water evaporation and pollutant degradation. Carbohydr. Polym. **360**, 123622 (2025)
85. Z. Sun, C. Han, S. Gao, Z. Li, M. Jing, H. Yu, Z. Wang, Achieving efficient power generation by designing bioinspired and multi-layered interfacial evaporator. Nat. Commun. **13**, 5077 (2022)
86. Z. Wei, Y. Wang, C. Cai, Y. Zhang, S. Guo, Y. Fu, S.C. Tan, Dual-network liquid metal hydrogel with integrated solar-driven evaporation, multi-sensory applications, and electricity generation via enhanced light absorption and Bénard–Marangoni effect. Adv. Funct. Mater. **32**, 2206287 (2022)
87. M. Liu, Y. Sun, K. Shao, N. Li, J. Li, P. Murto, Z. Wang, J. Chen, X. Xu, Synergistic solar-powered water-electricity generation: an integrated floating system on water. Nano Energy **119**, 109074 (2024)
88. Q. Ji, N. Li, S. Wang, S. Li, F. Li, L. Yu, P. Murto, X. Xu, Synergistic solar-powered water-electricity generation via rational integration of semitransparent photovoltaics and interfacial steam generators. J. Mater. Chem. A **9**, 21197–21208 (2021)
89. N.T.T. Nguyen, L.M. Nguyen, T.T.T. Nguyen, D.T.C. Nguyen, T.V. Tran, Synthesis strategies, regeneration, cost analysis, challenges and future prospects of bacterial cellulose-based aerogels for water treatment: a review. Chemosphere **362**, 142654 (2024)
90. W. Sakuma, S. Yamasaki, S. Fujisawa, T. Kodama, J. Shiomi, K. Kanamori, T. Saito, Mechanically strong, scalable, mesoporous xerogels of nanocellulose featuring light permeability, thermal insulation, and flame self-extinction. ACS Nano **15**, 1436–1444 (2021)

Chapter 16
Cellulose-Based Aerogels for Lightweight Structural Materials

Shuxue Wang, Lu Zong, and Jianming Zhang

Abstract Cellulose-based aerogels are a research hotspot as third-generation high-performance structural materials, prized for their ultra-lightweight nature, tunable mechanics, and eco-friendliness. This chapter systematically surveys recent advances in fabricating ultra-lightweight, high-strength CBAs through preparation strategies and structural design. Key techniques analyzed include: physical crosslinking (e.g., directional freezing), chemical cross-linking (e.g., esterification), and composite reinforcement (e.g., nanofillers based interfacial regulation), revealing microstructure-mechanical property relationships. It highlights structural designs (e.g., multiscale structure) and new technology for enhancing mechanics. Applications in energy-efficient buildings, aerospace, and electronics are detailed, emphasizing lightweight advantages. Finally, the review addresses scalability challenges, environmental stability limits, multifunctional synergy needs, and modeling gaps, proposing future directions involving microfluidics, machine learning, and renewable solvents to guide structural materials innovation.

Keywords Cellulose · Aerogels · Lightweight · High strength · Structural materials

16.1 Introduction

Cellulose, as the most abundant natural polymer on Earth with an annual global production exceeding 1.5×10^{12} tons, represents an ideal matrix for sustainable material design owing to its renewability [1], biodegradability, and versatile chemical modifiability. Cellulose-based aerogels (CBAs), recognized as the third-generation

S. Wang · L. Zong · J. Zhang (✉)
Key Laboratory of Rubber-Plastics, Ministry of Education/Shandong Provincial Key Laboratory of Rubber Plastics, Qingdao University of Science and Technology, Qingdao, China
e-mail: zjm@qust.edu.cn

L. Zong
e-mail: zonglu@qust.edu.cn

G. Wei and J. Zhang (eds.), *Cellulose-Based Hydrogels and Aerogels*, Lecture Notes in Nanoscale Science and Technology 41,
https://doi.org/10.1007/978-3-032-03008-5_16

aerogel materials following inorganic and synthetic polymer aerogels, were initially conceptualized by Kistler [2]. However, their scalable production was not realized until the early twenty-first century, facilitated by advancements in green solvent technologies. These materials are engineered by dissolving and regenerating cellulose or via nanoscale restructuring, thereby constructing novel lightweight solid materials featuring three-dimensional nanoporous networks. They uniquely combine the ultra-lightweight characteristics intrinsic to aerogels (with densities as low as 0.008 g/cm^3) and the inherent mechanical toughness of cellulose, thus overcoming the limitations of traditional aerogels, such as high brittleness and energy-intensive manufacturing processes [3].

The core advantages of ultra-lightweight, high-strength CBAs are manifested in three key aspects. First, their density ranges from merely 1/10–1/100 of traditional structural materials (0.008–0.25 g/cm^3) [4], enabling substantial reductions in transportation and operational energy consumption. Second, structural optimization endows them with exceptional mechanical properties, such as compressive strengths reaching 30 MPa and toughness exceeding 5 MJ/m^3, surpassing most polymer foams. Third, their tunable hierarchical pore architecture (porosity: 80–99%) confers outstanding thermal insulation (thermal conductivity: 0.023–0.035 W/(m · K)), adsorption capabilities, and damping characteristics. These combined attributes demonstrate considerable promise for applications in building, aerospace, electronic protection, and other fields demanding stringent lightweight and high-strength requirements [5].

In recent years, research in this field has exhibited three prominent trends. First, the advancement toward refined structural design, shifting from stochastic porous configurations to biomimetic oriented structures. Second, the pursuit of multifunctional integration, synergistically optimizing thermal insulation, flame retardancy, electromagnetic shielding, and mechanical properties. Third, innovations in green fabrication technologies, exemplified by closed-loop processes achieving ionic liquid recovery rates of up to 99% [6]. However, current studies predominantly concentrate on single-property optimization, with insufficient systematic analysis of the complete "structure–property–application" chain.

This chapter comprehensively addresses these aspects across four dimensions: preparation strategies, structural design, performance optimization, and application scenarios. Particular emphasis is placed on the application mechanisms of high-strength structural materials, aiming to provide theoretical guidance and technical roadmaps for designing next-generation sustainable structural materials.

16.2 Preparation Strategy: From Basic Molding to Performance Enhancement

The functional properties of porous structures assembled from cellulose are closely related to their structural design. The preparation of porous cellulose assemblies generally involves three steps: (i) dissolving or dispersing cellulose; (ii) assembling into layered structures using external stimuli; and (iii) controlled removal of liquid to obtain dry porous structures. During these processes, factors such as the size and surface chemistry of cellulose fibers, the concentration of cellulose suspension, and external stimuli and drying methods all affect the properties of the assembled structures, providing tunable functions.

16.2.1 Physical Cross-Linking

Physical cross-linking methods construct 3D networks through non-covalent interactions (Fig. 16.1a) [7], circumventing the environmental burden associated with chemical crosslinking agents, and have emerged as a mainstream green fabrication technology [8]. In physical cross-linking gelation, the individual gel components are incorporated into the CBA precursor through physical entanglement, hydrogen bonding or van der Waals interactions. This method mainly forms a gel by mixing functional pre-synthesized materials with cellulose in a solvent. The stability of the functional materials in the CBA structure prepared by physical cross-linking gelation is due to the physical entanglement or interaction between the functional materials and cellulose, forming a weak interaction force.

Among these, ice-templating enables the aligned arrangement of cellulose nanofibers (CNFs) by controlling the growth direction of ice crystals, yielding anisotropic structures [10]. A representative process involves placing a CNF suspension (concentration: 0.5 2 wt%) on a cold substrate (−20 to −196 °C). A temperature gradient induces vertically aligned ice crystal growth, extruding CNFs into aligned microchannels. Subsequent freeze-drying produces lamellar aerogels. Precise control of pore channel dimensions (10–200 μm) was achieved by regulating freezing rates (1–20 °C/min) [11]. When the freezing rate reaches 10 °C/min, the resulting aerogel exhibits over threefold greater axial compressive strength compared to disordered structures. For instance, Li et al. employed a stepwise cooling protocol (105–80 °C) combined with ultrasonic dispersion to fabricate an aerogel with oriented pores. This material demonstrated significantly lower axial thermal conductivity (0.028 W/(m · K)) than its radial counterpart (0.035 W/(m · K)), highlighting the anisotropy modulation advantages of structural design.

Solvent-induced phase separation represents another efficient physical crosslinking method. Its core principle exploits the solubility differences of cellulose in various solvents to induce network reconstruction. When using ionic liquids (e.g., [AMIM]Cl) as solvents, cellulose molecular chains fully extend. Upon adding

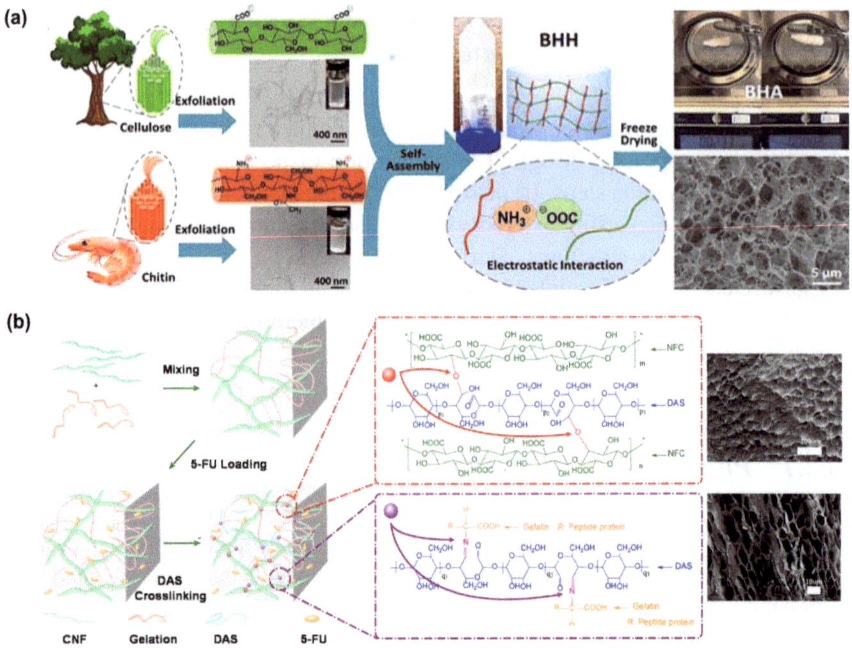

Fig. 16.1 **a** CNF/PDChNF aerogel via the electrostatic interaction between TOCNF and PDChNF. Reprinted from Ref. [7], Copyright 2019, American Chemical Society. **b** CNF/gelatin aerogel via hemiacetal/acetal and Schiff base reaction with dialdehyde starch. Reprinted from Ref. [9], Copyright 2019, American Chemical Society

a non-solvent (e.g., water), cellulose molecules undergo hydrogen-bond-driven self-assembly to form a nanofiber network [12]. Key control parameters include the solvent exchange rate (governing pore size distribution) and cellulose concentration (determining skeleton density). For instance, dissolving microcrystalline cellulose (8 wt%) in 1-butyl-3-methylimidazolium chloride ([BMIM]Cl), followed by gradient solvent exchange with ethanol, yields high-strength aerogels with specific surface areas up to 350 m^2/g. Notably, the integration of wet-spinning technology enables continuous production via physical crosslinking [13]. Ma's team dissolved cellulose (5 wt%) in an ionic liquid, extruded the solution through a spinneret into a water coagulation bath, where cellulose molecules instantaneously underwent in situ self-assembly and hydrogen-bond crosslinking. This process formed gel fibers with continuous 3D networks, which were subsequently converted via supercritical drying into aerogel fibers exhibiting toughness up to 5 MJ/m^3 [14]. This technique achieves integrated "dissolution-assembly-drying," producing fibers with controllable diameters (50–200 μm), tensile strengths of 17–30 MPa, and elongation at break of 82–110%, thereby establishing a foundation for textile-based structural materials.

16.2.2 Chemical Cross-Linking

Chemical cross-linking gelation involves chemical bonding interactions between gel functional components and cellulose. This method involves chemically cross-linking cellulose or a large number of surface-modified cellulose functional groups (such as hydroxyl, carboxyl, and methoxy) with functional components to form strong interactions [15]. This covalent cross-linking mechanism significantly improves the hydrothermal stability and mechanical properties of aerogels. Esterification crosslinking represents the most prevalent strategy, utilizing polycarboxylic acids (e.g., citric acid, 1,2,3,4-butanetetracarboxylic acid) to react with cellulose hydroxyl groups, forming ester linkages. Innovatively, Ma's team employed a citric acid (CA)-PVA hybrid system, wherein CA simultaneously acts as a crosslinker bridging cellulose molecules and forms an esterified network with PVA, establishing a ternary "cellulose-PVA-CA" crosslinked system [16]. At a CA loading of 20 wt% relative to cellulose, the aerogel exhibited a mere 14% stress loss after 500 compression cycles at 80% strain, substantially lower than uncross-linked samples (>50%) [17]. This crosslinking not only enhances mechanical performance but also confers pH responsiveness to the material, offering a new pathway for designing structurally intelligent materials.

Silane modification serves as a key technology for improving hydrophobicity and environmental stability. Methyltrimethoxysilane (MTMS) is introduced onto the aerogel surface via chemical vapor deposition (CVD), undergoing hydrolysis-condensation reactions to form a polysiloxane coating [18]. This process requires precise control of reaction humidity (40–60% RH) and duration (2–6 h)—insufficient time leads to incomplete coverage, while excessive time causes pore blockage. Optimized MTMS-modified aerogels achieve a water contact angle of 141.5° and maintain structural integrity after 24-h water immersion, with water absorption below 5% [19]. Recently, click chemistry has emerged as a novel crosslinking technique due to its efficiency and selectivity. For instance, thiol-ene reactions between cellulose derivatives containing double bonds and tetra-arm thiol-terminated poly(ethylene glycol) enable gelation within 5 min. The resulting aerogels combine high elasticity (95% resilience ratio) with self-healing capabilities, providing innovative solutions for dynamic load-bearing scenarios.

In a complementary strategy, Li and Wang et al. prepared nanocellulose–gelatin composite cryogels with tunable porosity by cross-linking with dialdehyde starch (DAS) (Fig. 16.1b) [9]. In this system, DAS's aldehyde groups react with cellulose hydroxyls to form hemiacetal and acetal bonds, and simultaneously form Schiff bases with gelatin's ε-amino groups. Compared to mere hydrogen bonding, these covalent linkages more effectively tailor the cryogel's pore architecture. Moreover, increasing the DAS content proportionally raises the density of hemiacetal/acetal and Schiff-base crosslinks, which in turn enhances the composite's water-uptake capacity.

16.2.3 Composite Enhancement Strategy: Multifunctional Synergy

Pure CBAs often fail to meet the demanding requirements of structural materials in extreme environments. Composite reinforcement strategies achieve performance breakthroughs through the incorporation of inorganic nanoparticles, polymers, or carbon materials, as shown in Fig. 16.2 [3].

Hydrogen Bonding: Cellulose nanomaterials (CNMs) feature abundant surface hydroxyl groups that form supramolecular hydrogen-bond networks, stabilizing β-1,4-glucan chains and enabling hierarchical assembly (Fig. 16.2a) [20]. These dynamic bonds enhance mechanical strength and reversibility in composites with polymers (e.g., lignin, chitosan, PVA). Three key strategies exploit multiple H-bonds: (i) Functionalized CNMs: Self-associative motifs (e.g., 2-ureido-4-pyrimidone quadruple H-bonds) enable UV-healable composites and sacrificial bonds in flexible chains for toughness [21]. (ii) External crosslinkers: Tannic acid clusters form robust supramolecular plastics via H-bond clustering. (iii) Chemical and mechanical treatments regulate bond density to achieve ordered structures. Hydrogen bond dynamics enable stimuli-responsive materials, such as hydro-setting bioplastics. Combining pH- and temperature-responsive polymers further yields smart nanocomposites with dual-responsive capabilities [22].

Electrostatic Interactions: Charged CNM surfaces enable colloidal stability and electrostatic assembly with oppositely charged polymers (e.g., chitin) or nanoparticles (Fig. 16.2b) [23]. Ionic bridging enhances toughness in CNF-polyacrylamide composites. As Pickering emulsifiers, low-charge CNCs stabilize non-spherical emulsions via interfacial jamming, while surfactant/protein adsorption (e.g., zein) improves tunability [24]. Stimuli-responsive emulsions (pH/redox-triggered) show promise in oil recovery and controlled delivery. Challenges include environmental sensitivity (pH/ionic strength) and toxicity concerns from charged polymer matrices.

Metal Coordination: Metal-carboxylate bonds (e.g., on TEMPO-CNFs) crosslink fibrils, with ion valence dictating properties (Fig. 16.2c) [25]: Trivalent ions (Fe^{3+}, Al^{3+}): Enhance stiffness/water retention. Divalent ions (Cu^{2+}, Zn^{2+}): Balance strength/swelling while adding functionalities (antimicrobial/catalytic)–Functional ligands (tannic acid) or MOFs integrate for advanced materials (filters, supercapacitors) [26]. CNM-templating addresses MOF processing limitations, while perovskite coordination boosts photovoltaic efficiency. Cost and recycling remain hurdles for industrial adoption.

Hydrophobic Interactions: Hydrophobization (acetylation/silylation) improves CNM compatibility with hydrophobic polymers (Fig. 16.2d) [27]. Optimal grafting density balances self-association (enabling shear-thinning gels) and matrix interaction. Amphiphilic CNMs bind 2D materials (graphene, MXene) via hydrophilic (110)/(1–10) planes and hydrophobic (200) planes [28], facilitating eco-friendly

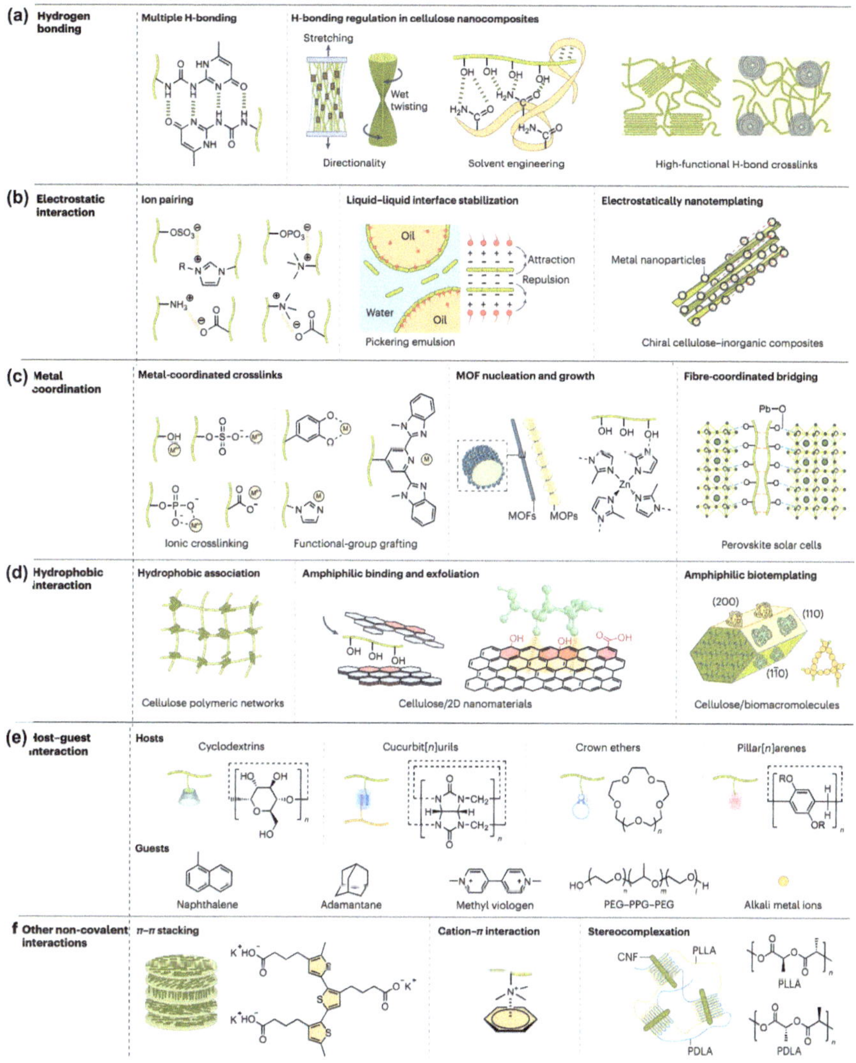

Fig. 16.2 Regulation strategies of cellulose materials: **a** hydrogen bonding, **b** electrostatic interactions, **c** chemical extraction or polymer grafting, **d** hydrophobically interaction, **e** Host–guest interactions, and **f** π–π stacking. Reprinted from Ref. [3], Copyright 2025, Springer

hydroplastics. Uniform modification without compromising biodegradability is challenging [29].

Host-Guest Interactions: Macrocyclic hosts enable dynamic composites: Cucurbit uril: Ternary complexes yield ultrahigh-water hydrogels (99.7 wt%) with multi-stimuli responsiveness and fiber processability (Fig. 16.2e) [30]. Crown ethers:

Facilitate heavy-metal adsorption; pillararenes enable pollutant sensing. Limitations include synthetic complexity and low guest-load capacity [31].

Other Non-covalent Interactions: π–π stacking: Aligns aromatic molecules for optoelectronics but risks phase separation (Fig. 16.2f). Cation–π interactions: Enhance rubber-like nanocomposites but face cation competition [32]. Stereo-complex formation: CNC-nucleated PLA improves melting temperature/packaging performance–Hydrogen bonding [33].

16.3 Structural Design: From Multi-level Construction to Bionic Innovation

16.3.1 Multi-level Structure Characterization and Performance Correlation

The exceptional properties of CBAs stem from their multi-scale structural characteristics, necessitating advanced characterization techniques to elucidate the "structure–property" relationships. In Fig. 16.3a, at the microscopic scale, scanning electron microscopy (SEM) and transmission electron microscopy (TEM) observations reveal that ideal high-strength aerogels possess continuous nanofiber skeletons (diameter: 5–20 nm) with uniform pore distribution [14, 34, 35]. The surface modification of the fibers introduced a significant difference in positive and negative charges between the two fibers, enabling the nanofibers to adsorb onto the microfiber surface and promoting the formation of high-density structures to fabricate biomimetic multi-scale cellulose-based structural materials (Fig. 16.3a, b) [36]. A team from North Carolina State University employed cryogenic electron microscopy (Cryo-EM) to capture, for the first time, the three-dimensional network reconstruction process of cellulose/graphene composite aerogels in the hydrated state (Fig. 16.3c, d) [11, 35, 37]. This revealed a graphene-mediated "bridging effect" between cellulose fibers, which enhances the material's fracture toughness by 55%. At the mesoscopic scale, small-angle X-ray scattering (SAXS) analysis demonstrates a strong correlation between pore size distribution concentration (quantified by the dispersity index D) and mechanical performance [38]. When $D < 0.3$ (indicating narrow distribution), the aerogel achieves a compressive modulus of up to 12 MPa—more than double that of broadly distributed samples.

Nanoindentation provides novel insights into micro-mechanical behavior. Ma's team utilized this technique to map the local modulus distribution of S-RCCP aerogels, identifying that silane-coated regions exhibit a significantly higher modulus (3.2 GPa) than the cellulose skeleton (1.8 GPa). This confirms the reinforcement effect of MTMS at fiber junctions. Advances in in situ characterization techniques enable dynamic performance studies [19]. For instance, in situ Raman spectroscopy monitoring during compression of cellulose/silica composite aerogels revealed a stress

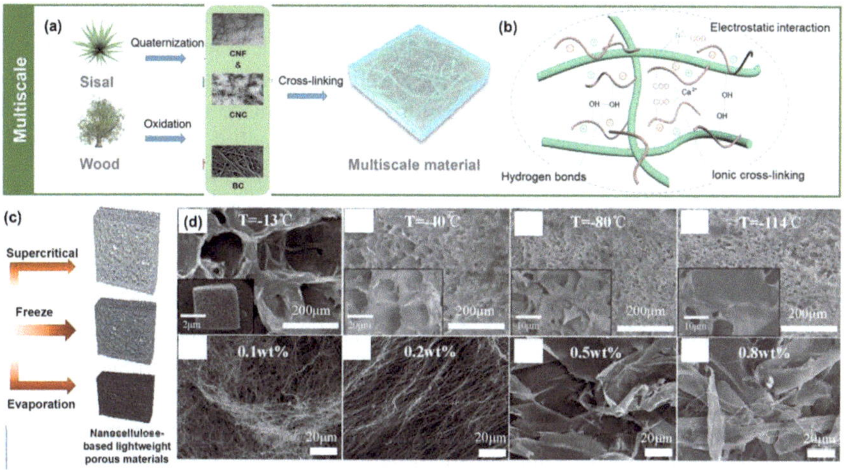

Fig. 16.3 **a** Synthesis of nanocellulose-based lightweight porous materials. Reprinted from Ref. [36], Copyright 2023, Wiley. Reprinted from Ref. [34], Copyright 2017, Springer. **b** Schematic illustration of chemical interactions between quaternized sisal (Q-sisal) microfibers and carboxylated nanofibers. Reprinted from Ref. [36], Copyright 2023, Wiley. **c** Reprinted from Ref. [35], Copyright 2020, Elsevier. **d** SEM images of freeze-dried CBAs frozen at different temperatures. Reprinted from Ref. [37], Copyright 2016, Elsevier. SEM images of freeze-dried aerogels with different CNF concentrations. Reprinted from Ref. [11], Copyright 2011, Royal Society of Chemistry

concentration factor of 2.5 around SiO_2 particles, elucidating their role as localized reinforcing points [18].

16.3.2 Interfacial Hydrogen Bonding and Interfacial Multiscale Mechanic Model

In recent years, the study of constitutive relationship and multiscale mechanical behavior of nanocellulose materials based on theoretical modeling has been gradually attracted the attention of domestic and foreign mechanics researchers. In this section, we will focus on the current status of the research on the interfacial hydrogen bonding behavior of nanocellulose, the interfacial mechanics of ontogeny and the multiscale mechanics model, and we will make a specific elaboration of the research on each of these three aspects in turn.

16.3.2.1 Interface Hydrogen Bonding Behavior

The abundant hydroxyl functional groups on the cellulose molecular chain are able to form an ordered hydrogen bonding network, which makes nanocellulose a typical material for hydrogen bonding system [39]. As a non-covalent interaction, hydrogen bonding has the property of breakable reconfiguration in deformation, and the understanding of the hydrogen bonding behavior of cellulose nanocrystals/nanofibers is the basis for studying the mechanical behavior of nanocellulose materials [40]. Intrachain hydrogen bonding supports the rigid skeleton of cellulose molecular chains and induces specific intrinsic deformations, such as bending, buckling, and torsion, under external loads [41–44]. Recent studies have shown that the cellulose hydrogen bonding conformation plays an important role in the local flexibility of the molecular chain, a finding verified by Kelvin et al. through nanoscale imaging of cellulose molecular chains [45]. Interchain hydrogen bonding, on the other hand, assembles parallel-arranged molecular chains into stable stacked structures and plays a key role in the multiscale interfacial deformation of nanocellulose materials [46]. Probing the interfacial hydrogen bonding behavior of nanocellulose from the molecular scale is a fundamental prerequisite for achieving bottom-up mechanical design of high-performance nanocellulose structured materials.

Hydrogen bonding predominantly affects the (100) crystalline plane of type-I nanocrystalline cellulose fibrils. Through molecular dynamics simulations (Fig. 16.4a), Sinko and Keten observed that adjacent 36-chain cellulose fiber models exhibit alternating alignment and misalignment of interfacial hydroxyl groups during shear slip along the (100) plane (Fig. 16.4b). Statistical analysis of hydrogen bond counts (Fig. 16.4c) demonstrates that when hydroxyl groups align, hydrogen bonds reach a maximum while the system's potential energy minimizes, enhancing structural stability; conversely, misalignment causes hydrogen bond rupture and stability degradation. The fracture-reformation nature of hydrogen bonds enables continuous interfacial slip to manifest pronounced stick–slip effects. The resulting energy dissipation provides interfacial toughness over extended slip distances [47].

Sinko and Keten further investigated interfacial water effects (Fig. 16.4d, e): water molecules on the (110) plane negligibly impact cellulose crystal separation adhesion energy. At hydrophobic (100) interfaces, however, water preferentially forms intermolecular hydrogen bonds, generating capillary bridging effects (Fig. 16.4f) that increase the adhesion energy required for separation. Under shear loading, water molecules mitigate interfacial barriers by reducing atomic friction, thereby diminishing inter-crystalline shear modulus. Collective simulations confirm that water molecules substantially alter interfacial hydrogen bonding behavior in nanocellulose, along with physical phenomena including adhesion, shear response, and stick–slip friction [50, 48].

Traditional material design often reduces component size to increase strength. For example, smaller grain diameters in metals inhibit dislocation expansion, raising strength but weakening crack-tip passivation, reducing fracture toughness. Bottom-up design using interfaces and intermolecular interactions offers a solution to this strength-toughness trade-off. Zhu et al. [49] revealed interfacial hydrogen bond fracture-remodeling in cellulose nanomaterials via MD simulations (Fig. 16.4g) and experiments. Nanocellulose film damage arises from fiber slippage. MD simulations of this slippage show a sawtooth-shaped potential energy curve (Fig. 16.4h): peaks correspond to unstable hydrogen bonds during slip, troughs to stable configurations

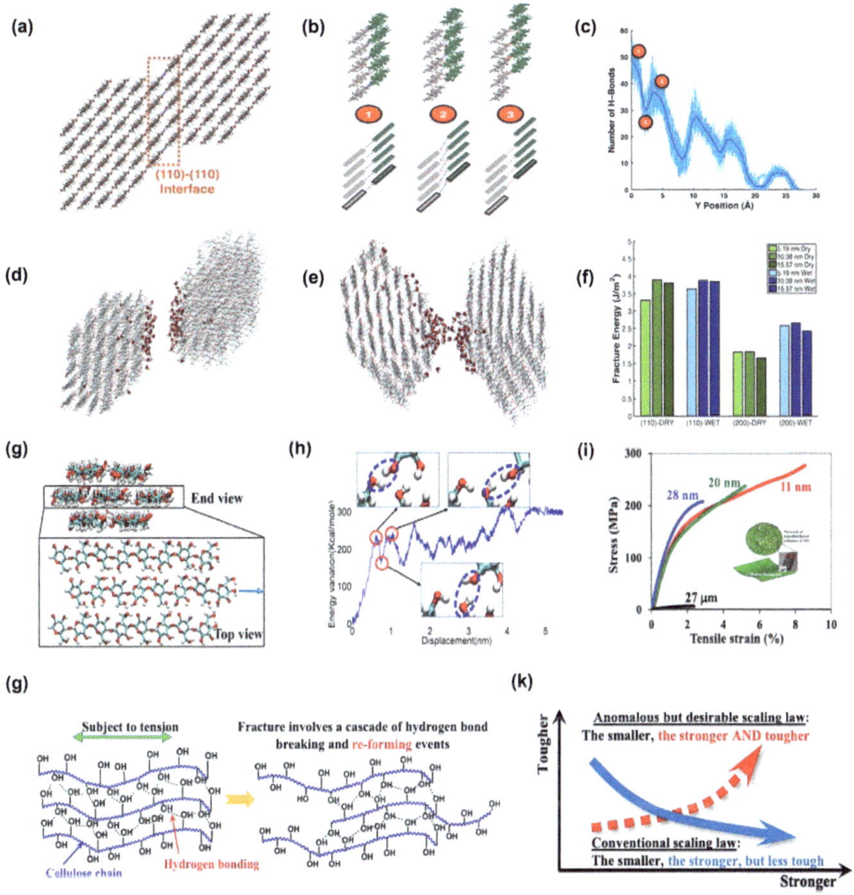

Fig. 16.4 **a** Schematic diagram of interfacial shear slip of type I β cellulose model; Evolution of hydrogen bond configuration and quantity at the interface during the **b** and **c** shearing process. Reprinted from Ref. [47], Copyright 2015, Elsevier. **d** and **e** Simulation snapshots of the cellulose model containing interfacial water molecules. **f** The influence of interfacial water molecules on interfacial separation energy. Reprinted from Ref. [48], Copyright 2014, AIP Publishing. **g** Simulation diagram of the molecular chain pull-out process in nanocellulose. **h** Evolution of interfacial hydrogen bond configuration and system potential energy during pull-out process. **i** Variation of mechanical properties of nanocellulose films with fiber diameter. **j** Schematic diagram of interfacial hydrogen bond breaking and reconstruction behavior. **k** Anomalous strength and toughness scale law of nanocellulose materials. Reprinted from Ref. [49], Copyright 2015, National Academy of Sciences

after bond breaking/reforming (Fig. 16.4j). This periodic process enhances fracture toughness through energy dissipation. Higher hydrogen bond density and more remodeling cycles further increase toughness.

Reducing nanocellulose fiber diameter from 27 μm to 11 nm increased film strength and toughness by 40× and 130×, respectively (Fig. 16.4i), due to increased fiber surface area and hydrogen bond density. Abundant interfacial hydrogen bonding thus establishes a new scaling law (Fig. 16.4k): smaller diameters simultaneously enhance strength and toughness, overcoming their traditional exclusivity.

16.3.2.2 Interface Mechanical Ontology

As noted, hydrogen bonds dominate cellulose nanomaterial interfaces. Their dynamic nature provides cohesive bonding between nanofibers while enabling efficient load transfer and energy dissipation via stick–slip behavior. To address this hydrogen-bond-mediated interfacial mechanics, researchers have developed intrinsic models to quantify multiscale deformation and toughening principles. Meng and Shi [51] proposed a shear-lag model for cellulose nanopaper. This model shows that macroscopic properties (strength, toughness, modulus) increase with smaller fiber diameter, greater length, and higher hydrogen-bond density. By linking interfacial behavior to hydrogen bond self-healing, it establishes a structure–property relationship using microstructural parameters, with predictions matching experimental data well. Furthermore, Meng et al. [52] investigated mode I fracture toughness using a modified crack-bridging model that accounts for nanofiber tilt near the crack surface. Applying cohesive laws, they revealed how fiber orientation affects toughness, establishing a quantitative angle-toughness relationship for designing strong, tough materials.

Meng et al. effectively model nonlinear deformation via hydrogen bond rearrangement, but neglect hydrogen bond spatial distribution effects on interface stacking. Metric interfaces inherently resist slip more than non-metric ones. Even with identical cross-linking, stacking differences yield divergent mechanical responses. Consequently, He et al. [53] performed MD shear simulations on cellulose nanocrystals and graphene oxide, correlating non-covalent interaction types, spatial distributions, and stacking configurations via intrinsic interface relationships. Cellulose nanocrystals exhibit a sawtooth shear profile, characteristic of functionalized metric interfaces. These display size-dependent shear lag: increasing fiber length triggers sequential deformation homogenization, localization, and kinking. He et al. adapted the Frenkel-Kontorova model to describe this size effect, revealing non-metric interfaces outperform metric ones in load transfer beyond critical fiber lengths with sufficient roughness. Using a "brick-and-mortar" shear-lag analysis, they defined a slip-cycle isomorphism. Integrating this with general shear-lag models established a multiscale theory for non-covalent interfaces. This framework describes nonlinear deformation in nanofibrillar cellulose, nacre, protein crystals, and graphene oxide, linking interface deformation modes, key characteristic parameters, and bulk mechanics to reveal a universal toughening mechanism in laminated nanocomposites.

16.3.2.3 Multiscale Mechanical Modeling

Resolving the strength-toughness trade-off remains a key challenge in advanced structural materials design. Zhu et al. proposed a bottom-up strategy for cellulose nanopaper, achieving simultaneous strength-toughness enhancement. Molecular dynamics (MD) simulations revealed an anomalous scaling law governed by hydrogen-bonded interfacial fracture-reconstruction. While they quantified toughness via energy absorption per unit volume (stress–strain integral), this metric fails to capture crack sensitivity and damage tolerance. A fracture mechanics approach is essential to elucidate toughening mechanisms and construct multiscale models.

Meng et al. [54] developed a multiscale crack-bridging model incorporating hydrogen-bond reconstruction. Their analysis showed that the stress intensity factor combines intrinsic fracture toughness with crack-tip bridging stress (for cracks perpendicular to nanofibers). Fracture toughness scales proportionally with fiber bridging length and interfacial shear strength, but inversely with fiber diameter. This model unifies microstructure-property relationships and aligns with experiments. Additionally, Chen et al. [55] measured fracture cohesion length in cellulose paper, finding that larger fiber length-to-diameter ratios reduce crack sensitivity. Zhang et al. [56, 57] analyzed moisture-induced anisotropic weakening in wood cell walls using molecular models.

Beyond hydrogen-bond-driven interfacial self-healing and damage mechanisms during cellulose nanofiber slip-pullout, microscale fiber motion under load also critically influences material mechanics. Cellulose nanomaterials' complex microstructure and size effects at the mesoscale challenge atomistic simulations [58]. Consequently, Qin et al. [59], Natarajan and Gilman [60] developed coarse-grained (CG) cellulose nanocrystal models, enabling efficient study of these microstructures. Shishehbor et al. [61] later used this CG approach to identify how fiber length, orientation, and interactions govern mechanics and nanocrystalline film failure. Ray et al. [62] employed stepwise multiscale CG modeling on random microfiber networks, revealing tensile-induced deformation localization via inter-fiber sliding causes cellulose paper damage. Zhang et al. [63] created a CG model of onion epidermal cell walls to probe cellulose-polysaccharide interactions underlying plant cell wall strength and ductility across scales. Their model's noncovalent bonds form bundled networks mimicking primary cell walls, where stress-dependent plasticity arises from oriented fiber motion (stretch, bend, slip, reorientation). This provides quantitative strategies for designing stiff, ductile, and elastic multifunctional fiber materials.

16.4 Applications of High-Strength Structural Materials

16.4.1 Building Energy Conservation and Safety Protection

In the field of building energy efficiency, CBAs represent a new generation of green building materials, distinguished by their ultra-low thermal conductivity (0.023–0.035 W/(m · K)) and tunable mechanical properties. Their primary application lies in exterior wall insulation systems, where they offer significant advantages over conventional materials. For instance, CNF/Fe_3O_4 structural materials—prepared by integrating CNF with 0D functional building blocks of Fe_3O_4 inorganic nanoparticles—effectively resolve impedance mismatch through gradient structural design (Fig. 16.5a) [64]. This design substantially reduces incident wave reflection at interfaces while preserving the material's high mechanical performance, achieving exceptional electromagnetic wave absorption ($\approx$59.5 dB absorption capacity, 5.20 GHz bandwidth) (Fig. 16.5b) [65].

Similarly, Li et al. [66] developed a novel radiative cooling material by incorporating inorganic microspheres, specifically integrating cellulose with SiO_2 particles. This material demonstrates radiative cooling effects capable of significantly reducing indoor temperatures (Fig. 16.5c). Integrating functional 1D materials like CNTs enables the balancing of contradictory properties via heterogeneous double-network structures [67]. In CNF/CNT systems, the CNT network confers excellent electrical conductivity but lacks mechanical strength due to weak interactions. Conversely, CNF forms a dense hydrogen-bond network providing high mechanical strength. Thus, constructing a heterogeneous double network of CNT and CNF achieves simultaneous high electrical and mechanical performance. Relative to other materials (Fig. 16.5d), these CNF/CNT structures exhibit superior electromagnetic interference (EMI) shielding efficiency (100 dB) at equivalent thicknesses, positioning them as competitive candidates for next-generation EMI shielding applications [64].

High flammability remains a challenge for CNF-based structural materials in construction. Combining CNF with 2D nanoclay to form a highly ordered brick-and-mortar structure effectively prevents oxygen diffusion, yielding nonflammable materials with 100% limiting oxygen index (LOI) (Fig. 16.5e) [64]. Comparative burning tests using butane blowtorch flames demonstrate this: a pine wood house model ignites upon contact, burns rapidly, and collapses within seconds. Conversely, the CNF/nanoclay composite structure remains non-combustible and retains structural integrity under identical conditions (Fig. 16.5a).

Overall, the bottom-up assembly strategy for integrating CNF with inorganic nanoscale building blocks demonstrates robustness and scalability. No fundamental scientific or technical barriers impede extending this approach to broader material systems. Rapid progress is anticipated in developing biomimetic CNF-based structural materials with integrated structure–function properties, thereby expanding CNF's application potential.

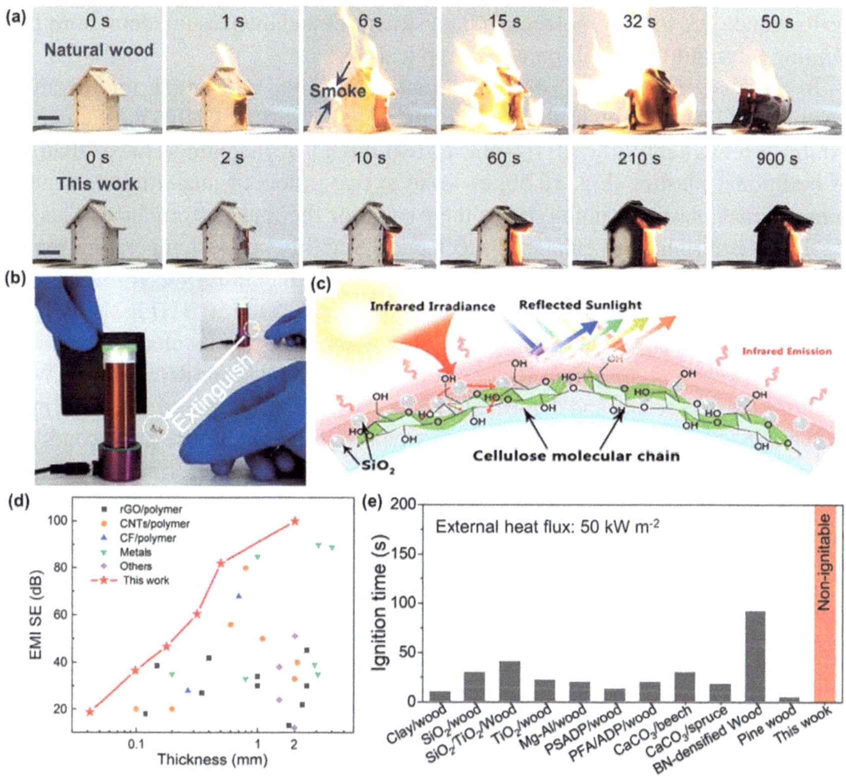

Fig. 16.5 **a** Combustion experiments on CNF/nanoclay structural materials and pine wood under a butane blowtorch flame. Reprinted from Ref. [64], Copyright 2023, Elsevier. **b** Demonstration of electromagnetic wave absorption in CNF-based structural materials. Reprinted from Ref. [65], Copyright 2024, American Chemical Society. **c** Schematic diagram of the radiative cooling principle of cellulose composite SiO_2 structural materials. Reprinted from Ref. [66], Copyright 2021, American Chemical Society. **d** Comparison of EMI SE between CNF/CNT and reported shielding materials. Reprinted from Ref. [64], Copyright 2023, Elsevier. **e** Comparison of the limit of the oxygen index (LOI) between CNF/nanoclay structural materials and different types of natural wood. Reprinted from Ref. [64], Copyright 2023, Elsevier

16.4.2 Aerospace Lightweight Structure

The aerospace sector imposes stringent requirements for lightweight design and multifunctional integration, areas where CBAs exhibit distinct advantages. A primary application lies in Thermal Protection Systems (TPS). Thermal dimensional stability—a critical parameter in material design and application—quantifies dimensional changes with temperature variations, typically measured by the coefficient of thermal expansion (CTE) [68]. This property is essential in fields demanding precise dimensional control, such as optical devices, electronic components, and aerospace engineering. However, petroleum-derived plastics often exhibit inadequate thermal performance due to their molecular chain structures and interchain interactions [69], limiting their utility in high-precision applications.

In contrast, CNF-based structural materials offer substantial improvements (Fig. 16.6a) [70]. Owing to the intrinsic thermal stability of CNFs, these materials exhibit a remarkably low CTE—one to two orders of magnitude lower than that of traditional plastics (Fig. 16.6b, c)—one to two orders of magnitude lower than conventional plastics—highlighting their superior thermal performance. This low CTE also contributes to enhanced mechanical stability across temperature ranges. Dynamic mechanical analysis reveals that the storage modulus of CNF-based materials changes minimally as temperature increases from 25 to 150 °C (Fig. 16.6d) [70]. Conversely, most petroleum-derived plastics undergo substantial modulus reduction with rising temperatures, often accompanied by significant shape deformation. While common plastics soften and deform markedly at 200 °C, CNF-based structural materials retain structural integrity and dimensional stability under the same conditions (Fig. 16.6e) [36]. This underscores their stable mechanical performance and durability across diverse temperatures, effectively overcoming the thermal limitations inherent in petroleum-derived plastics.

In addition to the thermal properties discussed above, CNF-based structural materials exhibit exceptional mechanical properties including high strength and toughness [71]. Owing to CNF's superior nanoscale mechanical performance and micro-nano structural design, these materials achieve strength and toughness values approximately twice and five times greater, respectively, than those of most plastics [36]. Ashby plots clearly demonstrate that CNF-based structural materials surpass conventional petrochemical-derived engineering plastics in these properties (Fig. 16.6f, g) [71]. Furthermore, CNF-based materials display excellent impact resistance and hardness, indicating greater robustness than typical plastics (Fig. 16.6h, i) [36].

Notably, CNF—whether plant-derived or bacterially synthesized—represents one of Earth's most abundant renewable resources and is soil-biodegradable. By crosslinking CNF hydrogels with calcium ions or incorporating biodegradable polymers to enhance bonding, the resulting structural materials retain their degradability [72]. Consequently, producing and implementing CNF-based structural materials shows significant promise for reducing the energy consumption and environmental impacts associated with traditional petrochemical plastics (Fig. 16.6a), thereby contributing to sustainable development challenges.

16.4.3 Electronic Equipment Protection and Electromagnetic Management

Driven by advancements in 5G/6G technology, electronic devices require enhanced electromagnetic compatibility (EMC) and thermal management solutions. Nanocellulose offers excellent dispersibility and facilitates efficient interactions with functional nanomaterials, enabling straightforward construction of macrostructures. Key

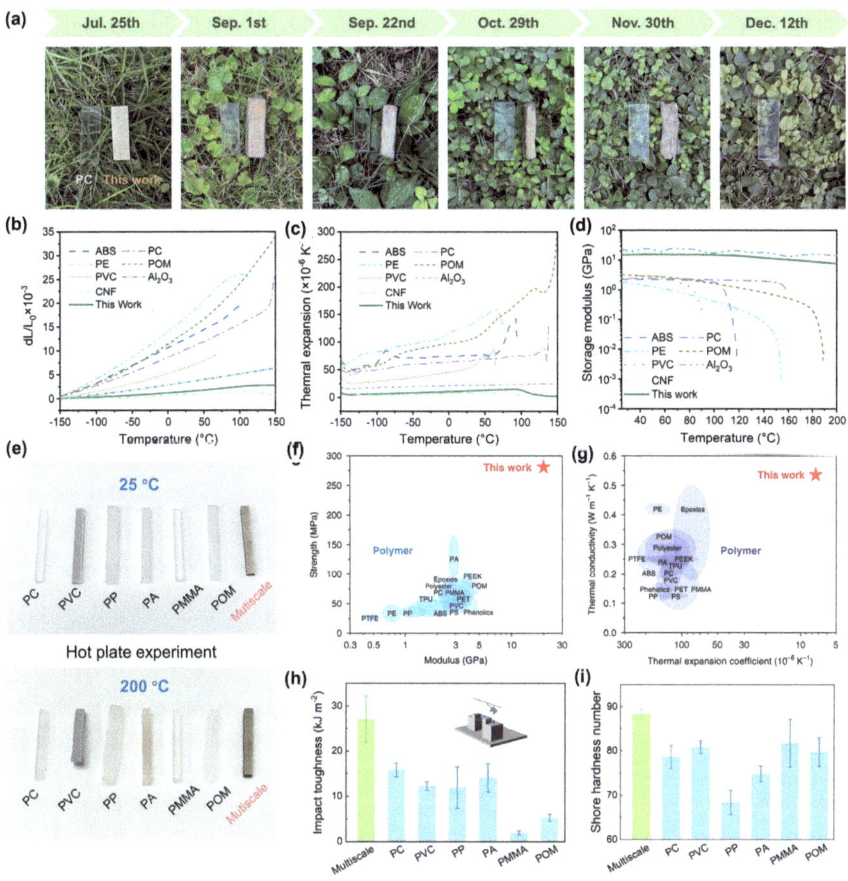

Fig. 16.6 **a** Degradation experiment of CNF-based structural materials and PC in the soil. Reprinted from Ref. [70], Copyright 2024, Wiley. **b–d** Comparison of the thermal expansion, coefficient of thermal expansion, and storage modulus between CNF based wave-absorbing structural materials and various plastics and Al_2O_3 ceramics. Reprinted from Ref. [70], Copyright 2024, Wiley. **e** Comparison of multiscale structural materials with plastics via thermal stability experiment. Reprinted from Ref. [36], Copyright, 2023 Wiley. **f, g** Comparison of the mechanical and thermal properties of CNF structural material with those of common polymers, metals, and ceramics. Reprinted from Ref. [71], Copyright 2024, Springer Nature. **h, i** Comparison of the impact toughness and Shore hardness of multiscale structural materials with those of various plastics. Reprinted from Ref. [36], Copyright, 2023 Wiley

properties include high strength-to-weight ratio, small diameter, large aspect ratio, high specific surface area, and abundant hydrophilic functional groups (such as –OH and –COOH). Chen et al. [73], Wu et al. [74] utilized one-dimensional CNFs to manufacture ultralow-density MXene aerogels via ice-templated freeze-casting (Fig. 16.7a, b). The longitudinal plane clearly shows oriented cell walls and unidirectional pore channels with ~20 μm spacing (Fig. 16.7c), while the transverse plane reveals an isotropic structure in the MXene/CNF hybrid aerogels (Fig. 16.7d). By adjusting the cell wall orientation relative to the electric field direction of incident electromagnetic waves (EMWs), the authors created specific angles (Fig. 16.7e, f). Here, 0° denotes alignment between the oriented cell walls/ pore channels and the EMW electric field. Figure 16.7g shows the measured EMI shielding effectiveness (SE) values for these hybrid aerogels (density: 4 mg/cm^3). These aerogels achieved outstanding performance: EMI SE up to 74.6 dB, SSE (specific SE) up to 30,660 $dB \cdot cm^3 \cdot g^{-1}$, and SSE/t (SSE normalized by thickness) up to 189,400 $dB \cdot cm^2 \cdot g^{-1}$ (Fig. 16.7h–j), surpassing previously reported MXene-based and other shielding designs [75]. These findings demonstrate that the highly oriented biomimetic cell walls enable precise tuning of EMI shielding efficacy through controlled angular orientation relative to the incident EMW field. This novel phenomenon agrees with finite element analysis modeling and allows optimization of EMI shielding performance without altering the aerogel framework materials.

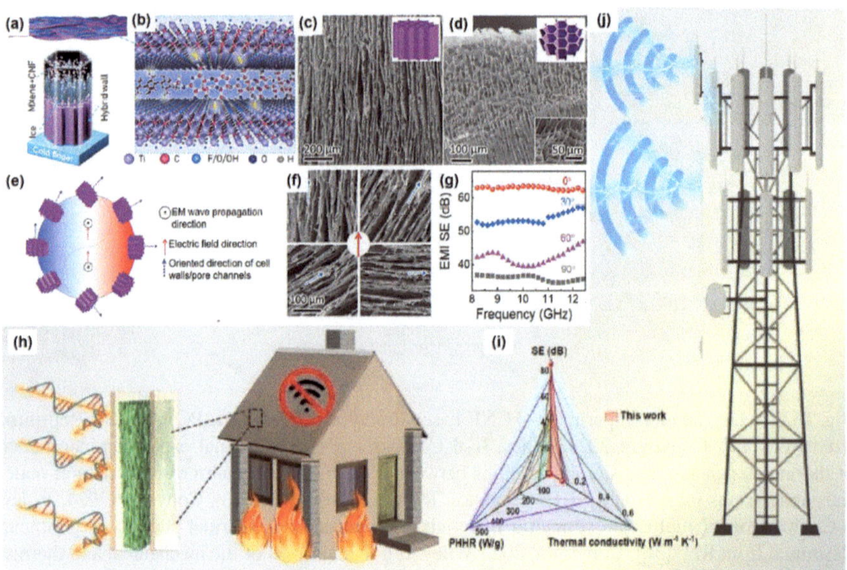

Fig. 16.7 **a** Schematic representation of the assembled MXene/CNF hybrid cell walls. **b** Diagram of the MXene, CNFs, and noncovalent interactions. SEM images of the MXene/CNF hybrid aerogels in the **c** longitudinal plane and **d** transverse plane. **e** The orientation-induced EMI shielding process. **f** SEM pictures of MXene/CNF hybrid aerogels at different angles (0°, 30°, 60°, and 90°). **g** EMI SE of MXene/CNF hybrid aerogels. Reprinted from Ref. [74], Copyright 2023, Elsevier. **h–j** Flame retardant and heat insulation multifunctional integrated EMI shielding material. Reprinted from Ref. [75], Copyright 2024, Springer Nature

16.5 Conclusion

The evolution of CBAs from ultra-light porous materials to high-performance structural materials has undergone a paradigm shift from "passive reinforcement" to "active design" [76]. This chapter systematically reviews advancements in preparation strategies, structural design, and applications, leading to the following key conclusions. Innovations in preparation technologies form the foundation for performance breakthroughs. Physical cross-linking methods (e.g., directional freezing, wet spinning) enable green fabrication; chemical cross-linking (esterification, silylation) enhances stability [77]; and composite reinforcement (inorganic hybridization, polymer interpenetration, carbonization) endows multifunctionality. Notably, wet spinning combined with in-situ self-assembly technology has successfully prepared aerogel fibers with toughness >5 MJ/m^3, opening new avenues for textile structural materials [65].

Advances in structural design lie at the core of high performance. Bio-inspired load-bearing column structures increase cyclic compressive performance to a loss rate of <10% after 500 cycles [78]; ligned pore channel design achieves an axial strength of 18 MPa (2.1 times that of the radial direction) [71]; and gradient structures satisfy thermo-mechanical synergistic protection under a 300 °C temperature difference. These structural innovations have pushed material performance close to theoretical limits [65].

Expansion of application scenarios validates their practical value. In the construction sector, dual-function materials integrating thermal insulation and self-heating have achieved an energy-saving rate >35% [79]; aerospace thermal protection systems withstand 800 °C high temperatures with a 60% weight reduction; and electronic protection materials combine electromagnetic interference shielding (SE > 45 dB) with impact protection (reducing drop damage rate by 70%) [78]. These applications demonstrate the maturity of CBAs as next-generation structural materials.

Future research could be focused on four key directions: microfluidic continuous production to address scale-up bottlenecks [80]; AI-driven design to accelerate material innovation; renewable solvent systems to achieve full-life cycle greening; and closed-loop recycling technologies to build a sustainable material ecosystem. With breakthroughs in these technologies, CBAs are poised to replace 20% of petroleum-based structural foams by 2030, providing material solutions for carbon neutrality goals [5].

References

1. F. Schué, *Biopolymers from renewable resources*, ed. by D.L. Kaplan (Springer, Heidelberg, 1998), pp. 417, Price DM278.00 ISBN 3-540-63567-X, Polym. Int. **49**, 472–473 (2000)
2. S. Kistler, Coherent expanded aerogels and jellies. Nature **127**, 741 (1931). https://doi.org/10.1038/127741a0

3. L. Chen, L. Yu, L. Qi, S.J. Eichhorn, A. Isogai, E. Lizundia, J.Y. Zhu, C. Chen, Cellulose nanocomposites by supramolecular chemistry engineering, Nat. Rev. Mater. (2025).
4. A. Isogai, Emerging nanocellulose technologies: recent developments. Adv. Mater. **33**, 2000630 (2021)
5. H.-B. Yang, X. Yue, Z.-X. Liu, Q.-F. Guan, S.-H. Yu, Emerging sustainable structural materials by assembling cellulose nanofibers. Adv. Mater. **37**, 2413564 (2025)
6. E.A. Appel, J. del Barrio, X.J. Loh, O.A. Scherman, Supramolecular polymeric hydrogels. Chem. Soc. Rev. **41**, 6195–6214 (2012)
7. J. Li, Y. Wang, L. Zhang, Z. Xu, H. Dai, W. Wu, Nanocellulose/gelatin composite cryogels for controlled drug release. ACS Sustain. Chem. Eng. **7**, 6381–6389 (2019)
8. C. Jiménez-Saelices, B. Seantier, B. Cathala, Y. Grohens, Spray freeze-dried nanofibrillated cellulose aerogels with thermal superinsulating properties. Carbohydr. Polym. **157**, 105–113 (2017)
9. X. Zhang, I. Elsayed, C. Navarathna, G.T. Schueneman, E.I.B. Hassan, Biohybrid hydrogel and aerogel from self-assembled nanocellulose and nanochitin as a high-efficiency adsorbent for water purification. ACS Appl. Mater. Interfaces. **11**, 46714–46725 (2019)
10. T. Zhang, Y. Zhang, X. Wang, S. Liu, Y. Yao, Characterization of the nano-cellulose aerogel from mixing CNF and CNC with different ratio. Mater. Lett. **229**, 103–106 (2018)
11. W. Chen, H. Yu, Q. Li, Y. Liu, J. Li, Ultralight and highly flexible aerogels with long cellulose I nanofibers. Soft Matter **7**, 10360–10368 (2011)
12. T. Saito, T. Uematsu, S. Kimura, T. Enomae, A. Isogai, Self-aligned integration of native cellulose nanofibrils towards producing diverse bulk materials. Soft Matter **7**, 8804–8809 (2011)
13. N. Lin, C. Bruzzese, A. Dufresne, TEMPO-oxidized nanocellulose participating as crosslinking aid for alginate-based sponges. ACS Appl. Mater. Interfaces **4**, 4948–4959 (2012)
14. D. Ciftci, A. Ubeyitogullari, R.R. Huerta, O.N. Ciftci, R.A. Flores, M.D.A. Saldaña, Lupin hull cellulose nanofiber aerogel preparation by supercritical CO_2 and freeze drying. J. Supercrit. Fluids **127**, 137–145 (2017)
15. J. Zhao, C. Lu, X. He, X. Zhang, W. Zhang, X. Zhang, Polyethylenimine-grafted cellulose nanofibril aerogels as versatile vehicles for drug delivery. ACS Appl. Mater. Interfaces **7**, 2607–2615 (2015)
16. H. Cheng, Y. Li, B. Wang, Z. Mao, H. Xu, L. Zhang, Y. Zhong, X. Sui, Chemical crosslinking reinforced flexible cellulose nanofiber-supported cryogel. Cellulose **25**, 573–582 (2018)
17. L. Guo, Z. Chen, S. Lyu, F. Fu, S. Wang, Highly flexible cross-linked cellulose nanofibril sponge-like aerogels with improved mechanical property and enhanced flame retardancy. Carbohydr. Polym. **179**, 333–340 (2018)
18. X. Yang, K. Shi, I. Zhitomirsky, E.D. Cranston, Cellulose nanocrystal aerogels as universal 3D lightweight substrates for supercapacitor materials. Adv. Mater. **27**, 6104–6109 (2015)
19. Z. Zhang, G. Sèbe, D. Rentsch, T. Zimmermann, P. Tingaut, Ultralightweight and flexible silylated nanocellulose sponges for the selective removal of oil from water. Chem. Mater. **26**, 2659–2668 (2014)
20. B.J.B. Folmer, R.P. Sijbesma, R.M. Versteegen, J.A.J. van der Rijt, E.W. Meijer, Supramolecular polymer materials: chain extension of telechelic polymers using a reactive hydrogen-bonding synthon. Adv. Mater. **12**, 874–878 (2000)
21. J.R. McKee, J. Huokuna, L. Martikainen, M. Karesoja, A. Nykänen, E. Kontturi, H. Tenhu, J. Ruokolainen, O. Ikkala, Molecular engineering of fracture energy dissipating sacrificial bonds into cellulose nanocrystal nanocomposites. Angew. Chem. Int. Ed. **53**, 5049–5053 (2014)
22. H. Sun, X. Fang, Y. Zhu, Z. Yu, X. Lu, J. Sun, Highly tough, degradable, and water-resistant bio-based supramolecular plastics comprised of cellulose and tannic acid. J. Mater. Chem. A **11**, 7193–7200 (2023)
23. J. Yang, X. Zhang, M. Ma, F. Xu, Modulation of assembly and dynamics in colloidal hydrogels via ionic bridge from cellulose nanofibrils and poly(ethylene glycol). ACS Macro Lett. **4**, 829–833 (2015)

24. L. Bai, W. Xiang, S. Huan, O.J. Rojas, Formulation and stabilization of concentrated edible oil-in-water emulsions based on electrostatic complexes of a food-grade cationic surfactant (ethyl lauroyl arginate) and cellulose nanocrystals. Biomacromolecules **19**, 1674–1685 (2018)
25. H. Dong, J.F. Snyder, K.S. Williams, J.W. Andzelm, Cation-induced hydrogels of cellulose nanofibrils with tunable moduli. Biomacromolecules **14**, 3338–3345 (2013)
26. J. Yan, A.M. Abdelgawad, M.E. El-Naggar, O.J. Rojas, Antibacterial activity of silver nanoparticles synthesized In-situ by solution spraying onto cellulose. Carbohydr. Polym. **147**, 500–508 (2016)
27. E.A. Appel, M.W. Tibbitt, M.J. Webber, B.A. Mattix, O. Veiseh, R. Langer, Self-assembled hydrogels utilizing polymer–nanoparticle interactions. Nat. Commun. **6**, 6295 (2015)
28. R. Xiong, H.S. Kim, S. Zhang, S. Kim, V.F. Korolovych, R. Ma, Y.G. Yingling, C. Lu, V.V. Tsukruk, Template-guided assembly of silk fibroin on cellulose nanofibers for robust nanostructures with ultrafast water transport. ACS Nano **11**, 12008–12019 (2017)
29. Y. Zhou, C. Chen, S. Zhu, C. Sui, C. Wang, Y. Kuang, U. Ray, D. Liu, A. Brozena, U.H. Leiste, N. Quispe, H. Guo, A. Vellore, H.A. Bruck, A. Martini, B. Foster, J. Lou, T. Li, L. Hu, A printed, recyclable, ultra-strong, and ultra-tough graphite structural material. Mater. Today **30**, 17–25 (2019)
30. Y. Wu, D.U. Shah, C. Liu, Z. Yu, J. Liu, X. Ren, M.J. Rowland, C. Abell, M.H. Ramage, O.A. Scherman, Bioinspired supramolecular fibers drawn from a multiphase self-assembled hydrogel. Proc. Natl. Acad. Sci. **114**, 8163–8168 (2017)
31. X. Tan, H. Yang, X. Ran, Z. Li, L. Zhang, W. Gao, X. Zhou, G. Du, L. Yang, Pillar[6]arene-modified gold nanoparticles grafted on cellulose nanocrystals for the electrochemical detection of bisphenol A. New J. Chem. **45**, 14126–14133 (2021)
32. B.E. Risteen, A. Blake, M.A. McBride, C. Rosu, J.O. Park, M. Srinivasarao, P.S. Russo, E. Reichmanis, Enhanced alignment of water-soluble polythiophene using cellulose nanocrystals as a liquid crystal template. Biomacromolecules **18**, 1556–1562 (2017)
33. R.K. Layek, V.S. Parihar, J. Seppälä, A. Efimov, S. Palola, M. Kanerva, S. Annurakshita, M. Kellomäki, E. Sarlin, Reduced graphene oxide integrated poly(ionic liquid) functionalized nano-fibrillated cellulose composite paper with improved toughness, ductility and hydrophobicity. Mater. Adv. **2**, 948–952 (2021)
34. Q. Yao, B. Fan, Y. Xiong, C. Jin, Q. Sun, C. Sheng, 3D assembly based on 2D structure of cellulose nanofibril/graphene oxide hybrid aerogel for adsorptive removal of antibiotics in water. Sci. Rep. **7**, 45914 (2017)
35. Y. Sun, Y. Chu, W. Wu, H. Xiao, Nanocellulose-based lightweight porous materials: a review. Carbohydr. Polym. **255**, 117489 (2021)
36. X. Yue, H.-B. Yang, Z.-M. Han, Y.-X. Lu, C.-H. Yin, X. Zhao, Z.-X. Liu, Q.-F. Guan, S.-H. Yu, Tough and moldable sustainable cellulose-based structural materials via multiscale interface. Engineering **36**, 2306451 (2024)
37. F. Martoïa, T. Cochereau, P.J.J. Dumont, L. Orgéas, M. Terrien, M.N. Belgacem, Cellulose nanofibril foams: links between ice-templating conditions, microstructures and mechanical properties. Mater. Des. **104**, 376–391 (2016)
38. G. Zhu, H. Xu, A. Dufresne, N. Lin, High-adsorption, self-extinguishing, thermal, and acoustic-resistance aerogels based on organic and inorganic waste valorization from cellulose nanocrystals and red mud. ACS Sustain. Chem. Eng. **6**, 7168–7180 (2018)
39. R.J. Moon, A. Martini, J. Nairn, J. Simonsen, J. Youngblood, Cellulose nanomaterials review: structure, properties and nanocomposites. Chem. Soc. Rev. **40**, 3941–3994 (2011)
40. T. Li, EML webinar overview: advanced materials toward a sustainable future—mechanics design. Extreme Mech. Lett. **42**, 101107 (2021)
41. L. Bu, M.E. Himmel, M.F. Crowley, The molecular origins of twist in cellulose I-beta. Carbohydr. Polym. **125**, 146–152 (2015)
42. T. Dumitrică, Intrinsic twist in Iβ cellulose microfibrils by tight-binding objective boundary calculations. Carbohydr. Polym. **230**, 115624 (2020)
43. P.N. Ciesielski, R. Wagner, V.S. Bharadwaj, J. Killgore, A. Mittal, G.T. Beckham, S.R. Decker, M.E. Himmel, M.F. Crowley, Nanomechanics of cellulose deformation reveal molecular defects that facilitate natural deconstruction. Proc. Natl. Acad. Sci. **116**, 9825–9830 (2019)

44. C.M. Altaner, L.H. Thomas, A.N. Fernandes, M.C. Jarvis, How cellulose stretches: synergism between covalent and hydrogen bonding. Biomacromolecules **15**, 791–798 (2014)
45. K. Anggara, Y. Zhu, G. Fittolani, Y. Yu, T. Tyrikos-Ergas, M. Delbianco, S. Rauschenbach, S. Abb, P.H. Seeberger, K. Kern, Identifying the origin of local flexibility in a carbohydrate polymer. Proc. Natl. Acad. Sci. **118**, e2102168118 (2021)
46. U. Ray, S. Zhu, Z. Pang, T. Li, Mechanics design in cellulose-enabled high-performance functional materials. Adv. Mater. **33**, 2002504 (2021)
47. R. Sinko, S. Keten, Traction–separation laws and stick–slip shear phenomenon of interfaces between cellulose nanocrystals. J. Mech. Phys. Solids **78**, 526–539 (2015)
48. R. Sinko, S. Keten, Effect of moisture on the traction-separation behavior of cellulose nanocrystal interfaces, Appl. Phys. Lett. **105** (2014)
49. H. Zhu, S. Zhu, Z. Jia, S. Parvinian, Y. Li, O. Vaaland, L. Hu, T. Li, Anomalous scaling law of strength and toughness of cellulose nanopaper. Proc. Natl. Acad. Sci. **112**, 8971–8976 (2015)
50. C. Zhang, S. Keten, D. Derome, J. Carmeliet, Hydrogen bonds dominated frictional stick-slip of cellulose nanocrystals. Carbohydr. Polym. **258**, 117682 (2021)
51. Q. Meng, X. Shi, A microstructure-based constitutive model of anisotropic cellulose nanopaper with aligned nanofibers. Extreme Mech. Lett. **43**, 101158 (2021)
52. Q. Meng, B. Li, T. Li, X.-Q. Feng, Effects of nanofiber orientations on the fracture toughness of cellulose nanopaper. Eng. Fract. Mech. **194**, 350–361 (2018)
53. Z. He, Y. Zhu, H. Wu, A universal mechanical framework for noncovalent interface in laminated nanocomposites. J. Mech. Phys. Solids **158**, 104560 (2022)
54. Q. Meng, B. Li, T. Li, X.-Q. Feng, A multiscale crack-bridging model of cellulose nanopaper. J. Mech. Phys. Solids **103**, 22–39 (2017)
55. Q. Chen, B. Chen, S. Jing, Y. Liu, T. Li, Flaw sensitivity of cellulose paper. Extreme Mech. Lett. **56**, 101865 (2022)
56. C. Zhang, M. Chen, S. Keten, B. Coasne, D. Derome, J. Carmeliet, Hygromechanical mechanisms of wood cell wall revealed by molecular modeling and mixture rule analysis. Sci. Adv. **7**, eabi8919 (2021)
57. K. Kulasinski, D. Derome, J. Carmeliet, Impact of hydration on the micromechanical properties of the polymer composite structure of wood investigated with atomistic simulations. J. Mech. Phys. Solids **103**, 221–235 (2017)
58. Q. Meng, T.J. Wang, Mechanics of strong and tough cellulose nanopaper. Appl. Mech. Rev. **71** (2019)
59. X. Qin, S. Feng, Z. Meng, S. Keten, Optimizing the mechanical properties of cellulose nanopaper through surface energy and critical length scale considerations. Cellulose **24**, 3289–3299 (2017)
60. B. Natarajan, J.W. Gilman, Bioinspired bouligand cellulose nanocrystal composites: a review of mechanical properties. Philos. Trans. Royal Soc. A: Math. Phys. Eng. Sci. **376**, 20170050 (2018)
61. M. Shishehbor, H. Son, M. Nuruddin, J.P. Youngblood, C. Davis, P.D. Zavattieri, Influence of alignment and microstructure features on the mechanical properties and failure mechanisms of cellulose nanocrystals (CNC) films. J. Mech. Behav. Biomed. Mater. **118**, 104399 (2021)
62. U. Ray, Z. Pang, T. Li, Mechanics of cellulose nanopaper using a scalable coarse-grained modeling scheme. Cellulose **28**, 3359–3372 (2021)
63. Y. Zhang, J. Yu, X. Wang, D.M. Durachko, S. Zhang, D.J. Cosgrove, Molecular insights into the complex mechanics of plant epidermal cell walls. Science **372**, 706–711 (2021)
64. H.B. Yang, Z.X. Liu, H. Chen, X. Yue, Z.C. Ling, Z.M. Han, C.H. Yin, Y.H. Ruan, X. Zhao, Z. Zhou, D.H. Li, S. Xie, K.P. Yang, Q.F. Guan, S.H. Yu, An all-natural fire-resistant bioinspired cellulose-based structural material by external force-induced assembly. Mater. Today Nano **23**, 100342 (2023)
65. Z.-X. Liu, H.-B. Yang, Z.-M. Han, W.-B. Sun, X.-X. Ge, J.-M. Huang, K.-P. Yang, D.-H. Li, Q.-F. Guan, S.-H. Yu, A bioinspired gradient design strategy for cellulose-based electromagnetic wave absorbing structural materials. Nano Lett. **24**, 881–889 (2024)

66. Y. Chen, B. Dang, J. Fu, C. Wang, C. Li, Q. Sun, H. Li, Cellulose-based hybrid structural material for radiative cooling. Nano Lett. **21**, 397–404 (2021)
67. Q.-F. Guan, Z.-M. Han, K.-P. Yang, H.-B. Yang, Z.-C. Ling, C.-H. Yin, S.-H. Yu, Sustainable double-network structural materials for electromagnetic shielding. Nano Lett. **21**, 2532–2537 (2021)
68. X. Pang, Y. Song, N. Shi, M. Xu, C. Zhou, J. Chen, Design of zero thermal expansion and high thermal conductivity in machinable xLFCS/Cu metal matrix composites. Compos. B Eng. **238**, 109883 (2022)
69. Y. Jia, J. Feng, Z. Feng, J. Liu, Y. Yang, X. Li, M. Lei, H. Guo, Z. Wei, Y. Lv, F. Xu, An endoscopically compatible fast-gelation powder forms Janus-adhesive hydrogel barrier to prevent postoperative adhesions. Proc. Natl. Acad. Sci. **120**, e2219024120 (2023)
70. H.-B. Yang, X. Zhao, Q. Wang, Y.-H. Ruan, Z.-X. Liu, X. Yue, Y.B. Zhu, H.A. Wu, Q.-F. Guan, S.-H. Yu, Simultaneously strengthening and toughening all-natural structural materials via 3D nanofiber network interfacial design, **63**, e202408458 (2024)
71. Q.-F. Guan, H.-B. Yang, Z.-M. Han, Z.-C. Ling, S.-H. Yu, An all-natural bioinspired structural material for plastic replacement. Nat. Commun. **11**, 5401 (2020)
72. Q.-F. Guan, H.-B. Yang, Z.-M. Han, Z.-C. Ling, K.-P. Yang, C.-H. Yin, S.-H. Yu, Plant cellulose nanofiber-derived structural material with high-density reversible interaction networks for plastic substitute. Nano Lett. **21**, 8999–9004 (2021)
73. L. Chen, S. Wang, S. Wang, C. Chen, L. Qi, L. Yu, Z. Lu, J. Huang, J. Chen, Z. Wang, X.-W. Shi, Z. Song, H. Liu, C. Chen, Scalable production of biodegradable, recyclable, sustainable cellulose-mineral foams via coordination interaction assisted ambient drying. ACS Nano **16**, 16414–16425 (2022)
74. Q. Wu, Z. Zeng, L. Xiao, From 2D graphene and MXene nanolayers to 3D biomimetic porous composite aerogels for electromagnetic interference shielding. Compos. A Appl. Sci. Manuf. **177**, 107939 (2024)
75. M. Chen, J. Zhu, K. Zhang, H. Zhou, Y. Gao, J. Fan, R. Chen, H.-L. Wang, Carbon nanofiber/polyaniline composite aerogel with excellent electromagnetic interference shielding, low thermal conductivity, and extremely low heat release. Nano-Micro Lett. **17**, 80 (2024)
76. D.-H. Li, Z.-M. Han, Q. He, K.-P. Yang, W.-B. Sun, H.-C. Liu, Y.-X. Zhao, Z.-X. Liu, C.-N.-Y. Zong, H.-B. Yang, Q.-F. Guan, S.-H. Yu, Ultrastrong, thermally stable, and food-safe seaweed-based structural material for tableware. Adv. Mater. **35**, 2208098 (2023)
77. Q.-F. Guan, H.-B. Yang, Z.-M. Han, L.-C. Zhou, Y.-B. Zhu, Z.-C. Ling, H.-B. Jiang, P.-F. Wang, T. Ma, H.-A. Wu, S.-H. Yu, Lightweight, tough, and sustainable cellulose nanofiber-derived bulk structural materials with low thermal expansion coefficient. Sci. Adv. **6**, eaaz1114 (2020)
78. Q.-F. Guan, H.-B. Yang, C.-H. Yin, Z.-M. Han, K.-P. Yang, Z.-C. Ling, S.-H. Yu, Nacre-inspired sustainable coatings with remarkable fire-retardant and energy-saving cooling performance. ACS Mater. Lett. **3**, 243–248 (2021)
79. S.-M. Chen, Z.-B. Zhang, H.-L. Gao, S.-H. Yu, Bottom-up film-to-bulk assembly toward bioinspired bulk structural nanocomposites. Adv. Mater. **36**, 2313443 (2024)
80. X. Yue, H.-B. Yang, Z.-M. Han, Y.-X. Lu, C.-H. Yin, X. Zhao, Z.-X. Liu, Q.-F. Guan, S.-H. Yu, Tough and moldable sustainable cellulose-based structural materials via multiscale interface engineering. Adv. Mater. **36**, 2306451 (2024)

Chapter 17
Cellulose-Based Aerogels for Energy Storage

Ting Xu and Chuanling Si

Abstract Cellulose, with the properties of wettability, abundant hydroxyl groups, eco-friendliness, and good mechanical performance, has been widely used in fabricating energy storage materials. Cellulose-based aerogel (CBA) has increasingly become a research hotspot in the field of energy storage due to its advantages such as abundant availability in nature, light weight, high porosity, and excellent stability in most solvents. Herein, the application progress of CBA in supercapacitors, zinc-ion batteries, lithium-ion batteries, and sodium-ion batteries are comprehensively discussed, based on the relationship of structure-performance and energy storage mechanism. Finally, the prospects and challenges are proposed for the problems faced in the large-scale application of CBAs, the coordination relationship between pore size regulation and electrical conductivity and mechanical properties, and the characterization of ion and electron conduction mechanisms in CBAs. It is hoped to provide new ideas for the material design and improving energy storage performance of CBAs.

Keywords Cellulose · Aerogels · Supercapacitors · Batteries · Energy storage

17.1 Introduction

Cellulose is one of the most abundant natural renewable resources in nature. It has the advantages of being renewable, inexpensive, environmental-friendly and biodegradable, and is widely used in energy storage fields [1–8]. Due to the poor electrical conductivity of cellulose, it usually needs to be compounded with functional materials such as conductive materials or to be carbonization for further application in energy storage [9–16]. The fabrication process of cellulose composite aerogels commonly

T. Xu · C. Si (✉)
State Key Laboratory of Biobased Fiber Materials, Tianjin Key Laboratory of Pulp and Paper, Tianjin University of Science and Technology, Tianjin, China
e-mail: sichli@tust.edu.cn

G. Wei and J. Zhang (eds.), *Cellulose-Based Hydrogels and Aerogels*, Lecture Notes in Nanoscale Science and Technology 41,
https://doi.org/10.1007/978-3-032-03008-5_17

requires the sol–gel of the dispersion and the drying process [17]. Currently, self-assembly induced by crosslinking agents, hydrothermal/solvothermal and cryogenic casting are the main methods of the gelation process. To maintain the internal three-dimensional network structure, drying methods include CO_2 supercritical drying and freeze-drying are needed. The prepared cellulose-based aerogel (CBA) has the advantages of light weight, high porosity and easy regulation of microstructure, and can be used as electrode or separator materials for energy storage devices [18].

CBA with tunable pore structure and high porosity has been used to the electrode of supercapacitors. CBAs possess significant hydrophilic surfaces, a hierarchical micro–/mesoporous structure for charge storage, rich adsorption sites to facilitate electrolyte ion absorption and transport, resulting in improved electrochemical performance of supercapacitors. Through carbonization, CBAs can be transformed to carbon aerogels, which enhance the electrical conductivity and contribute to enhanced electrochemical performance while retaining the original porous structure [19]. For batteries, the carbonized CBAs can well disperse metal oxides and can provide rich material/electrolyte contact area, thus enhancing the diffusion rates of ions. Otherwise, high porosity, structural tunability, and compatibility with active components of CBAs can inhibit metal dendrites, promote reaction kinetics and interfacial stability [20].

In conclusion, CBA can not only serve as a flexible framework to enhance the mechanical properties of aerogel, but also be used to construct a continuous conductive network through carbonization, providing an alternative for energy storage materials. This paper first exhibits the advantages and characteristics of CBA in energy storage. Then, from the aspects of energy storage mechanism and relationship of structure-performance, the application progress of CBA in supercapacitors, zinc-ion batteries, lithium-ion batteries, and sodium-ion batteries is comprehensively analyzed. Finally, prospects and challenges for the large-scale preparation of CBAs, the coordination relationship between structure and performance in energy storage applications, as well as the challenges and potential strategies faced in mechanism characterization are presented.

17.2 CBAs for Supercapacitors

Supercapacitors are a type of energy storage device that bridges the gap between traditional capacitors and rechargeable batteries. They combine the rapid charging and discharging characteristics of capacitors with the high energy storage capacity of batteries. Their core energy storage mechanism relies on the double-layer effect or pseudocapacitive response formed at the interface between the electrodes and electrolyte. CBAs possess significant hydrophilic surfaces, a hierarchical micro–/mesoporous structure for charge storage, rich adsorption sites to facilitate electrolyte ion absorption and transport, resulting in improved electrochemical performance of supercapacitors.

17.2.1 CBAs for Electrodes of Supercapacitors

Electrodes are the core components for capacity, requiring high specific surface area to provide sufficient active sites, high conductivity to ensure rapid electron transport, good electrochemical activity (contributing pseudocapacitance), and excellent structural stability to maintain long cycle life [21]. The three-dimensional nanoporous network structure of aerogels endows them with ultra-high porosity (>90%), extremely high specific surface area, and ultra-low density [22]. Cellulose, as the most abundant natural polymer on Earth, has extremely widespread and sustainable sources [23, 24]. It possesses high strength, high hydrophilicity, abundant hydroxyl functional groups that are easily chemically modified and functionalized, good thermal stability, and biodegradability. Its nanostructured forms are ideal building blocks for constructing high-performance aerogels.

Liao et al. assembled composite aerogels under mild conditions by two-step activation of Eucommia ulmoides tar-based activated carbon (ETAC), cellulose nanofibers (CNFs), and reduced graphene oxide (RGO) [25]. The doping of ETAC greatly increased the specific surface area of the aerogel, while CNFs extracted from Eucommia ulmoides enhanced the mechanical properties of the aerogel. Chen et al. proposed a simple "bottom-up" strategy using CNFs as the basic unit, introducing lignin (AL) to suppress severe volume shrinkage of CNFs during pyrolysis, and prepared a nanocellulose-based carbon aerogel with a tubular ordered structure [26]. The ordered tubular texture endowed the nanofiber cellulose-based carbon aerogel with high compressibility, elasticity, and excellent fatigue resistance (Fig. 17.1a), enabling it to serve as a self-supporting electrode for the fabrication of all-solid-state symmetrical supercapacitors with high electrochemical performance and superior mechanical flexibility.

Furthermore, Zhao et al. developed a high-load nitrogen-doped porous carbon electrode supported by a CNF/carbon nanotube (CNT) network [27], combining high capacity (133.4 mAh g^{-1}), long cycle life (95.4%@10,000 cycles), and commercial-grade loading (10 mg cm^{-2}). The CNF/CNT network was integrated with nitrogen-doped porous carbon (NPC) through vacuum filtration and freeze-drying processes, forming continuous ion/electron conduction pathways. As shown in Fig. 17.1b, this enables the electrode to achieve an areal capacity of 2.05 mAh cm^{-2} at a commercial-grade loading of 10 mg cm^{-2}, with a capacity retention rate of 86.2% after 3000 cycles, while mechanical strength is enhanced to 7.85 MPa. At low loading (2 mg cm^{-2}), the electrode exhibited a specific capacity of 133.4 mAh g^{-1} and long-term cycling stability of 95.4% (10,000 cycles), with an energy density of 113.4 Wh kg^{-1}, providing an efficient carbon-based electrode design strategy for the practical application of Zinc-ion hybrid supercapacitors.

Additionally, Liu et al. used a two-step freezing and carbonization process, based on hydrogen bonding, electrostatic interactions, and π-π interactions between the components, to construct a CNF/CNT/RGO composite carbon aerogel with a regular layered pore structure [28]. Benefiting from the pore structure and high surface roughness, the prepared CNF/CNT/RGO carbon aerogel exhibited ultra-low density

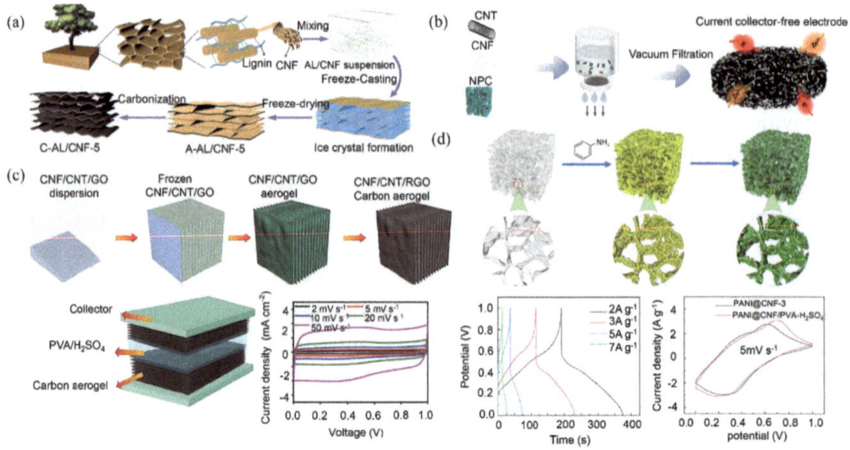

Fig. 17.1 **a** Illustration process of woody carbon aerogel with a biomimetic tubular structure [26]. Copyright 2020, Wiley. **b** Preparation of NPC-CNF/CNT electrodes [27]. Copyright 2024, Elsevier. **c** Fabrication process of CNF/CNT/RGO composite carbon aerogel and its application as a compressible supercapacitor electrode [28]. Copyright 2022, Wiley. **d** PANI@CNF composite material preparation process and its constant current charge–discharge (GCD) curve and cyclic voltammetry (CV) curve [29]. Copyright 2025, Elsevier

(2.64 mg cm^{-3}) and superhydrophilicity. As shown Fig. 17.1c, its honeycomb-like ordered porous structure can effectively transmit stress throughout the microstructure, endowing the carbon aerogel with high compressibility (capable of withstanding 80% strain) and excellent fatigue resistance (after 10,000 cycles of compression at 50% strain, the stress retention rate was 85.3%). Supercapacitors assembled using this aerogel as an electrode demonstrated excellent areal capacitance (109.4 mF cm^{-2}) and long-term compression cycling performance (88% capacitance retention after 5000 cycles at 50% strain). Xiong et al. used in situ chemical polymerization to graft conductive polyaniline (PANI) onto CNF aerogel with a high aspect ratio, then coated it with PVA-H_2SO_4 gel, successfully preparing a multifunctional composite material of PANI@CNF/PVA-H_2SO_4 [29], as shown in the Fig. 17.1d. The three-dimensional network of CNF served as a rigid scaffold, limiting the volume expansion of PANI. The capacitance retention rate of PANI@CNF/PVA-H_2SO_4 after cycling is >95%.

17.2.2 *CBAs for Separators of Supercapacitors*

The separator, its primary function is to physically isolate the positive and negative electrodes to prevent short circuits, while also allowing ions to move quickly and with low resistance. Additionally, it needs to possess good electrolyte wetting and retention properties, sufficient mechanical strength and flexibility, especially in flexible

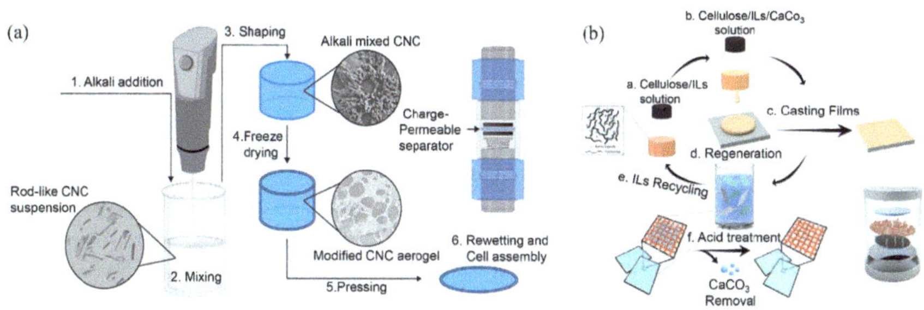

Fig. 17.2 **a** Preparation process of conductive separator material based on modified CNC [30]. Copyright 2025, Elsevier. **b** Illustration schematic of porous RC membranes [31]. Copyright 2023, Wiley

devices. CBAs with interconnected porous structure can facilitate rapid ion transport (low internal resistance). Their natural hydrophilicity/modifiability ensures good electrolyte wetting, their robust network provides mechanical support and flexibility, their thermal stability and flame-retardant potential enhance safety.

For example, Mijailović et al. regulated the structure of cellulose nanocrystal (CNC) gel through alkaline modification, followed by freeze-drying to form an aerogel, which was then compressed into a "cage-like" structure with uniform porosity [30]. This aerogel exhibited both high ionic permeability and mechanical strength. The ionic conductivity of the KOH-CNC separator reached 103.5 mS cm^{-1}, surpassing that of LiOH-CNC (68 mS cm^{-1}) and glass microfiber (92 mS cm^{-1}). Due to its high porosity and surface hydroxyl hydrophilicity, as shown in the Fig. 17.2a, the cyclic voltammetry (CV) curve of the symmetric supercapacitor is close to rectangular, indicating fast reversible capacitive behavior. Additionally, the charge transfer resistance (4.5 Ω) in the KOH-CNC impedance spectrum was lower than that of GM (10 Ω), demonstrating superior interfacial reaction kinetics.

Lou et al. used a chemical cross-linking method under alkaline conditions to induce ring-opening polymerization and cross-linking reactions between LNPs and polyethylene glycol diglycidyl ether, forming a sponge-like porous LF film with a porosity of 47.6%, an ionic conductivity of 3.26 mS cm^{-1}, and both high electrolyte absorption capacity (278%) and mechanical flexibility [32]. In addition, Wu et al. designed a high-performance regenerated cellulose (RC)-based membrane, as shown the Fig. 17.2b [31]. The membrane was prepared by effectively hydrolyzing and etching inorganic $CaCO_3$ nanoparticles from a filled RC membrane. The prepared RC membrane exhibits high tensile strength (75.83 MPa) and thermal stability (200 °C), outperforming commercial polypropylene-based membranes (Celgard 2500). The RC membrane exhibited rapid ion transport and permeability, as well as excellent ionic conductivity, due to its hydrophilicity, high porosity, and outstanding liquid absorption rate (208.5%). At a scan rate of 10 mV s^{-1}, the supercapacitor using the RC separator achieved a specific capacitance of 29.12 F g^{-1}, with a capacitance retention rate of nearly 100% after 10,000 cycles, and no significant damage to the separator structure. Its overall performance surpasses that of commercial polyolefin and non-woven separaors.

17.3 CBAs for Zinc-Ion Batteries

Zinc-ion batteries have gained great attention in the field of energy storage and conversion due to their high safety, low cost, and environmentally friendly properties [33]. Zn^{2+} is used to realize energy storage and release through the reversible reactions of embedding/de-embedding in the anode material and deposition/dissolution on the surface of the negative electrode during charging and discharging [34]. Currently, the core challenges are zinc dendrite growth and side reactions such as hydrogen precipitation and passivation [35]. Therefore, it further optimizes the ion transport channels and charge storage centers of the battery by utilizing CBA materials with unique three-dimensional porous networks [36], thus significantly improving the cycle life and safety of the battery. CBA, has become an effective method to solve the key bottleneck problems of zinc-ion batteries, such as dendrites and side reactions, by virtue of its dense and uniform nanopores, abundant hydroxyl groups, excellent ionic conductivity brought about by its excellent mechanical properties [37, 38].

17.3.1 CBAs for Electrodes of Zinc-Ion Batteries

CBA serves as a high-performance negative electrode substrate thereby accommodating zinc deposition. The huge internal space and uniform pore network of the aerogel provide rich nucleation sites and space for Zn^{2+} deposition, which can effectively inhibit the growth of zinc dendrites. Moreover, the uniformly distributed pores can help to guide the uniform deposition of zinc and avoid the dendrites from piercing the diaphragm due to the high local current density. Zhao et al. prepared highly conductive nanocellulose-based carbon aerogels with honeycomb-like porous structure and high specific surface area (Fig. 17.3a) [39]. And dendrite-free, nanocellulose-based carbon aerogel composite zinc anode (Zn@CA) as the backbone was synthesized by a facile electrodeposition method. As the carbon aerogel had a connected conductive network, the deposition position of Zn can be regulated, thus inhibiting the formation of dendrites (Fig. 17.3b). The assembled Zn@CA symmetric cell showed a stable voltage and can be cycled 13,000 times at 0.5 mA cm^{-2} with a Coulombic efficiency of about 100%. The high porosity and interconnected pores greatly reduce the ion transport resistance, optimize the ion transport channels, and improve the rate performance of the electrode.

Zhang et al. prepared CNT/CNF/CNC/Zn self-supported aerogel films (Fig. 17.3c). The pore size distribution of the self-supported film was adjusted by adding CNFs and CNCs [41]. Continuous interwoven network with dual transport channels for ions and electrons effectively reduced the growth of Zn dendrites by

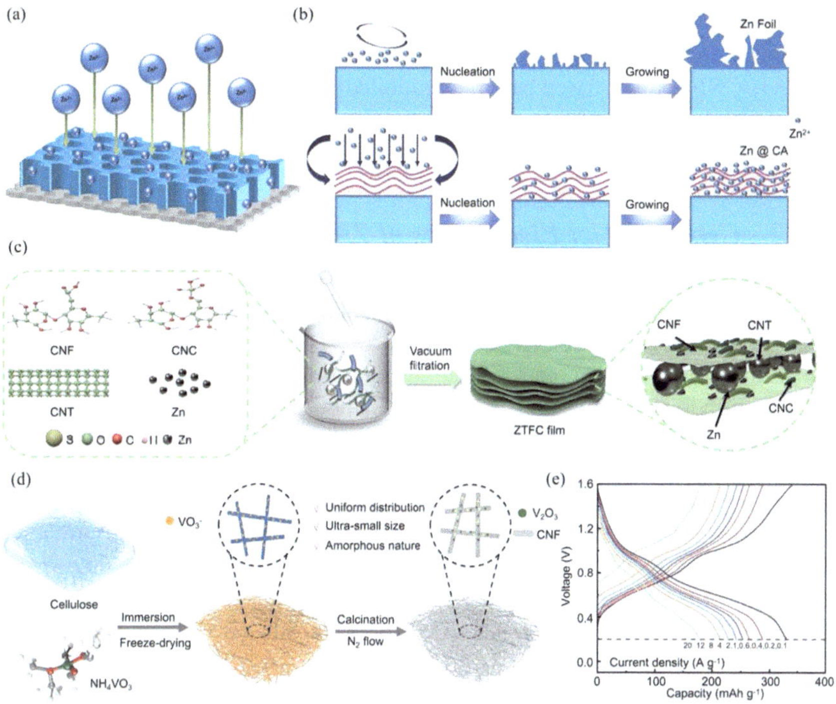

Fig. 17.3 **a** Schematic of deposition Zn@CA; **b** Schematic diagram of Zn deposition on Pristine Zn and Zn@CA [39]. Copyright 2023, Elsevier. **c** Fabrication illustration of the CNT/CNF/CNC/Zn self-supported aerogel films. Copyright 2025, Elsevier. **d** Schematic of the preparation of V_2O_3/CNF; **e** The corresponding GCD profiles at different current densities [40]. Copyright 2024, American Chemical Society

promoting ion diffusion kinetics. When CNF/CNC = 8:1, the pore size of the zinc ion anode membrane was concentrated in the range of 8–25 nm, with maximum electrical conductivity (1250 S m^{-1}) and excellent mechanical flexibility. The assembled symmetric and button batteries displayed excellent cycling performance, especially reflected in the 99% capacity retention of the button battery after 1500 cycles at 0.5 A g^{-1}.

CBAs can be used as a conductive backbone for cathode materials. A nanocellulose-based MnO_2-loaded composite carbon aerogel (CA@MnO_2) cathode was prepared by using a bi-directional freeze-carbonization and hydrothermal deposition method by Zhao and co-workers [42]. The ordered pore structure and high specific surface area effectively promoted the rapid migration of Zn^{2+} and provided an efficient electrode contact interface. In a two-electrode system, a reversible specific capacity of 480 mAh g^{-1} at 0.5 A g^{-1} was provided. Shi et al. synthesized V_2O_3 tightly coupled carbon aerogel by controlled ion impregnation and solid-phase reaction using bacterial cellulose and ammonium metavanadate as raw materials (Fig. 17.3d) [40]. Among them, cellulose aerogel plays the role of dispersion and supporting skeleton to achieve particle size reduction, uniform distribution and amorphous characteristics of V_2O_3. Charge–discharge cycling tests revealed that the V_2O_3/CNF aerogel-based batteries exhibited high multiplicity performance and excellent cycling stability (Fig. 17.3e).

17.3.2 CBAs for Gel Electrolytes of Zinc-Ion Batteries

CBA can act as the "reservoir" and "skeleton" of the liquid electrolyte, solidifying or semi-solidifying the liquid electrolyte to form a gel polymer electrolyte [43]. It is also a three-dimensional carrier platform that can guide the orderly deposition of zinc ions and limit the harmful side reactions. Zhou et al. filled the cellulose aerogel framework with gelatin to form an electrolyte film with a thickness of about 400 μm (Fig. 17.4a, b) [44], which was both highly ionic conductive (1.23×10^{-2} S cm^{-1}) and highly flexible (bending angle >120°). This solid electrolyte enabled the battery to achieve controlled degradation and excellent electrochemical performance and provided a specific capacity of 211.5 mAh g^{-1} at a current of 61.6 mA g^{-1} and a wide voltage range (0.85–1.9 V). Xu et al. prepared a guar gum cellulose aerogel membrane for electrolyte (Fig. 17.4c) [45]. The electrolyte membrane had a high ionic conductivity of about 4.73×10^{-2} S cm^{-1} and good mechanical stress characteristics. The cell exhibits a specific capacity of 309.1 mAh g^{-1} at a current density of 308 mA g^{-1} after 100 cycles and a stable cycling capability (100% capacity retention after 200 cycles) (Fig. 17.4d).

17.4 CBAs for Lithium-Ion Batteries

Lithium-ion batteries are secondary batteries that convert electrical energy and chemical energy through the reversible migration of lithium ions (Li^+) between the cathode and anode [46]. The core components include the cathode (e.g., $LiCoO_2$, $LiFePO_4$), the anode (e.g., graphite), the electrolyte (an organic solution containing lithium salts), and the separator. During charging, Li^+ is extracted from the cathode and intercalated into the anode (such as between graphite layers) through the electrolyte, while electrons flow to the anode via the external circuit. During discharge, Li^+ deintercalates from the anode and returns to the cathode, with electrons flowing through the external circuit to perform work. However, the safety, cycle life, and fast charging performance of lithium-ion batteries are limited by several key issues, including severe volume expansion and low conductivity of silicon anodes, as well as poor thermal stability and wettability of separators.

CBAs have demonstrated excellent structural tunability and performance advantages in the design of lithium-ion battery anode materials. Through innovative fabrication strategies, these materials effectively address key challenges such as high

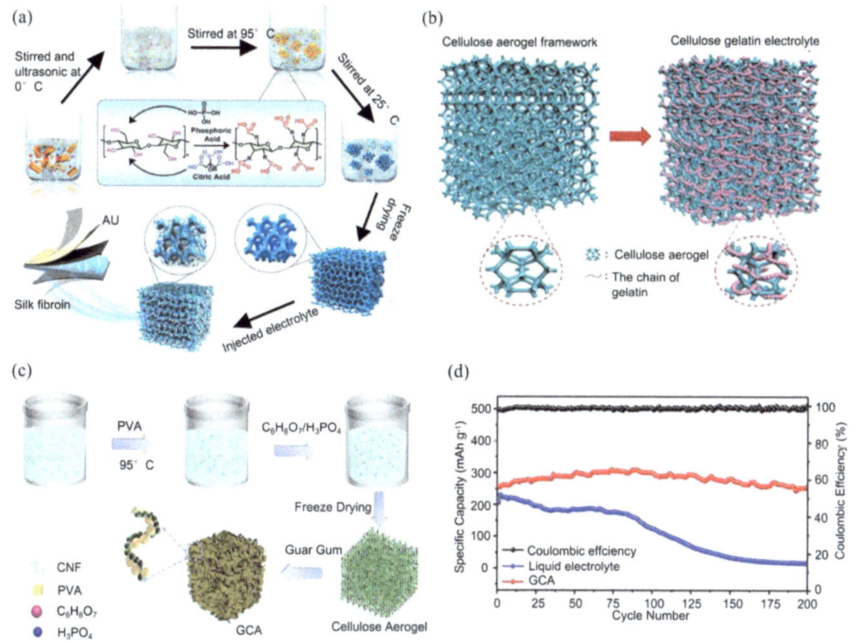

Fig. 17.4 **a** Schematic depicting the super-assembly synthesis route of the cellulose aerogel framework; **b** Structure and assembly of cellulose gelatin electrolyte [44]. Copyright 2022, Wiley. **c** Schematic of the synthesis route for guar gum cellulose aerogel; **d** Cycling stability of zinc ion batteries at 308 mA g^{-1} [45]. Copyright 2022, Royal Society of Chemistry

capacity, long cycle life, and sustainable development. For example, Liu et al. fabricated Fe_3O_4/carbon nanofiber aerogels (Fe_3O_4/CNFA) via a one-step pyrolysis of Fe^{3+}-crosslinked cellulose nanofibers [47]. Here, Fe^{3+} acts both as a crosslinking agent and iron source, forming uniformly distributed Fe_3O_4 nanoparticles (10–20 nm) embedded in a conductive carbon matrix. The synergistic effect between the conductive network and the high theoretical capacity of Fe_3O_4 enabled an ultrahigh specific capacity of 1635 mAh g^{-1} at 1 A g^{-1}, with the 3D network structure effectively buffering 200% volume expansion and maintaining over 90% capacity retention after 200 cycles. Addressing environmental concerns, Pham et al. extracted microfibrillated cellulose (MFC) from rice straw and produced nitrogen-doped carbon aerogels (MFC-CA) via urea-assisted crosslinking and carbonization [48]. The resulting material exhibited 10.49 wt% nitrogen doping (mainly pyridinic and graphitic nitrogen), offering abundant lithium storage sites and delivering a stable capacity of 348.6 mAh g^{-1} at 1C, with 97.9% capacity retention after 100 cycles, while reducing carbon emissions by 60% compared to conventional anodes.

Ye et al. developed nitrogen-doped carbon nanofiber aerogels (N-ACNF) with a high specific surface area of 696 m^2 g^{-1} by dissolving bamboo cellulose in ionic liquids and carbonizing under ammonia [49] (Fig. 17.5a). The hierarchical pore structure (micro-/mesopores) and uniform nitrogen doping (7.2 wt%) synergistically enhanced ion diffusion kinetics, achieving a capacity of 630.7 mAh g^{-1} at 1 A g^{-1} and retaining 430 mAh g^{-1} at 10 A g^{-1}, with over 90% capacity retention after 1000 cycles, significantly outperforming commercial carbon materials. For flexible battery applications, Wan et al. synthesized a self-supporting Fe_3O_4@BC-CNFs electrode by in situ growth of Fe_2O_3 nanosheets on bacterial cellulose (BC) followed by carbonization [50] (Fig. 17.5b). The unique 3D interconnected carbon fiber network provided excellent flexibility (bendable to 180°) and achieved a high capacity of 755 mAh g^{-1} at 0.1 A g^{-1}, due to the tight integration of Fe_3O_4 and graphitized carbon ($I_D/I_G = 0.89$). These studies collectively demonstrate that cellulose aerogels, through hierarchical pore structures, active site engineering, and integrated design, offer comprehensive solutions for high-performance lithium anodes, combining ultrahigh capacity (Fe_3O_4/CNFA), ultralong cycle life (N-ACNF), environmental friendliness (MFC-CA), and flexibility (Fe_3O_4@BC-CNFs).

Beyond anode materials, CBAs have also gained increasing attention in the design of electrolytes. Through diversified preparation strategies and structural designs, these materials have shown great potential in enhancing electrolyte performance. Li et al. developed a gel polymer electrolyte (GPE) composed of regenerated cellulose (RC) and polyethylene oxide (PEO) (Fig. 17.5c) [51]. Fabricated via regeneration and freeze-drying, the electrolyte exhibited a mechanical strength of 40 MPa (far exceeding the 1 MPa of conventional PEO-based electrolytes) and an ionic conductivity of 4.34×10^{-3} S cm^{-1} at room temperature. In $LiFePO_4$/Li cells, it delivered a discharge capacity of 149.3 mAh g^{-1} at 0.2C and retained 94.6% of its capacity after 100 cycles. The key innovation lies in the porous structure of cellulose enhancing the mechanical strength of PEO while maintaining high ionic conductivity. Wan et al. fabricated nanoporous cellulose aerogel membranes (CAMs) via ionic liquid regeneration and supercritical CO_2 drying [52]. With unique 20–50 nm pore structures, these membranes achieved an ionic conductivity of 2.81 mS cm^{-1} and excellent thermal stability at 120 °C. In $LiFePO_4$/Li batteries, a capacity of 138.1 mAh g^{-1} at 0.5C was achieved, with 98% retention after 100 cycles. This approach of using supercritical drying to precisely control pore architecture provides a new direction for high energy density batteries. These studies confirm that cellulose aerogels, through their porous networks, polar functional groups, and tunable chemical modifications, offer significant potential in enhancing ionic conductivity, mechanical strength, thermal stability, and interfacial compatibility of lithium-ion battery electrolytes.

In summary, CBAs, with their sustainable origin, abundant functional groups, and excellent porous structure tunability, are playing increasingly important roles in key components of lithium-ion batteries including anodes and separators. For anodes, strategies such as incorporation of metal oxides and nitrogen doping have achieved a synergy of high capacity, long cycle life, and mechanical flexibility. For separator design, the superior thermal resistance and electrolyte affinity of aerogel materials offer effective solutions for enhancing battery safety and long-term performance. However, despite their great potential demonstrated in laboratory settings, the

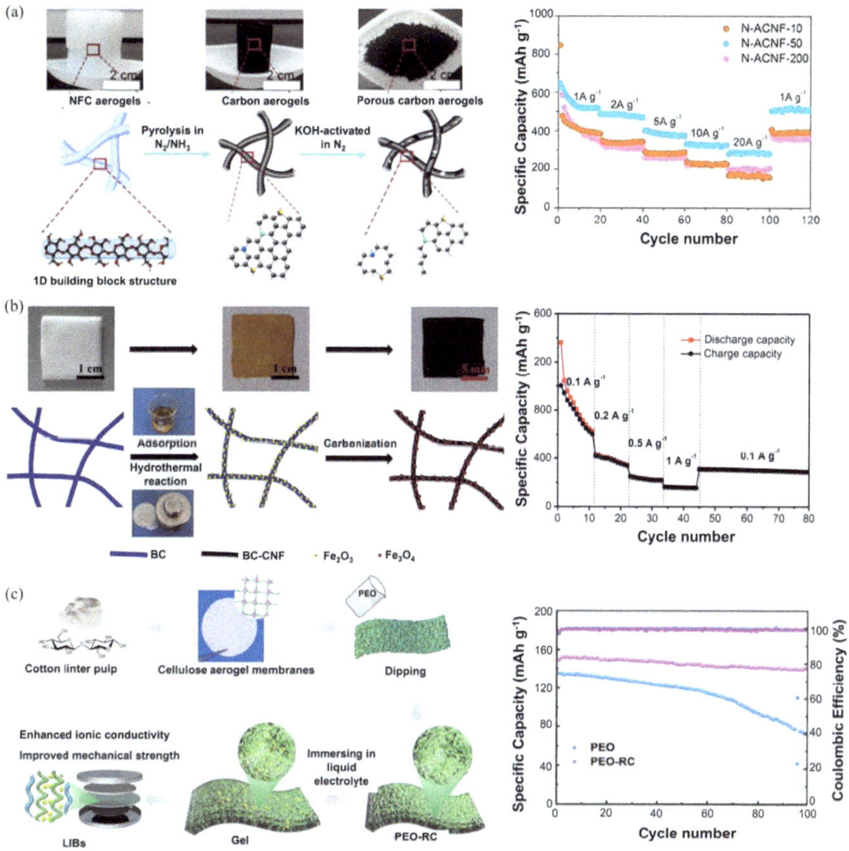

Fig. 17.5 Schematic diagram of CBA application in lithium-ion batteries. **a** Schematic diagram of the N-ACNF aerogel synthesis process and comparison of specific capacity at different current densities [49]. Copyright 2017, Royal Society of Chemistry. **b** Schematic diagram of the preparation process of Fe_3O_4@BC-CNFs and demonstration of their electrochemical performance [50]. Copyright 2015, Elsevier. **c** Schematic diagram of the preparation process of PEO-RC GPE and demonstration of its electrochemical performance in lithium-ion batteries [51]. Copyright 2023, Wiley

long-term stability of cellulose aerogels in real battery systems, scalable and consistent manufacturing processes, and compatibility with existing production lines still require further validation.

17.5 CBAs for Sodium-Ion Batteries

Sodium-ion batteries have gained increasing attention as an important source of energy storage devices, mainly due to the abundance and low cost of sodium, which is seen as a potential alternative to lithium-ion battery and is both cheap and environmentally friendly for large-scale energy storage [53]. However, the radius of Na^+ is much larger than that of Li^+ and most materials do not have body frames that provide a large interstitial space for Na^+ insertion/extraction. In addition, larger Na^+ radius is also associated with possible structural changes during insertion or extraction [54].

The 3D porous network of the aerogel can be used as a separator to adsorb electrolyte and form a continuous ion transport channel, which significantly reduces the interfacial impedance. In addition, the nanoscale pore size can selectively sieve ions and inhibit the generation of dendrites [55]. A CNF composite aerogel separator was prepared by Zhou et al. through solvent exchange and freeze-drying method [56]. Exhibited excellent pore structure, and the obtained diaphragms showed a reversible specific capacity of 320 mAh/g, with a capacity retention of 93.8% after 50 cycles at 50 mAh g^{-1} current density. Lu et al. prepared high nitrogen-doped carbon CNF aerogel with large-size pore structure and high nitrogen content derived from chitin [57]. The preparation process is shown in the Fig. 17.6a. When the aerogel was used as an anode material for sodium-ion batteries, the sample showed a high reversible capacity of 282 mAh g^{-1} at 0.1 A g^{-1} and more than 7000 cycles at 0.5 A g^{-1}, as shown in the Fig. 17.6b.

In addition, it is important to develop electrodes for application to high-performance sodium-ion batteries [60]. Huang et al. developed a BC composite aerogel based on three-dimensional nitrogen-doped carbon nanofibers [58], as shown in the Fig. 17.6c. The two key problems of insufficient electrical conductivity of carbon-based materials and limited sodium ion adsorption sites was solved. This nitrogen doping strategy not only significantly enhanced the electrical conductivity of the material, but its 3D interconnected nanofiber network (specific surface area of 469 m^2 g^{-1}) also effectively mitigated the volume expansion during charging and discharging. As an anode for sodium-ion batteries, the material maintains a stable capacity of 154 mAh g^{-1} after 1500 cycles at a high current density of 400 mA g^{-1}, as shown in the Fig. 17.6d. Directional freezing technology can also be applied to prepare cellulose-based composite aerogel, the hollow channels can effectively buffer the volume change of electrodes during charging and discharging and maintain the structural integrity, so as to improve the cycling stability of electrodes. Wang et al. prepared CNC/PEO-derived vertically aligned carbon aerogel anode materials for sodium-ion batteries via a unidirectional ice template strategy (Fig. 17.6e) [59]. The CNC/PEO composite aerogel had a high reversible capacity (about 298 mAh g^{-1} at 0.1 C) and good cycling stability (only 0.03% capacity decay with 2000 cycles at 0.5 C), as shown in Fig. 17.6f.

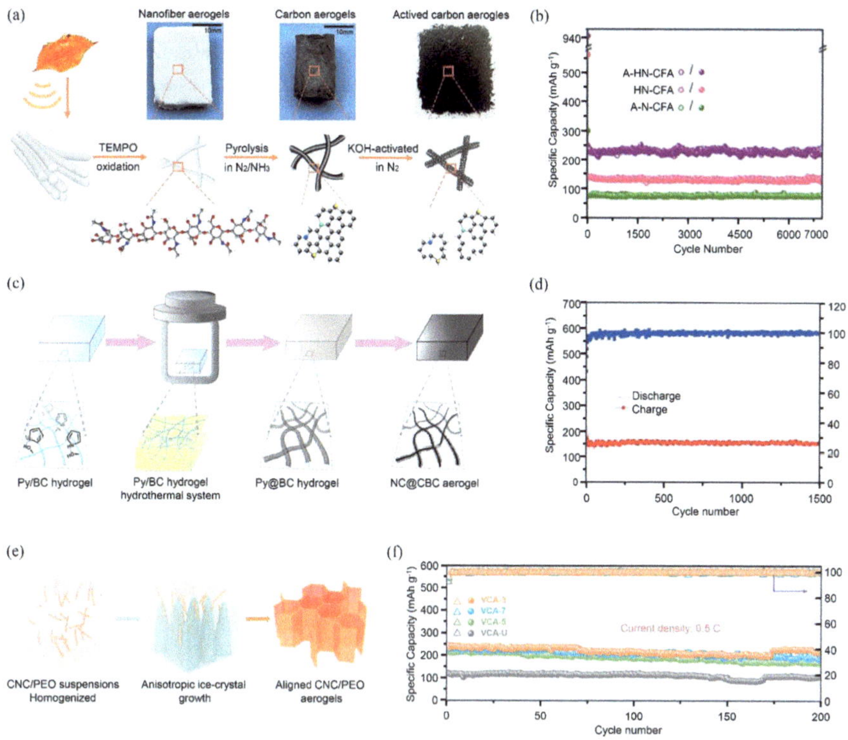

Fig. 17.6 **a** Fabrication of chitin-derived activated highly nitrogen-doped carbon fiber aerogel [57]. Copyright 2019, Elsevier. **b** Cycling performance at a current density of 0.5 A g^{-1}. **c** Design of the material preparation described in this work; **d** Long-term cycle performance as anode in SIBs [58]. Copyright 2018, Royal Society of Chemistry. **e** Schematic illustrating the formation mechanism of unidirectional ice-templated; **f** Cycling performance at a current density of 0.5 C [59]. Copyright 2022, Wiley

17.6 CBAs for Other Battery Systems

17.6.1 CBAs for Gel Electrolytes of Lithium-Sulfur Batteries

The application of CBAs in Li–S batteries provide a compelling route to overcome sulfur cathode challenges such as low conductivity, volume expansion, and polysulfide shuttle. The inherent characteristics of nanocellulose with renewability, and high surface area make it a suitable precursor for constructing conductive porous networks that facilitate sulfur loading and polysulfide immobilization.

Early breakthroughs leveraged bacterial cellulose-derived carbon nanofiber aerogels, offering low-density yet mechanically robust scaffolds with ample porosity. For instance, Li et al. constructed a CNFA from bacterial cellulose and achieved a high capacity of 1360 mAh · g^{-1} at 0.2 C [61]. The highly interconnected carbon network effectively trapped polysulfides while maintaining electrolyte permeability. Building on this, Li et al. further introduced a gel-based cathode composed of branched nanoribbon carbon aerogel with an ultra-high sulfur content of 90 wt% and mass loading of 6.4 mg · cm^{-2}, achieving a reversible capacity of 943 mAh · g^{-1} and areal capacity of 5.9 mAh · cm^{-2} [62]. To enhance redox kinetics, Lin et al. incorporated VO_2 onto nitrogen-doped carbonized bacterial cellulose [63]. After hydrogenation, the insulating VO_2 (M) was transformed into conductive VO_2 (R), reducing energy barriers for the sulfur reduction reaction. The optimized H-VO_2@N-CBC host achieved a high initial capacity of 1364.3 mAh · g^{-1} and retained 950.2 mAh · g^{-1} after 200 cycles at 0.2 C.

In parallel, 2D materials like MXene have been paired with TOCNF to form composite aerogels. The incorporation of $Ti_3C_2T_x$ MXene with oppositely charged CNFs creates a synergetic structure with improved electrical conductivity and strong polysulfide adsorption. Notably, Liu and Zhang et al. demonstrated that cationic CNF/MXene aerogels exhibited superior performance (1573.7 mAh · g^{-1} at 0.1 C), attributed to electrostatic interactions that reinforce structural integrity and ionic transport [64]. Further enhancing this strategy, Liu et al. carbonized MXene/CNF composites into a 3D conductive network with impressive stability over 200 cycles [65].

Expanding on the functionalization motif, recent efforts have explored single-atom and cluster catalysts within cellulose-derived carbon matrices. Lin et al. synthesized Fe single-atom catalysts anchored on nanocages within bacterial cellulose-derived carbon nanofiber aerogels [66]. This catalyst reduced the energy barrier for polysulfide conversion and achieved a reversible capacity of 1035 mAh · g^{-1} at 0.2 C, with 80.2% capacity retention after 300 cycles. Similarly, Cheng et al. designed cobalt cluster–carbon aerogels from cellulose nanocrystals, offering multiple active sites for Li_2S nucleation and improved reaction reversibility [67]. These findings indicate that cellulose aerogels are not merely passive hosts but can be tailored into catalytic frameworks by rationally embedding atomically dispersed metals. Another compelling route involves the construction of redox-active aerogels. Yu et al. embedded self-redox CeO_2 nanoparticles into a cellulose–graphene aerogel [68]. The resulting composite maintained a reversible capacity of 1010 mAh · g^{-1} at 0.1 C and demonstrated a stable performance of 805 mAh · g^{-1} after 400 cycles. The CeO_2 nanoparticles acted as redox mediators, reversibly interacting with soluble polysulfides while the graphene network provided efficient electron pathways.

Not all advancements require metallic additives. Chen et al. [69] developed carbon nanofibril aerogels from TOCNF with a high sulfur content of 78 wt% and a loading of 6.8 mg · cm^{-2}. The composite cathode delivered a high initial capacity of 1174 mAh · g^{-1} and maintained 680 mAh · g^{-1} over 200 cycles at 0.2 C. This suggests that well-designed carbon aerogels from pure nanocellulose can also meet the demands for high loading and long-term stability through tailored porosity and surface chemistry. Furthermore, Liu et al. [70] emphasized the importance of reaction kinetics by tailoring the nanostructure and surface functionality of TOCNF-based aerogels. Their work highlighted that not only the pore structure but also the surface chemistry plays a pivotal role in regulating the formation and conversion of lithium polysulfides, pointing to a delicate balance between adsorption strength and diffusion dynamics. Mao et al. [71] demonstrated a cotton cellulose-based graphene/carbon aerogel with a specific surface area of 657.85 $m^2 \cdot g^{-1}$ and porosity of 96%, enabling high sulfur loading of 9.11 mg $\cdot$ cm^{-2} and delivering an areal capacity of 8.60 $mAh \cdot cm^{-2}$ (High loading cotton CBA self-standing electrode for Li–S batteries). As illustrated in Fig. 17.7a and b, the aerogel was constructed through carbonization of cotton-derived cellulose, followed by graphene incorporation to form a self-standing, conductive framework. This design not only replaced traditional aluminum foil as the current collector, but also significantly improved sulfur utilization from 58 to 86%, as evidenced by the long-term cycling stability and high areal capacity retention shown in Fig. 17.7c.

Taken together, these studies highlight a clear progression in the design of CBAs for Li–S batteries, from passive porous hosts to active, multifunctional architectures.

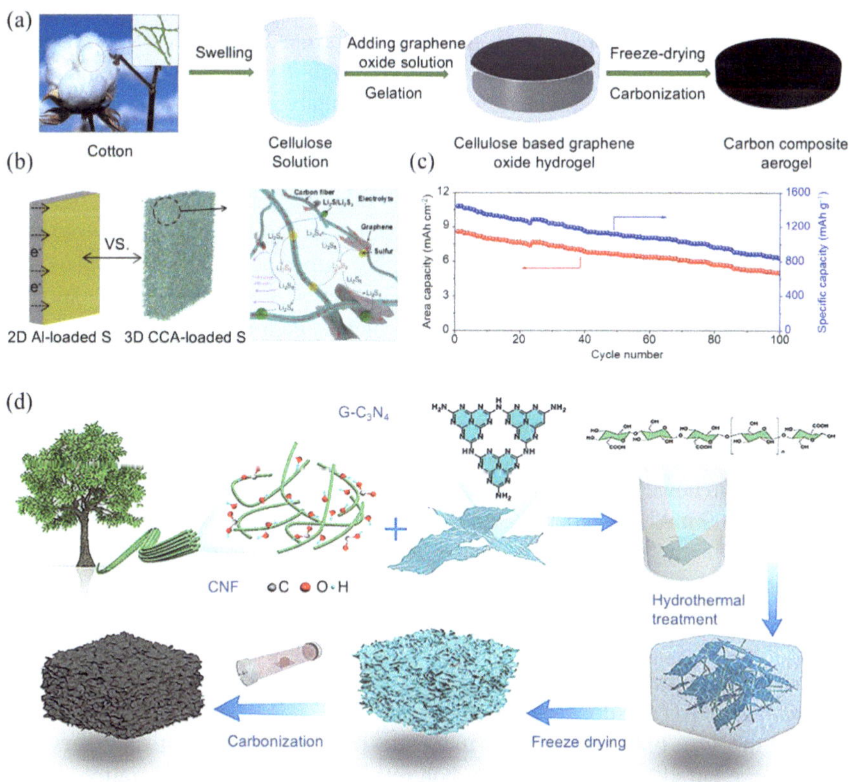

Fig. 17.7 **a** Schematic illustration of the preparation technology of CCA; **b** Contrast diagram of 2D Al and 3D CCA current collector and schematic illustration of the discharge process in CCA cathodes; **c** Cycling performance of high load S at 0.1 C [71] Copyright 2020, Elsevier. **d** Schematic illustrating the fabrication of the HNCA catalyst [72]. Copyright 2024, Royal Society of Chemistry

The early focus on structural integrity and sulfur encapsulation has evolved toward catalytic enhancement, interfacial engineering, and kinetic regulation. Cellulose-derived aerogels now stand at the intersection of sustainability and performance, providing a scalable and tunable platform for next-generation Li–S battery technologies. Their ability to integrate conductive, adsorptive, and catalytic functionalities within a bio-derived matrix opens new directions for both fundamental research and industrial application.

17.6.2 CBAs for Gel Electrolytes of Metal-Air Batteries

Metal-air batteries are energy storage devices that use metals as the negative electrode and oxygen in the air as the positive electrode active material. Metals such as zinc and aluminum release electrons when oxidized at the negative electrode, and oxygen is reduced to form hydroxide ions under the action of a catalyst at the positive electrode. CBA, as a three-dimensional porous material, can be used as an electrolyte carrier in metal-air batteries. It adsorbs the electrolyte through its porous structure, avoiding leakage and ensuring ion conduction, making it suitable for flexible devices. It can also load metal nanoparticles and other materials as catalyst carriers. The high specific surface area provides active sites, promoting oxygen reactions and material transport, and can also regulate the interface to suppress side reactions [73].

In response to these deficiencies, Shen's team prepared a single cobalt atom catalyst (Co SAs/NCNA) based on cellulose nanofiber carbon aerogel for efficient oxygen reduction reactions and zinc-air batteries. The preparation of cellulose nanofiber hydrogel precursors chelated by pyrolysis of metals, cadmium salts assist in cross-linking the aerogel, create micropores and defects, and isolate cobalt atoms. This catalyst features a high specific surface area and a three-dimensional porous structure, which promotes the accessibility of the catalytic center and the electron transport of substances. The half-wave potential of its oxygen reduction reaction is 0.855 V. The maximum power density of the zinc-air battery is 206 mW cm^{-2} and the specific capacity is 769 mAh · g^{-1}, demonstrating excellent catalytic performance and durability, providing a green and sustainable method for the large-scale production of single-atom electrocatalysts [74].

Liu et al. proposed an effective and sustainable strategy to construct TOCNF/ANF-Cd hydrogel/aerogel using TEMPO oxidized cellulose nanofibers (TOCNFs), thermally stable aramid nanofibers (ANFs), and low-boiling-point Cd^{2+} [75]. The hierarchical porous N-doped aerogel catalyst (N/CA-Cd) with TOCNF/ANF co-crosslinking was obtained through moderate pyrolysis. ANF not only serves as an N-source but also enhances the carbon retention rate and graphitization degree, and increases the electrical conductivity of carbon aerogel. The obtained $N/CA_{0.5}$-Cd catalyst has excellent electrocatalytic efficiency and a half-wave potential of 0.86 V. The peak power density of the zinc-air battery assembled by it reaches 186 mW·cm^{-2}, and the specific capacity is 730 mAh · g^{-1}, providing a sustainable method for the rational design of carbon aerogel. Besides, Zhang et al. [76] prepared a three-dimensional honeycomb-structured bifunction-functional oxygen catalyst based on cellulose nanofibers, composed of FeCo alloy and N and S dual-doped carbon aerogel (NSCA/FeCo), through directional freezing casting and annealing strategies. It combines the characteristics of metal-heteroatom-carbon bonding and graphitized carbon-coated alloy particles, is conductive and has a mesoporous structure, and shows excellent ORR/OER catalytic performance (ΔE is 0.75 V) under alkaline conditions. Based on this, the liquid flow and flexible rechargeable zinc-air battery respectively have high power density, excellent energy density and cycling stability, providing efficient new bifunction-functional electrocatalysts and feasible strategies for related fields.

Wei et al. [72] developed a nitrogen-doped carbon aerogel (HNCA) electrocatalyst based on nanocellulose and graphite-phase carbon nitride (g-C_3N_4) for zinc-air batteries. Figure 17.7d is a schematic diagram of the preparation of HNCA catalyst. Through hydrothermal pretreatment and carbonization process, with CNF as the framework, g-C_3N_4 as the nitrogen source and pore-forming agent, the prepared HNCA has a high specific surface area (785.4 $m^2 \cdot g^{-1}$) and abundant active sites of graphite nitrogen and pyridine nitrogen (69.1%). It demonstrates outstanding oxygen reduction (ORR) and oxygen precipitation (OER) performance. The zinc-air battery assembled by it revealed a specific capacity of 790 mAh $\cdot$ g^{-1}, an energy density of 945.6 Wh$\cdot kg^{-1}$, and a cycling stability of over 100 h, providing a new way for the conversion of biomass into a dual-functional air cathode.

CBA has broad prospects in metal-air batteries. Its high porosity and large specific surface area are conducive to gas diffusion and the storage of discharge products, and can optimize the three-phase interface of the electrode. The abundant hydroxyl and other functional groups on the surface, after modification, can enhance hydrophilicity, hydrophobicity and catalytic activity, and also load catalysts to improve reaction kinetics. In addition, its flexibility and mechanical stability are compatible with flexible devices, and the raw materials are green, renewable and low cost. In the future, multi-level pore structures can be constructed through bionic design, combined with conductive materials to enhance performance, surface modification can be optimized to improve interface stability, low-cost large-scale preparation processes can be developed, and combined with battery system collaborative design, problems such as electrolyte wettability and long-term cycle stability can be solved, breaking through material and engineering bottlenecks. It is expected to become a key material in Li-, Na- and Zn-air batteries, promoting the improvement of energy density and practicability of green high energy density batteries [77].

17.7 Conclusion

In summary, we focused on applications of CBA for energy storage devices including supercapacitors and rechargeable batteries. From the aspects of energy storage mechanism and relationship of structure-performance, the application progress of CBA in supercapacitors, zinc-ion batteries, lithium-ion batteries, and sodium-ion batteries is comprehensively analyzed. It demonstrated that CBA shows great promise for the development of next generation of flexible energy storage devices. However, cellulose foams and aerogels are very "young" materials, research on the applications of cellulose based composite foams and aerogels for flexible energy storage devices is still on the progress at fundamental level and laboratory work, much more efforts will be needed to realize their practical applications.

When employed as energy storage electrodes of CBA, typically demonstrate a "synergistic effect" among their constituent components. However, there remains a scarcity of studies that can qualitatively and quantitatively characterize the synergistic mechanisms between cellulose and active materials, as well as the electrochemical contributions of each component. Consequently, elucidating the ion diffusion and electron conduction processes between the active materials and nanocellulose at the electrode/electrolyte interface presents a significant challenge. In the future, it will be essential to corroborate findings through both theoretical modeling and experimental validation to fully understand the underlying mechanisms. Particularly, the charge storage mechanism of flexible cellulose composite aerogel materials in gel or solid-state flexible electrolytes is notably intricate and challenging to analyze due to ion capture within polymer chains and their interactions across multiple scales. Therefore, further advancements in experimental methodologies and in-situ characterization techniques are required.

References

1. T. Xu, H. Du, H. Liu, W. Liu, X. Zhang, C. Si, P. Liu, K. Zhang, Advanced nanocellulose-based composites for flexible functional energy storage devices. Adv. Mater. **33**, 2101368 (2021)
2. T. Xu, K. Liu, N. Sheng, M. Zhang, W. Liu, H. Liu, L. Dai, X. Zhang, C. Si, H. Du, K. Zhang, Biopolymer-based hydrogel electrolytes for advanced energy storage/conversion devices: properties, applications, and perspectives. Energy Storage Mater. **48**, 244–262 (2022)
3. W. Li, Y. Xu, G. Wang, T. Xu, C. Si, Design and functionalization of lignocellulose-derived silicon-carbon composites for rechargeable batteries. Adv. Energy Mater. **14**, 2403593 (2024)
4. H. Liu, H. Du, T. Zheng, K. Liu, X. Ji, T. Xu, X. Zhang, C. Si, Cellulose based composite foams and aerogels for advanced energy storage devices. Chem. Eng. J. **426**, 130817 (2021)
5. Q. Liang, K. Liu, T. Xu, Y. Wang, M. Zhang, Q. Zhao, W. Zhong, X.M. Cai, Z. Zhao, C. Si, Interfacial modulation of $Ti_3C_2T_x$ MXene by cellulose nanofibrils to construct hybrid fibers with high volumetric specific capacitance. Small **20**, 2307344 (2024)
6. Y. Wang, T. Xu, K. Liu, M. Zhang, X.M. Cai, C. Si, Biomass-based materials for advanced supercapacitor: principles, progress, and perspectives. Aggregate **5**, e428 (2024)
7. T. Xu, Y. Wang, K. Liu, Q. Zhao, Q. Liang, M. Zhang, C. Si, Ultralight mxene/carbon nanotube composite aerogel for high-performance flexible supercapacitor. Adv. Compos. Hybrid. Mater. **6**, 108 (2023)
8. Q. Zhao, T. Xu, K. Liu, H. Du, M. Zhang, Y. Wang, L. Yang, H. Zhang, X. Wang, C. Si, Biomass-based functional materials for rechargeable Zn-ion batteries. Energy Storage Mater. **71**, 103605 (2024)
9. Y. Wang, T. Xu, K. Liu, M. Zhang, Q. Zhao, Q. Liang, C. Si, Nanocellulose-based advanced materials for flexible supercapacitor electrodes. Ind. Crop Prod. **204**, 117378 (2023)

10. L. Zhu, Y. Cao, T. Xu, H. Yang, L. Wang, L. Dai, F. Pan, C. Chen, C. Si, Covalent organic framework membranes for energy storage and conversion. Energy Environ. Sci. **18**, 5675–5739 (2025)
11. J. Qi, M. Zhang, T. Xu, K. Liu, Y. Wang, H. Zhang, X. Wang, Z. Yuan, C. Si, Nanocellulose/metal-organic frameworks composites for advanced energy storage of electrodes and separators. Chem. Eng. J. **500**, 157318 (2024)
12. J. Qi, H. Wang, M. Zhang, T. Xu, X. Wang, H. Zhang, H. Du, J. Hu, K. Liu, C. Si, Simultaneous preparation of lignin-containing cellulose nanocrystals and lignin nanoparticles from wood sawdust by mixed organic acid hydrolysis. Ind. Crop Prod. **222**, 119658 (2024)
13. J. Yang, X. Han, W. Yang, J. Hu, C. Zhang, K. Liu, S. Jiang, Nanocellulose-based composite aerogels toward the environmental protection: preparation, modification and applications. Environ. Res. **236**, 116736 (2023)
14. M. Jaouahar, E.-H. Ablouh, Z. Hanani, M. Humar, M.E. Achaby, H. Sehaqui, Comprehensive characterization of hygroscopic, fire-retardant and thermally insulating sulfated nanocellulose aerogel. Chem. Eng. J. **506**, 160236 (2025)
15. Y. Mao, Y. Sheng, Z. Fan, J. Yang, J. Liu, C. Tang, S. Fu, Atmospheric pressure dried discontinuous pore gradient structured CNF-based aerogel for ultra-low reflection, broadband, and super-high EMI shielding. Adv. Funct. Mater. 2421492 (2025). https://doi.org/10.1002/adfm.202421492
16. H. Liu, T. Xu, Q. Liang, Q. Zhao, D. Zhao, C. Si, Compressible cellulose nanofibrils/reduced graphene oxide composite carbon aerogel for solid-state supercapacitor. Adv. Compos. Hybrid. Mater. **5**, 1168–1179 (2022)
17. H. Zhou, Y. Guan, T. Xu, D. Tang, X. Yu, L. Dai, W. Chen, C. Si, Revealing water-biopolymer interactions: Towards advanced water collection from air. The Innovation (2025). https://doi.org/10.1016/j.xinn.2025.101048
18. K. Du, P. Shi, D. Zhang, Y. Xiao, S. Zhang, Polydopamine-anchored cellulose nanofiber flexible aerogel with high charge transfer as a substrate for conductive materials. ACS Appl. Mater. Interfaces **16**, 30314–30323 (2024)
19. Y. Al Haj, A.B. Soliman, J. Vapaavuori, M. Elbahri, Carbon aerogels derived from anion-modified nanocellulose for adaptive supercapacitor performance. Adv. Funct. Mater. **34**, 2313117 (2024)
20. H. Al Abdallah, J.H. Tannous, B. Abu-Jdayil. Cellulose and nanocellulose aerogels, their preparation methods, and potential applications: a review. Cellulose **31**, 2001–2029 (2024)
21. A.K. Das, M.N. Islam, R.K. Ghosh, R. Maryana, Cellulose-based bionanocomposites in energy storage applications: a review. Heliyon **9**, e13028 (2023)
22. L.J. Andrew, E.R. Gillman, C.M. Walters, E. Lizundia, M.J. MacLachlan, Multi-responsive supercapacitors from chiral nematic cellulose nanocrystal-based activated carbon aerogels. Small **19**, 2301947 (2023)
23. W. Liu, S. Liu, T. Liu, T. Liu, J. Zhang, H. Liu, Eco-friendly post-consumer cotton waste recycling for regenerated cellulose fibers. Carbon. Polym. **206**, 141–148 (2019)
24. X. Yang, K. Shi, I. Zhitomirsky, E.D. Cranston, Cellulose nanocrystal aerogels as universal 3D lightweight substrates for supercapacitor materials. Adv. Mater. **27**, 6104–6109 (2015)
25. Z. Liao, J. Cheng, J.-H. Yu, X.-L. Tian, M.-Q. Zhu, Graphene aerogel with excellent property prepared by doping activated carbon and CNF for free-binder supercapacitor. Carbon. Polym. **286**, 119287 (2022)
26. Z. Chen, H. Zhuo, Y. Hu, H. Lai, L. Liu, L. Zhong, X. Peng, Wood-derived lightweight and elastic carbon aerogel for pressure sensing and energy storage. Adv. Funct. Mater. **30**, 1910292 (2020)
27. Z. Zhao, Y.-P. Hu, K.-Y. Liu, W. Yu, G.-X. Li, C.-Z. Meng, S.-J. Guo, Cellulose nanofiber enabled high-mass loading N-doped porous carbon electrode for zinc-ion supercapacitors. Sustain. Mater. Techno. **41**, e00982 (2024)
28. H. Liu, T. Xu, C. Cai, K. Liu, W. Liu, M. Zhang, H. Du, C. Si, K. Zhang, Multifunctional superelastic, superhydrophilic, and ultralight nanocellulose-based composite carbon aerogels for compressive supercapacitor and strain sensor. Adv. Func. Mater. **32**, 2113082 (2022)

29. C. Xiong, C. Zheng, Z. Zhang, Q. Xiong, Q. Zhou, D. Li, M. Shen, Y. Ni, Polyaniline@cellulose nanofibers multifunctional composite material for supercapacitors, electromagnetic interference shielding and sensing. J. Mater. **11**, 100841 (2025)
30. D.M. Mijailović, P.A.C. Gane, K. Dimić-Mišić, M. Imani, D.B. Stojanović, P.S. Uskoković, High-performance crystalline nanocellulose (CNC) supercapacitor separator: achieved by maximizing alkali-induced gel stress-response prior to aerogel compression forming. J. Power. Sour. **633**, 236476 (2025)
31. H. Wu, J. Mu, Y. Xu, F. Xu, S. Ramaswamy, X. Zhang, Heat-resistant, robust, and hydrophilic separators based on regenerated cellulose for advanced supercapacitors. Small **19**, 2205152 (2023)
32. R. Lou, T. Niu, F. Zhao, L. He, Y. Yuan, G. Wei, G. Lyu, Renewable symmetric supercapacitors assembled with lignin nanoparticles-based thin film electrolyte and carbon aerogel electrodes. Inter. J. Biol. Macromol. **277**, 134474 (2024)
33. C.L. Miao, X.X. Wang, D.H. Guan, J.X. Li, J.Y. Li, J.J. Xu, Spatially confined engineering toward deep eutectic electrolyte in metal-organic framework enabling solid-state zinc-ion batteries. Angew. Chem. **136**, e202410208 (2024)
34. H. Yan, S. Li, J. Zhong, B. Li, An electrochemical perspective of aqueous zinc metal anode. Nano Micro Lett. **16**, 15 (2024)
35. B. Zhang, J. Yao, C. Wu, Y. Li, J. Liu, J. Wang, T. Xiao, T. Zhang, D. Cai, J. Wu, Z.W. Seh, S. Xi, H. Wang, W. Sun, H. Wan, H.J. Fan, Electrolyte design for reversible zinc metal chemistry. Nat. Commun. **16**, 71 (2025)
36. B. Qin, Z.L. Yu, J. Huang, Y.F. Meng, R. Chen, Z. Chen, S.H. Yu, A petrochemical-free route to superelastic hierarchical cellulose aerogel. Angew. Chem. **135**, e202214809 (2023)
37. Y. Chen, L. Zhang, Y. Yang, B. Pang, W. Xu, G. Duan, S. Jiang, K. Zhang, Recent progress on nanocellulose aerogels: preparation, modification, composite fabrication, applications. Adv. Mater. **33**, 2005569 (2021)
38. W. Zhou, M. Chen, Q. Tian, J. Chen, X. Xu, C.-P. Wong, Cotton-derived cellulose film as a dendrite-inhibiting separator to stabilize the zinc metal anode of aqueous zinc ion batteries. Energy Storage Mater. **44**, 57–65 (2022)
39. Q. Zhao, T. Xu, M. Zhang, H. Liu, H. Du, C. Si, Zn@cellulose nanofibrils composite three-dimensional carbon framework for long-life Zn anode. Ind. Crop Prod. **194**, 116343 (2023)
40. H. Shi, J. Cao, W. Sun, G. Lu, H. Lan, L. Xu, Z.A. Ghazi, D. Fan, Z. Mao, D. Han, W. Liu, S. Niu, Ultrasmall, amorphous V_2O_3 intimately anchored on a carbon nanofiber aerogel toward high-rate zinc-ion batteries. ACS Appl. Mater. Interfaces **16**, 18812–18823 (2024)
41. H. Zhang, M. Zhang, T. Xu, X. Wang, J. Qi, Y. Wang, W. Liu, L. Zhu, Z. Yuan, C. Si, Ions and electrons dual transport channels regulated by nanocellulose for mitigating dendrite growth of zinc-ion batteries. Chem. Eng. J. **505**, 159476 (2025)
42. Q. Zhao, H. Zhang, X. Wang, T. Xu, M. Zhang, Y. Wang, L. Zhu, S. Tong, X. Zhou, J. Li, W. Huan, Z. Huang, C. Si, Highly reversible and rapid charge transfer Zn–MnO_2 battery by MnO_2 nanosheet arrays anchored nanocellulose-based carbon aerogel. Adv. Compos. Hybrid Mater. **7**, 90 (2024)
43. W. Yang, Y. Yang, H. Yang, H. Zhou, Regulating water activity for rechargeable zinc-ion batteries: progress and perspective. ACS Energy Lett. **7**, 2515–2530 (2022)
44. J. Zhou, R. Zhang, R. Xu, Y. Li, W. Tian, M. Gao, M. Wang, D. Li, X. Liang, L. Xie, K. Liang, P. Chen, B. Kong, Super-assembled hierarchical cellulose aerogel-gelatin solid electrolyte for implantable and biodegradable zinc ion battery. Adv. Funct. Mater. **32**, 2111406 (2022)
45. R. Xu, J. Zhou, H. Gong, L. Qiao, Y. Li, D. Li, M. Gao, G. Xu, M. Wang, X. Liang, X. Zhang, M. Luo, H. Qiu, K. Liang, Y. Li, Environment-friendly degradable zinc-ion battery based on guar gum-cellulose aerogel electrolyte. Biomater. Sci. **10**, 1476–1485 (2022)
46. H.N. Trinh, M. Eesaee, S. Shahgaldi, J. Singh, T.L.G. Hoang, P. Nguyen-Tri, Advancements and challenges in polymer-based separators for lithium-ion batteries. Energy Storage Mater. **77**, 104153 (2025)

47. Y. Liu, J. Chen, Z. Liu, H. Xu, Y. Zheng, J. Zhong, Q. Yang, H. Tian, Z. Shi, J. Yao, C. Xiong, Facile fabrication of Fe_3O_4 nanoparticle/carbon nanofiber aerogel from Fe-ion cross-linked cellulose nanofibrils as anode for lithium-ion battery with superhigh capacity. J. Alloys Compd. **829**, 154541 (2020)
48. C.D. Pham, K.D. Tran, T.M. Truong, P.K. Le, Cellulose-derived carbon aerogel from rice straw for high-performance lithium-ion battery anodes. Biomass Convers Biorefin. **15**, 4407–4415 (2024)
49. G. Ye, X. Zhu, S. Chen, D. Li, Y. Yin, Y. Lu, S. Komarneni, D. Yang, Nanoscale engineering of nitrogen-doped carbon nanofiber aerogels for enhanced lithium-ion storage. J. Mater. Chem. A. **5**, 8247–8254 (2017)
50. Y. Wan, Z. Yang, G. Xiong, R. Guo, Z. Liu, H. Luo, Anchoring Fe_3O_4 nanoparticles on three-dimensional carbon nanofibers toward flexible high-performance anodes for lithium-ion batteries. J. Power. Sour. **294**, 414–419 (2015)
51. C. Li, L. Ou, Y. Liu, L. Xu, S. Zhou, L. Guo, H. Liu, Z. Zhang, M. Cui, G. Chen, J. Huang, J. Tao, A cellulose/polyethylene oxide gel polymer electrolyte with enhanced mechanical strength and high ionic conductivity for lithium-ion batteries. Adv. Mater. Technol. **12**, 2202002 (2023)
52. J. Wan, J. Zhang, J. Yu, J. Zhang, Cellulose aerogel membranes with a tunable nanoporous network as a matrix of gel polymer electrolytes for safer lithium-ion batteries. ACS Appl. Mater. Interfaces **9**, 29 (2017)
53. Y. Zhao, Y. Kang, J. Wozny, J. Lu, H. Du, C. Li, T. Li, F. Kang, N. Tavajohi, B. Li, Recycling of sodium-ion batteries. Nat. Rev. Mater. **8**, 623–634 (2023)
54. W. Chen, H. Yu, S.-Y. Lee, T. Wei, J. Li, Z. Fan, Nanocellulose: a promising nanomaterial for advanced electrochemical energy storage. Chem. Soc. Rev. **47**, 2837–2872 (2018)
55. C. Li, Q. Fu, K. Zhao, Y. Wang, H. Tang, H. Li, H. Jiang, L. Chen, Nitrogen and phosphorous dual-doped graphene aerogel with rapid capacitive response for sodium-ion batteries. Carbon **139**, 1117–1125 (2018)
56. H. Zhou, J. Gu, W. Zhang, C. Hu, X. Lin, Rational design of cellulose nanofibrils separator for sodium-ion batteries. Molecules **26**, 5539 (2021)
57. Y. Lu, D. Li, C. Lyu, H. Liu, B. Liu, S. Lyu, T. Rosenau, D. Yang, High nitrogen-doped carbon nanofiber aerogels for sodium-ion batteries: synergy of vacancy defects to boost sodium ion storage. Appl. Surf. Sci. **496**, 143717 (2019)
58. Y. Huang, L. Wang, L. Lu, M. Fan, F. Yuan, B. Sun, J. Qian, Q. Hao, D. Sun, Preparation of bacterial cellulose based nitrogen-doped carbon nanofibers and their applications in the oxygen reduction reaction and sodium-ion battery. New J. Chem. **42**, 7407–7415 (2018)
59. J. Wang, Z. Xu, J.-C. Eloi, M.-M. Titirici, S.J. Eichhorn, Ice-templated, sustainable carbon aerogels with hierarchically tailored channels for sodium- and potassium-ion batteries. Adv. Funct. Mater. **32**, 2110862 (2022)
60. H. Zhu, F. Shen, W. Luo, S. Zhu, M. Zhao, B. Natarajan, J. Dai, L. Zhou, X. Ji, R.S. Yassar, T. Li, L. Hu, Low temperature carbonization of cellulose nanocrystals for high performance carbon anode of sodium-ion batteries. Nano Energy **33**, 37–44 (2017)
61. S. Li, J. Warzywoda, S. Wang, G. Ren, Z. Fan, Bacterial cellulose derived carbon nanofiber aerogel with lithium polysulfide catholyte for lithium-sulfur batteries. Carbon **124**, 212–218 (2017)
62. S. Li, T. Mou, G. Ren, J. Warzywoda, Z. Wei, B. Wang, Z. Fan, Gel-based sulfur cathodes with a high sulfur content and large mass loading for high-performance lithium-sulfur batteries. J. Mater. Chem. A. **5**, 1650–1657 (2017)
63. X. Lin, W. Li, X. Pan, S. Wang, Z. Fan, Electrocatalytic and conductive vanadium oxide on carbonized bacterial cellulose aerogel for the sulfur cathode in Li-S batteries. Batteries. **9**, 14 (2023)
64. Y.-E. Liu, M.-G. Zhang, Investigation of the effect of anion/cation-modified cellulose nanofibers/MXene composite aerogels on the high-performance lithium-sulfur batteries. Ionics **28**, 2805–2815 (2022)
65. Y. Liu, M. Zhang, J. Guo, High-performance lithium–sulfur battery based on carbonized 3D MXene/T-CNF aerogel composite membrane. Ionics **28**, 647–655 (2022)

66. X. Lin, W. Li, V. Nguyen, S. Wang, S. Yang, L. Ma, Y. Du, B. Wang, Z. Fan, Fe-single-atom catalyst nanocages linked by bacterial cellulose-derived carbon nanofiber aerogel for Li-S batteries. Chem. Eng. J. **477**, 146977 (2023)
67. M. Cheng, W. Xia, R. Wang, S. Pan, T. Shen, Q. Liu, J. Hu, T. Wei, Y. Ling, W. Li, B. Liu, Highly dispersed cobalt clusters/nanoparticles embedded in cellulose nanocrystal derived carbon aerogel as efficient sulfur host for high performance lithium-sulfur batteries. J. Power. Sour. **608**, 234645 (2024)
68. H. Yu, X. Zhou, P. Zeng, C. Guo, H. Liu, X. Guo, B. Chang, M. Chen, J. Su, X. Wang, 3D cellulose graphene aerogel with self-redox CEO_2 as Li_2S host for high-performance Li–S battery. Energy Technol. **10**, 2200616 (2022)
69. J. Chen, Y. Liu, Z. Liu, Y. Chen, C. Zhang, Y. Yin, Q. Yang, Z. Shi, C. Xiong. Carbon nanofibril composites with high sulfur loading fabricated from nanocellulose for high-performance lithium-sulfur batteries. Colloids Surf, A. **603**, 125249 (2020)
70. Y.-E. Liu, M.-G. Zhang, Y.-N. Gao, J. Guo, Regulate the reaction kinetic rate of lithium-sulfur battery by rational designing of TEMPO-oxidized cellulose nanofibers/RGO porous aerogel with monolayer MXene coating. J. Alloys Compd. **898**, 162821 (2022)
71. H. Mao, L. Liu, L. Shi, H. Wu, J. Lang, K. Wang, T. Zhu, Y. Gao, Z. Sun, J. Zhao, G. Gao, D. Zhang, W. Yan, S. Ding, High loading cotton cellulose-based aerogel self-standing electrode for Li-S batteries. Sci Bull. **65**, 803–811 (2020)
72. Y. Wei, Y. Huang, Y. Pi, W. Yang, L. Wu, Y. Luo, Z. Chen, Y. Chen, G. Shi, X. Chi, X. Peng, Preparation of nanocellulose-based nitrogen-doped carbon aerogel electrocatalysts through hydrothermal pretreatment for zinc–air batteries. New J. Chem. **48**, 5582–5588 (2024)
73. M. Barman, M. Pal, R. Biswas, A. Dutta, A comprehensive review of metal-air batteries: mechanistic aspects, advantages and challenges. Catal. Today **451**, 115229 (2025)
74. M. Shen, W. Hu, C. Duan, J. Li, S. Ding, L. Zhang, J. Zhu, Y. Ni, Cellulose nanofibers carbon aerogel based single-cobalt-atom catalyst for high-efficiency oxygen reduction and zinc-air battery. J. Colloid Interface Sci. **629**, 778–785 (2023)
75. Q. Liu, M. Shen, C. Duan, L. Zhang, J. Zhu, Y. Ni, 3d hierarchical porous carbon aerogel electrocatalysts based on cellulose/Aramid nanofibers and application in high-performance Zn–Air batteries. ACS Appl. Energy Mater. **5**, 15146–15154 (2022)
76. Y. Zhang, X. Zhang, Y. Li, J. Wang, S. Kawi, Q. Zhong, FeCo alloy/N, S Co-doped carbon aerogel derived from directional-casting cellulose nanofibers for rechargeable liquid flow and flexible Zn-Air batteries. Nano Res. **16**, 6870–6880 (2023)
77. Y. Chen, J. Xu, P. He, Y. Qiao, S. Guo, H. Yang, H. Zhou, Metal-air batteries: progress and perspective. Sci. Bull. **67**, 2449–2486 (2022)

Chapter 18
Cellulose-Based Aerogels for Bone Tissue Engineering

Hani Nasser Abdelhamid

Abstract Most reported methods or materials for producing porous three-dimensional (3D) scaffolds have considerable obstacles, such as insufficient mimicry of the natural extracellular matrix (ECM) structure, dependence on unsustainable production techniques, and restricted functional efficacy. Polysaccharides, such as cellulose, represent a viable alternative, characterized by their abundance, renewability, biodegradability, and biocompatibility, while also exhibiting structural attributes similar to those of the ECM. Cellulose-based materials, including cellulose nanofibrils (CNFs), nanocrystals (CNCs), and bacterial nanocellulose (BC), are distinguished by their capacity to create highly porous and interconnected 3D aerogels. These structures are suitable for biomedical applications and can be produced using sustainable, scalable, and economical freeze-drying methods. The properties of cellulose aerogels can be improved by including other materials such as nanoparticles and advanced materials. This book chapter highlights the current progress in creating 3D cellulose-based aerogels using advanced methods for tissue engineering applications, such as electrospinning, 3D printing, and freeze-drying. It discusses essential topics such as scaffold design, fabrication, and their applications from both in vitro and in vivo assessments. These advancements highlight the promise of cellulose-based 3D scaffolds as adaptable and efficient biomaterials for the next generation.

Keywords Cellulose · Aerogels · Tissue engineering · Scaffolds

H. N. Abdelhamid (✉)
Department of Chemistry, College of Science, Imam Mohammad Ibn Saud Islamic University (IMSIU), Riyadh, Saudi Arabia
e-mail: hnabdelhamid@imamu.edu.sa

G. Wei and J. Zhang (eds.), *Cellulose-Based Hydrogels and Aerogels*, Lecture Notes in Nanoscale Science and Technology 41,
https://doi.org/10.1007/978-3-032-03008-5_18

18.1 Introduction

Cellulose is a natural biopolymer composed of anhydroglucose monomers interconnected by β-(1–4) linkages [1, 2]. It was initially reported and used by human culture since the era of Egyptian papyri. It has been used a chemical raw material for several applications. Cellulose is frequently considered the most abundant polymer on the Earth. It is predominantly preserved inside plants and microbes. It can be produced in nature through photosynthesis. It has historically been seen as an abundant and sustainable ecological resource. Plants, seaweeds, sugarcane bagasse, tunicates, marine algae, and bacteria are the primary sources of cellulose. Cellulose demonstrates excellent mechanical, physical, and chemical attributes, including substantial stability in acidic environments, chirality, high tensile strength, good elastic properties, low density, biodegradability, and an abundance of hydroxyl functional groups on its surfaces, facilitating chemical modification with enhanced wettability [3]. The yearly output of cellulose exceeds multi-billion tons, with increasing market demand. Consequently, cellulose has progressed several fields, such as sensors [4, 5], environmental [6–10], and biomedical-based applications [11–14].

Human tissues may degenerate or sustain damage from multiple factors, including trauma, congenital anomalies, infections, and diseases [15]. In light of the rising incidence of such disorders, there is an escalating interest in devising techniques to repair or regenerate these tissues [16, 17]. Tissue loss or dysfunction has conventionally been addressed through transplantation methods, including autografts (tissues sourced from the same individual), allografts (from a different human donor), and xenografts (from other sources e.g., animal). Although these techniques have been crucial in preserving lives and restoring functionality, they are associated with constraints including donor site morbidity, immunological rejection, pathogen transmission risk, and substantial financial expense [18]. These obstacles have resulted in the emergence of tissue engineering as a promising alternative that seeks to replace and restore damaged tissues and organs [19]. Tissue engineering advanced the biomedical treatment, combining biology, materials science, and engineering to fabricate viable replacements for lost or damaged tissue. This strategy relies on scaffolds, porous, three-dimensional (3D) constructs replicating the extracellular matrix (ECM). These scaffolds offer a structural framework for cellular adhesion, migration, proliferation, and differentiation, while enabling nutrient delivery and waste disposal. Optimal scaffolds should emulate the mechanical strength and biological functionality of native ECM, including a highly porous and interconnected structure that facilitates tissue integration.

Several materials were reported for tissue engineering [20]. Advanced materials are extensively utilized in scaffold manufacturing; nevertheless, their utilization is frequently impeded by biocompatibility, biodegradability, and environmental sustainability concerns [21]. To achieve clinical and commercial success, tissue-engineered constructions must be biologically functional, cost-efficient, and scalable for mass production. An essential phase in scaffold manufacturing entails maintaining the structural integrity of porous materials, including aerogels, throughout solvent extraction operations, ensuring stability and usefulness. Various ways have been devised to create scaffolds with consistent and reproducible topologies utilizing diverse biomaterials. Aerogels, characterized by their nanoscale linked structures, serve as scaffolds for tissue engineering by providing adjustable porosity, customizable mechanical flexibility, and shape memory, attributes especially beneficial for repairing complex or irregular bone lesions [22]. New advancements and technologies offer high efficiencies in scaffold design, cell biology, materials engineering, tissue engineering, and transforming regenerative medicine. These achievements provide significant potential for fulfilling unmet clinical requirements. Among several biomaterials, cellulose-based materials are promising for tissue engineering [23–26].

18.2 Cellulose and Its Derivatives

Cellulose is a sugar polymer (polysaccharide, or biopolymer), of β-1,4-D-glucopyranose units, inherently establishing robust intramolecular hydrogen bonds that enhance interchain cohesion and facilitate crystalline domain development. It is composed of glucose monomers interconnected by β-(1–4) linkages $(C_6H_{10}O_5)_n$; where n denotes the degree of polymerization, ranging from 200 to 44,000. The degree of polymerization depends on several parameters including the cellulose extraction source. It has attracted considerable attention as a sustainable and biocompatible substitute for reinforcing composite materials at the microscale. The structural characteristics of cellulose confer remarkable mechanical strength and stability, while its non-toxic and biodegradable properties render it a highly desirable material for biomedical applications. Due to its significant water absorption capacity, porosity, flexibility, and adjustable mechanical features, cellulose obtained from natural and microbiological sources is progressively utilized to produce suspensions, hydrogels, aerogels, films, and drug delivery systems [27]. Cellulose can be processed into several derivatives, including carboxymethyl cellulose (CMC), microfibrillated cellulose (MFC), methyl cellulose (MC), microcrystalline cellulose (MCC), nanofibrillated cellulose (NFC), 2,2,6,6-tetramethyl-piperidine-1-oxyl (TEMPO)-mediated oxidized cellulose nanofibers (TOCNFs), cellulose nanocrystals (CNCs), cellulose nanowhiskers (CNW), and bacterial cellulose/nanocellulose (BC/BNC) depending on the preparation method (Fig. 18.1) [26, 28].

Cellulose is a linear, semi-crystalline organic polymer of β-1,4-linked D-glucopyranose units (Fig. 18.1). The structure comprises crystalline (ordered) and amorphous (disordered) parts, with crystallinity generally varying from 40 to 70%, depending on the source and processing technique. The polymer chains assume a chair conformation and establish substantial intra- and intermolecular hydrogen

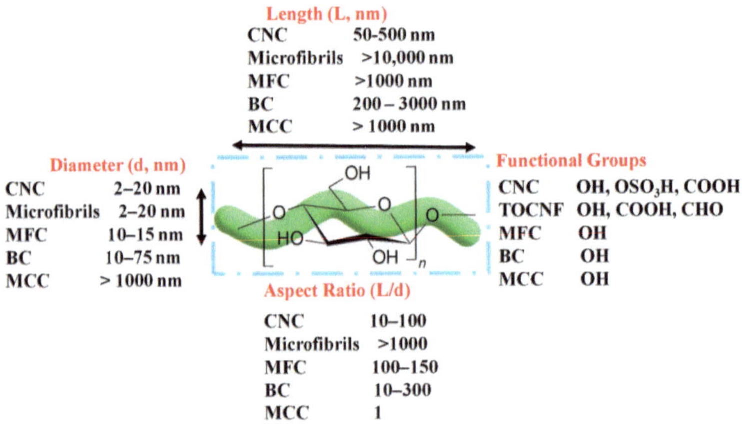

Fig. 18.1 Cellulose and its derivatives with properties such as functional groups and sizes. Figure printed from Open Access Ref. [29]

bonding through hydroxyl groups, leading to a highly cohesive microfibrillar network that resists disintegration in typical solvents. Cellulose contains a reducing end, a non-reducing end, and many glucose rings in its structure. The degree of polymerization (DP) exhibits considerable variation, ranging from roughly 200 in regenerated cellulose to exceeding 44,000 in Valonia.

Besides its structural characteristics, cellulose demonstrates hydrophilic properties owing to its hydroxyl groups, which also offer reactive sites for functionalization. These groups facilitate the creation of hydrogels and aerogels via physical procedures (e.g., blending, filler integration) or chemical approaches (e.g., cross-linking, grafting) [30]. Cationization, which entails incorporating positive or negative charges into the cellulose chain, improves water swelling and decreases the fiber cohesion, enhancing dispersion and facilitating the introduction of certain functions. These surface alterations are essential for improving performance in composite and biological applications.

Hemicellulose (HC) is the second most prevalent polysaccharide in lignocellulosic biomass, constituting approximately 20–35% [31]. It is a branching, heterogeneous polymer comprised of diverse sugar units, including xylose, mannose, glucose, and arabinose. In contrast to cellulose, HC possesses a reduced molecular weight and decreased structural regularity, constraining its commercial utilization in high-value applications. The variety in composition across various biomass sources further complicates its processing and incorporation into sophisticated materials.

Cellulose is available commercially, from microscale to nanoscale (Table 18.1). MCC comprises micron-sized cellulose particles and is extensively utilized in the pharmaceutical and food sectors. Microfibrillated cellulose (MFC), generally generated from mechanical processing of wood, contains a combination of crystalline and amorphous domains, establishing stable networks suitable for hydrogel synthesis.

Microfibrils are elongated cellulose fibers (diameter 2–20 μm, length exceeding 10 μm) utilized in structural composites.

CNCs and cellulose nanowhiskers (CNWs) are generated via acid hydrolysis at the nanoscale (Table 18.1) [32]. CNCs are rod-shaped entities (2–20 nm in diameter, extending up to several hundred nanometers in length) characterized by high crystallinity, low thermal expansion, and non-cytotoxicity, rendering them exceptional candidates for the reinforcement of biomaterials. CNWs display morphological similarities but are shorter and demonstrate somewhat increased crystallinity. They can infiltrate cells, showing potential for medication administration and bioimaging, but moderate cytotoxic effects have been observed in suspension.

CNFs are generated via mechanical defibrillation, frequently accompanied by chemical preparation. Their surface chemistry can be altered using techniques such as TEMPO-mediated oxidation, specifically targeting the C6 hydroxyl group of glucose,

Table 18.1 Summary of cellulose-based derivatives

Form	Preparation method	Size range	Diameter (nm/μm)	Length (nm/μm)	Functional groups	Description
MCC	Partial acid hydrolysis	Micro	>1 μm	<1 μm	Hydroxyl (−OH)	Micro-particles, used in pharma/food as filler/binder
MFC	Mechanical disintegration (e.g., refining, homogenization)	Micro	2–20 μm	>10 μm	Hydroxyl (−OH)	Fibrous with mixed crystalline/amorphous structure, for hydrogels and coatings
CNCs	Acid hydrolysis (e.g., hydrochloric acid)	Nano	2–20 nm	50–500 nm	Hydroxyl (−OH), sulfate half-esters ($-OSO_3^-$)	Rigid rods, high stiffness, used for reinforcement and optics
CNWs	Sulfuric acid hydrolysis	Nano	~5–20 nm	<200 nm	Hydroxyl (−OH), sulfate groups ($-OSO_3^-$)	Shorter CNCs with high crystallinity, drug delivery potential
CNFs	Mechanical fibrillation, TEMPO/carboxymethylation/oxidation	Nano to submicron	3–50 nm	>1 μm	Hydroxyl (−OH), carboxyl (−COOH), aldehyde (−CHO)	Flexible, high strength, tunable bioactivity, and porosity
BNC	Bacterial synthesis (*Acetobacter xylinum*)	Nano to micron	10–75 nm	200–3000 nm	Hydroxyl (−OH)	Moldable gel, pure and strong, for biomedical implants

or periodate oxidation to include aldehyde groups [10, 33]. CNFs have remarkable mechanical strength and adjustable surface charge, substantially affecting cell proliferation and scaffold rigidity. BC or BNC, produced by bacteria like Acetobacter xylinum, possesses excellent purity, crystallinity exceeding 85%, and mechanical strength. It possesses exceptional water retention capabilities and can be shaped into diverse forms, rendering it appropriate for biomedical applications. Nonetheless, its gradual breakdown rate following implantation poses a constraint for osseointegration, requiring additional modification via enzymatic, pharmacological, or genetic methods.

18.3 Tissue Engineering

Human tissues may deteriorate or be damaged for several reasons, including injuries and trauma. Consequently, increasing interest is in promoting repairing or regenerating such damaged tissues. Traditional therapies generally emphasize the transplantation of tissue from one location to another, whether sourced from the same individual (autograft), a donor (allograft), or an alternative species (xenograft). These treatments suffer from challenges such as donor site morbidity, pathogen infection during transmission, immune or body rejection, and costly. Thus, tissue engineering offers promising prospects for treating injured tissues and organs by facilitating their regeneration, rather than relying on donor transplantation.

Tissue engineering depends on porous structures, known as "scaffolds," which serve as templates for tissue creation and facilitate an environment for nutrient delivery and cell metabolite transport. The primary role of a scaffold is to provide alternative characteristics inherent in the ECM that are essential for the proper functioning of native cells. These scaffolds must exhibit structural parallels to the ECM by providing a highly porous, interconnected 3D structure with good mechanical properties, promoting cell migration, attachment, and differentiation. The fabrication of scaffolds can be achieved using several materials, i.e., synthetic polymers. Materials used for scaffold preparation play an essential role in the engineering of tissue applications. They should fulfill specific criteria, including biocompatibility, biodegradability, sustainability, and environmental friendliness. For a tissue-engineered construct to be commercially and therapeutically viable, it must be cost-effective and scalable from laboratory to batch production. Porous, structured materials can be derived from the wet gel by efficiently eliminating the solvent while preserving the structural framework without collapse or fracture. Consequently, numerous strategies facilitate the acquisition of scaffolds for tissue regeneration.

There are several methods for scaffold preparation. Table 18.2 summarizes most of these methods, including direct casting in models, 3D printing [34, 35], electrospinning [36–38], and freeze-drying (lyophilization) [39]. Free-drying stands out compared to other fabrication techniques due to its sustainability, scalability, and cost-effectiveness. It is particularly suitable for preparing scaffolds from heat-sensitive materials, enabling the formation of aerogels with extremely high porosity, a critical

Table 18.2 Summary of fabrication techniques for 3d scaffold development in tissue engineering

Fabrication technique	Advantages	Disadvantages	Notes on process/additional details
Solvent casting	-Tunable pore size -Simple -Cost-effective	-Limited pore interconnectivity -Require in some case volatile organic solvents -Few solvent options	Polymer is dissolved in a solvent and mixed with porogen; the solvent evaporates and leaches out
Phase separation	Good structural control	-Limited solvent choice -Complex operation	Phase separation between polymer and solvent at controlled temperatures creates micro- and macro-porous structures, which are not easily scalable
Hydrogel microfabrication	-Precise microstructural control -Tuneable mechanical properties	-Limited interconnectivity in 3DLimited materials	Used mainly for soft tissues, microfabrication methods include photolithography and molding. Limited for load-bearing scaffolds due to poor structure
Electrospinning	-Tunable nanofiber morphology -Broad material range	-Environmental sensitivity -Low fiber density and interconnectivity	High surface area fibers require stacking or additional steps for 3D structures. Often lacks pore interconnectivity needed for tissue ingrowth
Vacuum/ microwave drying	Generates meso- and micropores	-Risk of pore collapse -Pore size control is limited	Involves fast drying of wet gels. It is not ideal for maintaining structural fidelity; it can lead to inconsistent architectures
3D printing	-Highly controlled structure -SustainableGood mechanical properties	-Limited printable materials -Lower resolution for complex tissue types	CAD-based precision allows the production of tissue-specific architectures. Limited by available bioinks and often requires post-processing for bioactivity
Supercritical CO_2 drying ($ssCO_2$)	-High porosity and surface area -Prevents pore collapse	-Requires organic solvents -High-pressure system -Expensive	Aerogel formation: (1) Solution → (2) Gelation → (3) Supercritical drying Preserves fine structure and avoids capillary collapse. Ideal for temperature-sensitive biomolecules, but complex and high-cost

(continued)

Table 18.2 (continued)

Fabrication technique	Advantages	Disadvantages	Notes on process/additional details
Freeze-drying (Lyophilization)	-Sustainable and low-cost -Non-toxic, water-based -Mild, low-temperature drying process -Ideal for heat-sensitive biomaterials -Highly porous (up to 99%)	-Limited mechanical strength due to high porosity -Requires optimization of parameters	Aerogel Formation Steps: (1) Solution transition—polymer dissolved to form a hydrogel (2) Sol–gel transition—network formation via cross-linking (3) Freeze-drying—sublimation of ice directly to vapor creates porosity

requirement for tissue engineering, especially for vascularization and nutrient transport in vivo. Freeze-drying can be combined with several advanced methods, such as 3D printing, enabling drying of the scaffolds without destroying the materials' structures. It can also be used to develop aligned macroporous structures [40].

The structure and properties of scaffolds can be tuned by altering the rates of scaffold preparation, mold types, and cross-linking methods, making it a flexible and powerful technique. However, the main challenge is balancing high porosity and good mechanical properties. The solvent removal method can synthesize porous 3D-structured scaffolds derived from cellulose hydrogels for tissue engineering.

Maintaining the structural integrity of cellulose-based scaffolds is critical. There are several methods to preserve the scaffold's structure. One of those methods is cross-linking. Cross-linking is essential for preparing stable and functional cellulose aerogels, improving the aerogel network's structural integrity, mechanical strength, and durability [30]. Choosing a suitable cross-linking approach is crucial for enhancing the performance of cellulose-based aerogels. Cross-linking entails the establishment of chemical or physical links between polymer chains, resulting in a 3D network that preserves its structure throughout drying and subsequent application. Different cross-linking methodologies have been utilized in fabricating cellulose aerogels, each with unique mechanisms, benefits, and constraints. Physical cross-linking techniques depend on non-covalent interactions, including hydrogen bonding, Van der Waals forces, ionic interactions, coordination bonds, π-π stacking, electrostatic forces, dehydration heat treatment, and ultraviolet (UV) irradiation. These methods are typically regarded as secure, economical, and induce minimal tissue response, rendering them appealing for biomedical applications. Nonetheless, physical cross-linking frequently results in a comparatively low level of network stabilization, is labor-intensive, and is constrained by inadequate control over cross-linking density and homogeneity.

Chemical cross-linking establishes robust covalent connections between cellulose chains by the use of cross-linkers like calcium ions (Cd^{2+}), glutaraldehyde (GA), citric acid (CA), glyoxal (GP), hexamethylene diamine (HMDA), and tannic acid (TA). These chemical approaches enhance the resultant aerogels' mechanical durability and structural integrity. However, the cytotoxicity of some chemical agents, especially aldehydes, presents safety concerns for biomedical applications and requires post-processing or the selection of friendly alternatives.

Enzymatic cross-linking presents a more biologically advantageous method by employing an enzyme such as horseradish peroxidase, transglutaminase, and tyrosinase with oxidizing agents like hydrogen peroxide (H_2O_2). This technique facilitates exact regulation of the cross-linking reaction via parameters such as pH, temperature, and ionic strength, permitting the production of aerogels with customized characteristics. In general, enzymes are costly with high restriction for specific substrate.

18.4 Cellulose for Tissue Engineering

Renewable polymers, such as proteins and polysaccharides, have been investigated as alternatives for creating scaffolds using straightforward and economical methods. 3D porous cellulose scaffolds have garnered significant attention. Nanocellulose has become a compelling option for aerogel manufacturing due to its superior mechanical properties, biocompatibility, and porous 3D structure.

Mathew's group reported the effects of integrating chitosan into cellulose nanofibers to develop freeze-dried porous 3D scaffolds for cartilage tissue regeneration [41]. The addition of the second component reduced the mechanical characteristics of the composite. A low concentration of the CNF matrix phase is crucial for fiber adhesion, hence maintaining scaffold integrity and improving cellular interaction with the microenvironment. 3D porous scaffolds composed entirely of bio-based materials were fabricated utilizing freeze-dried CNFs (70–90 wt.%) and stabilized with a genipin cross-linked matrix of gelatin and chitosan. Scanning electron microscopy (SEM) images revealed that the scaffolds possess interconnected pores with average sizes ranging from 75 to 200 μm and nanoscale pore wall roughness conducive to cellular interactions in cartilage regeneration. X-ray tomography validated the 3D uniformity and interconnectedness of the pores, along with the fibrillar architecture of the scaffolds. The present study effectively tackled the difficulty of concurrently customizing the pore shape and mechanical properties for cartilage regeneration. Moreover, the scaffolds' high porosity ($\approx$95%) and favorable cytocompatibility with chondrocytes are advantageous for cellular adhesion and ECM synthesis [41].

A highly porous 3D scaffold was prepared utilizing a composite of pectin, CMC, and 0.1% MFC, cross-linked with calcium chloride ($CaCl_2$), by the lyophilization procedure [42]. The improved scaffold, designated C(0.1%), has shown considerable promise for tissue engineering applications due to its advantageous structural, mechanical, thermal, and biological characteristics [42]. Micro-computed tomography (micro-CT) study indicated that the C(0.1%) scaffold exhibited a pore size between 10 and 250 μm, aligning with the optimal range for tissue engineering, facilitating efficient nutrient flow and cellular infiltration. The scaffold's porosity was determined to be roughly 88%, facilitating improved mass transfer inside the structure. In contrast to the control scaffold (C(0%)), which exhibited a pore size range of 30–300 μm, the incorporation of MFC yielded a more consistent and regulated porosity. Mechanical evaluation revealed that C(0.1%) attained the greatest compression modulus of around 3.987 MPa, signifying enhanced load-bearing capacity. The C(0.1%) scaffold demonstrated a reduced degradation rate and regulated water absorption in phosphate-buffered saline (PBS), maintaining structural integrity for an extended period during in vitro evaluation. Cytocompatibility assays utilizing NIH3T3 fibroblast cells demonstrated the best cell viability in the C(0.1%) scaffold compared to all other compositions, signifying exceptional biocompatibility and appropriateness for regenerative medicine. The aerogel exhibited several advantages, such as high porosity, ideal pore dimensions, enhanced mechanical and thermal characteristics, and extraordinary biocompatibility, rendering the pectin/CMC/MFC (C(0.1%)) composite scaffold a compelling choice for tissue regeneration applications [42].

Cellulose-based tissue engineering exhibits good medical properties, such as high antibacterial activity. Thus, there is a high interest in developing adipose tissue for both in vitro and in vivo applications. Two-dimensional (2D) adipocyte culture systems are inadequate, necessitating the development of 3D culture methods to provide a more realistic model for investigating metabolic disorders such as obesity and diabetes. Porous BNC scaffolds were reported for 3D adipose tissue [43]. It was applied for the adipogenic development of murine mesenchymal stem cells derived from a cell line. To achieve a porous architecture, the 3D scaffolds were fabricated by cross-linking homogenized cellulose fibrils with alginate and freeze-drying the mixture. Rapidly cooling the scaffolds in liquid nitrogen produced smaller holes than gradually freezing with isopropanol. The cells were distributed on 2D surfaces and exhibited minimal lipid droplet production, whereas cells cultured in macroporous 3D scaffolds displayed increased clustering and contained larger lipid droplets. The scaffold types exhibited a substantial presence of adipocytes; however, scaffolds with smaller pores harbored larger cell clusters compared to those with larger pores, with viable adipocytes remaining even four weeks post-differentiation. Scaffolds with low alginate content maintained their pore integrity more effectively. BNC-based scaffolds exhibit macroporous scaffolds, representing a feasible technique for constructing adipose tissue as an in vitro model for adipose biology and metabolic disorders [43].

Pure CNC aerogels featuring regulated 3D structures and internal pore architecture are fabricated utilizing the direct ink writing (DIW) process and freeze-drying method (Fig. 18.2) [44]. Conventional cellulosic aerogel processing methods are limited in their capacity to construct comprehensive aerogel structures; however, DIW 3D printing combined with freeze drying can address this limitation, yielding CNC aerogels with negligible structural shrinkage or damage. CNC gels at 11.8–30 wt.% in water were synthesized and 3D printed into a basic 1 cm^3 cubic structure utilizing the DIW method (Fig. 18.2). Following freeze-drying, the density and porosity of the aerogels were assessed based on their weight and volume. As the weight percentage of CNC increased, the density increased while the porosity dropped linearly. The density and porosity can be extrapolated to the theoretical bulk cellulose values of 1600 mg/cm^3 and 0%, respectively. The linear reduction in porosity can be attributed to lower pore diameters within the aerogel. During freeze-drying, smaller ice crystals formed in structures with high CNC loadings due to reduced water content in the gel. At low freezing rates, the development of ice crystals was significantly more prevalent than their nucleation. As ice crystals developed, they further expanded and compelled CNC to aggregate, in addition to CNC's inherent propensity to create robust hydrogen bonds with itself. The obtained products possess significant promise for applications including tissue scaffold templates, drug delivery, and packaging, owing to their intrinsic sustainability, biocompatibility, and biodegradability. Diverse three-dimensional structures were successfully printed without support material, and print quality can be improved by augmenting CNC concentration and printing resolution. Dual-pore CNC aerogel scaffolds have been produced, enabling customizable 3D structures and internal pore topologies that may enhance sophisticated CNC scaffold designs suited to particular cell integration requirements [44].

TOCNFs were synthesized, with CNF1 exhibiting a higher carboxyl fraction (1.0613 mmol/g) and CNF2 displaying a lower carboxyl fraction (0.5543 mmol/g) [45]. These CNFs, characterized by varying carboxyl and aldehyde fractions, served as the primary aerogel matrix. TOCNF was composed of chitosan (Ch), gelatin (Gl), and alginate (Al) biopolymers in various proportions. Aerogels and their applicability in wound dressings were synthesized using a straightforward freezing and lyophilization process. Water-PBS (phosphate buffered saline) absorption rates, water vapor permeability (WVP), antibacterial activity, and biocompatibility were assessed alongside micro and morphological imaging. CNF/Ch and CNF/Gl bioaerogels exhibited a superior antibacterial effect against both bacterial strains. Conversely, CNF-Al-based aerogels exhibited little antibacterial efficacy. All aerogels exhibit

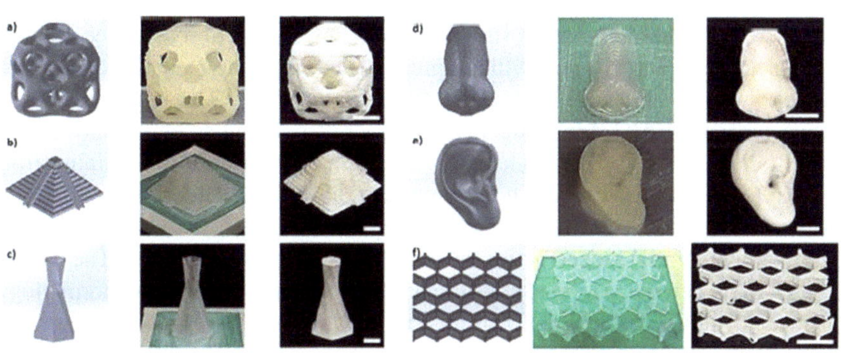

Fig. 18.2 DIW 3D printed octet cube (**a**), pyramid (**b**), hexagonally twisting vase (**c**), nose model (**d**), ear model (**e**), and honeycomb (**f**). Scale bars are 1 cm. Figure reprinted with permission from Ref. [44]. Copyright © 2017, The Author(s)

biocompatibility, facilitating cellular growth on the material's surface. Consequently, these aerogels may serve as wound dressings [45].

Assessment of bioactivity and initial hemocompatibility of dual-porous cellulose phosphate aerogels was reported and compared to a phosphate-free reference, produced through coagulation (water/ethanol), solvent exchange, and supercritical CO_2 drying [46]. The deposition of hydroxyapatite from simulated bodily fluid after 7 days of immersion demonstrated significant bioactivity, ranging from 1.5 to 2.2 mg Ca^{2+} per mg of scaffold. $ScCO_2$-dried and rehydrated scaffolding materials were incubated in heparin-anticoagulated human whole blood to examine specific parameters of hemostasis and the inflammatory response. The adhesion of leukocytes to the treated surfaces was evaluated using scanning electron microscopy and fluorescence microscopy. Data analysis indicates that phosphorylation at low DS does not enhance platelet activation. A high rise in platelet activation and thrombin generation was detected after Ca^{2+} ions neutralized a specific proportion of the negative surface charges. Integrating phosphorylation and calcification effectively regulated the inflammatory response [46].

3D printing provides an efficient technique for fabricating aerogels with hierarchical pore architectures, consisting of interconnected macropores and mesopores, essential for tissue engineering [47]. The process was applied for 3D-printed methylcellulose (MC) and methylcellulose doped with bacterial cellulose nanofiber (MCBCf) gels, cross-linked with an aliphatic polyurea, aliphatic triisocyanate Desmodur Ultra N3300, resulting in the corresponding aerogels (X-MC and X-MCBCf) after supercritical drying (Fig. 18.3). The cross-linking with polyurea results in the improved long-term stability of X-MC and X-MCBCf aerogels in water, unlike their native forms, and enables them to absorb water up to 1800% w/w within just 2 h. The initial biological assessment of the materials, encompassing in vitro (cell compatibility, hemolytic activity), in ovo (HET-CAM), and in vivo (A. salina model) evaluations, demonstrated favorable cell viability, blood compatibility, and safety for live creatures [47].

Cellulose aerogels with nanoparticles were also reported for tissue engineering. A dual architecture microporous structure, including a nano-fibrous linked network of cellulose aerogel reinforced with arginine-coated graphene oxide (CNF-GO-Arg aerogel), was investigated for cartilage regeneration [48]. The engineered dual-architectured CNF-GO-Arg aerogel, via dual ice templating assembly, exhibits an 80% strain recovery capacity under compression. The release of Arginine from CNF-GO-Arg aerogel resulted in a 41% reduction in intracellular reactive oxygen species (ROS) activity and facilitated chondrogenic development of hMSCs by redirecting mitochondrial bioenergetics towards oxidative phosphorylation. The created CNF-GO-Arg aerogel exhibited multifunctionality through biomimetic morphology, enhanced cellular bioenergetics, and reduced ROS generation to facilitate cartilage regeneration [48].

3D-printing and supercritical carbon dioxide ($scCO_2$) technologies were reported to produce highly porous aerogel scaffolds [49]. They were applied to fabricate cellulose aerogels containing superparamagnetic iron oxide nanoparticles

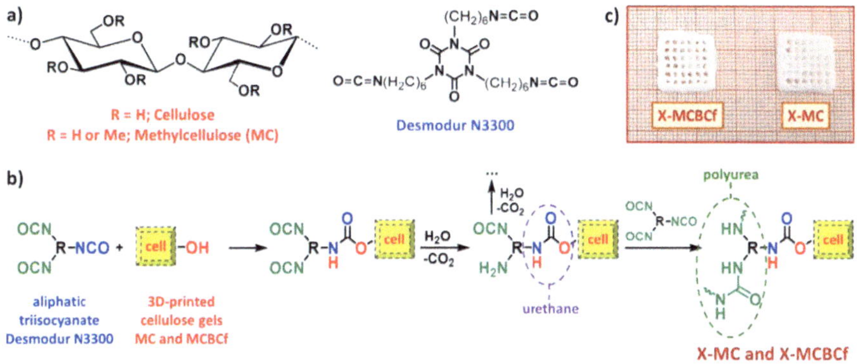

Fig. 18.3 **a** Chemical structures of cellulose, MC, and Desmodur N3300, **b** cross-linking of preformed cellulose gels with polyurea, and **c** Optical photograph of polyurea-cross-linked 3D-printed cellulose aerogels. Figure reprinted with permission from Ref. [47].

(SPIONs). Methylcellulose (MC) and BC are two adaptable cellulosic polysaccharides exhibiting exceptional physicochemical and biological properties, whereas SPIONs are frequently employed to functionalize biomaterials for tissue engineering applications. BC nanofibers doped with SPIONs were produced by combining never-dried BC with benzyl alcohol and iron (III) acetylacetonate ($Fe(acac)_3$) in a microwave vessel. After overnight incubation, the mixture was heated in a microwave oven to facilitate reagent homogenization. Ultimately, the BC/SPIONs were subjected to cooling, followed by filtration, and underwent two acetone washes and a single wash, before being autoclaved at 121 °C for 20 min. BC/SPIONs-doped aerogels were validated for the proper functionalization of the nanostructures. Biological data analysis using NIH/3T3 fibroblast cells revealed high hemocompatibility [49].

Hierarchical porous structure (Meso- and macroporous regime) of nanocrystalline cellulose (NCC)-based glycerogel with bioMOF system (CaSyr-1) consisting of calcium ions and syringic acid was proposed for musculoskeletal tissue regeneration [50]. The bioMOF framework uses a third bioactive component, ibuprofen (ibu), to create a multifaceted bioactive system. Initial in vitro findings indicate that the high glycerol content in the NCC scaffold may be cytotoxic upon direct exposure to human osteoblasts (HObs). Nonetheless, upon diluting the glycerol excess within the system (during the second day of the experiment), both direct and indirect assays validate complete biocompatibility and appropriateness for HOb proliferation [50].

18.5 Factors Affecting Tissue Engineering

The application of cellulose-based materials in tissue engineering is affected by various interconnected aspects, including material qualities, structural attributes, production methods, and biological interactions (Table 18.3). The type and composition of cellulose, e.g., MCC, CNC, CNF, BNC, and CMC, influence properties such as crystallinity, mechanical strength, porosity, and water absorption, thereby determining their appropriateness for diverse tissue types. The viscoelastic characteristics of plant-derived NFC hydrogel scaffolds significantly differ from those of the examined bacterial cellulose scaffolds [51]. The functionalization of cellulose with groups such as –OH, –COOH, –CHO, and $-NH_2$ improves cell adhesion, drug loading, and biodegradability, whereas chemically modified derivatives, including aminated or carboxymethylated cellulose, offer customized biological performance (Table 18.3). Additives such as MXene into cellulose-based aerogels increased water absorption by 852–1446% and retention (692–973%) ability [52]. Poly(lactic acid) improved and reinforced ethyl cellulose aerogels for bone regeneration [53].

Pure and cellulose-based materials were reported for scaffold fabrication. Composites have been thoroughly investigated for cellulose-based scaffold production. Additives are used for several purposes, including improving mechanical properties required for cell transporters for tissue growth and regeneration. They can also be added to decrease the cytotoxicity and immunogenicity by incorporating an additional biocompatible material. Biologically active compounds, can be integrated with cellulose to create cellulose-based scaffolds with enhanced functionality.

Fabrication techniques, e.g., freeze-drying, electrospinning, 3D printing, and solvent casting, are standard methods for controlling scaffold properties such as morphology, pore architecture, and alignment. These properties are vital for nutrient transport, vascularization, and cellular infiltration. Structural characteristics, including pore size, porosity, and mechanical stiffness, require optimization to replicate the physical environment of the target tissue. Cross-linking methods (physical, chemical, or enzymatic) and surface modifications (such as plasma treatment or peptide grafting) enhance mechanical stability and biofunctionality.

Furthermore, integrating other materials such as antibiotics, drugs, surfactants and nanoparticles (e.g., Ag [54], MXene [52], hydroxyapatite [55]) might confer supplementary attributes such as antibacterial efficacy, osteoinductivity, or electrical conductivity, enhancing the scaffold's suitability for particular applications like bone or nerve regeneration. From a biological standpoint, cellulose-derived materials are generally biocompatible and non-immunogenic; nonetheless, their degradation rate and capacity to facilitate cell signaling and tissue integration require meticulous regulation via chemical composition and structural design. Ultimately, choosing and altering cellulose materials according to these criteria facilitates the creation of scaffolds customized for various tissue engineering purposes.

Hydroxyl function groups in cellulose derivatives are advantageous for various chemical changes, which can enhance cell adhesion and proliferation [56]. It offers

Table 18.3 Summary of factors affecting the cellulose performance in tissue engineering

Item	Factor	Role in tissue engineering
Material	Cellulose type & composition	MCC, CNC, CNF, BNC, CMC, etc., impact crystallinity, mechanical properties, and processability
	Functional groups (–OH, –COOH, –CHO, –NH_2)	Allow cross-linking, drug loading, protein attachment, and enhanced bioactivity
	Functionalized derivatives	Aminated, carboxymethylated, and oxidized cellulose enhance cell adhesion, biodegradability, and biofunction
	Nanoparticle incorporation	Metal/metal oxide, MXene, graphene, and hydroxyapatite improve antimicrobial, osteoinductive, or electrical properties
Preparation	Fabrication method	Freeze-drying, electrospinning, 3D printing, and solvent casting define porosity, alignment, and structure
	Cross-linking technique	Physical (freezing), chemical (glutaraldehyde, genipin), and enzymatic; controls stability and degradation
	Surface modification	Grafting, plasma treatment, and coupling agents for tailored surface chemistry and cell interaction
Structural	Porosity & pore size	High porosity enhances vascularization; size must match specific tissue needs
	Pore architecture	Interconnectivity and orientation (e.g., unidirectional freezing) influence nutrient flow and cell migration
	Mechanical properties	Stiffness and strength must match native tissue; it is vital for bone, cartilage, etc.
	Water uptake/swelling behavior	Affects hydrogel behavior, cell viability, and drug diffusion
Biological	Biocompatibility	Essential for supporting cell adhesion, proliferation, and tissue integration
	Biodegradability	The rate must match tissue regeneration and be modified via cross-linking and polymer blending
	Immunogenicity	Low immune response preferred; cellulose is naturally non-toxic and inert
	Bioactivity/cell signaling	Enhanced by peptides, proteins, or nanoparticles incorporated into scaffolds
	Target cell type	Different cells (e.g., osteoblasts, chondrocytes, fibroblasts) require different conditions and environments

high surface wettability, improving cell adhesion and proliferation in vitro investigations. Birch wood pulp-derived CNF, changed through TEMPO-mediated oxidation, demonstrated a reduced pro-inflammatory response relative to unmodified CNF in human dermal fibroblasts in vivo studies [57]. Consequently, cellulose can be combined with several bioactive materials to improve biocompatibility while preserving biodegradability, mechanical characteristics, and structure.

Surface modification of cellulose may cause toxicity. The toxicity of the modified cellulose was reported by utilizing the cross-linking agent polyethyleneimine

(PEI) or the surfactant cetyl trimethylammonium bromide (CTAB) [58]. The sample treated with CTAB exhibited high toxic effects, adversely impacting cell survival, viability, and proliferation [58]. The toxicity of cellulose-modified CTAB was mainly due to the antibacterial activity of CTAB. The sample cross-linked with PEI exhibited a decrease in cell viability, indicated by a decline in DNA proliferation [58]. Amino-modified cellulose membrane was reported as non-toxic in human adipose-derived mesenchymal stem cells (hAMSC) [59]. Bacterial cellulose shows no toxicity toward human embryonic kidney 293 cells (HEK-293) [60]. Chemical modification of cellulose is critical and should be optimized.

The charged cellulose derivatives, especially in TOCNF, are crucial in modulating human mesenchymal stem cell activity, as reduced charge density enhances cell attachment and proliferation.

Mechanical properties of the prepared cellulose-based aerogels play an essential role in the material's performance. The compressive modulus of CNFs-based scaffolds was 1–3 MPa at ambient conditions, exceeding that of natural cartilage (about 1 MPa) [41]. The reduced compression modulus of 10–60 kPa in PBS at 37 °C was deemed advantageous for chondrogenesis [41].

18.6 Advantages of Cellulose-Based Tissue Engineering

Fabrication of cellulose-based scaffolds can be performed using several methods, offering the potential for materials optimization. The cellulose-based scaffolds can be easily prepared using several methods. It can be prepared via cross-linking MFC and Ca^{2+} via characteristic functional group interactions [42].

Cellulose-based materials exhibit high biocompatibility or minimal toxicity. Freeze-dried CNF demonstrates fibroblast cell adhesion on its surface, indicating the absence of cytotoxicity. Conversely, an aerogel composed of TOCNF cross-linked with polydopamine and tetracycline exhibited activity against *Staphylococcus aureus*, improving skin healing in rats in vivo [57]. The antibacterial activity is mainly due to the antibiotic, i.e., tetracycline. The cytokine behavior of a mechanically generated CNF suspension at several concentrations confirmed the absence of cytotoxic effects on cells after 6 and 24 h in culture conditions [61]. The cytotoxicity of TOCNF and unmodified CNF was investigated with no observation of toxicity to the cell membrane, mitochondrial function, or DNA proliferation [58]. This study utilized cellulose nanofibril structures derived from *Eucalyptus* and *Pinus radiata* pulp fibers. The nanofibrillated materials were produced using a homogenizer, both without pre-treatment and with TEMPO radical as pre-treatment, resulting in nanofibrils with low and high levels of anionic charge, respectively. Cytotoxicity assays were conducted on the samples, revealing that the nanofibrils do not induce acute harmful effects on the examined fibroblast cells (3T3 cells). The cell membrane, mitochondrial activity, and DNA proliferation were unaltered during the experiments, including direct and indirect contact between the materials and the 3T3 cells [58].

The nutrient diffusion rate of a CNF aerogel (1.7 wt.% CNF) was reported to be analogous to that of proteins found in the native extracellular matrix [51]. Plant-derived native NFC hydrogels serve as 3D cell culture scaffolds. The material is noteworthy for its distinct single-component structure, adjustable hydrogel characteristics, and accessibility. The structural properties of NFC hydrogel were examined, and its suitability as a cell culture scaffold was assessed utilizing hepatocyte and retinal epithelial cell lines. A singular component of the NFC scaffold facilitates hepatocyte three-dimensional cell culture without including supplementary bioactive elements. No cytotoxicity was detected, and mitochondrial activity was comparable in regular 2D and nanofibrillar cellulose cells. NFC was determined to be injectable owing to its fluid-like behavior under high stress, facilitated by reversible gelation and rheological properties that permitted the incorporation of cells into the gel. The spontaneously generated gel condition post-injection offered the necessary mechanical support for cellular growth and differentiation. The advantageous characteristics of NFC stem from its distinctive nanofibrillar architecture that emulates the properties of ECM [51]. CNF-based structures provide sufficient nutritional diffusivity for cells within 3D scaffolds.

Cellulose-based scaffolds exhibit good properties. Thermal study using differential scanning calorimetry (DSC) and dynamic mechanical analysis (DMA) indicated that the glass transition temperature (Tg) of mixed cellulose-based scaffolds increased to approximately 103 °C with the introduction of MFC, implying improved thermal stability [42]. X-ray diffraction (XRD) demonstrated distinct crystalline lattice planes, thereby corroborating the presence of MFC in the polymer matrix [42].

18.7 Challenges and Future Prospects

Despite cellulose being the most abundant natural polymer on Earth, present in plants, algae, and certain bacteria, the synthesis of nanocellulose continues to be expensive. This disparity stems from the technical difficulties in deconstructing bulk cellulose into nanoscale entities like CNCs, CNFs, or BNCs. The synthesis generally necessitates energy-demanding mechanical operations or severe chemical treatments, like acid hydrolysis or enzymatic digestion, substantially increasing production costs. Furthermore, these processes also require costly chemicals, intricate purification procedures, and significant waste management obligations. Attaining high yield and uniform quality for particle size, crystallinity, and surface attributes is challenging, constraining scalability. Moreover, nanocellulose is typically generated as a water-based solution, and drying it without inducing aggregation or compromising its nanostructure, particularly for applications like aerogels, necessitates sophisticated techniques such as freeze-drying or supercritical drying, which incur additional costs. The storage and reconstitution of dried nanocellulose pose further issues. The challenges are exacerbated by the fact that nanocellulose production has not attained full industrial scale, resulting in a lack of cost reductions typical of mass production.

Consequently, although cellulose is widely accessible, the complex and resource-demanding processes necessary for nanocellulose production result in elevated costs, highlighting the necessity for more efficient, sustainable, and scalable manufacturing techniques [62].

Although native cellulose exhibits favorable biocompatibility, it provides inadequate support for cell adhesion and proliferation, necessitating modifications and synthesizing derivatives to enhance its functioning in biological contexts. Modification of cellulose is critical and can lead to an increase in the material's toxicity.

Scaffolds of pure cellulose or nanocellulose can be achieved using advanced methods such as 3D printing. However, they are unsuitable for skin scaffolding due to their inadequate mechanical properties. Mechanical properties of cellulose-based scaffolds can be improved using post-treatment, such as cross-linking. For example, the mechanical properties of CNF-based skin scaffolds can be enhanced via cross-linking with Ca^{2+} and 1,4-butanediol diglycidyl ether (BDDE) [63].

18.8 Conclusions

In summary, cellulose is an abundant, renewable, and structurally resilient biopolymer with a historical usage that extends back to ancient civilizations, including the Egyptian era. Due to its natural origin, superior mechanical and chemical qualities, and inherent biocompatibility and biodegradability, cellulose has become essential in various scientific and industrial fields. The adaptability is further increased by the extensive array of derivatives that may be synthesized at the micro- and nanoscale, including MCC, MFC, CNFs, CNCs, and BNC, each presenting distinct advantages for biological applications. In tissue engineering, nanocellulose has advantageous characteristics, including high porosity, adjustable mechanical properties, and beneficial cellular interactions, making it an optimal choice for scaffolds replicating the original extracellular matrix. Nonetheless, despite the abundant availability of cellulose, the extraction and processing necessary to generate nanocellulose are intricate, energy-consuming, and expensive. The problems arise from the necessity for accurate mechanical, chemical, or enzymatic treatments and sophisticated drying methods such as freeze-drying or supercritical drying to preserve the integrity and functionality of nanostructures. Moreover, surface modification and cross-linking, crucial for improving performance in biological systems, introduce additional complexity and expense. The increasing demand for sustainable, scalable, and economical biomaterials in regenerative medicine necessitates innovating more straightforward and environmentally friendly nanocellulose manufacturing and scaffold fabrication methods. Surmounting these technical and economic obstacles could realize the complete promise of cellulose-based materials in advanced biomedical applications.

References

1. V. Girard, J. Chaussé, P. Vermette, Bacterial cellulose: a comprehensive review. J. Appl. Polym. Sci. 141 (2024). https://doi.org/10.1002/app.55163.
2. Q. Qin, S. Zeng, G. Duan, Y. Liu, X. Han, R. Yu, Y. Huang, C. Zhang, J. Han, S. Jiang, "Bottom-up" and "top-down" strategies toward strong cellulose-based materials. Chem. Soc. Rev. **53**, 9306–9343 (2024). https://doi.org/10.1039/D4CS00387J
3. S. Singh, S. Bhardwaj, P. Tiwari, K. Dev, K. Ghosh, P.K. Maji, Recent advances in cellulose nanocrystals-based sensors: a review. Mater. Adv. **5**, 2622–2654 (2024). https://doi.org/10.1039/D3MA00601H
4. Z. Sun, X. Wang, H. An, S. Liang, N. Li, A review on intelligence of cellulose based materials. Carbohydr. Polym. **338**, 122219 (2024). https://doi.org/10.1016/j.carbpol.2024.122219
5. H. Nasser Abdelhamid, D. Georgouvelas, U. Edlund, A.P. Mathew, CelloZIFPaper: cellulose-ZIF hybrid paper for heavy metal removal and electrochemical sensing. Chem. Eng. J. 136614 (2022). https://doi.org/10.1016/j.cej.2022.136614
6. A. Ashori, E. Chiani, S. Shokrollahzadeh, M. Madadi, F. Sun, X. Zhang, Cellulose-based aerogels for sustainable dye removal: advances and prospects. J. Polym. Environ. **32**, 6149–6181 (2024). https://doi.org/10.1007/s10924-024-03394-3
7. H. Nasser Abdelhamid, S. Sultan, A.P. Mathew, Binder-free three-dimensional (3D) printing of Cellulose-ZIF8 (CelloZIF-8) for water treatment and carbon dioxide (CO_2) adsorption. Chem. Eng. J. **468**, 143567 (2023). https://doi.org/10.1016/j.cej.2023.143567
8. H.N. Abdelhamid, MOFTextile: metal-organic frameworks nanosheets incorporated cotton textile for selective vapochromic sensing and capture of pyridine. Appl. Organomet. Chem. **37**(5) (2023). https://doi.org/10.1002/aoc.7078.10.1002/aoc.7078
9. H.N. Abdelhamid, S. Sultan, A.P. Mathew, 3D printing of cellulose/leaf-like zeolitic imidazolate frameworks (CelloZIF-L) for adsorption of carbon dioxide (CO_2) and heavy metal ions. Dalt. Trans. (2023). https://doi.org/10.1039/D2DT04168E
10. H.N. Abdelhamid, A.P. Mathew, Cellulose-based materials for water remediation: adsorption, catalysis, and antifouling. Front. Chem. Eng. **3**, 790314 (2021). https://doi.org/10.3389/fceng.2021.790314
11. H.P.S. Abdul Khalil, A.S. Adnan, E.B. Yahya, N.G. Olaiya, S. Safrida, M.S. Hossain, V. Balakrishnan, D.A. Gopakumar, C.K. Abdullah, A.A. Oyekanmi, D. Pasquini, A review on plant cellulose nanofibre-based aerogels for biomedical applications. Polymers (Basel). **12**, 1759 (2020). https://doi.org/10.3390/polym12081759
12. A. Jafari, A. Al-Ostaz, S. Nouranian, Biomedical potential of cellulose: current trends and future directions. Polym. Adv. Technol. **35** (2024). https://doi.org/10.1002/pat.6621.
13. H.N. Abdelhamid, Cellulose in drug delivery, in *Natural Biopolymers Drug Delivery* (Elsevier, 2025), pp. 105–126. https://doi.org/10.1016/B978-0-323-95367-2.00003-X
14. H.N. Abdelhamid, Polysaccharides for biomedical implants, in *Plant Polysaccharides as Pharmaceutical Excipients* (Elsevier, 2023), pp. 533–544. https://doi.org/10.1016/B978-0-323-90780-4.00015-2
15. M.U.A. Khan, G.M. Stojanović, M.F. Bin Abdullah, A. Dolatshahi-Pirouz, H.E. Marei, N. Ashammakhi, A. Hasan, Fundamental properties of smart hydrogels for tissue engineering applications: a review. Int. J. Biol. Macromol. **254**, 127882. https://doi.org/10.1016/j.ijbiomac.2023.127882
16. X. Huang, Y. Lou, Y. Duan, H. Liu, J. Tian, Y. Shen, X. Wei, Biomaterial scaffolds in maxillofacial bone tissue engineering: a review of recent advances. Bioact. Mater. **33**, 129–156 (2024). https://doi.org/10.1016/j.bioactmat.2023.10.031
17. K.H. Hussein, M. Soliman, M.S. Sabra, H.N. Abdelhamid, M. Abd-Elkareem, A.A. Sadek, Regenerative potential of graphene oxide-chitosan nanocomposite combined with fetal bovine serum on healing of full-thickness skin wound in rats. BMC Vet. Res. **21**, 324 (2025). https://doi.org/10.1186/s12917-025-04721-z

18. R.C. Nordberg, B.J. Bielajew, T. Takahashi, S. Dai, J.C. Hu, K.A. Athanasiou, Recent advancements in cartilage tissue engineering innovation and translation. Nat. Rev. Rheumatol. **20**, 323–346 (2024). https://doi.org/10.1038/s41584-024-01118-4
19. K. Melde, A.G. Athanassiadis, D. Missirlis, M. Shi, S. Seneca, P. Fischer, Ultrasound-assisted tissue engineering. Nat. Rev. Bioeng. **2**, 486–500 (2024). https://doi.org/10.1038/s44222-024-00166-5
20. M. Sreedharan, R. Vijayamma, E. Liyaskina, V. V. Revin, M.W. Ullah, Z. Shi, G. Yang, Y. Grohens, N. Kalarikkal, K. Ali Khan, S. Thomas, Nanocellulose-based hybrid scaffolds for skin and bone tissue engineering: a 10-year overview. Biomacromolecules **25**, 2136–2155 (2024). https://doi.org/10.1021/acs.biomac.3c00975
21. H.N. Abdelhamid, Degradable biopolymer-based nanocomposites in water treatment and air purification: a review. ACS Appl. Polym. Mater. **7**, 6576–6596 (2025). https://doi.org/10.1021/acsapm.5c00600
22. H. Liu, F. Xing, P. Yu, M. Zhe, S. Shakya, M. Liu, Z. Xiang, X. Duan, U. Ritz, Multifunctional aerogel: a unique and advanced biomaterial for tissue regeneration and repair. Mater. Des. **243**, 113091 (2024). https://doi.org/10.1016/j.matdes.2024.113091
23. Y. Zhang, S. Jiang, D. Xu, Z. Li, J. Guo, Z. Li, G. Cheng, Application of nanocellulose-based aerogels in bone tissue engineering: current trends and outlooks. Polymers (Basel) **15**, 2323 (2023). https://doi.org/10.3390/polym15102323
24. Z. Wei, C. Wu, R. Li, D. Yu, Q. Ding, Nanocellulose based hydrogel or aerogel scaffolds for tissue engineering. Cellulose **28**, 7497–7520 (2021). https://doi.org/10.1007/s10570-021-04021-3
25. C. Chen, Y. Xi, Y. Weng, Recent advances in cellulose-based hydrogels for tissue engineering applications. Polymers (Basel) **14**, 3335 (2022). https://doi.org/10.3390/polym14163335
26. R.R. Ujjwal, G. Slaughter, Advances in bacterial cellulose-based scaffolds for tissue engineering: review. J. Biomed. Mater. Res. Part A. **113** (2025). https://doi.org/10.1002/jbm.a.37912
27. J. Liesiene, S. Kiselioviene, A.S. Maruška, O. Baniukaitiene, Preparation and characterization of a highly porous, rigid cellulose-based hydrogel for biomedical and biotechnological applications. New J. Chem. **49**, 6603–6614 (2025). https://doi.org/10.1039/D5NJ00096C
28. E. Utoiu, V.S. Manoiu, E.I. Oprita, O. Craciunescu, Bacterial cellulose: a sustainable source for hydrogels and 3D-printed scaffolds for tissue engineering. Gels **10**, 387 (2024). https://doi.org/10.3390/gels10060387
29. H.N. Abdelhamid, A.P. Mathew, Cellulose-based nanomaterials advance biomedicine: a review. Int. J. Mol. Sci. **23**, 5405 (2022). https://doi.org/10.3390/ijms23105405
30. X. Shen, J.L. Shamshina, P. Berton, G. Gurau, R.D. Rogers, Hydrogels based on cellulose and chitin: fabrication, properties, and applications. Green Chem. **18**, 53–75 (2016). https://doi.org/10.1039/C5GC02396C
31. H.N. Abdelhamid, Nanocellulose-based materials for water pollutant removal: a review. Int. J. Mol. Sci. **25**, 8529 (2024). https://doi.org/10.3390/ijms25158529
32. R.M.A. Domingues, M.E. Gomes, R.L. Reis, The potential of cellulose nanocrystals in tissue engineering strategies. Biomacromolecules **15**, 2327–2346 (2014). https://doi.org/10.1021/bm500524s
33. S. Sultan, H.N. Abdelhamid, X. Zou, A.P. Mathew, CelloMOF: nanocellulose enabled 3D printing of metal-organic frameworks. Adv. Funct. Mater. **29**, 1–12 (2019). https://doi.org/10.1002/adfm.201805372
34. G. Siqueira, D. Kokkinis, R. Libanori, M.K. Hausmann, A.S. Gladman, A. Neels, P. Tingaut, T. Zimmermann, J.A. Lewis, A.R. Studart, Cellulose nanocrystal inks for 3D printing of textured cellular architectures. Adv. Funct. Mater. **27**, 1604619 (2017). https://doi.org/10.1002/adfm.201604619
35. K.M.O. Håkansson, I.C. Henriksson, C. de la Peña Vázquez, V. Kuzmenko, K. Markstedt, P. Enoksson, P. Gatenholm, Solidification of 3D printed nanofibril hydrogels into functional 3D cellulose structures. Adv. Mater. Technol. **1** (2016). https://doi.org/10.1002/admt.201600096

36. D. Atila, D. Keskin, A. Tezcaner, Cellulose acetate based 3-dimensional electrospun scaffolds for skin tissue engineering applications. Carbohydr. Polym. **133**, 251–261 (2015). https://doi.org/10.1016/j.carbpol.2015.06.109
37. X. He, Q. Xiao, C. Lu, Y. Wang, X. Zhang, J. Zhao, W. Zhang, X. Zhang, Y. Deng, Uniaxially aligned electrospun all-cellulose nanocomposite nanofibers reinforced with cellulose nanocrystals: scaffold for tissue engineering. Biomacromolecules **15**, 618–627 (2014). https://doi.org/10.1021/bm401656a
38. X. He, L. Cheng, X. Zhang, Q. Xiao, W. Zhang, C. Lu, Tissue engineering scaffolds electrospun from cotton cellulose. Carbohydr. Polym. **115**, 485–493 (2015). https://doi.org/10.1016/j.carbpol.2014.08.114
39. H. Cai, S. Sharma, W. Liu, W. Mu, W. Liu, X. Zhang, Y. Deng, Aerogel microspheres from natural cellulose nanofibrils and their application as cell culture scaffold. Biomacromolecules **15**, 2540–2547 (2014). https://doi.org/10.1021/bm5003976
40. R. Dash, Y. Li, A.J. Ragauskas, Cellulose nanowhisker foams by freeze casting. Carbohydr. Polym. **88**, 789–792 (2012). https://doi.org/10.1016/j.carbpol.2011.12.035
41. N. Naseri, J.-M. Poirier, L. Girandon, M. Fröhlich, K. Oksman, A.P. Mathew, 3-Dimensional porous nanocomposite scaffolds based on cellulose nanofibers for cartilage tissue engineering: tailoring of porosity and mechanical performance. RSC Adv. **6**, 5999–6007 (2016). https://doi.org/10.1039/C5RA27246G
42. N. Ninan, M. Muthiah, I.-K. Park, A. Elain, S. Thomas, Y. Grohens, Pectin/carboxymethyl cellulose/microfibrillated cellulose composite scaffolds for tissue engineering. Carbohydr. Polym. **98**, 877–885 (2013). https://doi.org/10.1016/j.carbpol.2013.06.067
43. P. Krontiras, P. Gatenholm, D.A. Hägg, Adipogenic differentiation of stem cells in three-dimensional porous bacterial nanocellulose scaffolds. J. Biomed. Mater. Res. Part B Appl. Biomater. **103**, 195–203 (2015). https://doi.org/10.1002/jbm.b.33198
44. V.C.-F. Li, C.K. Dunn, Z. Zhang, Y. Deng, H.J. Qi, Direct ink write (DIW) 3D printed cellulose nanocrystal aerogel structures. Sci. Rep. **7**, 8018 (2017). https://doi.org/10.1038/s41598-017-07771-y
45. O.Ç. Kurtuluş, S. Ondaral, N. Emin, A.E. Kadak, Bioaerogels produced from tempo oxidized nano cellulose with chitosan, gelatin, and alginate: general performances for wound dressing application. Cellulose **32**, 295–311 (2025). https://doi.org/10.1007/s10570-024-06282-0
46. C.B. Schimper, P. Pachschwöll, M.F. Maitz, C. Werner, T. Rosenau, F. Liebner, Hemocompatibility of cellulose phosphate aerogel membranes with potential use in bone tissue engineering, Front. Bioeng. Biotechnol. **11** (2023). https://doi.org/10.3389/fbioe.2023.1152577
47. A. Iglesias-Mejuto, G. Raptopoulos, N. Malandain, M. Neves Amaral, I. Ardao, M. Finšgar, A. Laromaine, A. Roig, C. Pinto Reis, C.A. García-González, P. Paraskevopoulou, 3D-printed cellulose aerogels minimally cross-linked with polyurea: a robust strategy for tissue engineering. ACS Appl. Mater. Interfaces **17**, 34444–34457 (2025). https://doi.org/10.1021/acsami.5c08389
48. P. Pareek, S. Chaudhary, S. Singh, A.T. Thodikayil, D. Kalyanasundaram, S. Kumar, Bridging biomimetic and bioenergetics scaffold: Cellulose-graphene oxide-arginine functionalized aerogel for stem cell-mediated cartilage repair. Int. J. Biol. Macromol. **278**, 134608 (2024). https://doi.org/10.1016/j.ijbiomac.2024.134608
49. A. Iglesias-Mejuto, N. Malandain, T. Ferreira-Gonçalves, I. Ardao, C.P. Reis, A. Laromaine, A. Roig, C.A. García-González, Cellulose-in-cellulose 3D-printed bioaerogels for bone tissue engineering. Cellulose **31**, 515–534 (2024). https://doi.org/10.1007/s10570-023-05601-1
50. A. Rosado, A. Borrás, M. Sánchez-Soto, M. Labíková, H. Hettegger, R.A. Ramírez-Jiménez, L. Rojo, L. García-Fernández, M.R. Aguilar, F. Liebner, A.M. López-Periago, J.A. Ayllón, C. Domingo, BioMOF@cellulose glycerogel scaffold with multifold bioactivity: perspective in bone tissue repair. Gels **10**, 631 (2024). https://doi.org/10.3390/gels10100631
51. M. Bhattacharya, M.M. Malinen, P. Lauren, Y.-R. Lou, S.W. Kuisma, L. Kanninen, M. Lille, A. Corlu, C. GuGuen-Guillouzo, O. Ikkala, A. Laukkanen, A. Urtti, M. Yliperttula, Nanofibrillar cellulose hydrogel promotes three-dimensional liver cell culture. J. Control. Release **164**, 291–298 (2012). https://doi.org/10.1016/j.jconrel.2012.06.039

52. T. Hu, P. Cai, C. Xia, MXene reinforced microporous bacterial cellulose/sodium alginate dual crosslinked cryogel for bone tissue engineering. Biomed. Mater. **19**, 055022 (2024). https://doi.org/10.1088/1748-605X/ad6520
53. G. Horvat, J. Rožanc, U. Maver, M. Finšgar, Ž Knez, Z. Novak, Reinforcing ethyl cellulose aerogels with poly(lactic acid) for enhanced bone regeneration. Cellulose **31**, 4421–4439 (2024). https://doi.org/10.1007/s10570-024-05905-w
54. M.P. Bernardo, M. Foschini, A.C.C. Santos, C. Ueira Vieira, N. Saito, M.E.C. Mundim, O.F. Lopes, D. Pasquini, Fabrication of antimicrobial cellulose and silver niobate aerogels for enhanced tissue regeneration. ACS Omega **10**, 15493–15502 (2025). https://doi.org/10.1021/acsomega.5c00351
55. C. Ao, Y. Niu, X. Zhang, X. He, W. Zhang, C. Lu, Fabrication and characterization of electrospun cellulose/nano-hydroxyapatite nanofibers for bone tissue engineering. Int. J. Biol. Macromol. **97**, 568–573 (2017). https://doi.org/10.1016/j.ijbiomac.2016.12.091
56. K. Syverud, *Tissue Engineering Using Plant-Derived Cellulose Nanofibrils (CNF) as Scaffold Material* (2017), pp. 171–189. https://doi.org/10.1021/bk-2017-1251.ch009
57. Y. Liu, Y. Sui, C. Liu, C. Liu, M. Wu, B. Li, Y. Li, A physically crosslinked polydopamine/nanocellulose hydrogel as potential versatile vehicles for drug delivery and wound healing. Carbohydr. Polym. **188**, 27–36 (2018). https://doi.org/10.1016/j.carbpol.2018.01.093
58. L. Alexandrescu, K. Syverud, A. Gatti, G. Chinga-Carrasco, Cytotoxicity tests of cellulose nanofibril-based structures. Cellulose **20**, 1765–1775 (2013). https://doi.org/10.1007/s10570-013-9948-9
59. S.C.M. Fernandes, P. Sadocco, A. Alonso-Varona, T. Palomares, A. Eceiza, A.J.D. Silvestre, I. Mondragon, C.S.R. Freire, Bioinspired antimicrobial and biocompatible bacterial cellulose membranes obtained by surface functionalization with aminoalkyl groups. ACS Appl. Mater. Interfaces **5**, 3290–3297 (2013). https://doi.org/10.1021/am400338n
60. W. Shao, J. Wu, H. Liu, S. Ye, L. Jiang, X. Liu, Novel bioactive surface functionalization of bacterial cellulose membrane. Carbohydr. Polym. **178**, 270–276 (2017). https://doi.org/10.1016/j.carbpol.2017.09.045
61. J. Vartiainen, T. Pöhler, K. Sirola, L. Pylkkänen, H. Alenius, J. Hokkinen, U. Tapper, P. Lahtinen, A. Kapanen, K. Putkisto, P. Hiekkataipale, P. Eronen, J. Ruokolainen, A. Laukkanen, Health and environmental safety aspects of friction grinding and spray drying of microfibrillated cellulose. Cellulose **18**, 775–786 (2011). https://doi.org/10.1007/s10570-011-9501-7
62. H.N. Abdelhamid, S. Norgren, A. Mathew, Scale-up processing of leaf-like zeolitic imidazolate frameworks (ZIF-L)/cellulose for water treatment. Int. J. Biol. Macromol. **315**, 144455 (2025). https://doi.org/10.1016/j.ijbiomac.2025.144455
63. C. Xu, B. Zhang Molino, X. Wang, F. Cheng, W. Xu, P. Molino, M. Bacher, D. Su, T. Rosenau, S. Willför, G. Wallace, 3D printing of nanocellulose hydrogel scaffolds with tunable mechanical strength towards wound healing application. J. Mater. Chem. B. **6**, 7066–7075 (2018). https://doi.org/10.1039/C8TB01757C

Chapter 19
Cellulose-Based Aerogels for Drug Delivery

Yuchen Jin and Zhiqiang Su

Abstract Cellulose-based aerogels (CBAs) have emerged as highly promising drug carriers due to their ultralow density, exceptional porosity (>99%), high specific surface area, and biocompatibility. They serve as versatile platforms for efficient drug loading, protection against degradation, and precise spatiotemporal control of drug release. This chapter reviews recent advances in the development and application of CBAs in diverse drug delivery scenarios, including antibacterial wound healing, tissue repair, hemostasis, and intelligent responsive release. Through various functionalization strategies, such as surface modification with bioactive groups, incorporation of nanoparticles, and design of hierarchical or core–shell structures, CBAs demonstrate enhanced drug loading capacity, tunable release profiles and multifunctionality. While in vitro and preclinical studies showcase significant potential, challenges such as precise morphology/size control, long-term in vivo performance validation, clinical scalability, and complex release mechanism understanding remain. Continued research is crucial to address these limitations and advance CBAs towards clinical translation. Overall, CBAs offer a sustainable, adaptable, and effective platform for next-generation intelligent and targeted drug delivery systems.

Keywords Cellulose-based aerogels · Drug delivery · Biomaterials · Intelligent drug release

19.1 Introduction

Cellulose-based aerogels (CBAs) have been widely applied in tumor treatment, wound healing, and oral drug delivery, among other fields. They exhibit revolutionary potential in drug delivery due to their ultralow density, high porosity (>99%) [1], and high specific surface area [2]. Porous aerogels with large pore volume and

Y. Jin · Z. Su (✉)
State Key Laboratory of Chemical Resource Engineering, Beijing Key Laboratory of Advanced Functional Polymer Composites, Beijing University of Chemical Technology, Beijing, China
e-mail: suzq@mail.buct.edu.cn

G. Wei and J. Zhang (eds.), *Cellulose-Based Hydrogels and Aerogels*, Lecture Notes in Nanoscale Science and Technology 41,
https://doi.org/10.1007/978-3-032-03008-5_19

tunable pore size are suitable for the storage of various drugs, preventing rapid degradation or premature release of the drugs in undesired areas [3]. Their unique physical structure not only supports efficient drug loading but also allows precise control of drug release behavior through biodegradability and chemical modifiability, significantly overcoming the limitations of traditional carriers [4]. Aerogels with special morphologies, such as spherical aerogels [5], which are endowed with core/shell structures, can subsequently produce reliable therapeutic effects when combined with drugs [6]. However, the biomedical applications of aerogel-treated materials are limited by the difficulty in simultaneously controlling the morphology and particle size of the aerogels, as well as the release patterns of the drug molecules [7]. Multilayered aerogel drug carriers have unique advantages such as stimulus-triggered release, adjustable drug release profiles, and the ability to load two different drugs in a core/shell structure, which are difficult to achieve in other types of aerogels [8, 9].

In this chapter, we review the transformative potential of CBAs as advanced drug delivery systems, highlighting their ultralight density, exceptional porosity (>99%), and tailorable surface functionality that enable high-efficiency drug loading and intelligent controlled release. We discuss innovative strategies—including pH/temperature dual-responsive modifications, core–shell structural designs, and mathematical release modelling—for achieving targeted spatiotemporal drug delivery in applications such as antibacterial wound healing, hemostasis, tissue regeneration, and tumor microenvironment-specific therapy. Furthermore, we analyze the multifunctional integration of CBAs (e.g., combining photothermal effects with coagulation promotion) and their superiority over traditional carriers in biodegradability, biocompatibility, and clinical efficacy, while addressing remaining challenges in scalability and long-term in vivo performance for future clinical translation.

19.2 Characteristics and Advantages of CBAs for Drug Delivery

CBAs, with their ultralight weight, high porosity, and high specific surface area, possess excellent performance advantages and broad application prospects in the field of drug delivery [10]. Their three-dimensional porous network structure enables efficient loading of active ingredients and allows for precise controlled drug release through pore structure regulation and modification methods, such as pH-responsive, temperature-responsive, and enzyme-triggered targeted release [11]. The hydrophilic cellulose fibers cross-linked with hydrophobic conjugated polymers via phosphorylation form an interpenetrating structure without distinct layer separation, optimizing the hydrophilic-hydrophobic balance and promoting blood flow and selective lipid adsorption [12].

Compared with traditional inorganic aerogels and polymer aerogels, CBAs exhibit better biodegradability and higher compressive strength. Moreover, the excellent biodegradability and biocompatibility of cellulose avoid the toxic residues of traditional carriers after drug delivery, and its chemical modifiability endows it with multifunctional integration capabilities (Fig. 19.1a and b). Unlike liposomes that are prone to leakage and polymer microspheres with low drug loading capacity, cellulose aerogels have higher drug loading efficiency [13]. Their mechanical stability and long circulation time significantly enhance the bioavailability of drugs, and no secondary intervention is required after secondary degradation in the body, making them an ideal choice for the next generation of intelligent drug carriers.

Carbon nanotube (CNT)-reinforced carbonized cellulose aerogels demonstrate excellent cell and blood compatibility, effectively absorbing red blood cells and platelets and activating platelets [14]. In in vivo experiments, their hemostatic effect is superior to that of gauze and gelatin sponge, and the photothermal effect can accelerate blood clot formation and improve its mechanical properties. Additionally, they can promote skin wound healing and represent a promising strategy for treating uncontrollable bleeding and promoting wound healing (Fig. 19.1c) [15].

19.3 Drug Loading Mechanism and Release Behavior

19.3.1 Drug Loading Mode

The drug loading mechanism of cellulose aerogel mainly relies on its unique three-dimensional porous structure and surface chemical properties [16]. Drug molecules can be encapsulated in nanoporous channels through physical adsorption, or functional groups can be introduced on the surface of cellulose through chemical modification to form covalent or non-covalent bonds with drugs [17]. Additionally, it can be combined with carriers such as liposomes and polymer microspheres to work synergistic with cellulose aerogel to enhance targeting and drug loading [18].

Drug molecules can be added to the reaction mixture before gel formation, after gel formation, and by immersing cellulose aerogel in a drug solution. The drug is directly added to the cellulose solution or suspension and then gelated. This drug loading method is simple to operate, but the drug may react with the gel precursor. Introducing the drug in the solvent exchange stage can promote the precipitation of the drug in the gel pores, and then enter the drying process, which can complete the drying and drug loading in one step [4]. However, the diffusion efficiency is limited by the molecular weight of the drug and the pore size. After the drying is completed, the dried unloaded aerogel can also be immersed in the drug solution, and the drug can be loaded through physical diffusion. The larger surface area of the cellulose aerogel drug carrier leads to faster diffusion of the drug and increased solubility. However, the relationship between the structure of the cellulose aerogel and the drug loading capacity has not been systematically studied [19]. Table 19.1 shows the advantages and disadvantages of the drug loading methods.

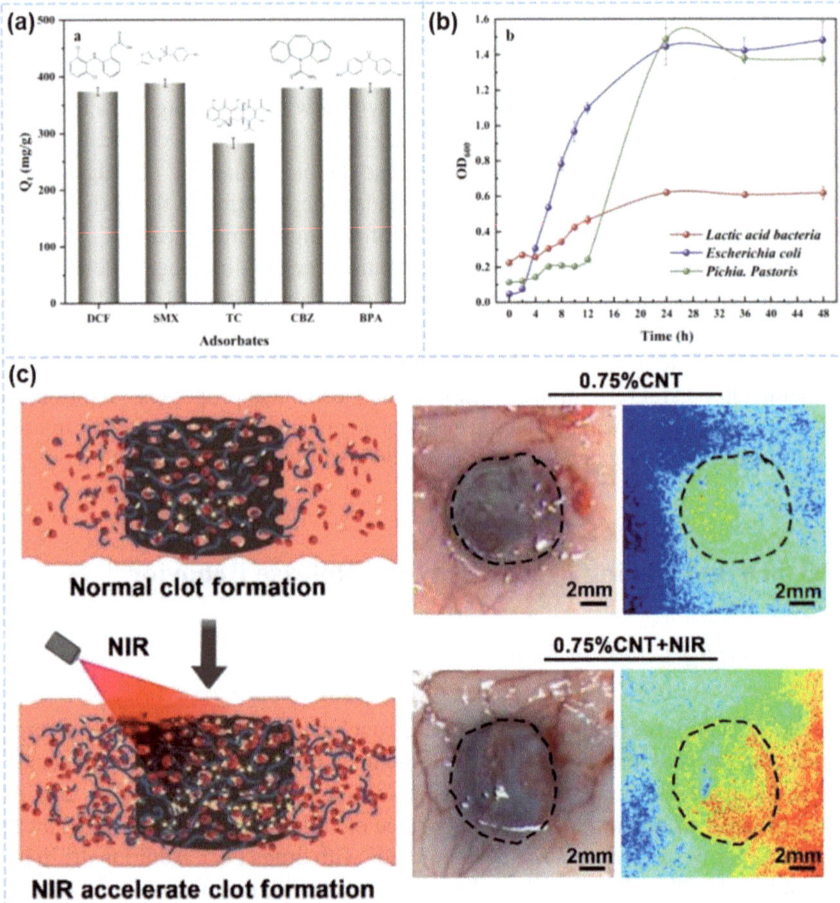

Fig. 19.1 **a** The biocompatibility and **b** biodegradability of cellulose and CBAs. Reproduced with the permission from Ref. [7], Copyright 2024, Elsevier Ltd. **c** Schematic diagram of photothermal effect of composite aerogel and the influence of photothermal effect on angiogenesis after subcutaneous implantation. Reproduced with the permission from Ref. [8], Copyright 2025, Wiley-VCH GmbH

Table 19.1 Advantages and disadvantages of drug loading methods

Drug loading timing	Merits	Drawbacks	Refernce
Before the wet gel forms	Simplicity of operator	The medicine will react in advance	[19]
After the wet gel is formed	Drying and drug loading are completed in one step	The diffusion efficiency is limited	
After the aerogel is formed	The drug spreads more quickly	Affected by hydrophilicity and hydrophobicity	

19.3.2 Controlled Release Mechanism

Drug release refers to the process in which drug molecules are transferred from the interior of the carrier to the outer surface of the carrier and eventually released into the surrounding environment [20]. This process is controlled by the random movement of drug molecules and driven by convection generated by the chemical potential gradient and osmotic pressure [21]. An important parameter for drug release is the release rate of drug molecules from the carrier. The release rate is affected by aspects such as the specific surface area, pore structure and chemical modification of the carrier [22]. Aerogel drug carriers have excellent specific surface area. Therefore, the drug molecules loaded with aerogel are expected to exhibit low solubility characteristics, which is beneficial for long-term drug delivery [23]. Nanopores are suitable for the diffusion of small-molecule drugs, while micropores (50–200 μm) are appropriate for the controlled release of large molecules [24]. The drug release from hydrophilic cellulose aerogels can be predicted by controlling the degradation rate of the carrier cellulose through which the drug molecules pass. Hydrophobic-loaded drug aerogels exhibit a slower drug release rate, primarily influenced by drug diffusion. To better investigate drug release systems, the release rate of the drug can be demonstrated through release profiles. Release profiles can be categorized into three types: burst release, sustained release, and controlled release [25].

For degradable drug carriers, the drug release rate is controlled by diffusion through the carrier's pore network. As the drug carrier degrades, the pores within the carrier gradually enlarge. Once the critical pore size is exceeded, it leads to drug release. In contrast, for non-degradable drug carriers, diffusion is the primary driving force for drug release [26]. The release rate is not affected by the concentration gradient, which in turn is influenced by the carrier's properties, such as permeability and thickness. Therefore, drug release can be predicted by controlling the diffusion of drug molecules through an insoluble polymer or simply by regulating the degradation rate of the carrier [27].

Mathematical modelling of drug release, through the comparative analysis of constructing theoretical models and experimental data, can accurately predict the diffusion rate and release kinetics characteristics of drugs, and clarify the mechanism of action of variables such as material properties and environmental parameters on the delivery process. The drug release curve of nanocellulose composite aerogel is related to pH. Based on classic models such as the first-order kinetic model, Higuchi equation and Korsmeyer-Peppas model, researchers can optimize the formulation process through parameter sensitivity analysis, significantly reducing the R&D cost and shortening the cycle compared with traditional experimental methods. Provide a quantitative research framework for the development of sustained-release and controlled-release preparations and personalized drug administration regimens. The drug release rate and cumulative release rate of nanocellulose composite aerogel under acidic conditions are both higher than those under neutral conditions. The drug release curve of the nanocellulose composite aerogel conforms to the first-order kinetic model and the Korsmeyer-Peppas model, and is controlled by Fick diffusion [28].

Cellulose nanocrystals (CNC) exhibit significant pH-responsive behavior in drug delivery. In acidic environments, the carboxyl groups grafted onto cellulose become protonated, enabling CNC-based stimuli-responsive nanoparticles to form stable emulsions and achieve slow drug release [29]. In neutral to alkaline environments, the grafted carboxyl groups on CNCs become deprotonated, leading to emulsion breaking and accelerated drug release (Fig. 19.2a). Through pH switching, the stabilization and breaking of emulsions can be repeatedly achieved, enabling intelligent control of drug release in the gastrointestinal tract, enhancing bioavailability, and providing a new strategy for the precise delivery of hydrophobic drugs. This approach is particularly valuable for improving the delivery efficiency and therapeutic efficacy of lipophilic drugs (Fig. 19.2b).

Furthermore, cellulose aerogels also exhibit pH-responsive behavior in drug delivery (Fig. 19.2c) and can combine with temperature to show dual pH-temperature

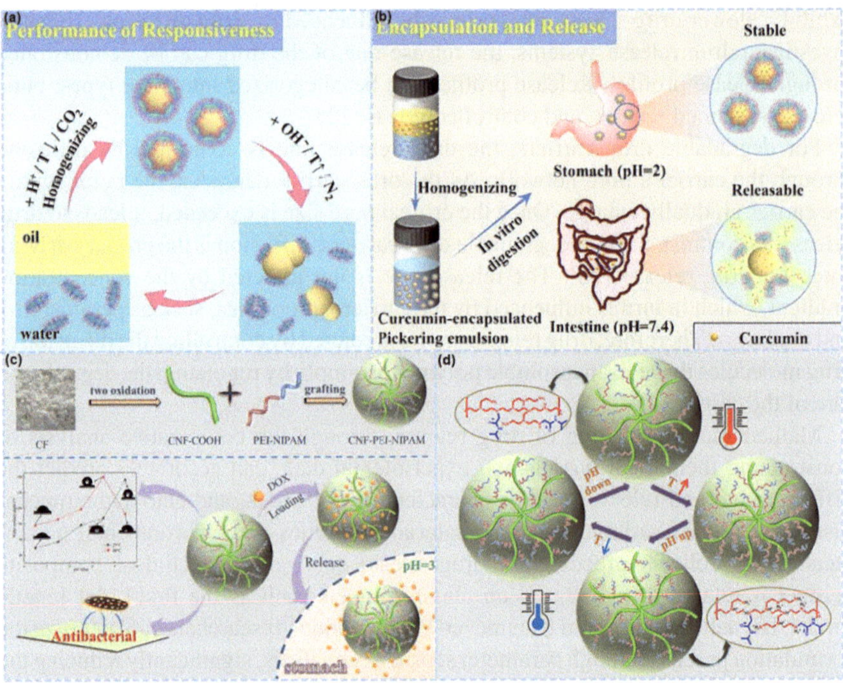

Fig. 19.2 **a** Schematic diagram of pH response of cellulose nanocrystalline multi-stimulus–response nanoparticles. **b** Schematic diagram of cellulose aerogel with multi-level pore structure and hemostatic function. Reproduced with the permission from Ref. [29], Copyright 2025, Elsevier Ltd. **c** The temperature and pH dual-stimulus response mechanism of CNF-PEI-NIPAM. **d** Schematic diagram of the structure of CNF/CNT aerogel after absorbing water. Reproduced with the permission from Ref. [30], Copyright 2020, Elsevier Ltd.

responsiveness (Fig. 19.2d). The pH responsiveness originates from the protonation/deprotonation of functional groups on the material surface under different pH conditions, which regulates the pore structure and thereby alters the drug release rate. For example, drug release is slower in acidic conditions and faster in alkaline conditions. Temperature responsiveness is due to the phase transition of thermosensitive polymers, where increased temperature leads to pore expansion and accelerated molecular motion, promoting drug release [30]. This dual-responsive mechanism enables precise drug release, such as rapid drug release in the tumor microenvironment (low pH and high temperature) to enhance therapeutic efficacy, while slow release in normal tissues reduces side effects, providing a technological basis for targeted delivery and personalized treatment.

19.4 Applications and Case Studies of CBAs for Drug Delivery

Composite nanocellulose aerogel shows great promise in the field of drug delivery due to its high porosity, large specific surface area, and modifiable surface hydroxyl groups. It can not only effectively encapsulate and protect drug molecules from premature degradation or release, but also achieve controllable, sustained, and pH/temperature-responsive release of drugs through chemical modification or morphological design. Research examples have demonstrated its potential in prolonging drug action and improving loading efficiency.

19.4.1 Antibacterial and Tissue Repair

The drug-loading mechanism of cellulose aerogel as an antibacterial and wound-healing material mainly relies on its unique 3D porous network structure. This high porosity and extremely large specific surface area provide ample space for loading various types of antibacterial agents. Natural components, such as curcumin, can be embedded into the cellulose framework through hydrophobic interactions [31].

Metal nanoparticles such as silver achieve stable anchoring with the help of hydroxyl groups on the surface of aerogel [32], Antibiotics such as amoxicillin are loaded through a combination of physical adsorption and pore encapsulation [6]. In terms of drug release behavior, cellulose aerogel exhibits intelligent and controllable characteristics (Fig. 19.3a). For instance, the release rate of the curcumin loading system significantly increases in the slightly acidic environment of simulated wounds, reaching 50% within 50 h, which is higher than 41% under neutral conditions, achieving targeted delivery to the infected site [31]. Amoxicillin drug-loaded aerogel presents a typical two-phase release pattern. The burst release within the initial 2 h can rapidly establish a local antibacterial concentration to inhibit early

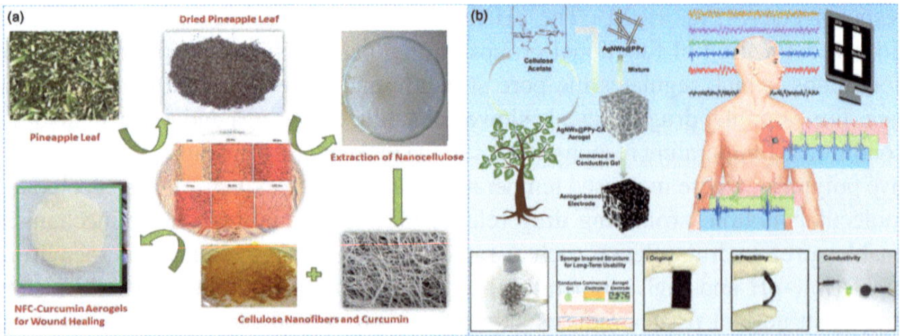

Fig. 19.3 (a) Schematic diagram of a new type of three-dimensional porous aerogel for the treatment of chronic wounds. Reproduced with the permission from Ref. [31], Copyright 2022, Elsevier Ltd. (b) Schematic diagram of antibacterial aerogel inspired by sponges, used for recording electrophysiological signals in human–machine interfaces. Reproduced with the permission from Ref. [33], Copyright 2024, Wiley-VCH GmbH

infection. Then it enters a sustained sustained-release stage, where the drug slowly diffuses through deep pores and maintains long-lasting antibacterial activity for more than 12 h [6].

At the same time, the composite drug-loading strategy further optimizes the release efficiency and antibacterial effect. Polypyrrole-coated silver nanowires (AgNWs@PPy) are synthesized in situ within the cellulose aerogel (Fig. 19.3b). The conductive polymer framework of polypyrrole not only promotes the uniform dispersion of silver ions, but also enhances the antibacterial effect through the positive charges of the molecular chains. The sustained-release characteristic of silver ions maintains the antibacterial rate at over 97.8% in the long term, effectively overcoming the disadvantage of traditional silver particles that are prone to aggregation and failure [33].

CBAs can also be applied in bone tissue repair. Liu et al. utilized bagasse nanofibrillated cellulose (NFC) combined with chitosan to construct an aerogel scaffold loaded with sodium risedronate, which possesses both high mechanical strength and drug sustained-release properties [34]. This scaffold significantly promotes the proliferation of osteoblasts and the expression of RUNX2 protein, while chitosan endows it with potential antibacterial properties.

It is worth noting that the structural designability of cellulose aerogels provides a key pathway for precise control of drug release. By introducing sodium alginate and performing calcium ion cross-linking, the wet-state stability of the aerogel can be significantly enhanced, preventing structural collapse during liquid absorption and thereby ensuring the reproducibility of the drug release profile [31]. The innovative design of Janus' asymmetric structure integrates the hydrophilic layer and the hydrophobic drug-carrying layer into one, which not only enables the unidirectional and rapid flow of exudate into the aerogel but also effectively prevents the reverse permeation of external contaminants. At the same time, it ensures the local high-concentration release of drugs carried in the hydrophobic layer at the wound interface, avoiding ineffective loss [35]. This structure–function integrated design concept enables cellulose aerogel to maintain stable and efficient drug loading and release performance in complex wound environments.

19.4.2 Intelligent Drug Release

CBA has shown significant advantages in the fields of antibacterial and wound repair due to its adjustable porous structure, high specific surface area and surface functionalization ability [18]. Its three-dimensional network structure provides an ideal carrier for the efficient loading of antibacterial drugs [36]. For example, the active sites introduced by carboxylating modified cellulose can electrostatically adsorb positively charged antibacterial molecules, significantly increasing the drug loading to 287.39 mg/g [37]. Meanwhile, the in-situ modification of the aerogel surface by copper nanoparticles (Cu NPs) not only forms a physical barrier to delay drug release, but also endows the material with its own broad-spectrum antibacterial properties [38].

In terms of the regulation of release behavior, cellulose aerogel achieves intelligent drug release in the wound microenvironment through the pH response mechanism. Under acidic conditions (such as in the early stage of an infected wound), the protonation of carboxyl groups causes the aerogel to shrink, and the drug release is slow, avoiding local toxicity caused by sudden release. When the wound tends to heal and the pH approaches neutral, the deprotonation of carboxyl groups causes gel swelling, accelerates the release of antibacterial drugs, and continuously inhibits pathogenic microorganisms. In addition, the calcium ion cross-linked pectin network or core–shell structure design can further extend the release time to more than 10 h, significantly enhancing the sustained-release performance [36].

It is particularly worth noting that the double-filled structure collaboratively optimizes the release kinetics. HAP fills the macropores to reduce the initial diffusion of the drug, and the Cu NPs layer enhances the physical barrier, enabling the composite aerogel to achieve two-stage sustained release for up to 34 h while maintaining a high drug loading capacity [38]. This design concept not only solves the problem of sudden release of traditional carriers, but also ADAPTS to the dynamic microenvironment of wounds through multiple response mechanisms, providing precise treatment strategies for chronic wound infections [18].

Therefore, CBA provides an efficient and safe solution for the delivery of antibacterial drugs and wound repair through the integration of functional modification, structural design and intelligent release mechanism. Future research needs to further optimize the drug loading efficiency and mechanical strength to promote their large-scale application in clinical infection management.

19.4.3 The Hemostatic Function of the Special Structure

The hemostatic material based on cellulose aerogel has the core advantage of a unique multi-level structure design. These structures not only endow the material with efficient hemostatic ability, but also provide an ideal platform for integrating drug loading and antibacterial functions [39]. The high porosity network composed of cellulose nanofibers, combined with in-situ anchored functional nanoparticles or synthetic fibers with special topological structures, jointly constructs an efficient hemostatic channel that physically absorbs blood, concentrates coagulation factors, and activates the coagulation cascade reaction. For instance, nanozeolites anchored on cellulose nanofibers achieve stable bonding at an ultra-low loading rate through multi-dimensional entanglement of hydrogen bonds. Their porous surface mimics the negative charge interface of platelets to accelerate thrombin generation. Meanwhile, the calcium ions released collaboratively significantly shorten the coagulation time, achieving a significant reduction in blood loss and extremely low temperature rise in animal models. It effectively avoided the risks of burns and thrombosis of traditional materials (Fig. 19.4a) [40] Specifically, through the hierarchical pores and microchannels constructed by bionic design, combined with the trench/pleated fiber topology structure, achieved rapid directional blood delivery and efficient blood cell enrichment, laying the foundation for the rapid distribution and targeting effect of drugs on the wound surface (Fig. 19.4b) [41].

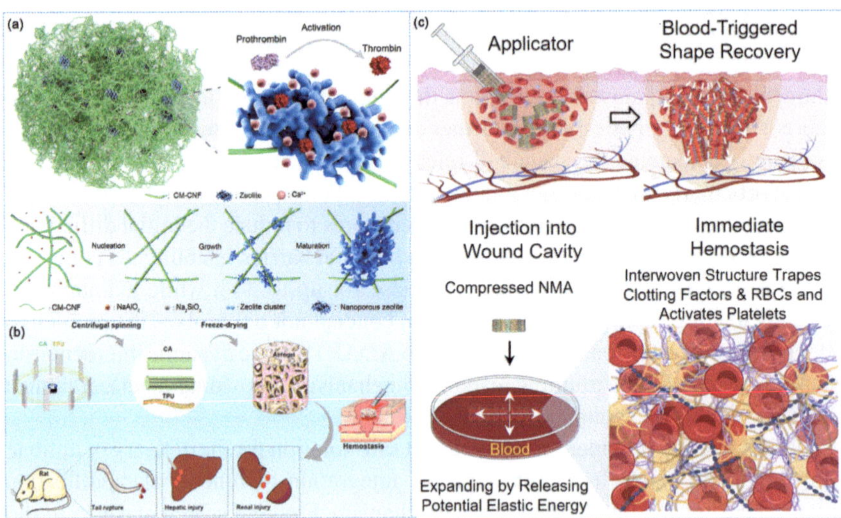

Fig. 19.4 **a** Schematic illustration of CNF aerogel structure and hemostatic mechanism Reproduced with the permission from Ref. [40], Copyright 2025, Wiley-VCH GmbH. **b** Schematic diagram of cellulose aerogel for rapid hemostasis. Reproduced with the permission from Ref. [41], Copyright 2025, American Chemical Society. **c** Schematic diagram and hemostatic mechanism of biomicrofiber aerogel as injectable hemostatic material. Reproduced with the permission from Ref. [39], Copyright 2025, The Author(s)

This high specific surface area and porous interconnected structure simultaneously creates superior conditions for the efficient loading and controllable release of antibacterial drugs. The active groups on the surface of the material can be loaded with broad-spectrum antibacterial agents through physical adsorption or chemical bonding [41]. For instance, in some studies, carbon nanotube-reinforced aerogels possess excellent photothermal conversion performance. Under near-infrared light irradiation, they can rapidly heat up to achieve physical sterilization, significantly reducing the risk of infection. The calcium ions loaded in the material themselves also have antibacterial effects, forming a chemical-physical synergistic antibacterial barrier [15]. The two-component nano-microfiber aerogels in other studies reserve space for integrating sustained-release antimicrobial peptides or growth factors through the characteristics of gelatin coatings and degradable polymer substrates. Their ultra-high porosity and rapid blood-sucking ability can ensure the efficient release of drugs on the wound surface and their action on enriched blood cells and microorganisms (Fig. 19.4c) [39].

In addition, the optimized mechanical properties and intelligent response characteristics of cellulose aerogel enable it to closely adhere to deep irregular wounds and maintain stable physical pressure, which provides a continuous working environment for the drug delivery system [41]. Animal experiments have confirmed that such materials possess excellent biocompatibility and potential for promoting healing. For instance, they can promote angiogenesis through photothermal effects and induce the polarization of repair immune cells to accelerate tissue repair, laying a solid foundation for the integrated application of drug-loaded antibacterial and regenerative medicine [15]. Therefore, CBA achieves multiple synergies of hemostasis, drug delivery, antibacterial and tissue repair through an integrated structure–function design, providing an innovative solution for the management of complex wounds.

19.5 Conclusion and Outlooks

CBAs demonstrate transformative potential as intelligent drug delivery platforms, leveraging their ultralight porous frameworks, modifiable surface chemistry, and inherent biocompatibility to achieve precision-controlled spatiotemporal release. Innovations in structural engineering and functional hybridization enable multi-functional synergy, simultaneously addressing hemostasis, antimicrobial activity, and tissue regeneration while outperforming conventional carriers in drug-loading efficiency and biodegradability.

Despite the notable advancements in functional regulation and application expansion achieved by CBAs, several key challenges remain to be addressed. Future research should prioritize scalable fabrication via low-cost, non-freeze-drying techniques for uniform morphology/size control, while rigorously validating long-term biosafety through in vivo studies of degradation kinetics, immunogenicity, and organ accumulation. Advancements must further leverage AI-driven design with computational models to predict release profiles and optimize formulations, expand functionality toward gene/cell delivery and novel stimuli, and accelerate clinical translation by establishing GMP-compliant production protocols and initiating human trials for chronic wound/oncology therapies.

References

1. M. Rostamitabar, R. Subrahmanyam, P. Gurikov, G. Seide, S. Jockenhoevel, S. Ghazanfari, Cellulose aerogel micro fibers for drug delivery applications. Mater. Sci. Eng. C **127**, 112196 (2021)
2. D. Xu, Y. Cheng, W. Lin, S. Han, S. Wu, A.K. Mondal, A. Li, F. Huang, Di-aldehyde tunicate cellulose nanocrystal (D-tCNC) aerogels for drug delivery: Effect of D-tCNC composition on aerogel structure and release properties. Int. J. Biol. Macromol. **256**, 128345 (2024)
3. Y. Sang, J. Gao, X. Han, T. Liang, T. Chen, Y. Zhao, Preparation and sustained release of diatomite incorporated and Eudragit L100 coated hydroxypropyl cellulose/chitosan aerogel microspheres. Int. J. Biol. Macromol. **267**, 131447 (2024)
4. T. Zhang, C. Cheng, G. Wei, X. Zhang, J. Zhang, Cellulose gels with controllable oriented porous structure by regulating regeneration process. Int. J. Biol. Macromol. **316**, 144753 (2025)
5. R. Wang, D. Shou, O. Lv, Y. Kong, L. Deng, J. Shen, PH-Controlled drug delivery with hybrid aerogel of chitosan, carboxymethyl cellulose and graphene oxide as the carrier. Int. J. Biol. Macromol. **103**, 248–253 (2017)
6. S. Ye, S. He, C. Su, L. Jiang, Y. Wen, Z. Zhu, W. Shao, Morphological, release and antibacterial performances of amoxicillin-loaded cellulose aerogels. Molecules **23** (2018)
7. T. Mohan, U. Ajdnik, C. Nagaraj, F. Lackner, A. Dobaj Štiglic, T. Palani, L. Amornkitbamrung, L. Gradišnik, U. Maver, R. Kargl, K. Stana Kleinschek, One-step fabrication of hollow spherical cellulose beads: application in pH-responsive therapeutic delivery. ACS Appl. Mater. Interfaces **14**, 3726–3739 (2022)
8. Q. Bi, Q. Zhao, Z. Qin, W. Gao, Z. Li, S. Deng, L. Mo, Lightweight, robust, and antimicrobial cellulose nanofibril aerogels decorated with TiO_2@PDA core–shell nanoparticles for highly efficient oil–water separation. Chem. Eng. J. **507**, 159989 (2025)
9. A. Zhang, Y. Zou, Y. Xi, P. Wang, Y. Zhang, L. Wu, H. Zhang, Fabrication and characterization of bamboo shoot cellulose/sodium alginate composite aerogels for sustained release of curcumin. Int. J. Biol. Macromol. **192**, 904–912 (2021)
10. Y.Y. Khine, M.H. Stenzel, Surface modified cellulose nanomaterials: a source of non-spherical nanoparticles for drug delivery. Mater. Horiz. **7**, 1727–1758 (2020)
11. A. Liu, H. Wu, A. Naeem, Q. Du, B. Ni, H. Liu, Z. Li, L. Ming, Cellulose nanocrystalline from biomass wastes: an overview of extraction, functionalization and applications in drug delivery. Int. J. Biol. Macromol. **241**, 124557 (2023)
12. H. Zheng, J. Chu, S. Li, X. Tang, S. Bi, K. Chen, A. Zhou, Y. Yang, X. Ning, Y. Xu, An optimized aerogel-based apheresis device for targeted lipid clearance in elderly hyperlipidemia patients. Adv. Mater. **n/a**, 2420091 (2025)
13. X. Hu, J. Zhou, Z. Deng, W.-X. Zhang, MIL-100-Fe self-assembled cellulose nanofibers sponge for Diclofenac cascade encapsulation. Carbohydr. Polym. **352**, 123182 (2025)
14. Y.-Y. Wang, Z.-H. Zhou, J.-L. Zhu, W.-J. Sun, D.-X. Yan, K. Dai, Z.-M. Li, Low-temperature carbonized carbon nanotube/cellulose aerogel for efficient microwave absorption. Compos. Part B: Eng. **220**, 108985 (2021)
15. J. Tan, Y. Fang, K. Wang, M. Wei, Z. Zhang, L. Wang, S. Chen, J. Pan, Carbon nanotube reinforced carbonized cellulose aerogels for enhanced hemostasis and accelerated skin wound healing. Adv. Funct. Mater. **n/a**, 2417798 (2025)

16. S. Yu, T. Budtova, Creating and exploring carboxymethyl cellulose aerogels as drug delivery devices. Carbohydr. Polym. **332**, 121925 (2024)
17. Z. Liu, S. Zhang, B. He, S. Wang, F. Kong, Synthesis of cellulose aerogels as promising carriers for drug delivery: a review. Cellulose **28**, 2697–2714 (2021)
18. W. Shi, Y.C. Ching, C.H. Chuah, Preparation of aerogel beads and microspheres based on chitosan and cellulose for drug delivery: a review. Int. J. Biol. Macromol. **170**, 751–767 (2021)
19. S. Wei, Y.C. Ching, C.H. Chuah, Synthesis of chitosan aerogels as promising carriers for drug delivery: a review. Carbohydr. Polym. **231**, 115744 (2020)
20. Y. Wang, R. Huang, S. Feng, R. Mo, Advances in nanocarriers for targeted drug delivery and controlled drug release. Chin. J. Chem. Eng. **23**, 513–528 (2025)
21. Q. Fu, Z. Gu, S. Shen, Y. Bai, X. Wang, M. Xu, P. Sun, J. Chen, D. Li, Z. Liu, Radiotherapy activates picolinium prodrugs in tumours. Nat. Chem. **16**, 1348–1356 (2024)
22. P. Tian, L. Zhao, J. Kim, X. Li, C. Liu, X. Cui, T. Liang, Y. Du, X. Chen, H. Pan, Dual stimulus responsive borosilicate glass (BSG) scaffolds promote diabetic alveolar bone defectsrepair by modulating macrophage phenotype. Bioact. Mater. **26**, 231–248 (2023)
23. J. Henschen, J. Illergård, P.A. Larsson, M. Ek, L. Wågberg, Contact-active antibacterial aerogels from cellulose nanofibrils. Colloid. Surface. B **146**, 415–422 (2016)
24. R. Sariboga, O.F. Sarioglu, Cellulolytic characterization of the rumen-isolated Acinetobacter pittii ROBY and design of a potential controlled-release drug delivery system. Eng. Microbiol. **4**, 100164 (2024)
25. H.M. Alkhalidi, A.A. Alahmadi, W.Y. Rizg, E.B. Yahya, A.K. Abdul Khalil, R.Y. Mushtaq, M.Y. Badr, A.Y. Safhi, K.M. Hosny, Revolutionizing cancer treatment: biopolymer-based aerogels as smart platforms for targeted drug delivery. Macromol. Rapid Commun. **45**, 2300687 (2024)
26. S. Zong, L. Zhu, F. Lei, J. Jiang, J. Duan, Low cost Non freeze-drying construction strategy of cellulose aerogels for efficient solar desalination. Chem. Eng. J. **500**, 156859 (2024)
27. H.-Y. Wang, W. Yang, Z. He, Y. Huang, G.-S. Wang, Y. Zhu, Construction of a partially rGO multifunctional aerogel with multi-stage pores for hydrophobicity, thermal insulation and efficient microwave absorption property. J. Mater. Chem. C (2025)
28. C. Zhang, S. Fan, J. Zhang, G. Yang, C. Cai, S. Chen, Y. Fang, W. Wan, A multifunctional trap-capture-kill antibacterial system for enhanced wound healing via modified decellularized mushroom aerogels. Bioact. Mater. **50**, 232–245 (2025)
29. K. Li, L. Peng, J. Li, X. Zhu, Multi-responsive Pickering emulsions stabilized by amphiphilic cellulose nanocrystals for building smart release systems of hydrophobic drugs. Carbohydr. Polym. **355**, 123348 (2025)
30. Y. Liang, H. Zhu, L. Wang, H. He, S. Wang, Biocompatible smart cellulose nanofibres for sustained drug release via pH and temperature dual-responsive mechanism. Carbohydr. Polym. **249**, 116876 (2020)
31. J. Jose, A.R. Pai, D.A. Gopakumar, Y. Dalvi, V. Ruby, S.G. Bhat, D. Pasquini, N. Kalarikkal, S. Thomas, Novel 3D porous aerogels engineered at nano scale from cellulose nano fibers and curcumin: An effective treatment for chronic wounds. Carbohydr. Polym. **287**, 119338 (2022)
32. C. Wan, J. Li, Cellulose aerogels functionalized with polypyrrole and silver nanoparticles: In-situ synthesis, characterization and antibacterial activity. Carbohydr. Polym. **146**, 362–367 (2016)
33. J. Wang, Q. Zhou, A. Wang, Z. Zhu, K. Moses, T. Wang, P. Li, W. Huang, Sponge inspired flexible, antibacterial aerogel electrode with long-term high-quality electrophysiological signal recording for human-machine interface. Adv. Funct. Mater. **34**, 2309704 (2024)
34. N. El-Wakil, K. Rabab, A.A. Mahmoud, D. Alain, R.E. Abouzeid, M.T. Abo El-Fadl, A. Maged, Risedronate-loaded aerogel scaffolds for bone regeneration. Drug Deliv. **30**, 51–63 (2023)
35. K. Zhang, X. Jiao, L. Zhou, J. Wang, C. Wang, Y. Qin, Y. Wen, Nanofibrous composite aerogel with multi-bioactive and fluid gating characteristics for promoting diabetic wound healing. Biomaterials **276**, 121040 (2021)
36. S. Groult, S. Buwalda, T. Budtova, Tuning bio-aerogel properties for controlling drug delivery. Part 2: Cellulose-pectin composite aerogels. Biomater. Adv. **135**, 212732 (2022)

37. J. Zhao, C. Lu, X. He, X. Zhang, W. Zhang, X. Zhang, Polyethylenimine-grafted cellulose nanofibril aerogels as versatile vehicles for drug delivery. ACS Appl. Mater. Interfaces **7**, 2607–2615 (2015)
38. L. Liu, T. Wang, M. Li, Y. Gao, L. Zhang, Carboxylated cellulose-based composite aerogel with double filling structure for sustained drug release. Ind. Crop Prod. **210**, 118126 (2024)
39. S.M.S. Shahriar, S.M. Andrabi, A.-M. Al-Gahmi, Z. Yan, A.D. McCarthy, C. Wang, Z.A. Yusuf, N.S. Sharma, M.E. Busquets, M.I. Nilles, C.P. Jara, K. Yang, M.A. Carlson, J. Xie, Bicomponent nano- and microfiber aerogels for effective management of junctional hemorrhage. Nat. Commun. **16**, 2403 (2025)
40. L. Chen, Z. Jin, T. Kamiya, X. Liu, Y. Tian, R. Xia, G. Wang, T. Li, Q. Zhang, Nanoporous zeolite anchored cellulose nanofiber aerogel for safe and efficient hemostasis. Small **21**, 2500696 (2025)
41. F. Fu, X. Zuo, Y. Wang, F. Zhao, C. Li, Y. Zeng, L. Wang, F. Wang, Centrifugal spinning-derived biomimetic aerogel for rapid hemostasis with minimal blood loss. Nano Lett. **25**, 6040–6050 (2025)

Chapter 20
Cellulose-Based Aerogels for Sensors and Electronics

Pinhong Chen, Shuibo Zheng, Zhipeng Sun, Lan Yang, and Haisong Qi

Abstract Cellulose-based aerogels (CBAs), characterized by ultrahigh porosity (>90%), ultralow density (0.01–0.05 g·cm^{-3}), and tunable functionality, exhibit transformative potential for advanced sensors and electronics. This review systematically examines their fabrication—including freeze-drying, supercritical drying, and 3D printing—alongside functionalization strategies (e.g., conductive nanofiller integration, crosslinking, and surface modification) that tailor mechanical, electrical, and interfacial properties. The hierarchical porous architecture enables exceptional performance in sensors: pressure/strain, humidity, gas, and biosensors. In electronics, they serve as flexible conductors, energy harvesters, supercapacitors, battery separators, electromagnetic interference shields and thermal insulators. Challenges persist in scalability, environmental stability, and interfacial compatibility, yet emerging solutions, such as in-situ functionalization and sustainable processing, promise to advance applications in wearable tech, IoT, and energy systems. Future directions prioritize multifunctional composites, AI-driven design, and scalable green synthesis to realize next-generation sustainable devices.

Keywords Cellulose aerogels · Preparation methods · Multi-purpose · Sensors · Electronics

20.1 Introduction

Conventional synthetic polymers have long dominated modern industrial applications, yet their limitations are increasingly evident as technological advancements and application demands evolve. In emerging fields such as wearable electronics, these materials often fall short of requirements for flexibility, lightweight properties, and biocompatibility. For instance, their inherent rigidity and weight compromise

P. Chen · S. Zheng · Z. Sun · L. Yang · H. Qi (✉)
State Key Laboratory of Advanced Papermaking and Paper-Based Materials, South China University of Technology, Guangzhou, China
e-mail: qihs@scut.edu.cn

G. Wei and J. Zhang (eds.), *Cellulose-Based Hydrogels and Aerogels*, Lecture Notes in Nanoscale Science and Technology 41,
https://doi.org/10.1007/978-3-032-03008-5_20

user comfort in wearable devices, where materials must seamlessly integrate with dynamic human motion [1]. Furthermore, traditional polymers frequently lack the multifunctionality required for advanced applications. In sensor technologies, for example, rapid responsiveness to diverse analytes and high selectivity are critical, yet conventional polymers typically exhibit narrow detection ranges and sluggish response kinetics, limiting their utility in complex environments [2]. Most critically, the environmental persistence of non-degradable synthetic polymers exacerbates global plastic pollution, contributing to long-term ecological harm [3]. Addressing these limitations necessitates the development of next-generation materials that combine tailored functionality, environmental compatibility, and adaptability to interdisciplinary challenges.

Cellulose-based aerogels (CBAs), an emerging class of bio-based materials, exhibit remarkable potential in flexible electronics and sensor technologies due to their unique physicochemical properties. For instance, their three-dimensional porous networks, fabricated via processes such as freeze-drying, achieve porosities exceeding 90% and ultralow densities of 0.01–0.05 $g \cdot cm^{-3}$ [4]. This hierarchical architecture not only minimizes device weight but also enhances user comfort and portability. The high porosity facilitates superior gas permeability and liquid absorption, enabling rapid analyte diffusion and response in sensing applications [5]. In gas sensors, for example, cellulose aerogels adsorb and transport target molecules with response times reduced by over 50% compared to conventional synthetic polymers [6]. Furthermore, the abundant hydroxyl (–OH) groups on cellulose surfaces permit tailored modifications, such as oxidation (e.g., sodium periodate treatment) or crosslinking (e.g., tannic acid) to precisely modulate conductivity, mechanical strength, and stability [7]. Surface functionalization via grafting reactions further expands their utility by introducing tunable electrical properties. Derived from natural plant fibers (e.g., softwood pulp), CBAs are inherently biocompatible and non-toxic, making them suitable for direct human contact [8]. Their environmental sustainability is underscored by microbial degradation into CO_2 and water, offering a biodegradable alternative to persistent synthetic polymers.

Owing to their exceptional properties, these materials have garnered significant attention in recent years, demonstrating broad application potential in sensors (e.g., environmental monitoring, biosensing) and electronic devices (e.g., flexible circuits, energy storage systems). In the field of sensing, such aerogels can serve as highly sensitive and selective sensing materials. For instance, when functionalized with specific chemical groups or nanoparticles, they enable rapid and precise detection of various analytes, including gases, humidity, and biomolecules. Gas sensors based on CBAs exhibit response times as short as a few seconds, a reduction of over 50% compared to conventional sensors fabricated from synthetic polymers [9]. For electronic applications, the ultralow density of cellulose aerogels offers distinct advantages in wearable and portable electronics, where they can function as lightweight, flexible substrates or components. Their porous architecture enhances device performance by improving heat dissipation and reducing overall weight [10]. Furthermore, these aerogels can be integrated into energy storage systems such as supercapacitors and batteries, delivering specific capacitances exceeding 200 $F \cdot g^{-1}$ and superior energy/power densities relative to traditional electrode materials. Notably, the inherent biocompatibility and environmental sustainability of CBAs make them uniquely suited for biomedical and biodegradable electronics. For example, in implantable medical devices, they can serve as biocompatible matrix materials, minimizing immune responses in host tissues while enabling natural degradation post-use, eliminating the need for secondary extraction surgeries.

Recent advances in CBAs, driven by in-depth research and refined functionalization strategies, have yielded unprecedented progress in sensor and electronic device applications, as shown in Fig. 20.1 [11–28]. These developments provide novel paradigms for engineering next-generation high-performance sensing and electronic systems. This chapter systematically summarizes the fabrication methodologies, functionalization approaches, and cutting-edge applications of CBAs in sensing and electronics, offering critical insights to advance their fundamental understanding and practical implementation. Our analysis not only informs optimization of aerogel synthesis for enhanced performance but also establishes theoretical and technical foundations for innovative device architectures, accelerating the integration of sustainable cellulose materials into advanced technologies.

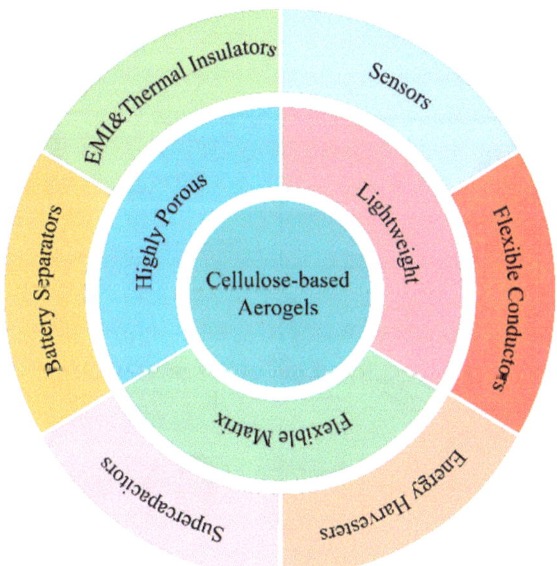

Fig. 20.1 CBAs: Platform for advanced sensors and electronics

20.2 Preparation and Functionalization Strategy of CBAs

20.2.1 Source of Cellulose

The fabrication of CBAs begins with strategic raw material selection. Cellulose sources, including wood, agricultural residues, and cotton fibers, exhibit distinct chemical structures and physical properties, necessitating source-specific choices aligned with target applications [29–31]. For instance, wood-derived cellulose possesses high crystallinity and mechanical strength, making it suitable for aerogels requiring structural robustness, while cotton cellulose offers superior flexibility and processability for deformable applications, as shown in Fig. 20.2a.

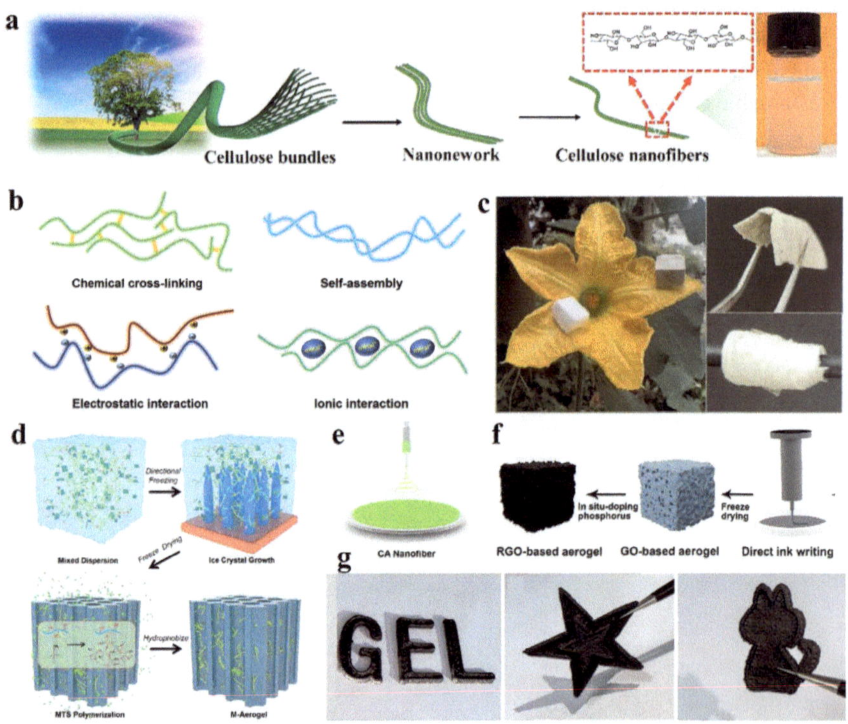

Fig. 20.2 Preparation and modification of CBAs. **a** Sources of cellulose. Reproduced with the permission from Ref. [38]. Copyright 2024, Elsevier. **b** Mode of cross-linking of cellulose gels. Reproduced with the permission from Ref. [39]. Copyright 2021, Wiley. **c** Physical drawing of the finished CBAs. Reproduced with the permission from Ref. [40]. Copyright 2021, Elsevier. **d** Freeze-drying process. Reproduced with the permission from Ref. [41]. Copyright 2023, Elsevier. **e** Preparation of cellulose nanofibers by electrospinning. Reproduced with the permission from Ref. [41]. Copyright 2023, Elsevier. **f** Preparation of CBAs by direct ink writing. Reproduced with the permission from Ref. [42]. Copyright 2024, Elsevier. **g** Preparation of CBAs by DIW. Reproduced with the permission from Ref. [42]. Copyright 2024, Elsevier

Following material selection, pretreatment emerges as a critical step to remove impurities (e.g., lignin, hemicellulose) and enhance cellulose purity and reactivity. Chemical methods employ acids, alkalis, or enzymes to hydrolyze or saponify contaminants [32, 33]. Alkaline pretreatment, for example, elevates cellulose purity to > 90%, significantly improving aerogel performance. Complementary physical approaches, such as mechanical grinding or ultrasonication, refine cellulose fibers from micrometer to nanometer scales. Ultrasonication increases specific surface area several-fold, creating optimal conditions for subsequent gelation.

20.2.2 Key Preparation Process

Gelation constitutes the defining stage in CBA synthesis, governing both the porous architecture and functional properties of the resultant material. This process typically proceeds through physical entanglement or chemical crosslinking pathways. Physical crosslinking relies on non-covalent interactions, including chain entanglement, hydrogen bonding, van der Waals forces, and ionic associations, to drive gel network formation [7]. While operationally straightforward and partially reversible (e.g., hydrogen bonds), this approach yields mechanically inferior gels with extended processing timelines, as shown in Fig. 20.2b. Conversely, chemical crosslinking exploits the abundant surface functionalities of cellulose (hydroxyl, carboxyl, or aldehyde groups) to form irreversible covalent bonds with agents such as glutaraldehyde or citric acid [16]. Chemically cross-linked gels exhibit superior mechanical strength, tunable cross-linking density, and enhanced stability, albeit at higher costs and with permanent network formation, as shown in Fig. 20.2b [34]. Critical processing parameters—such as the gelation time, temperature, and precursor concentration, demand precise optimization. Insufficient gelation time produces underdeveloped networks with heterogeneous pores, while excessive duration leads to over-condensed matrices and reduced porosity. Elevated temperatures accelerate kinetics but risk cellulose degradation at extremes. Precursor concentration directly dictates network density and ultimate porosity, necessitating application-specific calibration.

Drying constitutes the terminal phase in CBA fabrication, critically determining the structural fidelity and functional performance of the final material, as shown in Fig. 20.2c. Three primary drying methods are employed: ambient pressure drying, freeze-drying, and supercritical drying. Ambient pressure drying relies on solvent evaporation via heating under atmospheric conditions (e.g., oven drying) [35]. To mitigate capillary stress-induced structural collapse, low-surface-tension solvents (e.g., organic alternatives to water or alcohol) or oil-soluble surfactants are often introduced into the CBAs. This method offers mild processing conditions (atmospheric pressure, temperature control), cost-effectiveness, and scalability for industrial production. However, the inherent surface tension at liquid–gas interfaces during drying can collapse nanoscale pores, significantly reducing porosity.

Freeze-drying entails rapid freezing of the wet gel to immobilize solvents as ice crystals, followed by sublimation under vacuum to preserve the porous architecture, as shown in Fig. 20.2d [36, 37]. Controlled ice crystal growth (e.g., unidirectional freezing) enables the formation of vertically aligned macroporous or honeycomb-like structures, ideal for composite aerogels with uniform component distribution. However, this method requires expensive equipment (e.g., freeze dryers with cryogenic vacuum systems), high energy consumption, and is limited by sample capacity and prolonged drying cycles (ranging from hours to days).

Supercritical drying involves treating sol–gel-derived wet gels under supercritical conditions, where solvents are evaporated without liquid–gas interfaces, effectively eliminating capillary forces [35]. For instance, using CO_2 as the supercritical fluid, wet gels are placed in high-pressure reactors and heated to the critical point (31.1 °C and 7.38 MPa) of CO_2, followed by gradual depressurization to yield aerogels with ultrahigh porosity and ultralow density. While this approach optimally preserves pore integrity and specific surface area, it demands complex infrastructure, costly solvent exchange protocols, and stringent operational controls.

20.2.3 *Functional Modification*

Surface modification of CBAs proves instrumental in enhancing their functional performance and diversifying application horizons. Freeze-dried CBAs form sponge-like porous networks, yet often require further modification to enhance specific properties such as hydrophobicity, conductivity, and mechanical strength. Post-processing techniques including impregnation/coating, thermal treatment, surface modification, and cross-linking can enrich aerogel functionality while providing a foundation for advanced applications through the incorporation of functional materials. The inherent hydrophilicity of cellulose, arising from its abundant hydroxyl groups, poses challenges in humid environments. Water adsorption on surfaces or penetration into the matrix can disrupt conductive pathways and cause signal drift. More critically, moisture may dilute or disrupt the continuous distribution of conductive components (e.g., polypyrrole (PPy), silver particles), compromising electrical stability.

Surface modification strategies such as alkyl group introduction can effectively mitigate these effects while improving environmental stability. Rajabi-Abhari et al. developed silanized cellulose nanofiber (SCNF) aerogels for mechanically robust and reversibly compressible triboelectric nanogenerators (TENGs) capable of mechanical energy harvesting and self-powered sensing [22]. The SCNF aerogels were prepared via vinyltrimethoxysilane (VTMS) silanization of cellulose nanofibers (CNFs) followed by freeze-drying and post-thermal treatment, exhibiting optimized hydrophobicity, mechanical strength, flexibility, and compressibility. Functional materials (e.g., conductive polymers or hydrophobic agents) can be effectively incorporated through impregnation or spray coating to further enhance aerogel performance. Chen et al. fabricated compressible aerogel elastomers with superior electromechanical properties by freeze-drying silver nanowires and CoNi@carbon into magnetic conductive aerogels, followed by polydimethylsiloxane (PDMS) introduction via vacuum-assisted impregnation [37].

The inherent electrical insulation of cellulose restricts its capacity to transduce external stimuli into detectable electrical signals. Integrating conductive components into cellulose matrices, however, imparts tailored electrical properties that address application-specific requirements. A common strategy involves constructing conductive networks by integrating cellulose with conductive materials, such as metal ions/particles, carbon-based nanomaterials (e.g., graphene, carbon nanotubes), or conductive polymers (e.g., PPy, polyaniline (PANI)). For instance, Wang et al. developed a waterproof, breathable composite aerogel for wearable heaters using CNF aerogels, PPy, fluorinated finishing agents, and polyurethane (PU) [43]. The material exhibited ultralight density (0.14 $g{\cdot}cm^{-3}$), robust mechanical strength, and efficient Joule heating (reaching 173 °C at 4 V within ~70 s), alongside high air permeability (186.6 ± 22 $L{\cdot}s^{-1}{\cdot}m^{-2}$) and water vapor transmission (5.8 ± 0.8 $kg{\cdot}m^{-2}{\cdot}d^{-1}$). In a separate study, He et al. fabricated a cellulose/BNNSs-AgNP (CBAg) aerogel by hybridizing AgNPs, 1D cellulose, and 2D boron nitride nanosheets (BNNSs) [44]. The AgNPs bridged adjacent BNNSs, while cellulose stabilized the 3D framework. When incorporated into waterborne polyurethane (WPU), the CBAg-WPU composite showed a 69% enhancement in thermal conductivity and an 89% reduction in wear rate. Further advancing the field, Zheng et al. engineered an anisotropic conductive aerogel film via directional freeze-drying and controlled compression, combining aramid nanofibers (ANFs), CNFs, and silver nanowires (AgNWs) [34]. The resulting ACA film achieved a high electrical conductivity (10.81 $S{\cdot}m^{-1}$) and an electromagnetic interference shielding efficiency of 34.34 dB.

Furthermore, the inherent mechanical limitations of native cellulose, characterized by modest strength and toughness, can be overcome by incorporating reinforcing agents (e.g., CNFs, silica nanofibers) or implementing functional modifications such as chemical crosslinking or graft copolymerization, substantially enhancing its structural robustness. For instance, Chen et al. combined highly conductive nanocellulose with wood-derived CNFs to fabricate aerogels via directional ice-templating and freeze-drying [12]. These aerogels were subsequently employed as active layers in resistive pressure sensors on paper substrates. In another study, Wang et al. developed compressible carbon aerogels through sequential micro-fibrillation of cellulose fibers (CFs), periodate oxidation, and tannic acid (TA) cross-linking [33]. The resulting CF suspension was freeze-dried, and the aerogel was pyrolyzed at high temperatures. Micro-fibrillation imparted a hierarchical structure to the carbon aerogel, improving its binding strength and compressibility. Periodate oxidation and TA crosslinking enhanced mechanical performance, while pyrolysis conferred excellent conductivity, making the material well-suited for pressure sensing. When integrated into a pressure sensor, the device demonstrated a high sensitivity of 4 kPa^{-1}, rapid response/recovery times (32/30 ms), a low detection limit (0.8 Pa), and outstanding stability with excellent reproducibility (3000 cycles at 0.5 kPa).

20.2.4 Structural Design and Performance

Conventional freeze-drying of wet gels often results in disordered ice crystal formation, leading to porous architectures with compromised mechanical properties such as low strength, reduced elastic modulus, and structural collapse. To enhance mechanical resilience and structural stability, advanced strategies focus on precise regulation of micro/nanoscale assemblies, including the design of anisotropic frameworks, aligned molecular orientations, or chemical crosslinking to reinforce 3D network integrity and optimize physicochemical characteristics. Directional freeze-drying, an evolution of traditional methods, enables controlled ice-templating of pore structures through unidirectional or bidirectional freezing. Unlike isotropic freezing, temperature gradients are applied unidirectionally or bidirectionally, guiding ice crystal growth along defined axes to yield aerogels with aligned anisotropic porosity post-drying. Wang et al. fabricated nanocellulose/aramid nanofiber (CA) aerogels via unidirectional freeze-casting [45]. By utilizing the aligned CA aerogel as a template, they induced in-situ polymerization of conductive PPy within the matrix, producing CA/PPy (CAP) aerogels. These materials exhibited anisotropic mechanical properties and thermal conductivity while maintaining structural integrity under high-temperature conditions. Liu et al. developed conductive carbon aerogels incorporating CNFs, carbon nanotubes (CNTs), and reduced graphene oxide (RGO) through bidirectional freezing and annealing [16]. The honeycomb-like ordered porous structure, combined with ultralow density ($2.64\ mg \cdot cm^{-3}$) and superhydrophilicity (water contact angle $\approx 0°$ within 106 ms), enabled efficient stress distribution across the microstructure. This design conferred exceptional compressibility and fatigue resistance, sustaining 10,000 compression cycles at 50% strain without performance degradation.

Recent advances in 3D printing technology have enabled the fabrication of CBAs with precisely controlled architectures and internal pore structures, overcoming the limitations of conventional molding techniques, as shown in Fig. 20.2e–g. This approach allows for the creation of complex hierarchical or aligned porous structures, significantly expanding the application potential of nanocellulose aerogels, particularly in tissue engineering. Qin et al. developed conductive nanofiber aerogels through freeze-drying of in-situ polymerized PPy and electrospun cellulose acetate nanofibers, as shown in Fig. 20.2e [46]. The resulting aerogel exhibited an ultralow bulk density of $14.67\ mg \cdot cm^{-3}$, attributed to its nanofibrous skeleton and hierarchical porosity. The integration of direct ink writing (DIW) 3D printing with freeze-drying has emerged as a powerful strategy for producing cellulose aerogels with high-fidelity 3D porous architectures. Zhou et al. demonstrated this approach using an ink containing 20 wt% CNFs (GA-CNF-20), which exhibited a dynamic yield stress of 620 Pa due to intermolecular hydrogen bonding and optimal rheological properties, as shown in Fig. 20.2f and g [42]. The corresponding aerogels showed significantly enhanced mechanical performance, with an elastic modulus of 153 kPa and compressive stress of 400 kPa at 39.7% strain.

20.3 Applications of CBAs

CBAs have emerged as an ideal material for sensing and electronic devices, owing to their innate porous architecture, tunable chemical properties, exceptional compatibility with functional composites, and eco-friendly characteristics. These attributes are particularly instrumental in advancing flexible electronics, the Internet of Things (IoT), and wearable technologies. Future developments, integrating nanotechnology and artificial intelligence-driven design, are expected to further expand their performance limits.

20.3.1 Applications in Sensors

Sensors serve as transducers that convert external stimuli, including gases, solvents, acoustic waves, light, heat, mechanical force, and temperature, into electrical signals for subsequent processing and interpretation. High-performance sensing materials must exhibit exceptional sensitivity, stretchability, durability, linearity, selectivity, detection limits, and rapid response times. CBAs have emerged as promising sensing platforms due to their highly porous, continuous, and uniform architectures. Functionally, these can be categorized as strain/pressure sensors, temperature/humidity sensors, and biochemical sensors, with strain/pressure sensors typically operating through piezoresistive or piezoelectric mechanisms under mechanical loading.

20.3.1.1 Strain and Pressure Sensors

Recent advances demonstrate the versatility of CBAs sensors across applications. Chen et al. developed a magnetically responsive composite aerogel via freeze-drying silver nanowires and CoNi@carbon with ANFs/CNFs, functioning both as a piezoresistive sensor (for human motion monitoring, pressure array recognition) and triboelectric nanogenerator, as shown in Fig. 20.3a [37]. However, conventional cellulose-based electromechanical sensors often suffer from limited sensitivity and output due to inefficient stress transfer and weak electromechanical coupling. Li et al. addressed this through a "radial-longitudinal" biaxial compression technique that created anisotropic auxetic structures, amplifying piezoelectric output by 25.7-fold [47]. Structural engineering has enabled breakthrough performance metrics. Xu et al. designed a biomimetic hybrid aerogel using aligned CNF/CNT "mortar" with tubular MXene 'bricks'via bidirectional freezing, achieving ultralow density (7.48 mg·cm^{-3}) and exceptional conductivity (~2400 S·m^{-1}), as shown in Fig. 20.3b [48]. As a pressure sensor, it demonstrated record linear sensitivity (817.3 kPa^{-1}). Similarly, Cheng et al. produced AgNWs/NFC aerogels with ordered porosity through unidirectional freeze-drying, yielding sensors with 3.86 kPa^{-1}

sensitivity, 180 ms response, and >10,000-cycle durability [40]. Tian et al. engineered a layered porous aerogel integrating CNT/MXene conductors with cellulose nanofibers, enabling simultaneous temperature (33.5 $\mu V \cdot K^{-1}$) and pressure (−45.2% kPa^{-1}) detection without signal interference [49]. The dual-mode sensor achieved 0.03 K temperature resolution and 0.3 Pa pressure detection limits with outstanding stability.

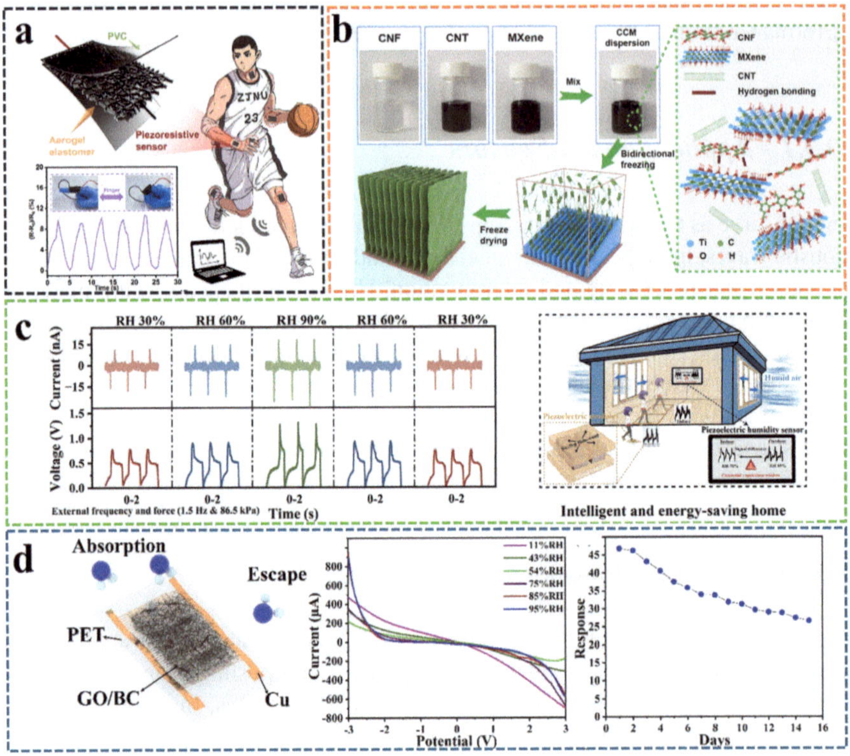

Fig. 20.3 CBAs for sensor applications. **a** Composite aerogel elastomers as piezoresistive sensors to monitor changes in relative resistance for various body movements. Reproduced with the permission from Ref. [37]. Copyright 2024, Elsevier. **b** Schematic diagram of the preparation process of CNF/CNT/MXene aerogel used as a pressure sensor. Reproduced with the permission from Ref. [48]. Copyright 2024, Springer. **c** The humidity-sensitive piezoelectric outputs of Swollen Holowood under an external frequency and force of 1.5 Hz and 86.5 kPa, and Schematic illustration for its application for intelligent and self-powered indoor humidity control. Reproduced with the permission from Ref. [52]. Copyright 2025, Wiley. **d** I-V curves of GO/BC humidity sensor under different humidity conditions and two-week stability at 54% RH. Reproduced with the permission from Ref. [21]. Copyright 2023, Elsevier

20.3.1.2 Humidity Sensor

Humidity sensing, which involves the detection of ambient moisture content (relative humidity, RH%), finds broad applications in meteorological monitoring, industrial control, healthcare, agriculture, and smart home systems.

Jiang et al. demonstrated a reversible hemicellulose supramolecular modulation strategy to achieve in situ nanofibrillation of wood cell walls, forming a "core (cellulose)-shell (hemicellulose)" nanostructure that mimics natural configurations, as shown in Fig. 20.3c [52]. This design exhibits intriguing humidity-responsive mechanical behavior, unlocking significant potential for applications in sensing, process monitoring, and energy management systems. Sun et al. reported a multifunctional, integrated, and low-power wireless flexible sensing platform based on biocompatible bacterial cellulose (BC) and graphene hybrids, as shown in Fig. 20.3d [21]. This platform not only enables real-time monitoring of physiological signals and respiration but also supports Morse code-based wireless communication. The system exhibits exceptional humidity-sensing performance, facilitating continuous tracking of respiratory waveforms and frequency. When worn during sleep, the wireless flexible pressure and humidity sensor provides synchronous diagnostic data for sleep apnea syndrome (SAS). Jin et al. developed an efficient method to fabricate aligned CNF/multiwalled CNT (CNF/MWCNT) aerogel films (AGFs), yielding a humidity sensor with high sensitivity and rapid response [20]. By optimizing water molecule exchange channels and modulating electrical conductivity, the sensor achieves superior performance across a 0–75% RH range.

20.3.1.3 Gas Sensor

Gas detection has become a critical focus in laboratories, industrial facilities, and households. Gas sensors, which measure the concentration of specific gases (e.g., CO_2, NH_3, and VOCs) in the environment, are widely employed in environmental monitoring, industrial safety, medical diagnostics, and smart home applications.

Wei et al. modified bacterial cellulose (BC) aerogels with polyethylene glycol (PEG) and incorporated graphene along with ZnO or TiO_2 to develop a novel VOC-sensing platform, as shown in Fig. 20.4a [50]. The material demonstrated real-time monitoring capability for multiple VOCs, achieving a detection limit as low as 1.43 ppm, including formaldehyde at the threshold limit value (TLV) concentration (0.75 ppm). Fu et al. engineered conductive CA/Ni-HITP composites through in situ growth of Ni-HITP frameworks on CBAs, subsequently integrating these with PVDF films to develop a respiratory TENG (R-TENG), as shown in Fig. 20.4b [14]. This device enables real-time monitoring of wearer respiration, distinguishing between breathing patterns through frequency and intensity analysis. Gao et al. engineered a compressible cellulose-based triboelectric aerogel with a hierarchically porous structure via heterointerface modulation [10]. The material exhibited high sensitivity to ammonia (NH_3), accurately detecting concentration variations in the range of 20–150 ppm with an excellent linear response ($R^2 = 0.996$).

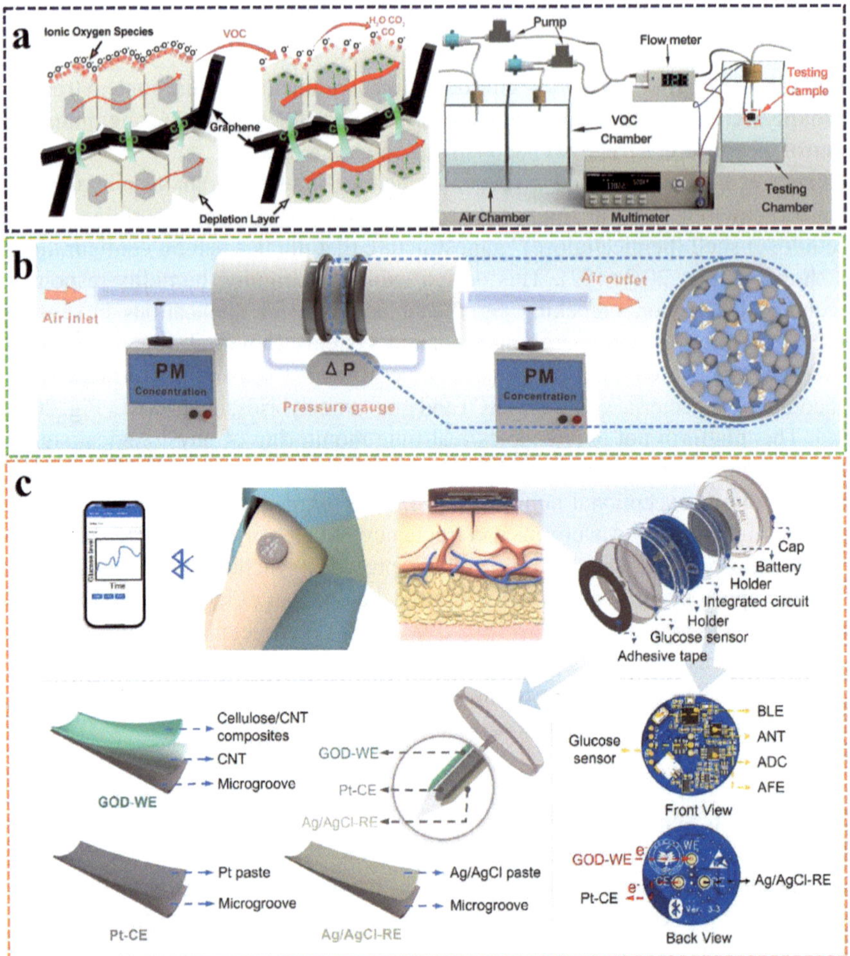

Fig. 20.4 **a** Representative physical model of the mechanism of BC/G/PEG/MOs with graphene and metal oxides added and the gas sensing performance for different VOCs. Reproduced with the permission from Ref. [50]. Copyright 2024, Elsevier. **b** Schematic diagram of pressure drops and submicron particles concentration measurement. Reproduced with the permission from Ref. [14]. Copyright 2022, Elsevier. **c** Highly sensitive continuous glucose sensor (CGS) with porous 3D cellulose/CNT network for real-time continuous glucose monitoring. Reproduced with the permission from Ref. [51]. Copyright 2025, Elsevier

20.3.1.4 Biosensors

CBA-based biosensing platforms leverage biorecognition elements (e.g., enzymes, antibodies, DNA, or cells) to achieve highly sensitive detection of target biomolecules through physicochemical signal transduction (electrical, optical, or thermal). These systems are widely applied in disease diagnostics, environmental monitoring, food safety, and pharmaceutical development.

Wang et al. reported a high-sensitivity continuous glucose sensor (CGS) featuring a porous 3D cellulose/ CNT network on the working electrode, as shown in Fig. 20.4c [51]. This design enhances enzyme-loading capacity for subcutaneous deployment, enabling continuous glucose monitoring in interstitial fluid with a sensitivity of 9.15 $\mu A \cdot mM \cdot cm^{-2}$ while maintaining functionality in live rats for up to 9 days. Wu et al. developed a chemically crosslinked, fluorescent composite film via a one-pot Schiff base reaction between aldehyde-functionalized cellulose nanofibers (DNF) and amino-modified carbon dots [53]. The film exhibits high transparency and mechanical strength, serving as a fluorescent probe for Fe^{3+} quantification with a linear response range of 0–100 μM. To address nonspecific adsorption of macromolecules, which causes surface fouling and pore clogging, limiting selective enrichment and surface-enhanced Raman scattering (SERS) detection. Wang et al. engineered an innovative aerogel-embedded hydrogel composite (CH@S-CNF/SA/Ag NPs) [24]. The aerogel component, with its high surface area and electrophilic properties, functions as a SERS-active substrate for target molecule adsorption, demonstrating utility in urine analysis for both healthy individuals and stage-5 chronic kidney disease (CKD5) patients.

20.3.2 Applications in Electronics

CBAs demonstrate significant potential across multiple electronic domains, including flexible electronics, energy harvesting, supercapacitors, battery separators, electromagnetic shielding, and thermal management, where their distinct physicochemical characteristics enable novel technological paradigms.

20.3.2.1 Flexible Conductors and Electrodes

CBAs, enriched with highly reactive hydroxyl (−OH) groups, enable facile surface modification and robust integration with conductive fillers via covalent or hydrogen bonding, positioning them as promising candidates for high-performance electrodes. Their porous architecture further facilitates the loading and uniform dispersion of carbon-based conductive materials, ensuring structural stability in electrode fabrication. Notably, the inherent flexibility and mechanical resilience of cellulose aerogels allow their use as substrates for flexible circuits, enduring repeated bending and folding without degradation.

Wang et al. developed a sponge-inspired flexible electrode based on cellulose aerogels through thermally induced phase separation, as shown in Fig. 20.5a [1]. This electrode conforms seamlessly to skin surfaces, achieving a signal-to-noise ratio (SNR) ~15% higher than commercial rigid electrodes in 24-h electrocardiogram (ECG) monitoring. Remarkably, the aerogel electrode exhibits exceptional liquid absorption capacity, retaining solutions up to eight times its own weight (≈800%). Liu et al. engineered a permeable, stretchable, and recyclable CBA-based electronic system by sandwiching a screen-printed silver sensing layer between a biocompatible CNF/HPC/PVA aerogel sublayer and a permeable polyurethane top layer, as shown in Fig. 20.5b [26]. This design achieves notable sensing performance, including high sensitivity (gauge factor ≈ 238), ultralow detection limits (0.1% strain), and rapid response times (18 ms). In another case, Lin et al. achieved a breakthrough in high-resolution QLEDs using nanoimprinting and surface modification [54]. The optimized devices exhibited a record resolution of 9072 PPI, peak external quantum efficiency (EQE) of 15.7%, and maximum luminance of 15,163 $cd{\cdot}m^{-2}$. By tuning the elastic modulus of patterned films, the QLEDs retained 91% of initial luminance after 400 bending cycles, underscoring exceptional mechanical durability.

20.3.2.2 Energy Harvesting

Beyond sensing, CBAs serve dual roles as structural supports and triboelectric layers in applications such as triboelectric nanogenerators (TENGs) and moisture-enabled power generation.

Zhang et al. synthesized a 3D cadmium metal–organic framework (Cd-MOF) as a filler for ethyl cellulose (EC), producing a mechanically durable and flexible Cd-MOF@EC composite film [55]. A contact-separation mode TENG (FCEC-TENG) fabricated with 10 wt.% Cd-MOF@EC demonstrated exceptional energy-harvesting efficiency, powering 14 commercial white LEDs and enabling visible-light-driven bromination reactions with high yield and tolerance. Similarly, Mo et al. developed a nanocellulose-based hydrogel moisture-enabled generator (N-HMEG) by combining sulfated cellulose nanofibrils (SCNFs), polyvinyl alcohol (PVA), and a glycerol-water binary solvent, as shown in Fig. 20.5c [56]. The SCNFs enhanced ion diffusion within the hydrogel, while glycerol improved hygroscopicity, enabling the N-HMEG to generate an open-circuit voltage of 0.9 V and a short-circuit current density of 92 $\mu A{\cdot}cm^{-2}$ at 80% relative humidity. In a parallel innovation, Lin et al. engineered a temperature-gated nanochannel membrane using functionalized Nostoc-derived nanocellulose for controllable osmotic energy harvesting, as shown in Fig. 20.5d [57]. Under a 50-fold salinity gradient at 50 °C, the system achieved a maximum power density of ~10.1 $W{\cdot}m^{-2}$, showcasing its potential for scalable blue energy applications.

20.3.2.3 Supercapacitors

Supercapacitors, emerging as a hybrid energy storage system bridging conventional capacitors and rechargeable batteries, have garnered significant attention for their high power density, exceptional rate capability, prolonged cycle life, and rapid charge/discharge kinetics. CBAs, with their ultrahigh specific surface areas (up to 500 m^2/g) and hierarchical porosity, are ideal electrode materials for supercapacitors, offering

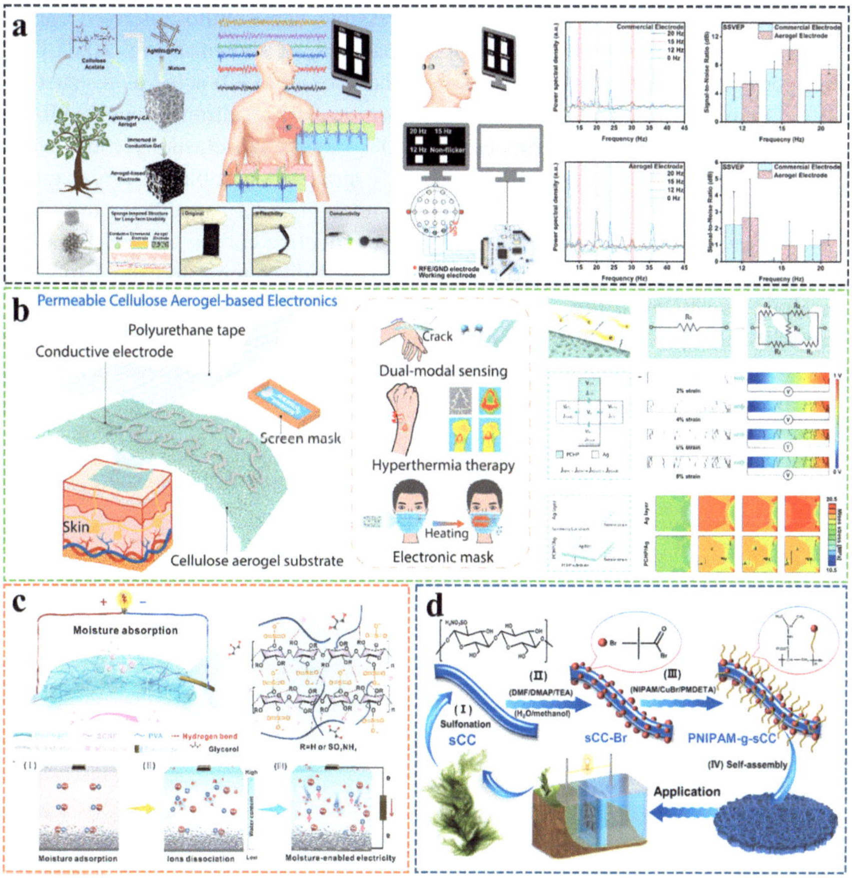

Fig. 20.5 Applications of CBAs in electronic devices. **a** AgNWs@PPy-CA aerogel sponge-like flexible electrode for 24 h electrocardiogram (ECG) recording. Reproduced with the permission from Ref. [1]. Copyright 2024, Wiley. **b** CNF/HPC/PVA aerogel for easy strain/moisture sensing and thermal healthcare. Reproduced with the permission from Ref. [26]. Copyright 2025, American Chemical Society. **c** SCNF-PVA aerogel is used to collect air moisture and for power generation. Reproduced with the permission from Ref. [56]. Copyright 2024, Elsevier. **d** PNIPAM-g-sCC can regulate and boost the output of osmotic energy harvesting systems by reversibly and stably switching temperatures. Reproduced with the permission from Ref. [57]. Copyright 2023, Elsevier

abundant active sites for charge storage. Their intrinsic flexibility further ensures stable electrochemical performance under mechanical deformation, making them suitable for energy storage in flexible electronics.

Haj et al. addressed the long-standing challenge of poor wettability in carbon-based supercapacitor electrodes by strategically modulating carbon aerogel (CA) properties via conjugated anions, a previously underexplored design parameter [58]. This approach yielded remarkable capacitance retention of 120.29% and 134.42% after 5000 and 10,000 cycles, respectively, surpassing conventional performance benchmarks. Liu et al. engineered a compressible nanocellulose-derived carbon aerogel for composite electrodes in supercapacitors, as shown in Fig. 20.6a [59]. CNFs formed an elastic network, while graphene oxide (GO) acted as a structural scaffold to mitigate shrinkage during carbonization. The resulting aerogel exhibited exceptional compressibility (withstanding 80% strain) and elasticity (96% stress retention after 2000 compression cycles at 30% strain). Assembled supercapacitors demonstrated robust electrochemical stability, retaining 82% capacitance over 5000 cycles. In another study, Liu et al. fabricated conductive carbon aerogels incorporating CNFs, CNTs, and reduced GO (RGO) via bidirectional freeze-casting and annealing, as shown in Fig. 20.6b [16]. The composite demonstrates exceptional electrochemical performance, including an areal capacitance of 109.4 mF·cm^{-1} at 0.4 mA·cm^{-2}, and remarkable mechanical resilience, retaining 88% capacitance after 5000 compression cycles at 50% strain.

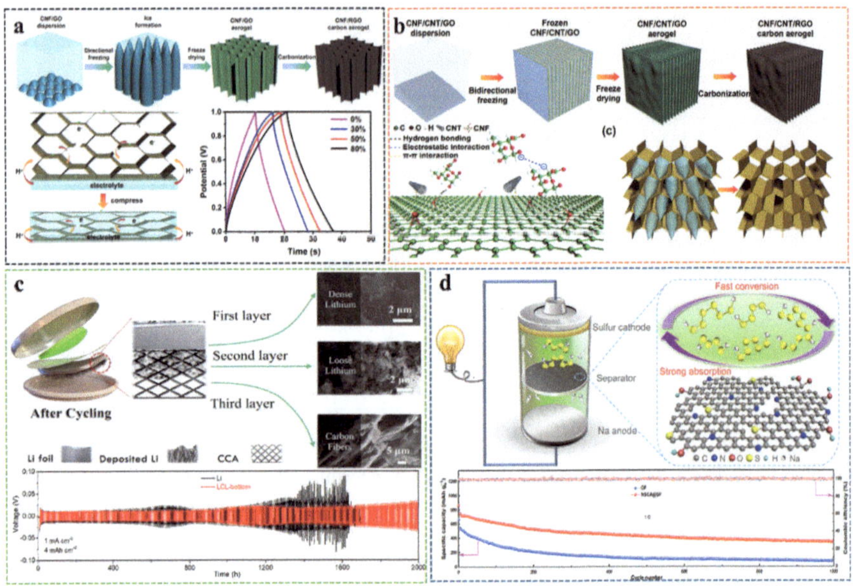

Fig. 20.6 **a** 3D porous structure carbon aerogels of nanocellulose for the preparation of high-performance compressible supercapacitor composite electrodes. Reproduced with the permission from Ref. [59]. Copyright 2024, Springer. **b** CNF/CNT/RGO carbon aerogels for highly compressible supercapacitors and strain sensors. Reproduced with the permission from Ref. [16]. Copyright 2022, Wiley. (d) Cellulose/graphene carbon composite aerogel to inhibit upward dendrite growth through the diaphragm for lithium metal batteries. Reproduced with the permission from Ref. [60]. Copyright 2021, Wiley. **c** Lightweight 3D cellulose nanofiber-derived carbon aerogel (NSCA@GF) coated on a glass fiber diaphragm to inhibit the shuttle effect for room temperature sodium-sulfur (RT/Na-S) batteries. Reproduced with the permission from Ref. [17]. Copyright 2023, Wiley

20.3.2.4 Battery Separators

CBAs exhibit dual utility in battery technologies, functioning both as separators and electrode scaffolds. When utilized as separator materials, their porous architecture enhances electrolyte permeability and ion transport efficiency, thereby improving charge–discharge kinetics.

For instance, lithium-ion batteries incorporating cellulose aerogel separators demonstrate markedly enhanced cycling stability, with capacity retention exceeding 80% after 1000 cycles. As electrode scaffolds, these aerogels bolster structural integrity and electrical conductivity of electrodes, further enhancing overall battery performance. Yang et al. developed a multifunctional separator by coating a glass fiber membrane (NSCA@GF) with a lightweight 3D nanocellulose-derived carbon aerogel, effectively mitigating polysulfide shuttle effects in sodium-sulfur (Na-S) batteries, as shown in Fig. 20.6c [17]. The hierarchical porous structure, high electronic conductivity, and 3D network of the N, S co-doped CA endowed the separator with robust polysulfide anchoring capacity and accelerated redox kinetics for polysulfide conversion. Heteroatomic N/S sites further enhanced chemical adsorption and catalytic activity. Na-S cells equipped with the NSCA@GF separator demonstrated a high reversible capacity of 788.8 $mAh \cdot g^{-1}$ at 0.1 C after 100 cycles and exceptional cyclability, with a minimal capacity decay of 0.059% per cycle over 1000 cycles at 1 C.

Mao et al. proposed a novel "directional growth" strategy by relocating the primary deposition interface from the anode/separator to the anode/current collector, as shown in Fig. 20.6d [60]. A cellulose/graphene carbon composite aerogel (CCA) interlayer, positioned between the current collector and lithium anode (LCL bottom), acted as a charge organizer to homogenize current density and guide lithium deposition. At an ultrahigh current density of 25 $mA \cdot cm^{-2}$, the LCL bottom electrode achieved a stripping/plating capacity of 13 $mAh \cdot cm^{-2}$, maintaining stable overpotentials below 50 mV for over 300 h—significantly outperforming bare lithium (1 V overpotential at ~50 h). Paired with lithium iron phosphate (LFP) cathodes, full cells retained 94% of initial capacity after 1000 cycles. Bharti et al. leveraged carbonized bacterial cellulose (CBC) as a sulfur host and pristine BC as a separator (CBC@BC) in lithium-sulfur (Li–S) batteries [61]. The CBC@BC configuration delivered a remarkable capacity of 740 $mAh \cdot g^{-1}$ at 1 C and retained 310 $mAh \cdot g^{-1}$ even at 4 C, highlighting its potential for high-rate applications.

20.3.2.5 Electromagnetic Interference (EMI)

CBAs demonstrate emerging potential in EMI shielding applications. Although their intrinsic EMI shielding efficiency (SE) is limited, their performance can be significantly enhanced through surface metallization (e.g., copper or silver coatings) or integration with conductive materials. Cheng et al. engineered MXene $Ti_3C_2T_x$/carboxymethyl cellulose (MXene/CMC) aerogels as multifunctional energy harvesters capable of scavenging mechanical energy while mitigating electromagnetic radiation exposure [62]. These aerogels achieved exceptional EMI shielding efficiencies of 52.15 dB (X-band), 60.31 dB (Ku-band), and 80.36 dB (K-band), highlighting their dual functionality in energy conversion and health protection. Mai et al. fabricated a novel bilayer aerogel composed of zeolitic imidazolate framework-derived hollow CoNi carbon nanocages, MXene $Ti_3C_2T_x$, and nanocellulose (BZMN) via a two-step freeze-casting method, as shown in Fig. 20.7a [63]. The BZMN aerogel exhibited balanced EMI shielding (35.1 dB) and low reflection coefficients (R = 0.28 at 35.1 dB; R = 0.25 at 21.2 dB), suggesting superior absorption-dominated shielding mechanisms. Similarly, Li et al. developed MXene/sodium carboxymethyl cellulose composite aerogels using unidirectional freezing and dual silane crosslinking [27]. With 90 wt.% MXene loading, the aerogel demonstrated an ultrahigh specific SE of 45.8 dB at a remarkably low density (19.75 $mg{\cdot}cm^{-3}$), achieving a normalized SE of 11,595 $dB{\cdot}cm^2{\cdot}g^{-1}$, a benchmark for lightweight EMI shielding materials. Wu et al. fabricated anisotropic cellulose/polyacrylamide/graphene nanoplatelet/AgNWs (CPGxAy) hybrid aerogels via directional freeze-drying, followed by vacuum-assisted impregnation with polyethylene glycol (PEG), as shown in Fig. 20.7b [64]. The synergistic effect of graphene nanoplatelets (GNPs) and AgNWs endowed PEG@CPGxAy superhydrophobic phase-change materials (ss-CPCMs) with a thermal conductivity of 0.84 $W{\cdot}m^{-1}{\cdot}K^{-1}$ (200% higher than pure PEG) and anisotropic electromagnetic interference shielding efficiencies of 71.08 dB (transverse) and 35.21 dB (longitudinal), while maintaining high melting/crystallization enthalpy efficiencies (93.47% and 93.08%, respectively).

20.3.2.6 Thermal Insulation

Electronic devices frequently generate substantial and sustained heat during operation, which can lead to catastrophic failure under extreme conditions. CBAs, with their 3D nanoporous networks (porosity up to 99.8%), effectively suppress thermal conduction, achieving ultralow thermal conductivities as low as 0.02984 $W{\cdot}m^{-1}{\cdot}K^{-1}$, surpassing conventional insulation materials. These attributes, combined with their lightweight structure and sustainability, have positioned cellulose aerogels as a focal material for advanced thermal management in electronics.

Zhou et al. engineered a continuous heat-transfer network by functionalizing liquid metal (LM) surfaces with alkoxy and carboxyl groups to enhance interfacial interactions with hydroxyl groups on CBA walls, as shown in Fig. 20.7c [23]. This strategy anchored LM onto CBA, yielding a thermal conductivity of 7.421 $W{\cdot}m^{-1}{\cdot}K^{-1}$ and a thermal anisotropy ratio of 23–1.35-fold higher than unmodified LM-CA composites. Huang et al. developed regenerated cellulose aerogel fibers (CAFs) through wet spinning and gradient solvent exchange [28]. CAF textiles exhibited superior thermal insulation (39.31 $mW{\cdot}m^{-1}{\cdot}K^{-1}$), maintaining a 45 °C temperature differential at 100 °C, with potential applications in heat-resistant and cold-weather apparel.

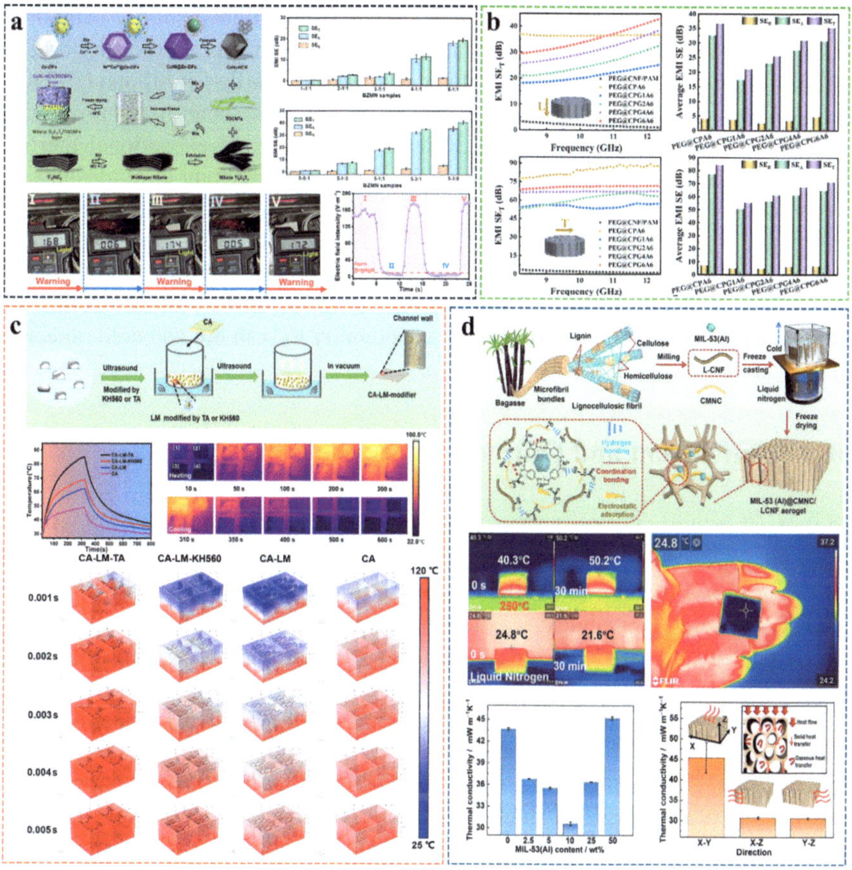

Fig. 20.7 **a** Bilayer aerogels consisting of hollow CoNi carbon nanocages/MXene $Ti_3C_2T_x$/ cellulose nanoparticles (BZMN) derived from zeolitic imidazolium ester frameworks are candidates for low-reflection electromagnetic shielding devices, which can be used to protect future communication and electronic devices. Reproduced with the permission from Ref. [63]. Copyright 2025, Wiley. **b** Anisotropic PEG@CPGxAy ss-CPCMs aerogels were successfully prepared by directional freeze-drying method for heat transfer and electromagnetic interference shielding of modern electronic devices. Reproduced with the permission from Ref. [64]. Copyright 2023, Springer. **c** CA-LM-TA composites were constructed by surface functionalization on liquid metal (LM) surfaces with the introduction of alkoxy and carboxyl groups to build continuous heat transfer networks with high thermal conductivity and excellent EMI shielding properties. Reproduced with the permission from Ref. [23]. Copyright 2025, Springer. **d** MIL-53(Al)@CMNC/LCNF aerogels (M@CLA) with anisotropic structures were prepared by unidirectional freezing to exhibit excellent thermal insulation at 250 °C and −196 °C. Reproduced with the permission from Ref. [65]. Copyright 2024, Springer

Li et al. integrated MIL-53(Al) nanoparticles into lignocellulose nanofibril (LCNF) aerogels using carboxylated nanocellulose (CMNC) as a molecular bridge, as shown in Fig. 20.7d [65]. Robust interactions between MIL-53(Al) and CMNC's hydroxyl/carboxyl groups imparted flame retardancy via char layer formation, achieving a thermal conductivity of 30.6 $mW{\cdot}m^{-1}{\cdot}K^{-1}$ and exceptional thermal insulation at extreme temperatures (250 °C to −196 °C). Zhong et al. designed a passive isothermal flexible sensor using hierarchical cellulose aerogel (HCA) as a tri-directional radiative cooling layer [25]. The HCA's hollow microfibers provided ultralow thermal conductivity (reducing winter heat loss) while enabling radiative cooling (summer temperature reduction). This sensor maintained operational stability across 0–100 °C, demonstrating adaptability to both hot and cold objects.

20.4 Conclusion and Outlook

CBAs have emerged as a highly promising class of sustainable materials for advanced sensor and electronic applications, owing to their ultrahigh porosity (>90%), ultralow density (0.01–0.05 $g{\cdot}cm^{-3}$), tunable surface chemistry, biocompatibility, and biodegradability. Key fabrication methods—including freeze-drying, supercritical drying, and emerging 3D printing—enable precise control over hierarchical porous structures. Functionalization strategies (e.g., crosslinking, conductive nanomaterial integration, hydrophobic modification) overcome inherent limitations like electrical insulation and mechanical fragility, unlocking exceptional performance in gas/humidity/pressure sensors, biosensors, flexible electrodes, supercapacitors, battery separators, EMI shields, and thermal insulators. Innovations in structural design (e.g., anisotropic aerogels, biomimetic architectures) further enhance sensitivity, stability, and multifunctionality.

Despite significant progress, challenges remain in scalability, environmental stability, interfacial compatibility in composites, and long-term durability under operational stresses. Future research could be focused on these aspects.

(1) Green manufacturing: Developing low-cost, energy-efficient processes (e.g., ambient-pressure drying, microwave-assisted techniques) to replace energy-intensive supercritical/freeze-drying.
(2) Stability enhancement: Engineering moisture-resistant (e.g., via hydrophobic coatings) and mechanically robust aerogels through advanced crosslinking or nanocomposite designs.
(3) Multifunctional integration: Leveraging nanotechnology and AI-driven design to create "smart" aerogels with synchronized sensing, energy harvesting, and self-healing capabilities.
(4) Scalability & standardization: Optimizing raw material sourcing (e.g., agricultural waste) and establishing standardized protocols for industrial translation.
(5) Application diversification: Expanding into emerging fields such as implantable bioelectronics, neuromorphic computing, and sustainable IoT devices.

Bridging fundamental research with scalable engineering will accelerate the adoption of CBAs in next-generation, eco-friendly electronics.

Acknowledgements This work was supported by Guangdong Basic and Applied Basic Research Foundation (2024A1515012143) and State Key Laboratory of Pulp and Paper Engineering (2024QN02).

References

1. J. Wang, Q. Zhou, A. Wang, Z. Zhu, K. Moses, T. Wang, P. Li, W. Huang, Sponge inspired flexible, antibacterial aerogel electrode with long-term high-quality electrophysiological signal recording for human-machine interface. Adv. Funct. Mater. **34**, 2309704 (2024)
2. M. Wang, R. Li, X. Feng, C. Dang, F. Dai, X. Yin, M. He, D. Liu, H. Qi, Cellulose nanofiber-reinforced ionic conductors for multifunctional sensors and devices. ACS Appl. Mater. Interfaces **12**, 27545–27554 (2020)
3. Z. Zeng, C. Wang, G. Siqueira, D. Han, A. Huch, S. Abdolhosseinzadeh, J. Heier, F. Nüesch, C.J. Zhang, G. Nyström, Nanocellulose-MXene biomimetic aerogels with orientation-tunable electromagnetic interference shielding performance. Adv. Sci. **7**, 2000979 (2020)
4. X. Feng, X. Wang, C. Zhang, C. Dang, Y. Chen, H. Qi, Highly conductive and multifunctional nanocomposites based on sulfated nanocellulose-assisted high dispersion limit of single-walled carbon nanotubes. Carbon **183**, 187–195 (2021)
5. X. Feng, X. Wang, M. Wang, S. Zhou, C. Dang, C. Zhang, Y. Chen, H. Qi, Novel PEDOT dispersion by in-situ polymerization based on sulfated nanocellulose. Chem. Eng. J. **418**, 129533 (2021)
6. M. Wang, Z. Lai, X. Jin, T. Sun, H. Liu, H. Qi, Multifunctional liquid-free ionic conductive elastomer fabricated by liquid metal induced polymerization. Adv. Funct. Mater. **31**, 2101957 (2021)
7. V. Rahmanian, T. Pirzada, S. Wang, S.A. Khan, Cellulose-based hybrid aerogels: strategies toward design and functionality. Adv. Mater. **33**, 2102892 (2021)
8. Y. Chen, Y. Liu, Y. Li, H. Qi, Highly sensitive, flexible, stable, and hydrophobic biofoam based on wheat flour for multifunctional sensor and adjustable EMI shielding applications. ACS Appl. Mater. Interfaces **13**, 30020–30029 (2021)
9. H. Zhi, X. Zhang, F. Wang, P. Wan, L. Feng, Flexible Ti_3C_2 T_x MXene/PANI/bacterial cellulose aerogel for e-skins and gas sensing. ACS Appl. Mater. Interfaces **13**, 45987–45994 (2021)
10. C. Gao, W. Zhang, T. Liu, B. Luo, C. Cai, M. Chi, S. Zhang, Y. Liu, J. Wang, J. Zhao, C. Qin, S. Nie, Hierarchical porous triboelectric aerogels enabled by heterointerface engineering. Nano Energy **121**, 109223 (2024)
11. S. Wei, X. Qiu, J. An, Z. Chen, X. Zhang, Highly sensitive, flexible, green synthesized graphene/biomass aerogels for pressure sensing application. Compos. Sci. Technol. **207**, 108730 (2021)
12. S. Chen, Y. Chen, D. Li, Y. Xu, F. Xu, Flexible and sensitivity-adjustable pressure sensors based on carbonized bacterial nanocellulose/wood-derived cellulose nanofibril composite aerogels. ACS Appl. Mater. Interfaces **13**, 8754–8763 (2021)
13. R. Chen, X. Li, Q. Huang, H. Ling, Y. Yang, X. Wang, Self-assembled porous biomass carbon/RGO/nanocellulose hybrid aerogels for self-supporting supercapacitor electrodes. Chem. Eng. J. **412**, 128755 (2021)
14. Q. Fu, Y. Liu, T. Liu, J. Mo, W. Zhang, S. Zhang, B. Luo, J. Wang, Y. Qin, S. Wang, S. Nie, Air-permeable cellulosic triboelectric materials for self-powered healthcare products. Nano Energy **102**, 107739 (2022)
15. D. Liao, Y. Wang, P. Xie, C. Zhang, M. Li, H. Liu, L. Zhou, C. Wei, C. Yu, Y. Chen, A resilient and lightweight cellulose/graphene oxide/polymer-derived multifunctional carbon

aerogel generated from Pickering emulsion toward a wearable pressure sensor. J. Colloid Interface Sci. **628**, 574–587 (2022)
16. H. Liu, T. Xu, C. Cai, K. Liu, W. Liu, M. Zhang, H. Du, C. Si, K. Zhang, Multifunctional superelastic, superhydrophilic, and ultralight nanocellulose-based composite carbon aerogels for compressive supercapacitor and strain sensor. Adv. Funct. Mater. **32**, 2113082 (2022)
17. W. Yang, W. Yang, R. Zou, Y. Huang, H. Lai, Z. Chen, X. Peng, Cellulose nanofiber-derived carbon aerogel for advanced room-temperature sodium–sulfur batteries. Carbon Energy **5**, e203 (2023)
18. B. Luo, C. Cai, T. Liu, X. Meng, X. Zhuang, Y. Liu, C. Gao, M. Chi, S. Zhang, J. Wang, Y. Bai, S. Wang, S. Nie, Multiscale structural nanocellulosic triboelectric aerogels induced by hofmeister effect. Adv. Funct. Mater. **33**, 2306810 (2023)
19. M.U. Khan, Y. Abbas, H. Abunahla, M. Rezeq, A. Alazzam, N. Alamoodi, B. Mohammad, Biocompatible humidity sensor using paper cellulose fiber/GO matrix for human health and environment monitoring. Sens. Actuat, B Chem. **393**, 134188 (2023)
20. G. Jin, M. Sun, Y. Gao, X. Zhou, C. Wei, L. Lyu, A CNFs/MWCNTs aerogel film-based humidity sensor with directionally aligned porous structure for substantially enhancing both sensitivity and response speed. Chem. Eng. J. **473**, 145304 (2023)
21. J. Sun, K. Xiu, Z. Wang, N. Hu, L. Zhao, H. Zhu, F. Kong, J. Xiao, L. Cheng, X. Bi, Multifunctional wearable humidity and pressure sensors based on biocompatible graphene/bacterial cellulose bioaerogel for wireless monitoring and early warning of sleep apnea syndrome. Nano Energy **108**, 108215 (2023)
22. A. Rajabi Abhari, H. Yoo, J.S. Kim, H.Y. Yoon, J.Y. Park, N. Yan, R. Tabassian, I.K. Oh, Reversibly compressible silanated cellulose nanofibril aerogel for triboelectric taekwondo scoring sensors. Small **20**, 2405664 (2024)
23. X. Zhou, L. Dong, S. Zheng, D. Cao, J. Chen, X. Feng, J. Zhu, X. Lu, L. Mu, Enhancing effective anisotropic thermal conductivity and electromagnetic interference shielding via interface engineering of natural cellular channels in wood and liquid metal/cellulose aerogel. Adv. Compos. Hybrid Mater. **8**, 82 (2025)
24. H. Wang, P. Xu, J. Wei, L. Qiu, J. Zou, C. Lin, R. You, Y. Hu, L. Zhang, Y. Lu, Urine collection by sodium alginate/CMC composite self-calibrating aerogel SERS platform for accurate screening and staging of chronic kidney disease. Int. J. Biol. Macromol. **302**, 140520 (2025)
25. S. Zhong, B. Lu, D.C. Wang, B. Arianpour, S. Wang, H. Han, J. Yin, H. Bao, Y. Liu, Z. Wen, Y. Zhou, Passive isothermal flexible sensor enabled by smart thermal-regulating aerogels. Adv. Mater. **37**, 2415386 (2025)
26. S. Liu, W. Li, X. Wang, L. Lu, Y. Yao, S. Lai, Y. Xu, J. Yang, Z. Hu, X. Gong, K.C. Leung, S. Xuan, Permeable, stretchable, and recyclable cellulose aerogel on-skin electronics for dual-modal sensing and personal healthcare. ACS Nano **19**, 3531–3548 (2025)
27. J. Li, C. Liang, C. Lin, Y. Wang, Y. Liang, D. Dong, Lightweight, superhydrophobic, and superelastic MXene/carboxymethylcellulose sodium composite aerogels for efficient and multifunctional electromagnetic interference shielding. Chem. Eng. J. **506**, 160274 (2025)
28. Z. Huang, B. Wang, Z. Shi, S. Qiao, A. Tong, J. Wang, J. He, A. He, X. Chen, P. Hu, W. Ke, N. Yao, W. Xu, F. Chen, Regenerated cellulose aerogel fibers with lightweight and exceptional mechanical performance for thermal insulation. Small, 2501154 (2025)
29. M. Wang, O.J. Rojas, L. Ning, Y. Li, X. Niu, X. Shi, H. Qi, Liquid metal and Mxene enable self-healing soft electronics based on double networks of bacterial cellulose hydrogels. Carbohydr. Polym. **301**, 120330 (2023)
30. J. Mo, Y. Liu, Q. Fu, C. Cai, Y. Lu, W. Wu, Z. Zhao, H. Song, S. Wang, S. Nie, Triboelectric nanogenerators for enhanced degradation of antibiotics via external electric field. Nano Energy **93**, 106842 (2022)
31. J. Mo, C. Zhang, Y. Lu, Y. Liu, N. Zhang, S. Wang, S. Nie, Radial piston triboelectric nanogenerator-enhanced cellulose fiber air filter for self-powered particulate matter removal. Nano Energy **78**, 105357 (2020)
32. C. Dang, C. Shao, H. Liu, Y. Chen, H. Qi, Cellulose melt processing assisted by small biomass molecule to fabricate recyclable ionogels for versatile stretchable triboelectric nanogenerators. Nano Energy **90**, 106619 (2021)

33. G. Wang, X. Liu, Z. Song, D. Yu, G. Li, H. Wang, S. Ge, W. Liu, Multifunctional flexible pressure sensor based on a cellulose fiber-derived hierarchical carbon aerogel. ACS Appl. Electron. Mater. **5**, 1581–1591 (2023)
34. J. Zheng, T. Hang, Z. Li, W. He, S. Jiang, X. Li, Y. Chen, Z. Wu, High-performance and multi-functional conductive aerogel films for outstanding electromagnetic interference shielding, Joule heating and energy harvesting. Chem. Eng. J. **471**, 144548 (2023)
35. T. Mai, P. Wang, M. Ma, Promising cellulose-based aerogel composites: preparation methods and advanced applications. Compos. B Eng. **281**, 111559 (2024)
36. B. Wang, X. Yin, R. Cheng, J. Li, G. Ying, K. Chen, Compressible, superelastic and fatigue resistant carbon nanofiber aerogels derived from bacterial cellulose for multifunctional piezoresistive sensors. Carbon **199**, 318–328 (2022)
37. Y. Chen, W. He, H. Zhou, J. Shen, X. Li, J. Zheng, Z. Wu, Compressible and conductive multi-scale composite aerogel elastomers for electromagnetic wave absorption, energy harvesting, and piezoresistive sensing. Nano Energy **119**, 109100 (2024)
38. L. Zhu, H. Zhang, Z. Hu, J. Yu, Y. Wang, Mesoporous MXene nanosheets/CNF composite aerogels for electromagnetic wave absorption and multifunctional response. Chem. Eng. J. **502**, 157770 (2024)
39. Y. Chen, L. Zhang, Y. Yang, B. Pang, W. Xu, G. Duan, S. Jiang, K. Zhang, Recent progress on nanocellulose aerogels: preparation, modification, composite fabrication, applications. Adv. Mater. **33**, 2005569 (2021)
40. R. Cheng, J. Zeng, B. Wang, J. Li, Z. Cheng, J. Xu, W. Gao, K. Chen, Ultralight, flexible and conductive silver nanowire/nanofibrillated cellulose aerogel for multifunctional strain sensor. Chem. Eng. J. **424**, 130565 (2021)
41. Z. Qin, X. Chen, Y. Lv, B. Zhao, X. Fang, K. Pan, Wearable and high-performance piezoresistive sensor based on nanofiber/sodium alginate synergistically enhanced MXene composite aerogel. Chem. Eng. J. **451**, 138586 (2023)
42. Q. Zhou, W. Hong, K. Sheng, M. Pan, C. Liu, C. Mei, T. Ma, Direct ink writing 3D printed graphene oxide nanocomposite aerogel for intelligent fire-warning and exceptional fire-shielding. Chem. Eng. J. **499**, 156325 (2024)
43. Y. Wang, L. Chen, H. Cheng, B. Wang, X. Feng, Z. Mao, X. Sui, Mechanically flexible, water-proof, breathable cellulose/polypyrrole/polyurethane composite aerogels as wearable heaters for personal thermal management. Chem. Eng. J. **402**, 126222 (2020)
44. Y. He, Y. Zhang, Y. Zhou, J. Yuan, X. Men, Construction of 3D aerogels consisting of cellu-lose and BNNSs bridged by AgNPs for enhancing thermal and tribological properties of polyurethane composites. Nano Mater. Sci. **6**, 752–763 (2024)
45. S. Wang, W. Meng, H. Lv, Z. Wang, J. Pu, Thermal insulating, light-weight and conductive cellulose/aramid nanofibers composite aerogel for pressure sensing. Carbohydr. Polym. **270**, 118414 (2021)
46. Z. Qin, Y. Lv, X. Fang, B. Zhao, F. Niu, L. Min, K. Pan, Ultralight polypyrrole crosslinked nanofiber aerogel for highly sensitive piezoresistive sensor. Chem. Eng. J. **427**, 131650 (2022)
47. S. Li, C. Liu, W. Chen, J. Huang, L. Gan, Negative Poisson's ratio in cellulose nanocrystal aerogels for enhanced sports piezoelectric sensing. Adv. Funct. Mater., 2418425 (2025)
48. T. Xu, Q. Song, K. Liu, H. Liu, J. Pan, W. Liu, L. Dai, M. Zhang, Y. Wang, C. Si, H. Du, K. Zhang, Nanocellulose-assisted construction of multifunctional MXene-based aerogels with engineering biomimetic texture for pressure sensor and compressible electrode. Nano-Micro Lett. **15**, 98 (2023)
49. L. Tian, F.L. Gao, Y.X. Li, Z.Y. Yang, X. Xu, Z.Z. Yu, J. Shang, R.W. Li, X. Li, High-performance bimodal temperature/pressure tactile sensor based on lamellar CNT/MXene/cellulose nanofibers aerogel with enhanced multifunctionality. Adv. Funct. Mater. **35**, 2418988 (2025)
50. W. Wei, S. Song, C. Meng, R. Li, Y. Feng, X. Chen, J. Chang, B. Fei, W. Yang, J. Li, Bacterial cellulose/polyethylene glycol composite aerogel with incorporated graphene and metal oxides for VOCs detection. Chem. Eng. J. **499**, 156510 (2024)

51. X. Wang, Z. Dong, W. Li, D. Xiao, G. Liu, Z. Yu, S. Yin, M. Liang, A high-sensitivity continuous glucose sensor using porous 3D cellulose/ carbon nanotube network. Talanta **283**, 127201 (2025)
52. Y. Jiang, L. Cheng, M. Yang, L. Xiao, S. Wang, X. Liu, Ultrastrong, Highly resilient, and humidity-sensitive wood nano-aerogel composed of resembling native-state fibrils. Adv. Funct. Mater., 2419155 (2025)
53. J. Wu, Q. Lu, H. Wang, B. Huang, Passivator-free microwave-hydrothermal synthesis of high quantum yield carbon dots for all-carbon fluorescent nanocomposite films. Nanomaterials **12**, 2624 (2022)
54. L. Lin, Z. Dong, J. Wang, H. Hu, W. Chen, T. Guo, F. Li, Flexible ultrahigh-resolution quantum-dot light-emitting diodes. Adv. Funct. Mater. **34**, 2408604 (2024)
55. Q. Zhang, C. Huang, Y. Tao, Y. Zhang, J. Cui, D. Wang, P. Wang, Y.Y. Zhang, Flexible triboelectric nanogenerators based on cadmium metal-organic framework/eethyl cellulose composites as energy harvesters for selective photoinduced bromination reaction. Small **21**, 2406623 (2025)
56. J. Mo, X. Wang, X. Lin, X. Feng, C. Qiu, S. Tao, P. Chen, K. Zhu, H. Qi, Sulfated cellulose nanofibrils-based hydrogel moist-electric generator for energy harvesting. Chem. Eng. J. **491**, 152055 (2024)
57. X. Lin, Y. Dong, S. Tao, X. Feng, X. Wang, T. Song, J. Liu, Z. Zhong, Y. Wang, H. Qi, Temperature-gated nanocellulose membrane for enhanced and controllable osmotic energy harvesting. Nano Energy **107**, 108156 (2023)
58. Y. Al Haj, A.B. Soliman, J. Vapaavuori, M. Elbahri, Carbon aerogels derived from anion-modified nanocellulose for adaptive supercapacitor performance. Adv. Funct. Mater. **34**, 2313117 (2024)
59. H. Liu, T. Xu, Q. Liang, Q. Zhao, D. Zhao, C. Si, Compressible cellulose nanofibrils/reduced graphene oxide composite carbon aerogel for solid-state supercapacitor. Adv. Compos. Hybrid Mater. **5**, 1168–1179 (2022)
60. H. Mao, W. Yu, Z. Cai, G. Liu, L. Liu, R. Wen, Y. Su, H. Kou, K. Xi, B. Li, H. Zhao, X. Da, H. Wu, W. Yan, S. Ding, Current-density regulating lithium metal directional deposition for long cycle-life Li metal batteries. Angew. Chem. Int. Ed. **60**, 19306–19313 (2021)
61. V.K. Bharti, A.D. Pathak, C.S. Sharma, M. Khandelwal, Flexible and free-standing bacterial cellulose derived cathode host and separator for lithium-sulfur batteries. Carbohydr. Polym. **293**, 119731 (2022)
62. Y. Cheng, W. Zhu, X. Lu, C. Wang, Lightweight and flexible MXene/carboxymethyl cellulose aerogel for electromagnetic shielding, energy harvest and self-powered sensing. Nano Energy **98**, 107229 (2022)
63. T. Mai, L. Chen, Q. Liu, Z.H. Guo, M.G. Ma, Zeolitic imidazolate frameworks derived magnetic Nanocage/MXene/Nanocellulose bilayer aerogels for low reflection electromagnetic interference shielding and light-to-heat conversion. Adv. Funct. Mater. **35**, 2417947 (2025)
64. C. Wu, L. Zeng, G. Chang, Y. Zhou, K. Yan, L. Xie, B. Xue, Q. Zheng, Composite phase change materials embedded into cellulose/polyacrylamide/graphene nanosheets/silver nanowire hybrid aerogels simultaneously with effective thermal management and anisotropic electromagnetic interference shielding. Adv. Compos. Hybrid Mater. **6**, 31 (2023)
65. Z. Li, Z. Chen, Q. Huang, S. Zhang, W. Wang, W. Li, MOF@lignocellulosic nanofibril aerogel designed by carboxymethylated nanocellulose bridging for thermal insulation and fire retardancy. Adv. Compos. Hybrid Mater. **7**, 28 (2024)

MIX
Papier aus verantwortungsvollen Quellen
Paper from responsible sources
FSC® C105338

If you have any concerns about our products,
you can contact us on
ProductSafety@springernature.com

In case Publisher is established outside the EU,
the EU authorized representative is:
Springer Nature Customer Service Center GmbH
Europaplatz 3, 69115 Heidelberg, Germany

Printed by Libri Plureos GmbH
in Hamburg, Germany